Views from the Alps

Views from the Alps

Regional Perspectives on Climate Change

edited by Peter Cebon, Urs Dahinden, Huw Davies, Dieter M. Imboden, and Carlo C. Jaeger

The MIT Press
Cambridge, Massachusetts
London, England

This book was set in Palatino on the Monotype "Prism Plus" PostScript Imagesetter by Asco Trade Typesetting Ltd., Hong Kong.

Printed and bound in the United States of America.

Library of Congress Cataloging-in-Publication Data

Views from the Alps : regional perspectives on climate change / edited
  by Peter Cebon ... [et al.].
      p.   cm. — (Politics, science, and the environment)
  Includes bibliographical references and index.
  ISBN 0-262-03252-X (alk. paper)
  1. Alps Region—Climate. 2. Climatic changes—Alps Region.
I. Cebon, Peter. II. Series.
QC989.A43V54   1998
551.69494'7—DC21   98-3946
                                                          CIP

Contents

Series Foreword

As our understanding of environmental threats deepens and broadens, it is increasingly clear that many environmental issues cannot be simply understood, analyzed, or acted upon. The multifaceted relationships between human beings, social and political institutions, and the physical environment in which they are situated extend across disciplinary as well as geopolitical confines, and cannot be analyzed or resolved in isolation.

The purpose of this series is to address the increasingly complex questions of how societies come to understand, confront, and cope with both the sources and the manifestations of present and potential environmental threats. Works in the series may focus on matters political, scientific, technical, social, or economic. What they share is attention to the intertwined roles of politics, science, and technology in the recognition, framing, analysis, and management of environmentally related contemporary issues, and a manifest relevance to the increasingly difficult problems of identifying and forging environmentally sound public policy.

Peter M. Haas
Sheila Jasanoff
Gene Rochlin

Preface

In the preface to *Inadvertent Climate Modification* (MIT Press, 1971) it was noted presciently that "the implications of inadvertent climate modification, both in terms of direct impact on man and the biosphere and of the hard choices that societies might face to prevent such impacts, are profound. Should preventive or remedial action be necessary, it will almost certainly require effective cooperation among the nations of the world." The intervening quarter of a century has seen the issue of human-induced climate change perplex politicians, challenge scientists, galvanize pressure groups, and affirm the mission of environmental organizations. At stake is not only the future of the natural environment but also the nature of future society.

One major political achievement has been the United Nations' establishment in 1992 of the "Framework Convention on Climate Change." This treaty seeks commitments from sovereign states that would limit the causes of, and prompt mitigating responses to, climate change. The scientific basis for such action has been set out in the regular assessments provided by the Intergovernmental Panel on Climate Change (IPCC). These assessments conclude that significant change is possible and that the associated signal is now discernible in the climate record.

Crucial to the further consideration of the diverse aspects of climate change are the two related issues of the uncertainties linked to climate prediction and the centrality of the regional dimension. First, prediction of the physical component of the change is beset with uncertainties linked to its timing, magnitude and spatial pattern, and the myriad possible future societal activities that could impact upon and modify that change compound these uncertainties. Assessments of and strategies for responding to and mitigating putative change need to take into account these complex interrelated uncertainties. Second, the distribution of the trace gases that induce climate change and the concerted political action required for its amelioration have an indubitable global dimension, whereas the realized patterns both of physical and ecological changes and of socioeconomic impacts and responses are expected to exhibit strong regional variations. Estimates of the anticipated mean global change can perforce neither embody the multifaceted aspects of the change nor adequately convey its potential scale and nature. There is, in effect, a pressing need to explore the subcontinental and regional dimensions of climate change.

This book is built around the two foregoing issues. It deals with regional aspects of climate change in the specific context of the alpine region of Europe, and the issue of uncertainty is a recurring underlying theme. Overviews are presented of the region's past, present, and possible future climate and ecology. These studies complement IPCC's current orientation toward elucidating the regional aspects of climate change. An examination of the possible impact of regionally instigated, specific, restrictive governmental measures and innovative technological initiatives follows. These studies contribute directly to the knowledge base required to construct an integrated environmental assessment (IEA) for the alpine region. Assessments of this kind should help in the generation of reasonable and realizable responses to the putative threat of climate change. The design of an appropriate framework for a regional IEA is the theme of the book's last chapter. On the premise that the processes of social learning should be central to the design of an IEA, we argue that it should be a dynamic adaptive process that acknowledges the spectrum of available information and accommodates the multiplicity of different perspectives.

Cross-disciplinary research is but one component of the "ideal" IEA process, and such research entails confronting different perspectives. The contributors to this book, drawn from the disparate disciplines of the physical, earth, biological, and social sciences, will readily testify that such research is challenging. It requires at the very least commitment, scientific acumen, interpersonal sensitivity, and an abundance of patience.

The Swiss Environmental Priority Programme (SPPU), one component of the Swiss National Science Foundation, funded the research for this book. The SPPU, and in particular its directors, are to be thanked for their early and enlightened recognition that appropriate cross-disciplinary research is fundamental to tackling many of today's pressing environment-related problems.

Contributors

Brigitta Ammann
Geobotanical Institute, University of
Bern, Switzerland

Daniel Ariztegui
Geological Institute ETH
Zürich, Switzerland

Harald Bugmann
PIK—Potsdam Institute for Climate
Impact Research
Potsdam, Germany

Peter Cebon
Melbourne Business School,
University of Melbourne
Australia

Huw C. Davies
Atmospheric Science ETH
Zürich, Switzerland

Trevor D. Davies
Climatic Research Unit, University
of East Anglia
Norwich, England

Gregor Dürrenberger
Swiss Federal Institute of
Environmental Science and
Technology (EAWAG)
Dübendorf, Switzerland

François Felber
Botanic Institute, University of
Neuchâtel, Switzerland

Marlyse Fierz
Laboratoire d'Ecologie végétale et
de Phytosociologie
Neuchâtel, Switzerland

Andreas Fischlin
Systems Ecology, Institute of
Terrestrial Ecology, ETH
Schlieren, Switzerland

Christoph Frei
Atmospheric Science, ETH
Zürich, Switzerland

Patricia Geissler
Botanical Garden
Geneva, Switzerland

Jean-Michel Gobat
Laboratoire d'Ecologie végétale et
de Phytosociologie
Neuchâtel, Switzerland

Antoine Guisan
Systems Ecology, Institute of
Terrestrial Ecology, ETH
Schlieren, Switzerland

Dimitrios Gyalistras
Institute of Terrestrial Ecology, ETH
Schlieren, Switzerland

Wilfried Haeberli
Institute of Hydraulics, Hydrology
and Glaciology (VAW), ETH
Zürich, Switzerland

D. M. Imboden
Swiss Federal Institute of
Environmental Science and
Technology (EAWAG)
Dübendorf, Switzerland
and Environmental Sciences ETH
Zürich, Switzerland

Carlo C. Jaeger
Swiss Federal Institute of
Environmental Science and
Technology (EAWAG)
Dübendorf, Switzerland
and Darmstadt University of
Technology
Germany

Philippe Küpfer
Botanic Institute, University of
Neuchâtel, Switzerland

Heike Lischke
Systems Ecology, Institute of
Terrestrial Ecology, ETH
Schlieren, Switzerland
and National Forest Inventory, Swiss
Federal Institute for Forest, Snow,
and Landscape Research
Birmensdorf, Switzerland

Guy S. Lister
Geological Institute, ETH
Zürich, Switzerland

David M. Livingstone
Swiss Federal Institute of
Environmental Science and
Technology (EAWAG)
Dübendorf, Switzerland

André F. Lotter
Swiss Federal Institute of
Environmental Science and
Technology (EAWAG)
Dübendorf, Switzerland
and Geobotanical Institute,
University of Bern
Switzerland

Christian Ohlendorf
Geological Institute, ETH
Zürich, Switzerland

Claudia Pahl-Wostl
Swiss Federal Institute of
Environmental Science and
Technology (EAWAG)
Dübendorf, Switzerland
and Environmental Sciences, ETH
Zürich, Switzerland

Christian Pfister
Historical Institute
University of Bern
Switzerland

S. Rayner
Battelle/Pacific Northwest
Laboratories
Washington, D.C., U.S.A.

Silvia Rothen
Rothen Ecotronics
Bern, Switzerland

Roman Rudel
IRE
Bellinzona, Switzerland

Christoph Schär
Atmospheric Science, ETH
Zürich, Switzerland

André Schlüssel
Botanical Garden
Geneva, Switzerland

Jakob Schwander
Institute of Physics, University of
Bern, Switzerland

Fritz Schweingruber
Forest, Snow, and Landscape
Institute, ETH
Birmensdorf, Switzerland

Bernard Stauffer
Institute of Physics, University of
Bern, Switzerland

Michael Sturm
Swiss Federal Institute of
Environmental Science and
Technology (EAWAG)
Dübendorf, Switzerland

Jean-Paul Theurillat
Centre alpien de Phytogéographie
Champex, Switzerland
and Botanical Garden
Geneva, Switzerland

Bernhard Truffer
Swiss Federal Institute of
Environmental Science and
Technology (EAWAG)
Dübendorf, Switzerland

M. van Asselt
International Centre for Integrative
Studies (ICIS) Maastricht,
Netherlands

A. Vckovski
Department of Geography,
University of Zürich
Switzerland

Caterina Velluti
Botanical Garden
Geneva, Switzerland
and Botanic Institute, University of
Pavia, Italy

Heinz Wanner
Institute of Geography
University of Bern
Switzerland

Martin Widmann
Atmospheric Science, ETH
Zürich, Switzerland

Martin Wild
Geography, ETH
Zürich, Switzerland

Jann Williams
School of Environmental and
Information Science
Charles Sturt University
Albury, Australia

Nicolas Wyler
Botanical Center of the University of
Geneva, Switzerland

Gui-Fang Zhao
Botanic Institute
University of Neuchâtel
Switzerland
and Department of Biology
Northwest University
Xi'an, China

1 Introduction

Dieter M. Imboden

1.1 GLOBAL AND REGIONAL WEATHER AND CLIMATE

In the last hundred years, a new chapter has opened in the history of the atmosphere's chemical composition—a chapter written by humankind. Atmospheric carbon dioxide (CO_2) concentrations measured at the summit of the volcano Mauna Loa in Hawaii gave the first intimations of an anthropogenic influence on the climate system (Keeling et al. 1976). Data from numerous other sources later confirmed these intimations, which finally connected to data on earlier climates extracted from ice cores (figure 1.1).

At present we are not certain of all the ramifications these changes may have for the climate in the next few centuries. However, the data accumulating indicate that climate has become warmer during the last hundred years (see box 1.1). Furthermore, there is growing evidence that changes in the atmosphere's chemical composition can influence global climate to an extent comparable to or exceeding that associated with natural climate variability. This effect is attributed to the increase in the atmospheric greenhouse gases (GHGs; see box 1.2). Some of these GHGs are purely of anthropogenic origin, whereas the emission of others—especially CO_2—has increased significantly during the last hundred years (figure 1.2). In contrast to the loading of the atmosphere with relatively short-lived pollutants that have a merely local impact, the changes in the atmosphere brought about by the fairly stable GHGs are cumulative and have a global impact. Potential consequences and appropriate response strategies therefore involve our planet as a whole. These inferences have led to the concept of global climate change.

This book is one product of an ongoing research program, CLEAR[1], which is concerned with climate dynamics in the various subregions of the European Alps[2] and with its related regional and global ramifications. The adoption of this particular approach implies that it is possible to gain insight into the problem of global change by looking at an individual region: specifically, the European Alps. Indeed, this book is geared to demonstrating that a regional focus represents an indispensable complement to the standard global analysis of climate change.

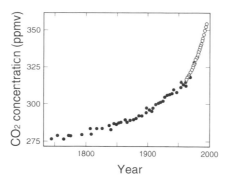

Figure 1.1 Variation in atmospheric CO_2 concentration from 1750 to the present. Black dots show values between 1750 and 1989 measured in air bubbles enclosed in an ice core from Siple Station (Antarctica); open circles give values measured in air at Mauna Loa station (Hawaii). ppmv = parts per million by volume. From Friedli et al. (1986).

Box 1.1   Global Climate during the Last Hundred Years

In July 1995 at the General Assembly of the International Union of Geodesy and Geophysics (IUGG) in Boulder, Colorado, evidence from different observational techniques, model calculations and analyses of natural archives were brought together to provide the message that our present climate is anomalously warm. Observations on land and in the ocean show a global warming of 0.3 to 0.6 Kelvin[3] over the last hundred years. This warming trend's causes are not yet fully known, but evidence for the importance of the role played by anthropogenic greenhouse gases (see box 1.2) and sulphate aerosols is becoming stronger.

Part of the uncertainty regarding the cause of global warming is linked to the fact that data from the last 1,000 years demonstrate that significant climate fluctuations have occurred as a result of natural phenomena such as volcanic eruptions, variations in solar irradiance, changes in the earth's orbit around the sun, and internal fluctuations of the combined ocean/atmosphere system (e.g., the El Niño/Southern Oscillation phenomenon). Model calculations of increasing complexity show that the superposition of natural and anthropogenic factors can explain why the global temperature trend may have a different sign in some parts of the world (e.g., cooling in the North Atlantic and in some parts of China and North America), but that natural factors alone cannot readily explain the observations made during the last hundred years.

Pointing in this direction is the fact that between 1951 and 1990 the average daily surface air temperature over land has increased by only 0.28 K, whereas the minimum daily temperature has increased on average by 0.84 K (Harvey 1995). Such behavior can only be explained by the influence of warming due to anthropogenic greenhouse gases combined with cooling due to sulphate aerosols, which increase the reflection of solar radiation out to space. This latter effect results in a weakening of heating during the day but not during the night. This kind of detailed understanding of the observations lends credence to the case for greenhouse gases' and aerosols' being responsible for global climate change. (See MacCracken 1995 for more details.)

Box 1.2 Greenhouse Gases and Climate Change

The earth's climate system is driven primarily by the influx of (shortwave) solar radiation. Although thermal energy can be stored in the ocean, on the continents, and in the atmosphere (mainly as water vapor), on a long-term basis the earth's heat budget is in a state of quasibalance, so that the outgoing radiation (both shortwave and infrared) is in equilibrium with the incoming solar insolation. The atmosphere's chemical composition influences the outgoing infrared radiation. The greenhouse gases (GHGs) absorb part of the infrared radiation the earth emits to space and reemit it earthward. As a consequence, low-level atmospheric temperatures are maintained at 15° Celsius, that is, at a level 33 Kelvin higher than would be the case were the earth without GHGs. The most important natural GHGs are water vapor and carbon dioxide (CO_2).

The term "man-made GHGs" (mGHGs) refers to the increase in natural GHGs brought about by human activities and to new GHGs of purely anthropogenic origin. Atmospheric concentrations of CO_2 have increased by more than 25 percent from pre-industrial values. This increase accounts for 55 percent of the anthropogenic greenhouse effect. Although absolute concentrations of other mGHGs are still much smaller, they contribute significantly to the greenhouse effect because they have a stronger effect on atmospheric radiation. Examples of such mGHGs are methane (its concentration increase is responsible for 15 percent of the greenhouse effect), nitrogen oxides (6 percent) and chlorofluorocarbons (CFCs, 24 percent). Agriculture is the major cause of the concentration increase of nitrogen oxides and methane in the atmosphere.

The IPCC found on the basis of the results from global climate models that, if nothing is undertaken to halt the input of mGHGs, the atmosphere may warm on average by 1.5 to 4.5 Kelvin by the end of the next century. Recently, the IPCC characterized a careful analysis of the global warming measured in this century (see box 1.1) as "unlikely to be entirely due to natural causes" (Kerr 1995, p. 1167). This was the first unequivocal affirmation of global warming due to anthropogenic causes by this international body of climate experts.

The authors of this book acknowledge the global dimension of climate dynamics. However, they also recognize that neither individuals, families, communities, nations, nor ecosystems experience either global mean surface air temperature, global mean precipitation or the average frequency of occurrence of hurricanes, but rather the local manifestations of the global climate system: Although climate is controlled globally, it is endured locally. In one of its original meanings, the word climate describes a property of the local and regional environment, something that matters for the plants, animals, and people living at a specific place (see box 1.3). In contrast, global climate, a relatively recent scientific concept, has remained for most people an abstract idea that is not easily accepted as a motivation for drastic changes in behavior. The word "global" reflects a new kind of understanding of the interrelationships among different phenomena observed at different locations on the globe. One of the best known examples of this is the relationship between the occurrence of so-called El Niño events (involving the non-appearance of the cold, nutrient-rich Corriente del Niño from the southeast at the Pacific coasts of Ecuador and Peru during Christmas time) and a shift in the major atmospheric pressure systems in the South Pacific known as the

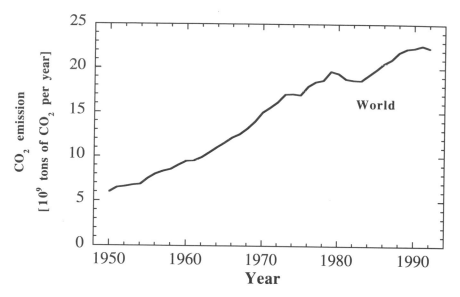

Figure 1.2 Global anthropogenic emission of carbon dioxide (CO_2) due to the burning of fossil fuels increased about fourfold between 1950 and 1991. Note the influence of the oil price crises in 1973 and in the early 1980s.

Southern Oscillation (SO). Bjerknes was one of the first to show that the occurrences of El Niño events are closely related to an unusually small SO index. He used the word "teleconnection" to describe the seemingly simultaneous appearance of these anomalies at locations many thousands of kilometers apart (Bjerknes 1969). Today, the El Niño/Southern Oscillation (ENSO) phenomenon is just one of many known examples of teleconnection.

The classification of climate phenomena according to their spatial dimension and coherence helps the scientist to analyze and understand the underlying causes of climate changes and to build mathematical models of the global climate. Beyond the realm of scientific interest, however, the perception of the global dimension of climate and its relation to human activities is also of utmost relevance to human society. Like no other human activity, the input into the atmosphere of long-lived substances such as GHGs or the radionuclides produced during nuclear bomb tests in the 1960s has evoked the perception that certain (local) environmental occurrences can have an impact far from the local source. In April 1986, when a nuclear reactor was burning at Chernobyl, people in Europe were shocked when they became aware that a catastrophe in an unknown, distant city could affect their well-being within a few days. Similar lessons from the global climate are usually less spectacular and thus less effective, although some cases, such as the problem of skin cancer due to the ozone hole over the southern hemisphere or the 1992 hurricane Andrew, are beginning to make their presence felt in the international media.

A book focusing on the regional perspective of global climate change confronts several challenges. There is only one climate system on earth, the

Box 1.3 Weather, Climate, and the Climate System

Weather refers to the atmosphere's current state as people feel it. With the invention of more and more sophisticated instrumentation, a more objective, quantitative description based on the measurement of solar radiation, precipitation, air pressure, wind speed and other variables replaced the once rather qualitative characterization of weather (sunshine, cloudiness, rain, storms, etc.). With increased understanding of the basic principles that determine the atmosphere's physics, it became possible to predict the weather for the next one or two days. One important condition for this was the establishment of a worldwide observational network (including weather ships on the ocean) and efficient dissemination of the data obtained. In spite of the significant increase in data availability over the last twenty years, accurate weather forecasts are still not normally possible for longer than about a week to ten days. Responsible for this limit is the unstable and chaotic nature of the solution to the atmosphere's governing physical equations.

Climate is different from weather. According to the *Oxford English Dictionary*, the English word *climate* was borrowed from the French *climat*, which in turn is derived from the ancient Greek word *klima* with the root *klinein*, "to slope." The word was adopted in late Latin, where it was used in the sense of region, district or neighborhood. This meaning has survived in the more poetic word *clime*. In the sixteenth century and later climate became more and more associated with phenomena related to the weather. For instance, it was used to describe a belt of the earth's surface contained between two given parallels of latitude, or, closer to its present meaning, to characterize a region with respect to its atmospheric conditions and its weather as they affect animal or vegetable life.

According to Hantel, (Kraus, and Schönwiese (1986), in modern science the concept of climate cannot be separated from the climate system, which comprises all climatically relevant parts of the earth. These authors define climate as being concerned with an object (the climate system), a method (proper statistics) and an effect (the impact of climate).

To summarize, *climate* in its original meaning described a local or regional property of the environment, a set of conditions difficult to define precisely and completely but considered to be of paramount importance for life. Later the word was also used to describe the system causing these conditions. Finally, the word *climate* also became popular beyond its scientific context to describe the imponderable factors that make up people's cultural, economical, and social environment. The following example to illustrate the difference between climate and weather is based on this metaphorical meaning. The climate of a marriage is determined by numerous and often subtle habits of communication between the spouses that are difficult to define precisely. However, this climate may occasionally manifest itself in an extreme "weather event" such as, for instance, marvelous holidays. The climate then corresponds to the pattern and frequency of these events, the weather to the actual events themselves.

global one, and every attempt to separate it into independent regional pieces seems destined ultimately to fail. Thus when dealing with climate, it seems reasonable to adopt a top-down approach; that is, an approach in which an understanding of local climate follows as a by-product of analyzing global climate. Climate first of all depends on the composition and dynamics of the atmosphere, which neither knows regional boundaries nor respects national borders. In fact, during the last twenty years we have learned from studies of various environmental archives—ice cores, tree rings, and the like (see chapter 3)—that on geological time scales, climate has always varied, and

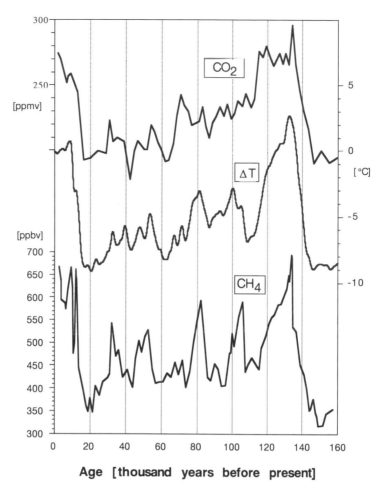

Figure 1.3 Natural variations in the concentrations of two important greenhouse gases, carbon dioxide (CO_2) and methane (CH_4), during the last 160,000 years, as inferred from the composition of air bubbles enclosed in the Vostok ice core (Antarctica). ΔT is the deviation from the present global mean air temperature, derived from stable oxygen isotope ratios. Note the sharp increase in ΔT and in the concentrations of CO_2 and CH_4 at the end of the last ice age, about 10,000–15,000 years ago. (The anthropogenically induced increase in CO_2 and CH_4 is not shown.) ppmv = parts per million, ppbv = parts per billion (i.e., 10^9) by volume.

that most changes, especially the major ones, have been of global nature (figure 1.3). Although the processes causing climate changes are still only poorly understood, we realize that the dynamics of the atmosphere alone are inadequate to explain them and that other components of the climate system, such as the oceans, the biosphere, and the polar ice caps (the so-called cryosphere) must be included in the analysis of climate variability.

Why, therefore, look at local or regional climate only? What are the advantages and the limitations of such an approach? What motivates scientists to participate in a research project on climate change focused on the European Alpine region? What is so special about this part of central Europe that it deserves the attention of scientist, politician, or concerned citizen? Why should global change be of particular importance for mountain areas, especially the Alps? Does not the very concept global indicate that climate change is a problem that spans the whole earth and requires a global perspective for both scientific analysis and action?

Three distinct aspects to the approach adopted here render the present book special and meaningful: the regional perspective, the specific choice of the Alpine region, and the research's interdisciplinary focus. These three aspects are the foundation on which the book's content rests and are the key to interpreting its contents. The reader might therefore find it helpful to review these points before he or she begins studying this book. The remainder of this chapter describes how these three aspects are related, why we consider them to be important, and how they relate to the debate on global climate change.

1.2   REGIONAL PERSPECTIVES ON GLOBAL CLIMATE CHANGE

Climate is a central factor for plants, animals, and humans. Together with the soil's physical and chemical properties, it is one of the essential nonbiological conditions that determine the evolution of life at a given location on earth. At a time when humankind's capability to shape its own environment appears unlimited, we tend to forget that the amount of rain that falls on a particular piece of land, the humidity and temperature of the air, and the force of the wind can make the difference between paradise and hell, between life and death. In fact, these factors set the stage for the spreading out of the human race and for the evolution of agriculture, on which all human cultures are based—in short, for the history of the species *Homo sapiens*. Recently, Richard Potts, an archaeologist at the Smithsonian Institution in Washington, D.C., suggested that the variability of climate, and thus of the physical and ecological environment, has been a major contributor to the significant brain evolution of the genus *Homo* over the past two million years (see Bower 1995).

In spite of climate's vital importance, for a long time we had little hope of gaining a deep understanding of its dynamics. The development of the weather seemed to obey a different kind of law from that governing, for instance, the rigid and thus predictable motion of the sun, moon, and planets. Certain patterns in weather, such as seasonal variability or the occurrence of the monsoon, must have caught the attention of early man, especially the farmer. Yet these phenomena could vary strongly from year to year, perhaps causing famines and leading ultimately to the collapse of whole cultures. Climate catastrophes could be neither prevented nor predicted. Humankind could only prepare for the unexpected and observe the weather closely to develop a certain "statistical experience" of the local or regional climate. Since in most cases some deity or deities were made responsible for the weather and related phenomena, the observation of key factors such as the onset of the monsoon or the height of the annual flood of the Nile in ancient Egypt was considered a religious or royal act. It was unthinkable to the ancient Egyptians that the Nile flood could have such a solid cause as the amount of precipitation falling in eastern Africa during the preceding few months (which in turn seems to be related to the ENSO phenomenon even further away) and could therefore be predictable.

Modern science has greatly changed our understanding of how the weather functions. Most societies have discharged their deities from their climatic responsibilities, although divine weather makers have not yet retired completely—especially during droughts. Today we know the principal laws that govern the atmosphere's physics. We can express them as mathematical equations and solve these equations on computers. There was even a time when scientists extrapolated their achievements so far as to dream of total control of the environment. Suddenly, the ability to master weather and climate seemed just on the horizon. But before this could happen, hope turned to irritation or even fear when it became clear that not only is climate more difficult to understand than anticipated, let alone to control, but humankind may already have inadvertently triggered climate changes, and not necessarily to our benefit.

Recognition of the possibility of inadvertent climate change gave climate research both a tremendous boost and unprecedented importance. In the last twenty years, atmospheric scientists have developed a series of climate models of increasing complexity, the General Circulation Models (GCMs). These have increased progressively in the complexity of their representations of the climate system by modeling more dimensions, resolving smaller spatial scales, describing more physical processes, and treating interfaces and boundary conditions in a more sophisticated way. Because the ocean carries a long-term climate memory by storing heat, advanced GCMs also include ocean models.

Essentially, climate models are forms of weather models. Weather models generally fail for a time span of more than five to ten days because of the atmospheric system's nonlinear nature. Since atmospheric flow is mostly unstable or even chaotic, flow patterns that develop from slightly different initial states often evolve into very different states over a time period on the order of ten days. This is the basis for the often cited metaphor that a tiny event such as the flap of a butterfly's wings might ultimately be responsible for engendering a hurricane.

A second consequence of the nonlinearity of the governing equations is that no time and space scales exist that would allow a natural separation of small- from large-scale phenomena. In other words, for weather and climate, everything ultimately depends on everything else. The regional climate in the Alps, for instance, depends on the global state of the ocean and atmosphere, which weather events in the Alpine region, in turn, partly influence. In essence this is why science emphasizes the global aspects of climate dynamics.

A third factor also makes climate forecasting difficult: The earth's climate system is not autonomous, that is, not independent of changes occurring external to the earth. The sun is obviously the engine driving the weather, but only recently was it discovered that changes in this engine's setup may be the main cause of past climate changes and even of part of the changes in climate that we have been experiencing this century (see box 1.1). These climate changes may be directly related to the sun's properties, such as variations in the solar constant (i.e., in the amount of energy the sun emits per unit area per unit time), or to the geometry of the earth-sun system. Milutin Milankovic, the father of climate modeling and of the astronomical theory of paleoclimates, used the astronomical parameters of the earth's orbit[4] to explain the periods of roughly 22,000, 40,000, and 100,000 years observed in long-term climate records extending over the past period of changing glaciation. However, only some of these external factors are known and are predictable[5], whereas others, such as the short-term fluctuation in solar radiation and the accompanying variation of sunspot abundance, are still only partially understood.

These, then, are three major factors that complicate climate prediction. Two originate from the nonlinearity of the governing equations. The first is related to the limitations imposed upon the weather's predictability and the second to its global nature. Climate scientists tackle these challenges using the enormous power of modern computers. Most climate assessments are produced by simulating the global weather over an extended period (years, decades or even longer), but rather than analyzing the individual courses of weather events, the results are examined in a statistical sense. In effect, the issue addressed is not whether it will be warm or rain in Zurich, say, on August 22, 2023, but the future mean values and ranges of variation of air temperature or precipitation in a particular month, season, year, or decade.

Calculating climatic averages rather than examining specific daily forecasts is the generally accepted method of assessing climate changes. Nevertheless, this has intrinsic and endemic problems, as Wigley (1995) has recently discussed. Today's computers are still much too slow to allow calculations to be performed with the spatial resolution necessary to resolve regional features. For example, a small country like Switzerland would be represented by only about, say, four points in the spatial grid used to calculate the atmosphere's physical state (i.e., air temperature, humidity, cloud cover, wind speed, etc.). Likewise, in a conventional GCM, the altitude of only four points would represent the country's complex topography. In such a representation, the majestic Alpine peaks shrink to harmless round hills of less than a thousand meters in altitude. This has two key impacts. First, by shortening and smoothing the peaks, we may well be introducing modeling distortions that affect vast regions of the earth, if not the entire model. For example, the flow regimes over and around the Alps discussed in chapters 2 and 8 have impacts as far away as Siberia. Second, Alpine climate can vary drastically from one locality or valley to another. Global models do not reflect this fine structure. Thus important questions such as the mean frequency of heavy rain events at a specific Alpine locality or the anticipated mean snow cover cannot be answered with the global models currently in use. Flood conditions in Alpine rivers do not result from the yearly mean precipitation over the Alps but from the rainfall's temporal and local distribution. It is this latter information that people living in the Alpine region need to know to prepare for possible climate changes.

It could be argued that as larger and faster computers continue to develop, the above-mentioned limitation due to computer capacity will rapidly disappear. In fact, this is not the case. If we make the (implausible) assumption that computers will continue to double in speed at their historical rate of once every two years, it will still be several decades before we have machines fast enough to run GCMs with spatial resolutions of a few kilometers. Furthermore, for any finite representation of the climate system, the need would remain to represent the effect of unresolved processes.

A related problem of climate modeling is the need for input data on the initial state of the climate's subsystems (atmosphere, ocean, etc.) For complex climate models, this need is substantial and is exacerbated by higher spatial resolution. At present, we simply lack an adequate data set to be able to calibrate and run a high-resolution global model, and we have no immediate hope that this situation will change in the near future.

As a complement to global climate model–based analysis, regional analysis follows a different path and can produce results fundamentally different from those yielded by GCMs. Global and regional approaches are like the two sides of a coin; although they give independent pictures, they cannot be separated. With respect to physical climate modeling, Giorgi and Mearns (1991) distinguish between three different categories of regional analysis:

1. In the *purely empirical approach*, regional climate scenarios are constructed using instrumental data records or paleoclimatic analogues. One reason why the European Alpine region turns out to be ideal for analyzing regional climate change is the rather large number of available data sets (both instrumental data and paleodata).

2. The *semi-empirical approach* uses GCMs to describe the atmospheric response to large-scale forcing (e.g., GHGs, solar forcing, etc.) and employs empirical techniques to account for the so-called mesoscale forcing (i.e., local topography, the presence of a large inland water body, etc.), which can modify the local circulation. Again, such local features are expected to be important in the Alps.

3. The *modeling approach* describes mesoscale features by increasing the model resolution only over the area of interest, either by using a grid of variable resolution or by nesting techniques, where a higher-resolution region is imbedded in a lower-resolution global model.

Obviously, the modeling approach is purely hierarchical (top-down) if information from the high-resolution subregion does not feed back into the low-resolution domain. To benefit fully from the global and regional approaches' complementarity, two-way nested models are needed, where information flows in both directions between the global and regional sub-models, that is, both top-down and bottom-up. The latter gives a fundamentally different role to regional studies: Instead of merely being a blown-up regional view of the global problem, regional studies may capture aspects that trigger a global phenomenon in much the same way as a local disturbance in atmospheric flow, such as that a mountain generates, may develop into a much larger atmospheric structure.

Until now we have discussed the interrelationship between global and regional features by using examples from atmospheric physics. In fact, as exemplified in other chapters, a similar dualism also exists for biological as well as for socioeconomic aspects of the climate change problem. However, because causal processes are so radically different in physical, biological, and social systems, we don't expect exactly the same relationships between global and regional phenomena. For instance, the response of the biosphere to enhanced atmospheric CO_2 concentrations may be very different on a regional level compared to a global one. Nonlinearities in both the regional climate response to global climate change (see chapters 2 and 4) and ecosystems' response to climate forcing (chapters 5 and 6), when combined with the ecosystems' complexity, mean that it is effectively impossible to employ a completely top-down approach to understanding biotic responses. Regional ecological niches may become important for the global biosphere under climatic stress. As discussed in section 5.7, high mountains may take over the role of "sanctuaries" in which biodiversity is preserved.

The top-down/bottom-up dualism is even more relevant for socioeconomic aspects. As explained above, it is difficult to make people act on such general statements as "The global mean temperature will rise by 1.5–4.5°C by the end of next century." Chapter 7, "Innovative Social Responses in the Face of Global Climate Change," discusses problems related to local and regional concerns. We argue that the complex interaction between technology and society demands the emergence of new socioeconomic patterns to achieve the policy goals stated in global treaties such as the Framework Convention on Climate Change (FCCC) prepared at the 1992 Rio de Janeiro conference, which has now been signed by more than 150 nations. For the necessary processes of social learning to be successful, social contact networks are essential for the creation and diffusion of radical innovations. Often these contact networks have a regional dimension. Thus, the regional analysis of sociotechnical transformation processes in combination with regional climate scenarios provides an approach to the problem of climate change that cannot easily be extrapolated from a global approach.

The fact that the regional and the global are complements and not substitutes means that it is impossible to create a unique coherent view of climate change in a region. That the various scientific disciplines are far from distinct strongly reinforces this notion. Just as the global and the local must be seen as complements, so must the contributions from various scientific fields. As such, there is no such thing as one single regional "view" of climate change.

To summarize, we believe that investigations on global climate change are a vital complement to global studies because they enable us to capture local and complementary aspects of the phenomenon that would otherwise be missed. Notwithstanding, a regional approach will not converge on a singular view of the climate change phenomenon. The best we can hope for is a set of complementary (but often contradictory) views that when juxtaposed provide a relatively comprehensive picture. Chapter 8 looks at the problem of creating that juxtaposition by considering how various views of the Alps presented in chapters 2 through 7 can be integrated to facilitate better policy making. It emphasizes the need for incorporating local knowledge into regionally relevant techniques of integrated assessment.

1.3   THE ALPS AS A KEY REGION FOR THE STUDY OF GLOBAL CHANGE

The term "regional" lies somewhere between the terms "global" and "local." A region is more than a single city or mountain valley but less than a continent or the whole globe. A region may be defined by some special geomorphologic characteristics or by a common sociocultural structure. Since the former often determines the development of the latter, a region typically exhibits both physical and social similarities.

This certainly applies to the Alpine region, where, in spite of obvious dissimilarities such as the various languages spoken, the special topographic conditions have triggered the evolution of rather similar socioeconomic structures. By the Alpine region, we mean the area bordered roughly by the cities of Stuttgart (Germany), Vienna (Austria), Milan (Italy), and Marseilles (France) (see figure 2.1). The Alps are a major mountain range extending along an 800-kilometer arc from the Mediterranean in the southwest to the edge of the central European plain on the northeast. The range has a mean width of about 200 km, an average ridge height of about 2,500 m a.s.l. and a maximum elevation of more than 4,800 m a.s.l. (Mont Blanc).

The Alpine region has had a very dynamic history. Continual confrontation with the vagaries of an unpredictable climate coupled with the vulnerability of both man and animals in an area where natural catastrophes are frequent and often severe has traditionally stimulated innovative action. Seven hundred years ago, one of the first democratic systems emerged in the mountainous central cantons of Switzerland, triggered by the need for wise self-organization by people living under extreme conditions and granted by the then-powerful European monarchs to guarantee free access to the important transit routes across the Alps. Alpine streams and rivers, fed during the whole year by snow and ice in the high mountains, provided an important source of mechanical energy. In those valleys, following the example of

Great Britain, the industrial revolution first took hold on the European continent. Even now, the Alpine region is still responsible for many technical innovations. Its relative wealth and independence serves as a lifestyle model for much of the less developed world.

For the climatologist, the Alps are also of great interest. Meteorological stations, several at high altitude, have accumulated long time series of weather data that have proved very valuable in the present debate on climate change. These data show that the Alpine weather and climate is influenced not only by large-scale global forcing, but also by local topographical details. This demonstrates the importance of a regional approach to the problem of global climate variation. The Alpine trends in mean air temperature during the last hundred years are qualitatively similar to those observed globally (see chapter 2). Aspects of the Alpine regions' sensitivity to global climate change are explored in chapters 2, 3, and 4.

As shown in chapter 3, for the paleoscientist, the Alps are an extremely interesting source of data. Because of the existence of long weather records and the abundance of traces the weather has left on natural objects (glaciers and moraines, lakes, permafrost and block flow, the altitude of the tree line, etc.) paleoecological studies in this region are among the most sophisticated on Earth. In this connection we should not forget that the many observations of fossils in the Alps helped lay the foundation for the development of the theory of evolution. Similarly, paleo-observations made scientists in the last century realize that climate has not been constant in the past but has a spectacular history, with a major event (the last Ice Age) having occurred about 10,000 years ago. In fact, the Swiss scientist Louis Agassiz was among the first to postulate the existence of the Ice Ages.

In addition to considering its direct impact on society, climate change's possible consequences on ecosystems are also examined. According to our current understanding of ecosystem adaptation mechanisms, the speed of change rather than its magnitude determines ecosystem survival. From paleo-studies we know that not all climate changes must result in high rates of extinction and a drastic collapse of biodiversity. On the contrary, climate changes have often triggered a new evolutionary thrust.

In principle, ecosystems are capable of moving geographically, provided that the driving force, climate change, does not exceed a certain critical speed. The global climate policy debate would benefit if a specific upper limit for a noncatastrophic ecological adaptation to warming could be defined. In fact, during the Rio conference, a limit of 0.1 K per decade was under discussion, but this limit never made its way into the final documents, because too many open questions remained regarding the different response patterns of the Earth's various ecosystems.

In most cases ecosystems' response to change results in a spatial redistribution of different types of ecosystems. Commonly these shifts occur horizontally, that is, from north to south or vice versa. For instance, the transition zone from forest to tundra has moved back and forth considerably during the last million years. In the Alps such movements occur mainly along the vertical axis, that is, between low and high altitudes. Thus, the dislocation of habitats found over several hundreds or thousands of kilometers along the horizontal axis may be studied in the Alps within a range of some hundreds of meters in altitude. During this process, the form of the local topography can result in isolated vegetational islands or even in the local disappearance of a given plant community, in spite of the fact that at some distance, higher mountains might yield the right conditions for the community. As shown in chapters 5 and 6, the Alps serve as ideal test areas for investigations on ecological changes and adaptation mechanisms evoked by climate change.

We have mentioned the intimate relationship between the geographical and social characteristics typical of the Alpine region. The socioeconomic case studies chapter 7 presents are examples of social learning processes and potentially important for a reduction of greenhouse emissions both in the Alpine region and globally. The first case deals with the problem of transalpine freight traffic, an issue that, ironically, was of central importance when the first Swiss cantons became independent about 700 years ago. This example demonstrates how conventional policy can run into serious problems if the process of social learning is ignored. The second case concentrates on the problems related to energy-efficient individual transport using lightweight vehicles. Here the different stages of the innovation process and the role of the regional contact networks are analyzed. Both examples are typical for the Alpine region, where there is a long history of finding collective solutions to key problems, and where problems related to economic restructuring, environmental policy and climate change are becoming more acute. However, the case studies can also serve as examples for the mechanisms of innovative change in other regions.

Finally, the selection of the Alps as a test region also has another rather basic (but essential) reason. As discussed in section 1.5, most of the ideas and data reported in this book have resulted from the interdisciplinary research project CLEAR, which started in 1992 and was financed mainly by the Swiss National Science Foundation within the framework of its Priority Programme Environment (see box 1.4). This leads to this book's third special aspect: its interdisciplinary approach.

1.4   AN EXPLORATORY APPROACH TO GLOBAL CHANGE

In the past, the classical scientific paradigm, according to which physical observations such as the measurement of air temperature, humidity, solar radiation and wind speed can be explained in terms of natural laws, has dominated research on global climate change. The construction of ever larger climate models is based on the conviction that these laws will eventually be able to depict the essential aspects of future climate, that is, that computers will take over the role of the former weather gods. Tacitly, this approach

Box 1.4 The CLEAR Process

CLEAR (CLimate and Environment in Alpine Regions) is a new research program within which coordinated, multifaceted research on climate change issues relating to Alpine regions is carried out in the natural and social sciences. Within this program, climate models are used more as an explanatory than as a forecasting tool. Using climate models, we try to enhance our understanding of key processes, emphasizing levels of uncertainty and studying processes of risk communication referring to environmental dynamics. CLEAR is based on a multilevel research strategy that combines the analysis of climate dynamics with the study of how ecosystems may respond to varying climatic conditions and with the investigation of how human society can cope with the expectation of anthropogenically induced climatic change. A numerical climate model for the Alpine region enables researchers to make detailed state-of-the-art analyses of key climate processes. The program investigates the integration of data and processes on different spatial scales, including interactions between regional and global modeling. Ecosystem models are elaborated to study the dynamics of Alpine ecosystems under changing climatic conditions with the help of paleoecological data. Communication between science and society concerning climatic risks is analyzed, and the possibility of innovative regional milieus to take action with regard to global change are investigated.

CLEAR was initiated in 1992 and is supported financially by the Swiss Priority Programme Environment, a program of the Swiss National Science Foundation. In 1996, financial support for a second investigation period (1996–99) was granted.

also implies that, given a "correct" scientific prediction of climate change, society will accept the new gods' verdict and respond rationally to the given recommendations.

This may sound simple, but it is not. As chapter 2 will demonstrate, the consequences of the few basic physical laws that are relevant for describing the energy balance and water mass balance in the atmosphere are extremely complex and far from being completely understood in all their ramifications. It is by no means clear whether they will ever be applicable to such basic questions as, for instance, how much snow will fall in the Alps next century. However, most climate researchers would agree that the best strategy for shifting the present limits of understanding consists of two approaches: (1) more and better observations and (2) better and larger models.

The natural science approach has been extremely successful in the last 200 years. In a simplified manner, scientific progress can be described as a continuous process of hypothesis building, followed by hypothesis testing by well-defined experiments until the hypothesis eventually takes on some kind of final form and becomes a fixed rule or even a law, in the sense of Newton's Laws. Because such rules and laws must be mutually consistent (meaning they should not lead to contradictions), mathematics offers itself as the ideal language for formulating these rules and deriving new rules by combining old ones. A set of rules—or a set of mathematical equations—is the basis of what today is called a mathematical model.

Today we are well aware that the paradigm of the mathematical description of the world (the physics paradigm)—like every concept—has its limits.

With respect to the problem of climate change, two different kinds of possible limits have to be considered: the technical limitations of the physics paradigm applied to a specific aspect of climate, for example, to climate change in the Alpine region; and the restricted role scientific information (even if it is of the most complete kind imaginable) plays within the context of society's perception of climate change and the design of a possible strategy for action.

Regarding the first limit, the great success the scientific method has achieved tends to conceal the fact that its power unfolds mainly when tackling a rather special category of questions related either to the very small (molecular level) or to the purely physical (nonliving matter, astrophysics, etc.). Problems related to the environment usually do not belong to this category. For instance, it is difficult, if not impossible, to design experiments related to environmental problems that affect the dynamics of the globe as a whole. In the case of climate, only one atmosphere exists, in which humankind has already inadvertently initiated many simultaneous experiments, the outcomes of which are difficult to interpret. Thus, it is often impossible to test hypotheses by well-defined experiments, a procedure that according to Popper (1959) separates science from metaphysics.

In this context, mathematical models can take on a new role. Remember that climate models' mathematical structure is extremely complex. Their solutions may become unstable and chaotic, and thus the results of all model calculations are subject to uncertainty. Individual model runs' *predictive* power may be small. However, in combination with the modern computers' enormous capacity, the models can substitute for real climate experiments. By repeating model calculations with slightly different assumptions as well as by comparing the results from differently structured models, a feeling can be developed for the models' overall behavior. For instance, more robust features can be separated from more arbitrary ones. We call this the *exploratory* use of models.

The difference between the predictive and exploratory use of models can be illustrated by using the metaphor of a popular computer game, the flight simulator (FS). The game's two key components are (1) a mathematical model of a given airplane that describes the plane's reaction to the manipulation of the control stick and other elements of the keyboard, and a selected landscape of mountains, rivers, and shorelines as well as cities, roads, and airfields. In our metaphor, the airplane represents the weather/climate system (or rather, a mathematical model of it, e.g., a GCM). The control stick for the rudder, elevator, engine thrust, and so forth stands for the forces external to the climate system such as the sun, but also for anthropogenic factors such as the concentration of GHGs. In contrast, the landscape represents the potential global climatic states; flying over it is like passing through a series of weather and climate situations.

Using the FS predictively, the operator wants to fly from A to B on a preselected route. The pilot anticipates, for a well-defined operation of the control stick, the plane's course and the series of landscape segments that will appear on the screen. However, the FS's real strength is that it can be used to perform experiments that in the real world are either too expensive, too dangerous, or not at all feasible. For instance, the simulator pilot can explore the plane's limits before it crashes. She or he can also explore new parts of the unknown landscape and thus acquire a certain familiarity with its geography. As in the case of weather forecasts, the plane's response or the landscape that is going to appear on the screen are fairly well determined in the short term but open if the pilot is trying out new maneuvers or flying far into new territory.

In a game, as in real life, it is important to be prepared to make decisions based on premises whose validity is uncertain. For this purpose, exploratory models are of great value. They can produce alternative scenarios that can then be analyzed for their impact on the plane's course (i.e., on climate). Doing so may reveal strategies for the future that are robust with respect to details still unknown. For instance, if one must fly from A to B and knows neither the exact location of B's airport nor the direction from which the runway must be approached, a first robust decision would be to direct the plane in B's general direction. To bring this metaphor back to the problem of climate change: whatever the value of the global mean temperature in the year 2050 may be, we know that reducing our energy dependence must be a move in the right direction, since it improves our flexibility with respect to future energy strategies.

Yet, the above discussion on mathematical models' strengths and limitations misses the essential role such tools play for social processes. In fact, the real issue is neither the FS nor the climate model alone, but the interaction of model and operator (pilot or humankind). The peculiar combination of a mathematical tool and human beings represents a system able to learn by assembling virtual experience. Mathematical models are learning tools that can be important in overcoming the physics paradigm's second limitation, which refers to the role scientific information can really play in determining society's perception of climate change and in promoting societal changes.

Although precise climate forecasting is relevant, people also have other gods. Perhaps they are more concerned with health, wealth, or prestige, or they struggle with short-term problems like finding a job or—even more basically—getting enough food and water to satisfy their immediate survival needs. In addition, the climate modelers' message may not be as straightforward and unequivocal as one could hope. As discussed in chapter 8, there are clearly doubts concerning the scientific community's assertion that a mathematical model can ultimately be built that will be capable of predicting future climate, at least with the spatial resolution many politicians require to justify action. However, this would ignore the need to implement short- and medium-term remedial action to forestall irreversible long-term changes. In this context, we note that the Intergovernmental Panel on Climate Change (IPCC) has linked the scientific community directly with the policy debate, and the IPCC's activities have placed climate change high on the agenda of national governments and international organizations. However, it is imperative that scientists acknowledge the intrinsic limitations of their often quite mechanistic approach and allow the debate to be opened up to encompass the panoply of wider issues. Taking no action while waiting for better models should not become the tacit policy either of the scientists, who may be tempted by financial support for their work, or of the politicians, who could postpone implementing costly and unpopular measures.

Betting everything on the one card of better climate prediction may have other consequences for the socioeconomic and sociopolitical system. Let us make the unrealistic assumption for a moment that we are able to predict the future climate and its impact on the hydrological cycle to any desired degree of precision. Would political decisions be simpler and easier to make under these circumstances? Such predictions would make known not only the losers of climate change, but also the winners. Is it realistic to assume that solidarity for a global reduction of CO_2 emissions would then survive? We believe that the present situation, in which anthropogenic climate change becomes more probable every day, but uncertainty remains concerning the further evolution of these changes, could be used fruitfully to trigger a process of social discourse and learning.

A successful climate policy must communicate an uncertain message to a society that may still have other priorities. As chapter 8 discusses, because of this problem, the old idea of "speaking truth to power" no longer works. As an alternative, *integrated assessment* (IA) is an attempt to combine scientific and social expertise in an exploratory manner to find new political and technical strategies for the future. Those strategies are aimed simultaneously at reducing the potential for global change (e.g., by reducing the emission of GHGs) as well as at increasing society's capability of dealing with anticipated changes. According to Rotmans and Dowlatabadi (1997), IA can be characterized as "an interdisciplinary process of combining, interpreting and communicating knowledge from diverse scientific disciplines in such a way that the whole cause-effect chain of a problem can be evaluated from a synoptic perspective with two characteristics: (1) it should have added value compared to a single disciplinary oriented assessment; (2) it should provide useful information to decision makers." The application of IA within a regional context can lead to a different, more effective and concrete dialogue between science and society than within a global context (Pahl-Wostl 1997). The global perspective is necessarily more complete with respect to causes (the changes in the atmosphere's chemical composition and the climate changes this is inducing are occurring on a global scale) but less binding with respect to social learning, which often occurs as a local or regional process.

1.5   THE CLEAR PROCESS AND SCIENCE FOR POLICYMAKERS

In 1992 a group of scientists of different disciplines submitted a set of proposals to the then-new Swiss Priority Programme Environment, a program financed by the Swiss National Science Foundation. They had coordinated their projects within the CLEAR umbrella project. CLEAR's overall aims were to use climate models less for forecasting purposes and more in an exploratory way, trying to enhance our understanding of key processes, emphasizing levels of uncertainty and studying processes of risk communication referring to environmental dynamics (see box 1.4).

CLEAR was funded by the priority program and viewed as an ongoing process of learning and cross-fertilization among different disciplines. This book reports on the first three years of this process. Thus the book does not primarily focus on issues of global change (although these are addressed where appropriate), and the authors do not attempt to provide a final recipe for dealing with global change in the Alpine region. In effect, the book attempts to carry the debate on global change from the scientific into the political community. It is meant as science for policymakers, but not according to the outdated recipe of hierarchical knowledge transfer ("We scientists tell you politicians the right solution to the problem"). There is no "right" solution, there are just options for the next couple of steps, and these can only be found if science, politics, and economy cooperate.

At the book's core is the mutual relationship between the science community and its object (the Alpine climate). The former started to unravel the secrets of the latter while the latter, due to its complexity, continuously changed the former's perceptions, methods, and expectations. Global change as felt on a regional level will continue to be with us, scientists, politicians, and citizens, for quite some time. To respond appropriately, we must develop a new culture that confronts the unwarranted hope that technology will "fix it," and at the same time we must acquire the requisite insight and promote the necessary action to avoid the fatalist's trap *"Après nous, le déluge!"*

NOTES

1. CLEAR is financially supported by the Swiss National Science Foundation "Priority Programme Environment."

2. In this book, the term "Alpine region" refers to the European Alps including their foreland; that is, it is understood to include the various subregions such as the Swiss Plateau or the large inner valleys like the Swiss Engadine or the Italian Valtellina.

3. The degree Kelvin (K) is the official temperature unit. It is based on the absolute zero point of temperature at -273.2 degrees Celsius (°C). A temperature step of 1 K corresponds to a step of 1 °C. Thus, a temperature expressed in Kelvin is equal to the temperature expressed in degrees Celsius plus 273.2. Whenever information is given on temperature changes, the Kelvin scale should be used.

4. The Earth's orbital elements around the sun are characterized by (1) precession (periodicity about 22,000 years), (2) obliquity (inclination of the equator to the ecliptic, about 40,000 years),

and (3) eccentricity of the orbit (about 100,000 years) (Berger 1977). According to Mitchell (1976), the relative variance for climate forcing is given by annual = 130, daily = 100, precession = 10, obliquity = 5, eccentricity = 12.

5. Our understanding of the physical processes occurring in the sun predicts that for the next few billion years the solar irradiance will steadily increase by about 1 percent per 100 million years. In fact, about 4.5 billion years ago, the solar output was approximately 25 percent less than today (Gough 1981). It is still an unsolved puzzle how our planet's climate system has adapted to this variation without a significant change in the global mean temperature.

REFERENCES

Berger, A. L. 1977. Support for the astronomical theory of climate change. *Nature* 269:44–5.

Bjerknes, J. 1969. Atmospheric teleconnections from the Equatorial Pacific. *Mon. Weather Rev.* 97:163–72.

Bower, B. 1995. Brain evolution: Climate shifts into gear. *Science News* 148:359.

Friedli, H., H. Lötscher, H. Oeschger, U. Siegenthaler, and B. Stauffer. 1986. Ice core record of the $^{13}C/^{12}C$ ratio of atmospheric CO_2 in the past two centuries. *Nature* 234:237–8.

Giorgi, F., and L. O. Mearns. 1991. Approaches to the simulation of regional climate change: A review. *Reviews of Geophysics* 29:191–216.

Gough, D. O. 1981. Solar interior structure and luminosity variations. *Solar Phys.* 74:21–34.

Hantel, M., H. Kraus, and C.-D. Schönwiese. 1987. Climate definition. *In Landolt-Börnstein: Zahlenwerte und Funktionen aus Naturwissenschaften und Technik*, eds. K.-H. Hellwege and O. Madelung, Neue Serie, Vol. 4, Chap. 11. p. 1–28.

Harvey, L. D. D. 1995. Warm days, hot nights. Nature 377:15–16.

Keeling, C. D., R. B. Bacastow, A. E. Bainbridge, C. A. Ekdahl Jr., P. R. Gunther, L. S. Waterman, and J. F. S. Chin. 1976. Atmospheric carbon dioxide variations at Mauna Loa Observatory, Hawaii. *Tellus* 28:538–51.

Kerr, R. A. 1995. Global Change—Scientists see greenhouse, semiofficially. *Science* 269:1667.

Mitchell, J. M. Jr. 1976. An overview of climate variability and its causal mechanisms. *Quaternary Research* 6:481–93.

MacCracken, M. 1995. The evidence mounts up. *Nature* 376:645–6.

Pahl-Wostl, C. 1997. Integrated assessment of regional climate change and a new role for computer models at the interface between science and society. In *Prospects for Integrated Assessment and Climate Change*, eds. A. Sors, A. Liberatore, S. Funtowicz, J. C. Hourcade, J. L. Fellons, European Commission DG XII, Report No. EUR 17639, pp. 156–160.

Popper, K. R. 1959. *The Logic of Scientific Discovery*. London: Hutchinson.

Rotmans, J., and H. Dowlatabadi. 1998. Integrated assessment noclelius. In *Human Choice and Climate Change, Vol. 3: The Tools of Policy Analysis*, eds. S. Rayner and E. Malone, Columbus, Ohio: Battelle Press.

Wigley, T. M. L. 1995. A successful prediction? *Nature* 376:463–4.

2 Current Alpine Climate

Christoph Schär, Trevor D. Davies, Christoph Frei,
Heinz Wanner, Martin Widmann, Martin Wild, and
Huw C. Davies

2.1 INTRODUCTION

In the past, the Alpine climate has exhibited large variations on a wide range of timescales. On the 100,000 year timescale, the advance and retreat of Alpine glaciers that accompanied the onset and decay of the Ice Ages left indelible marks on the landscape. On century and multidecadal timescales, significant climate variations have served, for example, to modify agricultural practice. Again, shorter-term climate variations have also had notable impacts. For instance "the year without a summer," 1816, resulted in a severe famine in the Alpine region. These climate signals experienced in the Alpine region were the local manifestation of three forms of natural changes in the planet's climate system linked respectively to (1) variations in the earth's orbit around the sun, (2) the redistribution of energy between the atmosphere and ocean, and (3) volcanic eruptions that substantially but transiently changed the atmosphere's aerosol composition.

Anthropogenic effects also contribute to global climate change. Incomplete observations and inadequate understanding of the climate system currently limit estimates of their impact, but their amplitude could rival that of natural variations. Recent assessments suggest that the increasing atmospheric concentration of specific long-lived trace gases (carbon dioxide, methane, nitrous oxide, and chlorofluorocarbons), although ameliorated somewhat by the countereffect of anthropogenic sulphate aerosols, could result in an unprecedentedly rapid (on the order of 0.3 K per decade) global mean temperature change. Such a change in the global mean would be expected to exhibit significant regional variations with accompanying ecological and socioeconomic effects.

This study focuses on aspects of Alpine climate and climate change. The Alps are a major mountain range that exerts a strong influence upon the in situ weather and climate. Their height is such that from the foothills to the Alpine crest the mean temperature decreases by typically some 15–20 K, with accompanying large spatial gradients in the soil type and cover (e.g., glaciers and snow cover) and in the types of ecosystems. The mountain range's scale and geometry enables it both to modify and to trigger weather systems and thereby to establish distinct climatic characteristics. Again the Alps demark the boundary between two major climate zones: the mid-latitude temperate and the Mediterranean type. The topographic factors also influence the nature of the Alpine response to global climate change and indeed suggest that the potential for complex (and significant) climate variations is enhanced in this region. Moreover, the resulting impacts could be substantial, since the population distribution, the broad range of ecosystems, agricultural activities and tourism are all shaped by, and adapted to, the physico-climatic topography of the region.

The foregoing remarks emphasize the need to consolidate our understanding of Alpine climate, to examine the nature and amplitude of possible climate change in the region, and to assess the form and extent of the impacts that could ensue from such a change. The undertaking of such a program is constrained, however, by the climate system's complexity and nonlinearity. The complexity is reflected in the myriad of interacting physical, chemical, and biological processes that render unattainable both precise specification and comprehensive understanding and explicit representation of the system. Likewise, the nonlinearity betokens an intrinsic limit to the system's predictability. These limitations to our observational base, scientific understanding, and ability to model, together with the inevitable uncertainty attached to any forecast of the climate's evolution, serve both to color the approach to and qualify the results of climate and climate change studies.

The present chapter is predominantly a synthesis of the extant Alpine climate studies, and our interrelated objectives are to describe the current Alpine climate, devoting particular attention to the climatic elements' geographical distribution and variability, and to outline the nature and physics of the processes that contribute decisively to establishing the particular features of the Alpine climate.

2.2 ALPINE CLIMATOLOGY: WHAT ARE THE GEOGRAPHICAL DISTRIBUTION AND VARIABILITY OF THE CLIMATIC ELEMENTS?

The Alps are an 800-kilometer, arc-shaped mountain range with a mean width of approximately 200 kilometers and an average ridge height of about 2.5 kilometers. Additional distinctive features of the range (see figure 2.1) are the major valleys that run predominantly north or south onto the foreland and several east-west inner-alpine valleys aligned along the main ridge.

The severity of Alpine weather and the occurrence of several distinctive orographically related atmospheric flow phenomena (e.g., the wind systems of the Bise, Bora, Föhn, and Mistral; Alpine lee cyclogenesis events; and orographic precipitation enhancement) has long attracted the interest of natural scientists and stimulated scientific investigation. Already in the eighteenth century Horace Bénédict de Saussure, with his combination of instrument design, field measurements, and physical considerations, laid the foundations for the discipline of mountain meteorology. This early interest has three sig-

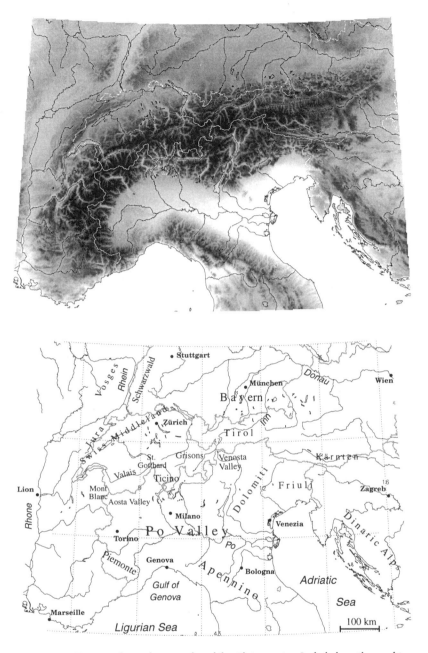

Figure 2.1 Topography and geography of the Alpine region. Included are the areal terms used in the text. (Courtesy of Dr. Daniel Lüthi, Atmospheric Science ETH.)

nificant repercussions for the present study. First, it led to the early establishment of meteorological observational stations, including several at high elevations. For example, of the principal mountain observatories Barry (1992) lists, almost one-third are Alpine stations. Observations at these stations represent an invaluable climate resource in the form of long time series for various climatic elements. Second, the systematic study of these data has provided the rudiments for an Alpine statistical climatology (for example, detailed documentation and cartographic displays of mean state climatic elements and their associated variance and trends), and the study of the spatial organization, dynamical characteristics and associated weather elements of the various Alpine-related flow systems forms the basis for a synoptic-dynamic climatology of the region (for example, the development and compilation of refined classifications of the prevailing weather types). Third, the early Alpine glaciologists provided both the first observational evidence for climate change in the form of ice ages (Agassiz 1840), and outlined the linkage of climate and its variation to the abundance of atmospheric greenhouse gases (Tyndall 1863).

In the present section we draw on and synthesize the available climatological data and the associated studies to provide an overview of Alpine climate in terms of the distribution of the climatic elements. In doing so, we restrict attention mostly to the past 100 years, for which modern instrumental records are available. For information on longer-term variability and paleoclimatic proxy data, the reader is referred to chapter 3.

2.2.1 Temperature

Our current knowledge of the climatic distribution of the air temperature is based on regular observations undertaken at the networks of surface stations and upper-air balloon soundings operated in the Alpine countries. In addition, special field campaigns during which observations were conducted at higher spatial and temporal resolution yielded information on the thermal conditions that prevail on scales ranging from that of the entire ridge down to that of specific valleys and slopes. In the following, we discuss the picture of the distribution that emerges from the analyses of these observations and proceed from the overall spatial distribution, including its seasonal and diurnal patterns, to the interannual variations since the beginning of regular observations.

Maps of the yearly and seasonal mean surface air temperature are available for individual Alpine countries and provide much local detail. An Alpine-wide overview with a coarser resolution is included in the Atlas of the World Meteorological Organization (WMO 1970). These maps' salient feature is a general temperature decrease with elevation, typically 0.65 K per 100 meters. This feature is a characteristic of the troposphere and is not directly related to specific topographic effects. Some of the topography-related flow systems, however, are linked to pronounced spatial, seasonal

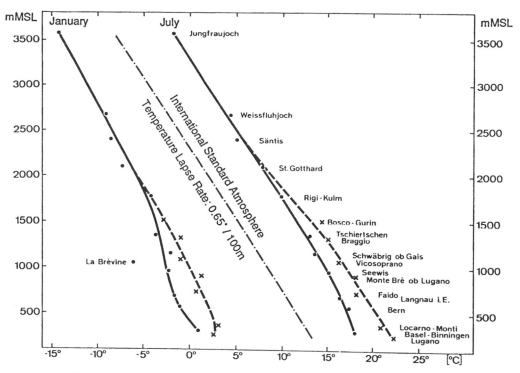

Figure 2.2 Vertical profiles of mean January and July temperature as given by surface observations in the Swiss Alps. Dots and crosses signify stations located to the north and south, respetively, of the main ridge crest. (After Schüepp et al. 1978; figure taken from Wanner 1991.)

and diurnal variations of the temperature and its vertical gradient. Effects of this kind are best illustrated by comparing observations on the same elevation, for example, between mountain tops or slopes and the free atmosphere, or between the valley floor and the adjacent flat land.

Figure 2.2 shows the vertical profile of mean January and July temperatures drawn from surface observations at various elevations in the Swiss Alps. In both seasons, distinct thermal regimes can be identified between the north and south sides of the main Alpine crest. Below 1500 m MSL, the southern Alpine area is about 2 to 4 K warmer than the north. This spatial variation over a horizontal distance of only about 200 kilomters exceeds the typical latitudinal temperature gradients in Central Europe (see, e.g., Schönwiese et al. 1993) and is indicative of topographic effects. The shielding of the south side against cold air advances from the north and northwest (cf. section 2.2.2) as well as influences on the distribution of cloud cover and hence solar insolation (cf. section 2.2.5) contribute to the regional enhancement of the temperature gradient across the Alps.

Horizontal temperature gradients are also observed between the Alpine ridge and the conditions in the free atmosphere over the adjacent foreland, both on seasonal and diurnal timescales. Based on frequent upper-air balloon soundings and summit observations during three summer months, Richner

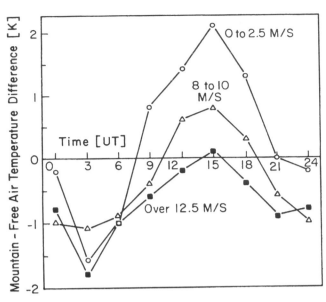

Figure 2.3 Mean summit minus free air temperature difference (K) in the Alps as a function of the time of day (universal time) and wind speed in summer. (After Richner and Phillips 1984; figure taken from Barry 1992.)

and Phillips (1984) describe the summit minus free-air temperature difference (measured at the same altitude) as a function of the time of day and wind speed (see figure 2.3). The thermal difference has a pronounced diurnal cycle, especially during calm situations, with relatively warmer (colder) summits during daytime (nighttime). During episodes of strong flow, mean summit air temperatures are colder than free-air temperatures at the same altitude, a property that derives from the adiabatic cooling associated with flow over the mountains (see section 2.3.2). Observations of the diurnal temperature variations during fair weather conditions (Wagner 1932; Phillips 1984) reveal a larger amplitude of the temperature cycle at Alpine valley floor stations as compared to the foreland regions, reflecting thermal gradients between the mountains and the foreland. Although such behavior can be expected primarily for summer high-pressure situations, it is also evident in the multiyear climatological mean (Schüepp and Schirmer 1977). Section 2.3.2.4 discusses some of the flow systems and meteorological phenomena associated with the inner-Alpine region's relative daytime warmth and nighttime coolness.

Motivated in part by the current discussion of climate change, a number of research groups have examined the interannual and longer-term temperature fluctuations in the Alpine region (von Rudloff 1971, 1986; Müller-Westermeier 1992; Pasquale, Flocchini, and Russo 1992; Sneyers, Böhm, and Vannitsem 1992; Auer, Böhm and Mohnl 1993; Böhm 1992, 1993; Weber, Talkner, and Stefanicki 1994; Beniston et al. 1994). Such investigations require a homogenization of the time series to correct for artificial variations introduced by frequent changes in observational techniques or in local envi-

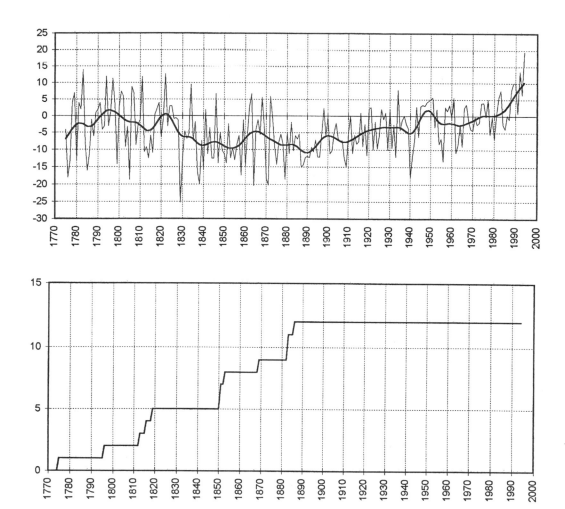

Figure 2.4 Surface air temperature in Austria 1775–1991. Top panel: Deviations of annual means from the thirty-year mean (1961–90) in units of 0.1 K, and low-pass filtered temperature (smooth curve; filter width twenty years). Bottom panel: Number of stations involved in the calculation of the annual means. (Courtesy of Dr. Reinhard Böhm, ZAMG Wien; see also Böhm 1993.)

ronmental conditions such as urbanization, agricultural changes, and deforestation. The reconstructed time series provide reliable records. For some of the Alpine stations, these date back to the beginning of the instrumental period in the late eighteenth century.

Figure 2.4 depicts the variations of annual mean temperature for Austria (Böhm 1993). The time series is based on a maximum of twelve single-station temperature records and is found to be representative for the northeastern Alpine region. A substantial year-to-year variability on the order of 1 K is evident. The longer-term variation, illustrated by the low-pass filtered (smooth) curve, shows warm periods at the beginning of the instrumental record until about 1820, and during the most recent fifty years. Between these two warm periods, the low-pass filtered temperature is substantially lower, with two major minima around 1850 and 1890. The warming since the beginning of this century was found to be stronger in autumn and winter than in spring and summer.

The salient characteristics of the temperature evolution in Austria are in qualitative agreement with other Alpine observations of similar duration (e.g., von Rudloff 1986; Müller-Westermeier 1992). Moreover, the temperature increase between the late nineteenth century and the 1950s is documented for many surface observing sites all over Central Europe (see, e.g., Schönwiese et al. 1993; Heinemann 1994). Concomitant with this warming, major recessions have been registered for many Alpine glaciers (cf. section 2.2.4).

The observed warming in the Alpine region during this century is in accord with (although slightly larger than) the increase in Northern Hemisphere temperatures (Houghton, Jenkins, and Ephraums 1990; Houghton et al. 1995) and accompanies the increase in atmospheric CO_2. It does not necessarily follow from these observations, however, that a causal relation exists between the regional warming and anthropogenic GHG emissions. For example, the warm phase around 1800 indicates that regional long-term temperature fluctuations on the order of 1 K can also result from natural variations of the climate system without simultaneous variations in CO_2. Nevertheless, the observed warming during this century is roughly consistent with the expected magnitude of GHG effects, and currently available climate change scenarios for the Alpine region suggest that an accelerated warming could take place in the near future (see chapter 4).

2.2.2 Pressure and the *Grosswetterlagen*

The pressure value at a particular location has comparatively little direct meaning for local weather and climate. The pressure's horizontal gradient, however, is closely related to the larger-scale horizontal flow field, and the three-dimensional pressure distribution is an indicator of the associated vertical velocity field. Thus the pressure pattern itself is related directly to the type of weather experienced in a region. Hence a customary approach to the study of weather and climate is to adopt a synoptic-climatological method and to characterize the weather in the European sector in terms of the occurrence of distinct weather regimes: the *Grosswetterlagen* types. Each regime corresponds to the persistence of a certain typical circulation pattern for a period comparable to or greater than the lifetime of individual synoptic systems.

Several such classification schemes have been developed, and among the most widely used are those of Lamb (1972) and Hess and Brezowsky (1952; see also references in Gerstengarbe and Werner 1993). These schemes categorize the pattern in terms of the ambient wind direction and introduce a further subdivision in terms of the dominant pattern of the pressure systems (high or low). For example, a southwesterly flow with a dominant low located west of Ireland with cyclonic curvature over central Europe (the *SWz*

Table 2.1 Relative frequencies of the Alpine weather types between 1945 and 1991

Weather type	Spring [%]	Summer [%]	Autumn [%]	Winter [%]	Year [%]
Convective					
Anticyclonic	12	20	18	14	16.0
Weak pressure gradients	30	36	28	16	27.5
Cyclonic	10	8	6	6	7.5
Advective					
West	10	10	10	14	11.0
North	16	12	16	24	17.0
East	6	2	4	8	5.0
South	12	6	12	10	10.0
Eddy over the Alps	6	6	6	6	6.0
Total					100.0

Note: Categorizations after Schüepp 1968.

category in the Hess and Brezowsky scheme) is consistent with cyclones tracking toward Scandinavia and the passage of fronts southward toward the Alps.

Further variants of the synoptic-climatological approach are worth noting. First, the method has been refined to apply specifically to the Alps by directly classifying weather types based on the pressure pattern in the immediate vicinity of the region itself. The two most frequently used schemes (cf. Wanner 1980; Fliri 1984) are those of Lauscher (1958) and Schüepp (1968). Table 2.1 illustrates a climatology in these terms and lists the relative frequencies of the Schüepp weather types in the seasons. Note, for instance, the predominence of northerly flows in winter and the two maxima of southerly flows (which include Föhn) during autumn and spring. Second, in a seminal paper, Kirchhofer (1974) examined the statistical relationship between European-scale weather types and the meteorological parameters at specific Alpine locations and in so doing pioneered the present downscaling studies (see chapter 4). Third, modern statistical techniques are increasingly being used to deduce fully objective synoptic classification schemes (see, e.g., Cacciamani, Nanni, and Tibaldi 1995).

It follows that distinctive aspects of the climate and its variability can be inferred from the statistics for the occurrence, persistence, and alternation of the contrasting weather regimes. The statistics for the last decades show several significant features (Bardossy and Caspary 1990; Gerstengarbe and Werner 1993; Murray 1993). For the annual mean, Bardossy and Caspary detected an increase in the frequency of southerly circulation types (from approximately 5 percent to approximately 10 percent) between 1880 and 1990. The most substantial changes in the last few decades occurred in winter, when the frequency of zonal (westerly) circulation patterns increased substantially at the expense of easterly and northerly configurations. These trends are consistent with the increased frequency of wet and warm winters during the last decades. We return to the link between the foregoing European-scale variations and larger-scale features of the atmospheric circulation in section 2.3.1.

2.2.3 Precipitation

Precipitation in mountainous regions often differs considerably from that in the surrounding lowlands, and section 2.3.2 dicusses some of the meteorological mechanisms responsible for these differences. The Alpine precipitation signal reveals a great spatial variability from the scale of the whole Alps to that of single slopes, and this concerns both the long-term mean as well as the occurrence of strong precipitation. To describe the main features of the Alpine precipitation climate, we proceed by discussing in turn the long-term mean distribution, the seasonal variability, and finally the long-term changes.

Today there are approximately 8,000 rain gauges in use in the Alps belonging to different national networks, and only a small fraction of their observations are distributed internationally. Collection and homogenization of the data is a demanding logistical and statistical task, and a large number of studies are based merely on national rather than Alpine-wide data sets. Precipitation analysis is particularly hampered by observational errors that can attain 15 percent for rain and up to 50 percent for snow (Sevruk 1985). The principal error source is induced by wind field deformations above the gauge, and the high and exposed Alpine stations are especially subject to this problem. Measurements are sensitive to the ambient wind speed, and this introduces an additional error source associated with changes in the surroundings of the gauges, such as tree growth, erection of buildings, and relocation of the gauges. These factors add to data inhomogeneity.

Long-term means based on national data sets can be found in Steinhauser 1953 for Austria, in the *Atlas Climatique de la France* (see Direction de la Meteorologie Nationale 1988, 1989) for France, Schirmer and Vent-Schmidt 1979 for Germany, Touring Club Italiano 1989 for Italy, and for Switzerland in Uttinger 1949 or in the Hydrological Atlas (Landeshydrologie und Geologie 1992). Long-term means of precipitation for the whole Alpine region are provided by Fliri (1974); Baumgartner Reichel, and Weber (1983); and Frei and Schär (1998). Figure 2.5 shows the yearly precipitation totals of the period 1931–60. The map is based on data from about 1,000 rain gauges, most of which are located at altitudes lower than 2000 meters. A relatively dry zone can be noted along the main crest of the Alps surrounded by wet zones to the north and south, and only at a few locations, for example, the Mont Blanc or St. Gotthard regions, do the two wet regions merge.

Not apparent in the smoothed distribution of figure 2.5 are the local extremes found at the driest spots located in the Aosta Valley, with annual

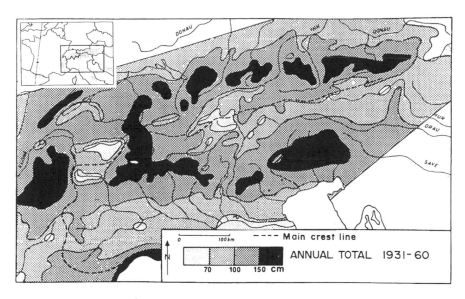

Figure 2.5 Mean annual precipitation sums, 1931–60. (From Fliri 1974.)

means below 550 millimeters per year; in the Venosta Valley and the Valais, with values around 600 millimeters per year; and in the dry northern Alpine valleys of the Grisons and northern Tirol. In the northern wet zone, annual means exceed 2,000 millimeters per year at several locations; in the southern wet zone in the Friuli region, precipitation values reach 3,000 millimeters per year at a few spots, whereas about 2,500 millimeters per year is observed in the Centovalli valley in Ticino. In general, the precipitation totals increase with altitude, but this relation shows strong variations with respect to location and season (Uttinger 1951; Lauscher 1976a; Lang 1985; Blumer and Spiess 1990; Frei and Schär 1998). Figure 2.6 displays an indication of the seasonal cycle of precipitation, based upon the same data as figure 2.5. In the northeastern Alps there is one maximum in summer, whereas the northern and north-western Alps show a second, somewhat weaker maximum in winter. In the southern regions, two maxima arise in spring and autumn.

Particular interest in Alpine precipitation also arises from the frequent occurrence of strong precipitation events in the region and their attendant adverse effects. Either directly, by flooding populated valley floors, or indirectly, by triggering landslides and avalanches, severe precipitation can cause catastrophic damage to agriculture and human infrastructure. A sequence of flooding events during the last few years (Piedmont, November 5–7, 1994; Brig, September 23, 1993; Vaison-la-Romaine, September 23, 1992), some involving loss of human lives, has tragically demonstrated the threat of extreme precipitation and the vulnerability of the Alpine region. In addition, during winter and spring, rapid snow melt during episodes of sudden warming can enhance normal runoff, and the resulting water discharge through the major rivers can affect areas far remote from the Alps.

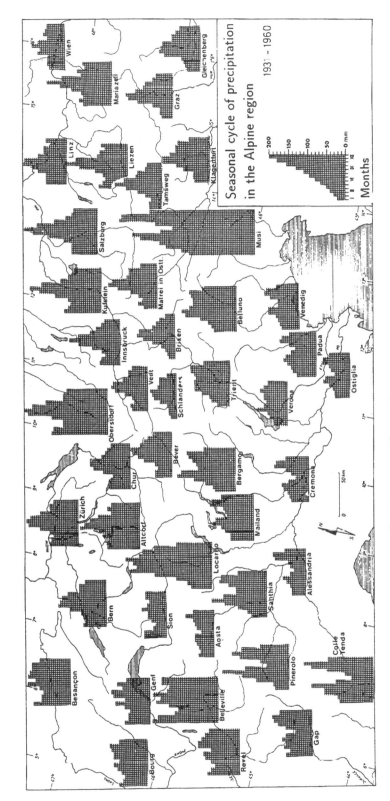

Figure 2.6 Mean monthly precipitation, 1931–60, at selected sites. (From Fliri 1974.)

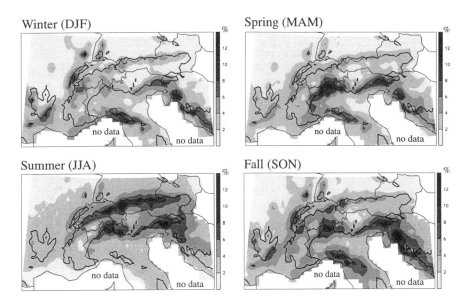

Winter (DJF) Spring (MAM)

Summer (JJA) Fall (SON)

Figure 2.7 Frequency (in percent) of days with strong precipitation (daily total ≥ 20 millimeters) in the Alpine region for the seasons of the year. The bold contour indicates the topography at a height of 800 meters. (Based on data described in Frei and Schär 1998.)

The significance of strong precipitation in the Alpine region also arises from the fact that relatively rare intense events contribute a substantial amount to the long-term mean. For instance, precipitation during the wettest 4 percent of days contributes about 40 percent to Alpine precipitation totals. Statistical analyses of extreme precipitation in some areas of the Alpine region can be found, for example, in Geiger, Zeller, and Röthlisberger 1991; Nobilis, Haiden, and Kerschbaum 1991; and Bonelli and Pelosini 1992. Figure 2.7 displays an Alpine-wide analysis of the frequency of strong (but not necessarily extreme) cases based on twenty years of data at about 3,000 rain gauge stations. Regions with frequent occurrence of strong precipitation (daily values of at least 20 millimeters) can be identified along the northern and southern rim of the Alpine ridge, whereas inner-alpine valleys and the adjacent flatland areas appear less affected. Strong precipitation is most frequent during summer, and during this season it mainly originates from severe thunderstorms (heavy, organized, convective systems; cf. section 2.3.2.5). In spring and fall, high activity is found in particular to the south of the main Alpine crest, that is, in the northwestern Po valley and the Ticino and Friuli areas. Here, strong precipitation usually relates to the advection of warm and moist air from the Mediterranean toward the Alps, induced by the approach of a cold front from the west or the development of mesoscale cyclones over the western Mediterranean (cf. section 2.3.2.2). Finally, during the winter season the frequency of strong precipitation is generally lower.

Figure 2.8 shows the time series for the annual precipitation totals and the number of wet days (with at least 0.1 millimeter) at the Sonnblick Observatory (3106 meters MSL) in Austria. This remote site, unaffected by direct

Precipitation-frequency (at least 0.1 mm, days)

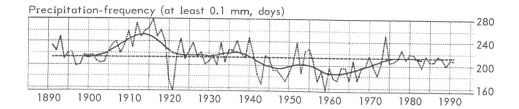

Precipitation-totals (totalizer, wind-shielded, mm)

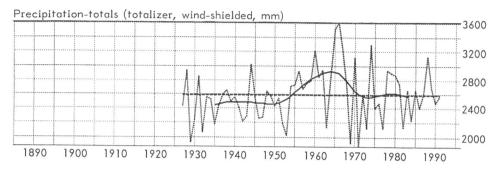

Figure 2.8 Annual precipitation frequency and annual precipitation sums at Sonnblick Observatory. Smoothed curves obtained by twenty-year Gaussian low pass filtering. (From Auer, Böhm, and Mohnl 1993.)

human activities, provides a very homogeneous data set appropriate for the study of interannual variability. Both series reveal large year-to-year variability. Furthermore, the twenty-year low pass filtered data show fluctuations on longer time scales, like the precipitation maximum in the 1960s or the decrease in the frequency of wet days from 1915 to 1960. The low correlation between the amount of precipitation and the frequency of wet days indicates the complexity of precipitation processes. Since the maximum in the totals coincides with the frequency minimum, the increase in the precipitation totals during the 1960s is attributable to an increase in intermediate- and high-intensity events.

The most comprehensive study of precipitation trends in Europe is the *Klimatrend-Atlas Europa 1891–1990* (Schönwiese et al. 1993, 1994). The study revealed a long-term increase of winter precipitation over the whole Alpine region, in its western part by up to 20 percent, during the last 100 years. This trend is consistent with a recent trend analysis conducted for Switzerland based on a relatively dense precipitation network of 113 rain gauge stations with daily resolution (Widmann and Schär 1997). This study detected an increase of winter precipitation in the northern and western parts of Switzerland by as much as 30 percent, and smaller but still positive trends for the southeastern parts of the country. The statistical significance of the trend at single stations using the Mann-Kendall test (Sneyers 1990; Denhard and Schönwiese 1992) is rather low, and the significance level is about 70 percent (Schönwiese et al. 1993). However, the notion of significance in climatological applications is dependent on the spatial resolution. If a region with a common long-term trend is considered, the investigation of the area mean rather than single station records can substantially increase the significance level. Using such (weighted) area means, the increase of winter precipitation during this century can be demonstrated to be statistically significant at a level of up to 90 percent and furthermore to possess some regional variations of similar significance (Widmann and Schär 1997). On the other hand, the decrease of autumn precipitation during this century by 20 percent reported by Schönwiese et al. is of substantially lower statistical significance over Switzerland.

In the European-scale analysis of Schönwiese et al., trends have also been detected for the shorter period 1961–90. An increase of spring precipitation by 30 percent is observed in the southwestern regions of the Alps, whereas the northeastern regions show a decrease of 25 percent. During summer, no significant trends can be detected. The northern and southern regions of the Alps behave differently in the autumn season, with an increase (up to 25 percent) to the north and a 25 percent decrease to the south. As for the centennial period, an Alpine-wide increase of precipitation (25 percent) occurs during winter (Schönwiese et al. 1993, 1994; Widmann and Schär 1997).

The occurrence of warm winters with a substantial surplus of precipitation represents an anomaly on a timescale of many hundreds of years. Already Pfister (1984; see also Pfister 1992) has noted in his analysis of proxy and early instrumental records starting in the sixteenth century that the winters in 1965–80 were warmer (by 1.3 K) and moister (by 25 percent) than the 1901–60 mean, and that such pronounced deviations had never occurred in the previous 500 years. As indicated above, this trend toward warmer, wetter winters has continued into the 1990s, and may constitute one of the first signals of global climate change in the Alpine region.

2.2.4 Snow and Ice

The seasonal snow cover, permanent high-Alpine snow fields, glaciers, and permafrost regions are characteristic features of the Alpine landscape and constitute an integral part of the regional climate system. The significance of the Alpine cryosphere components relates to their sensitivity to the distribution and variation of primary climatic elements (temperature and precipitation), to their effects on atmospheric climate processes, and to their relevance for ecosystems and human infrastructure in the region (agriculture, tourism, water resources, hydropower, etc.). In this section, we discuss the spatial and temporal variations in the Alpine snow cover and report on signatures of past climate variations inferred from glacier observations during the last few hundred years. For longer-term variations, including the ice ages, see chapter 3.

The duration, depth, and water equivalent of the snow cover in the Alps depends primarily on altitude (Schüepp, Gensler, and Bouët 1980; Witmer et

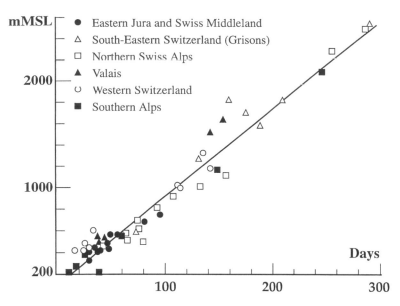

Figure 2.9 Altitude dependence of the average number of days with snow cover at Swiss observing stations (1959/60–1978/79). Distinct symbols are used for various subregions. (After Schüepp, Gensler, and Bouet 1980.)

al. 1986; Fliri 1991) and is controlled by the temperature decrease and precipitation increase with altitude (cf. sections 2.1 and 2.3). Figure 2.9, based on Swiss observations, shows the average number of days with snow cover per year as a function of elevation. For the northern and southern alpine foreland (200–500 meters MSL), continuous snow cover is found rarely for more than two consecutive weeks. At these elevations, snow depths during snowfall episodes typically reach only several to a few tens of centimeters, and intermittent melting periods often lead to widespread disappearance of the snow even deep in the winter. For Alpine areas at elevations between 1,000 and 3,000 meters MSL, however, the snow cover is predominantly seasonal. It accumulates during winter (typically starting in November) and then melts off during spring and early summer (see figure 2.10). Finally, above the climatological snowline, which varies from 2,400 to 3,400 meters MSL (Stone 1992), extended snow fields remain throughout the year. At these elevations, mean summer temperatures are still near freezing, and a substantial fraction of summer precipitation falls as snow (Lauscher 1976b).

Consider now further temporal and spatial features of the snow cover. Its seasonal character, with accumulation of winter snow falls, and its reduction during the melting period are responsible for the summer maximum in the annual runoff pattern of major Alpine rivers (see, e.g., Martinec 1987). Spatial variations also modulate the cover's primary altitude dependence. On the scale of the entire ridge, observations indicate a larger snow-water equivalent, a longer-lasting seasonal cover, and a lower elevation of the snowline along the northern and southern flanks of the Alps as opposed to the inner-

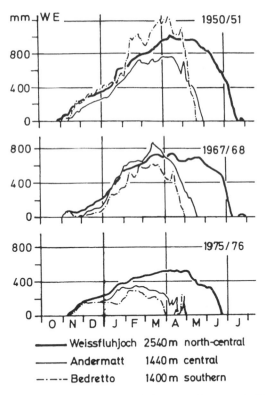

—— Weissfluhjoch 2540 m north-central
—— Andermatt 1440 m central
—·—·— Bedretto 1400 m southern

Figure 2.10 Time trace of the accumulation and ablation of the seasonal snow cover water equivalent (WE) at a north central, a central, and a southern site in the Swiss Alps. The three panels represent winter periods with a relatively thick, an average, and a relatively thin snow cover, respectively. (From Lang and Rohrer 1987.)

Alpine region (Lang and Rohrer 1987; Fliri 1991; Stone 1992). This feature is attributed to the partial shielding of deep central Alpine valleys during major snowfall events (Lang and Rohrer 1987) and correlates with the regional distribution of mean winter precipitation (Fliri 1974; see also section 2.3).

The amount, extent, and distribution of the snow cover in the Alps is subject to large year-to-year variations. Conditions between extreme winters can vary by more than a factor of 2 for the maximum snow-water equivalent and by one month for the spring termination of the snow cover (see figure 2.10). Föhn (1990) discusses longer-term variations and trends inferred from snow cover observations in the Swiss Alps. For the last fifty years, no significant trends are found for deep-winter snow conditions (i.e., snowfall, depths, profiles and avalanche activity; see also De Quervain and Meister 1987). There are some indications, however, of recent changes in the early winter situation, characterized by the snow depth at January 1. Since about 1982, a pronounced decrease is evident at a number of western and southern alpine sites (Föhn and Plüss 1989). As a result of the pronounced warming observed in the Alpine region (see section 2.2.1), winter precipitation is increasingly falling as rain rather than snow. Simple estimates taking into account this and other factors suggest that a warming of 1 K can lead to a rise in the snowline of 100 to 200 m (see, e.g., Abegg and Froesch 1994).

The large year-to-year fluctuations of primary climatic elements (cf. sections 2.2.1 and 2.2.3 and the discussion earlier in this section) often make it difficult to discern slowly varying climate signals such as long-term trends. In this respect the behavior of mountain glaciers and their long-term monitoring in the Alps offer valuable additional information. Mountain glaciers are sensitive to changes in external climatic conditions. In contrast to the seasonal response, which is more immediate, glaciers react with some delay to climatic variations on annual or secular timescales (Patzelt and Aellen 1990) and exhibit a favorable "signal-to-noise ratio" for the observation of slowly varying climate signals.

The growth and retreat of glaciers is associated with snow accumulation and with ice ablation by melt and sublimation. Whereas snowfall during the winter half year predominantly controls accumulation, ice ablation is particularly sensitive to summer temperature and sunshine and also depends on the deposition of pollutants and dust, which influences the radiative characteristic of the glacier surface. An imbalance between accumulation and ablation is accompanied by a reaction or adaptation of the glacier's ice flow characteristics and typically results in an advance (retreat) of the glacier terminus in case of a mass increase (decrease). The type, magnitude, and lapse time of this reaction can vary greatly, however, between individual glaciers (see, e.g., Kuhn 1990) and makes it necessary to consider an ensemble of glaciers for an indirect assessment of climate variations.

Patzelt and Aellen (1990) discuss the evolution of Alpine glaciers since the modern glacial maximum around 1850. Analyses of available observations indicate a major reduction of glacier area by 46 percent in the Austrian Alps and by about 30 percent in the Swiss Alps. The general tendency is clearly evident in the evolution of length changes (see figure 2.11). A majority of Swiss glaciers have been retreating during most of the period. In accord with similar observations for the Austrian Alps, the retreat is interrupted by short episodes of recovery, which are attributed to cool, cloudy summers (in the 1890s; see also section 2.2.1) and a number of winters with heavy snowfall (1915–20). Increased temperatures and sunshine duration as well as reduced precipitation in the high-Alpine area were found to accompany the major retreat phase between 1930 and 1950 (see also von Rudloff 1962). The brief readvances during the late 1970s and early 1980s, which affected predominantly smaller glaciers, largely ceased and the general retreat of Alpine glaciers continued during the last decade of this century.

2.2.5 Surface Energy Balance

Classical climatological parameters like temperature measure the underlying physical processes indirectly and provide little indication of why a particular climate prevails at a specific location. The local climate depends both on

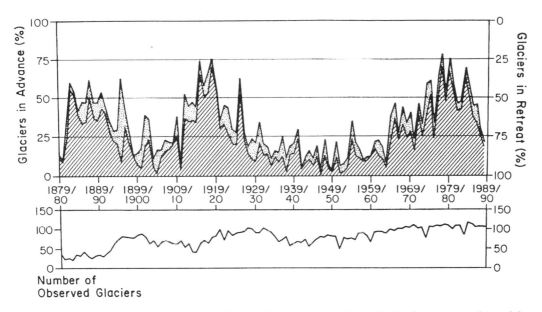

Figure 2.11 Percentage of Swiss glaciers showing advance (hatched) or retreat (white) of the terminus. The stippled portion denotes stationary glaciers. The bottom panel gives the number of observed glaciers in each year. Data from the World Glacier Monitoring Service, Zürich. (See also Aellen 1994; figure taken from Barry 1992.)

regional processes (like the synoptic setting; see section 2.2.2) and also on point processes (discussed in this section). Energy transformations at the earth's surface heavily affect the evolution of the near-surface temperature and humidity field. The associated processes (see figure 2.12) include the absorption of solar direct and diffusive radiation (together referred to as global radiation), the emission and absorption of longwave infrared radiation, and the turbulent exchange of energy in the atmosphere's surface layer through vertical fluxes of sensible and latent heat.

The net radiative energy flux (defined as the surplus of absorption over emission) drives the near-surface processes, and the "surface energy balance" describes how this energy input is redistributed into the turbulent fluxes of sensible and latent heat, the heating of the soil, and the melting of snow. The relative importance of the sensible and latent heat fluxes depends on the properties of the underlying surface and the local atmospheric conditions. Over dry soil, evaporation plays a minor role and the net radiative flux is primarily converted into turbulent fluxes of sensible heat and the heating or cooling of the soil. In contrast, over wet soils and the sea, evaporation and transpiration can dominate the turbulent energy flux by transporting the latent (condensational) heat. As with sensible energy fluxes, its effect is to cool the surface (by evaporation) and ultimately heat the atmosphere (by condensation).

Mountains are instrumental in affecting the shortwave radiation budget. Analyses of clear-sky mountain observatories' data in the Alps show that

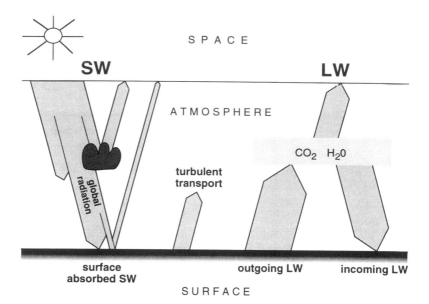

Figure 2.12 The atmosphere's global energy balance and its major contributions: shortwave (visible) radiative fluxes, longwave (infrared) radiative fluxes, and turbulent atmospheric transport of sensible and latent heat.

the global radiation increases in tandem with the altitude by approximately 0.8 Wm^{-2} per 100 meters (Müller 1984, Barry 1992). This effect is related to the overlying atmosphere's optical depth, which determines the amount of shortwave absorption and scattering. In the climatological mean (which includes clear-sky and overcast conditions), the vertical gradient of the global radiation is less firmly established. Data from sites at different altitudes in Switzerland (Ohmura et al. 1990) indicate an increase of insolation with altitude during winter and a decrease during summer. The decrease in summer is related to the shading associated with topographically generated convective clouds, whereas in winter lowland areas are frequently beneath a stratiform cloud cover. Observations of global radiation furthermore indicate that sites on the Alps' southern slope receive substantially more insolation than those on the northern slope (see figure 2.13), an effect associated with the smaller mean cloud amounts to the south of the Alps.

The albedo of the underlying surface controls the absorption of incoming global radiation. The persistence of snow covers in elevated areas substantially increases the albedo and leads to increased reflection and reduced absorption of global radiation (Barry 1992). On the scale of individual valleys, the valley slopes' orientation with respect to the sun as well as shading effects by neighboring mountains also control the absorption of shortwave radiation. Over complex terrain, the resulting surface temperature contrast drives a range of mountain wind systems (see section 2.3.2.4).

The incoming longwave radiation emitted from the atmosphere and the outgoing longwave emission from the surface determine the longwave-

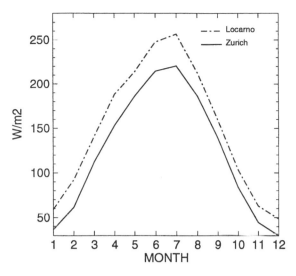

Figure 2.13 Mean annual cycles of global radiation at Zurich (north of the Alps) and Locarno (south of the Alps) in W/m². (From Wild et al. 1995, based on data from Ohmura, Gilgen, and Wild 1989.)

radiation balance at the surface. Both the incoming and outgoing fluxes decrease with altitude because of the decreasing water vapor content of the overlying air, and the decreasing soil temperature, respectively. The observed vertical gradient of incoming longwave radiation at different altitudes in the Alps is estimated to be close to -3 Wm^{-2} per 100 meters, whereas the vertical gradient of outgoing longwave radiation is estimated to be close to -2 Wm^{-2} per 100 meters. This yields a slight increase of the net longwave cooling with altitude.

The sum of the absorbed short- and longwave radiative exchanges between the surface and the atmosphere form the surface net radiation. This quantity determines the amount of energy available for the nonradiative components of the surface energy balance. The mean net radiation tends to decrease with altitude because of the longer duration of snow cover and the increase in net longwave cooling (Barry 1992). Accordingly, turbulent fluxes transfer less energy from the surface into the atmosphere at higher elevations. Quantitative estimates of the reduction of evapotranspiration with height show a large spread from -7 to -36 mm y^{-1} per 100 meters, as reviewed by Lang (1981).

Time series that allow an estimation of the temporal variability of the components of the surface energy balance over many years are very limited and largely restricted to global radiation. Figure 2.14 shows time series of annual and seasonal global radiation for the region of Zurich (average of three stations). In this area, the global radiation decreased from 1959 to the end of the 1970s by as much as 20 percent, and tended to recover thereafter, a trend found in winter and summer as well as in the annual mean. A similar trend was detected at other monitoring stations in the interior of Europe (including Salzburg and Potsdam). A substantial number of these decreases

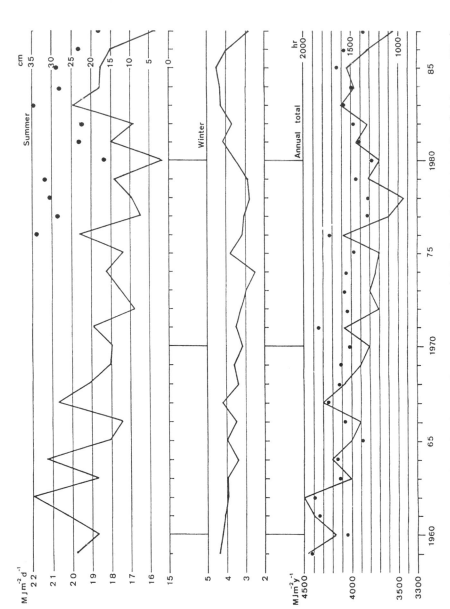

Figure 2.14 Time series of annual and seasonal (summer, winter) global radiation for Zurich (1 MJ m^{-2} d^{-1} = 11.6 W/m^2; 1000 MJ m^{-2} year^{-1} = 31.7 W/m^2). Dots in the upper right are summer evapotranspiration. Dots in the bottom parel are annual total sunshine duration for Zurich. (From Ohmura and Lang 1989.)

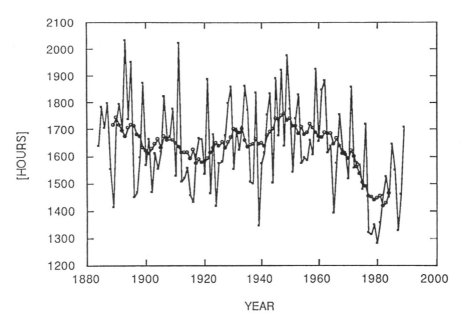

Figure 2.15 Annual total sunshine duration hours for 1884–1985 with the eleven-year running means for Zurich. (From Ohmura and Lang 1989.)

resulted from changes in cloud cover. Global radiation's variability significantly affects terrestrial processes and is considered the primary reason for the fluctuations in the time series of evapotranspiration measured at the experimental hydrological basin Rietholzbach, 30 kilometers east of Zurich (see figure 2.14).

The monthly global radiation correlates well with the monthly sunshine duration, as can be seen from figure 2.14. The analysis of the global radiation's variability can thus be extended back to the time when sunshine monitoring was initiated. Figure 2.15 shows such a 102-year time series of annual total sunshine duration measured at Zurich. A power spectral analysis of this time series (see Ohmura and Lang 1989) suggests that the global radiation has changed with a periodicity of twelve years and an overlying decreasing trend.

Changes in the radiative fluxes are an important aspect of climatic changes. The most direct effect of an increased concentration of carbon dioxide in the atmosphere is an increase of the longwave atmospheric radiation directed to the surface. It is a fundamental question in climate change studies how this additional energy will be redistributed within the components of the surface energy balance.

2.3 CLIMATE PROCESSES: WHAT DETERMINES THE ALPINE CLIMATE?

The climate experienced in the Alpine region is linked to a long chain of processes that operate on a wide range of spatial scales. On the global scale, the incident solar radiation and the atmosphere's composition are the main contributors to setting the global mean temperature and establishing the (seasonally) varying pole-to-equator temperature differences. Again, the longitudinal variations in the planet's distribution of oceans and continents and the presence of major mountain ranges contribute to establishing the atmosphere's continental-scale time mean atmospheric circulation pattern. Within this large-scale pattern, smaller (synoptic-scale), transient mid-latitude weather systems form, propagate, and decay, and to a great measure the passage of these systems—fronts, cyclones and anticyclones—determines the day-to-day weather variations in the extratropics. In turn, Alpine climate is a register of the ensemble effect of the synoptic-scale weather systems, including their orographic modification.

Four aspects of the foregoing chain of spatial scales merit further comment. First, the influence is not merely down-scale. For example, the genesis, track, and strength of the synoptic-scale weather systems not only is strongly dependent upon, but also significantly influences the pole-to-equator thermal difference and the large-scale circulation pattern. Second, each of these scales shows significant temporal variability. For example, the planetary waves exhibit considerable fluctuations with respect to the long-term time mean and thereby induce important anomalies in the occurrence of the different weather types (*Grosswetterlagen*) on time periods of between a week and a season. Third, any change in the global climate setting is also transmitted along this chain. In particular, because Alpine climate is strongly influenced by the eastward passage of fronts and cyclones from their genesis region in the Atlantic, changes in the strength, frequency, location and track length of Atlantic cyclones substantially influence the region's climate variations. Fourth, phenomena resulting from the Alps' modifications of weather systems have considerable influence at large distances downstream along the (variable) storm track, affecting at times practically the entire eastern Mediterranean region. A program to establish the causal connection between global and alpine climate change therefore needs to examine the possible multiple linkage(s) in the aforementioned chain.

In this section we consider, in the context of climate, the form of the large- and synoptic-scale circulation patterns in the European sector and some of their principal variations; the dynamics of the Alpine modification of the incident flow; and the form of the subalpine-scale systems.

2.3.1 Large-Scale Setting and Variability

2.3.1.1 The Large-Scale Mean Flow and the Atlantic Storm Track Figure 2.16 shows the winter and summer seasonal mean sea-level pressure (SLP) pattern and depicts a measure of the day-to-day variability of the mid-tropospheric flow in the east Atlantic–European sector. In winter, (see panel (a) of the figure) the Alps are at the center of the deformation-like pattern formed by the Icelandic Low to the northwest, the Azores High to the

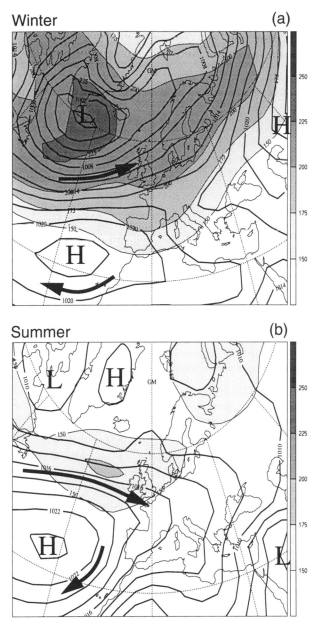

Figure 2.16 The time mean sea-level pressure pattern (contour interval 2 hPa) over the eastern Atlantic and the European sector in (a) winter and (b) summer, shown by the contours. The shading depicts a measure of the day-to-day variability of the upper tropospheric flow (the standard deviation of the 250 hPa surface in meters). (Based on data from Hoskins et al. 1989.)

southwest, the cold continental Siberian High to the east, and a weak Mediterranean Low feature to the south. In effect, the Alps are embedded in one of the slackest pressure gradient fields of the extratropical northern hemisphere. In summer (panel (b) of the figure), the Azores High extends northeastward to become the major pressure influence in the vicinity of the Alps.

The shaded patterns in figure 2.16 indicate the mean storm track. In winter, the majority of cyclones that influence Europe traverse the western Atlantic into northeastern Europe. Thus, the Alps are located south of the tail end of this storm track, and for the most part the weather is linked at the surface only to the cyclones' southernmost frontal features. The southward passage of these fronts around the Alpine massif itself is often a precursor of cyclogenesis to the lee, and thereby the central Mediterranean marks the entrance region of a weaker secondary storm track (Whittaker and Horn 1984; Buzzi and Tosi 1989). These cyclones' low-level orographic initiation results in their vertical structure's differing radically from that they exhibit elsewhere (von Ficker 1920; Tibaldi, Buzzi, and Speranza 1990). In summer, the Alps are further removed from the weaker and shortened Atlantic storm track, and in consequence fronts and their associated rain bands pass less frequently, and lee cyclogenesis is less frequent. Note also that changes in the incident fronts' frequency and strength would influence Alpine lee cyclogenesis and thereby exert a significant influence downstream affecting the entire Mediterranean region of southeastern Europe.

2.3.1.2 Variations on the Large-Scale Pattern

North Atlantic Patterns The North Atlantic Oscillation (NAO) relates to changes in the strength and location of the Icelandic Low and the Azores High on interannual and multidecadal time scales. One of the NAO's central features is that the changes in the two systems' intensity tend to be anticorrelated. A crude, one-parameter index for the change is the normalized mean surface pressure difference between Ponta Delgadas in the Azores and Akureyri in Iceland (see Rogers 1984), and its time series for the winter period shows an amplitude increase from 1890 to 1920, a subsequent decrease to around 1963, and thereafter a strong positive trend (see figure 2.17). Likewise, the Alpine surface pressure signal and that for the Icelandic region show a distinct inverse correspondence (see, e.g., Exner 1913, 1924).

Statistical studies examining the spatial structure of the anomalies in the large-scale sea-level pressure distribution indicate that the wintertime NAO signal is one of the two most recurrent low-frequency teleconnections—simultaneous variations at geographically separate regions—in the Northern Hemisphere. The NAO's north-south dipolar signal is present in the seasonal mean SLP at all seasons (Glowienka-Hense 1990; Rogers 1990) but it undergoes a seasonal variation in intensity, phase, and alignment. Other patterns identified applying the empirical orthogonal functions (EOF) technique to the surface pressure signal of the Atlantic-European sector include an east-

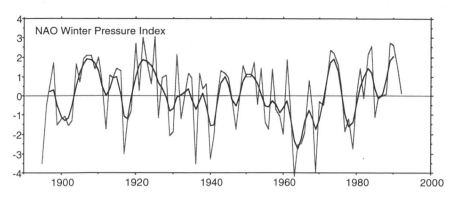

Figure 2.17 Winter index of the North Atlantic Oscillation (1895–1992) based on the mean normalized pressure difference between Ponta Delgadas, the Azores, and Akureyri, Iceland. The solid line represents smoothed data using a 1-4-6-4-1 low-pass filter. (Based on Rogers 1984; Bresch 1995; and data from Hoskins et al. 1989.)

west dipole and a dipole with centers located east of Iceland and in southern Europe (Rogers 1990). The NAO signal's amplitude in the Alpine region is rather weak, whereas those of the other two patterns are, respectively, about 4hPa and 8 hPa. These latter two patterns also resemble those derived using data for the eastern Atlantic and Europe (Glowienka-Hense 1990; Hagen and Schmager 1991; Fraedrich, Bantzer, and Burckardt 1993).

Consider now the climate imprint of the weather associated with the prevalence of these three surface signatures. For the NAO, it has been shown that high index values, and hence strong mean surface westerlies in the central Atlantic, tend to be accompanied by storm tracks extending further into northern Europe and cold (warm) thermal anomalies over Greenland (Scandinavia) and vice versa. Again an extreme, but not uncommon, flow setting is a strong negative anomaly of the NAO index associated with a persistent (over, say, a few weeks) blocking event in the North Atlantic resulting in a reversal of the latitudinal surface pressure gradient and an anomalous north-south high-low dipole. In winter, notable below normal temperatures in Scandinavia and enhanced precipitation over the Mediterranean usually accompany such an event (cf. Rex 1950, 1951; Moses et al. 1987; Lamb and Peppler 1987; Hurrell 1995). In contrast, the persistence of the other two patterns is not necessarily reflected strongly in the NAO index and yet can exert a significant impact on the in situ weather. Positive values of these signatures combine to yield stronger southwesterlies over northeast Europe, higher pressure over central Europe, a concomitant northward shift of the cyclone track, and a prediction of drier conditions. Contrary negative values would herald more westerly wind regimes penetrating into central Europe, a higher frequency of cyclone and frontal passages, and increased precipitation.

In terms of their cause and dynamics these variations of the atmospheric SLP signal have been linked to changes in sea surface temperature (SST) distribution of the North Atlantic (Deser and Blackmon 1993; Kushnir 1994).

On the interannual scale, the SST anomaly patterns are consistent with wind-driven atmospheric forcing. There is also some evidence, however, that SST anomalies in the western North Atlantic influence European weather on monthly timescales (Ratcliffe and Murray 1970; Palmer and Sun 1985). On decadal and longer timescales, there is a geographical and scale mismatch between the SST and SLP patterns. It has therefore been argued that the longer-term variations indicate a basinwide dynamical interaction between the atmosphere and ocean, with the ocean providing the principal forcing. On these long timescales, the SST signal has undergone a marked fluctuation during the last century, with negative anomalies in the 1920s, positive anomalies in the 1960s, and a return to colder conditions thereafter. The positive trend in the NAO index is consistent with the increase in the frequency of weather types with a southerly flow component, as noted in section 2.2.2. Again, on the associated timescales, an integral feature of the flow in the Atlantic is the overturning component, with dense water formation in the northern Atlantic being fed at the surface by water from the other oceanic basins and disgorging of the North Atlantic deep water to those basins at depth (see, e.g., Held 1993).

Clearly there is a complex chain of linkages to consider in this case. In effect the sequence takes the form global climate change ⇔ the interocean basin exchange ⇔ the Atlantic SST pattern ⇔ the atmospheric mean pattern in the North Atlantic ⇔ the characteristics of the in situ storm track ⇔ the frequency of the *Grosswetterlagen* ⇔ the response in the Alps. For instance, the positive trend of the winter NAO index noted above is consistent with the increased frequency of westerly circulation patterns (cf. section 2.2.2) and the increased frequency of warm and wet winters (cf. sections 2.2.1 and 2.2.3).

The El Niño/Southern Oscillation The El Niño/Southern Oscillation (ENSO) phenomenon is the dominant mode of interannual variability of the tropical ocean–atmosphere system. It is linked with anomalous SST patterns over a substantial portion of the tropical Pacific and with a shift in the main region of deep cloud-diabatic heating of the tropical atmosphere. ENSO induces strong, well-defined SLP and SST signals and anomalous weather across a broad belt of the tropics (Ropelewski and Halpert 1987; Halpert and Ropelewski 1992), and in addition, a teleconnection pattern extends as an arc-shaped, quasi stationary wave train over North America that tapers out over the southeastern United States.

The Atlantic-European sector is somewhat removed from both the forcing region for the El Niño and the tail end of the aforementioned wave train. However, perturbations of the large-scale pattern in the genesis region of the Atlantic storm track could, albeit intricately, influence the track's downstream features and thereby European weather. The nature of the influence(s) is related to the characteristics of the cyclones comprising the storm track and can vary with the strength and location of the tropical SST anomaly, the intensity of the midlatitude westerlies, and the prevailing settings of the SST and NAO in the Atlantic. Thus detecting an ENSO-European link and determining its dynamical nature is a challenging task. Nevertheless, there are indications of such a link (Van Loon and Madden 1981; Hamilton 1988; Kiladis and Diaz 1989; Fraedrich 1990; Fraedrich and Müller 1992; Wilby 1993; see also the overviews of Palmer and Anderson 1994; Fraedrich 1994). The effects include the forcing of anomalous winter weather patterns; modification in the frequency of cyclonic and anticyclonic *Grosswetterlagen*; signatures in the surface pressure (~ 1 hPa), temperature ($\sim 0.2^0$K) and precipitation (~ 10 mm); and changes in the intensity and location of the Atlantic storm track.

2.3.2 Alpine Effects on Weather and Climate

Alpine topography influences the atmospheric circulation by deflecting the flow horizontally and vertically, by introducing elevated sources and sinks of sensible and latent heat, and by inducing waves that propagate into the free atmosphere. These processes are mediated on a wide range of scales. On the meso-α-scale (horizontal scales in the range 200–2,000 kilometers), they include the retardation and modification of approaching synoptic systems and the formation of lee cyclones. On the meso-β-scale (20–200 kilometers), many of the flow phenomena (such as Föhn and valley winds) are associated with the generation and propagation of so-called internal gravity waves— which in turn are associated with the restoring effects of the atmosphere's stable density stratification. On the meso-γ-scale (2–20 kilometers), topographic effects include stationary mountain lee waves, deep convective clouds, and planetary boundary layer effects.

This section discusses some of the Alpine effects by first giving consideration to archetypal flow configurations, then in turn to the processes on the synoptic scale, the Alpine scale, the valley scale, and finally to aspects of orographic precipitation.

2.3.2.1 Archetypal Flow Configurations Although the real atmosphere is rarely so simple, consider a steady and uniform flow incident on an isolated topographic obstacle. This idealized setting illustrates some key aspects of flow past topography. For mountains with horizontal scales in the range of a few to about fifty kilometers, effects associated with the atmosphere's stable stratification dominate the dynamical response. The stratification tends to inhibit vertical displacement and thereby reduces the air parcels' ability to flow over the mountain. This feature gives rise to two distinct flow regimes, which figure 2.18 depicts. In the first regime (panels (a) and (b)), the flow is over the obstacle, and a pattern of vertically propagating gravity waves is established (see Smith 1979; Durran 1990). Even at upper levels, air parcels experience vertical excursions over a depth that is comparable to the mountain height.

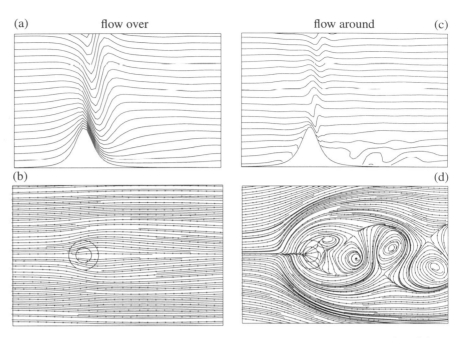

Figure 2.18 Nonrotating flow past an idealized isolated mountain. The flow is from left to right, and the figure shows both the flow-over regime (panels (a) and (b)) and the flow-around regime (panels (c) and (d)). Panels (a) and (c) show the instantaneous streamlines in a vertical section across the mountain top, and panels (b) and (d) show the same features on the surface. (From Schär and Durran 1997.)

If stratification effects become stronger, or if the upstream wind speed is smaller, the air parcels are unable to surmount the orography (panels (c) and (d) of figure 2.18). The flow splits upstream, and the air parcels make broad, quasi horizontal excursions as they tend to circumscribe the mountain rather than traverse it. For this configuration, there is still some gravity wave activity in the upper portion of the obstacle, but the vertical extent of the associated oscillations is significantly reduced compared to those of the flow-over regime. To the lee, a wake is established that can become unstable and undergo a transition toward vortex shedding (Schär and Smith 1993; Schär and Durran 1997). In the case of isolated topographic obstacles (such as mountainous islands), these effects can imprint on satellite pictures as spectacular Kármán vortex streets (Etling 1989). In the Alpine context, vortex shedding is not observed, but edge vortices have been identified during periods of strong flow, both in observations and numerical experiments (Steinacker 1984b; Thorpe, Volkert, and Heimann 1993; Aebischer and Schär 1998). There are also indications that these features can contribute to the formation of Alpine lee cyclones.

The transition between the two archetypal flow regimes shows some of the characteristics of a bifurcation (Smith and Grønås 1993; Smolarkiewicz and Rotunno 1989) and can result from small changes in the upstream conditions. Here the governing control parameter is related to the mountain height and geometry, the static stability, and the approaching airstream's speed. On the scale of the Alpine massif, the effects of the earth's rotation must in addition be accounted for. Numerical experiments have revealed that rotational effects relevant for large obstacles tend to favor the flow-over regime (Trüb and Davies 1995). The flow past the Alps is hence less likely to be around than the flow past a smaller-scale topographic feature of the same height. It then follows that the flow past the Alps can be over on the alpine scale but at the same time around on the scale of individual massifs and mountains.

An evaluation based on typical Alpine upstream parameters also indicates, in agreement with observational studies (Binder, Davies, and Horn 1989; Chen and Smith 1987), that both the flow regimes of figure 2.18 can occur on the Alpine scale. It is important to represent this feature in studying the Alpine climate, since flow-over and flow-around result in very different distributions of key climatic variables such as temperature and precipitation. In this regard, it should be noted that the implicit smoothing of the topography in low-resolution numerical models (such as global climate models) strongly shifts the response toward the flow-over regime.

2.3.2.2 Effects on Approaching Synoptic Systems and Lee Cyclogenesis The Alps are most effective in modifying the ambient circulation when the flow is roughly perpendicular to the main Alpine ridge, that is, from the north or south. These flow situations usually occur in conjunction with the approach of a trough or low-pressure system from the west or northwest, toward central Europe or Scandinavia, a configuration that is particularly common during autumn, winter, and spring (cf. section 2.3.1). Many of the approaching systems are close enough that their fronts come into direct contact with the Alps. When a frontal system impinges on the Alps, the dynamical processes can no longer be viewed as the modification of a quasi steady airstream, but rather represent a complicated three-dimensional and time-dependent flow evolution (for a concise review, see Egger and Hoinka 1992), which often induces transitions from flow-over to flow-around and vice versa. Changes in the wind speed and stratification of the incoming air mass trigger such transitions, consistent with the discussion of the archetypal flow regimes in the previous section.

Ahead of approaching cold fronts there is a strong southwesterly flow toward the Alps (see figure 2.19a). This airstream consists of warm, moist air, is comparatively weakly stratified, and has a high wind speed. It is thus able to pass over the Alpine ridge, leading to South Föhn in the north. In contrast, the cold air behind the front is stably stratified and often capped by a pronounced inversion. It is thus unable to climb the Alps but rather deflected laterally. In effect this leads to retardation and deformation of the incident low-level cold front, and the lateral deflection of the northerly flow induces Mistral and a pronounced cold-air outbreak to the west (figure 2.19b) and Bora to the east of the Alps (figure 2.19c). At the same time, a lee cyclone

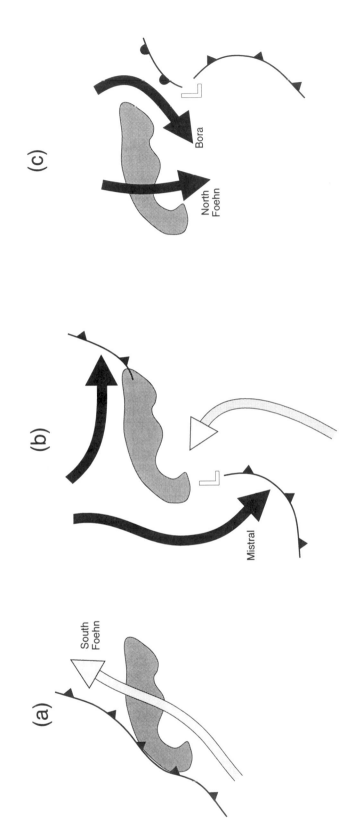

Figure 2.19 Schematics of the interception of a cold front by the Alps. The three phases correspond to (a) deformation of cold front and onset of South Föhn, (b) cold-air outbreak into the western Mediterranean (Mistral) and formation of lee cyclone, (c) eastward progression of lee cyclone and onset of Bora and North Föhn (adapted from Smith 1986.)

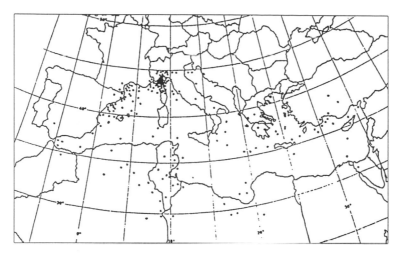

Figure 2.20 Climatological distribution of winter cyclogenesis in the Mediterranean region. (From Reiter 1975.)

may form over the Gulf of Genoa. Finally, as cold air piles up to the north, the associated cold-frontal inversion is lifted, and the cold air underneath spills over the mountain passes, inducing North Föhn. The circulation associated with the eastwards propagating lee cyclone also supports the formation of North Föhn and Bora winds.

Synoptic conditions are most suitable for lee cyclogenesis when a deep upper-level trough and its associated surface cold-front enter the Alpine region (Bleck and Mattocks 1984; Tafferner 1990; Tibaldi, Buzzi, and Speranza 1990). On the average, about thirty Alpine lee cyclones form per year, most frequently during the spring and the autumn. Figure 2.20 shows the climatological distribution of lee cyclogenesis. The concentration of events over the Gulf of Genoa near the southwestern tip of the Alps betokens an orographic influence. Once a lee cyclone has formed, it becomes an important governor of the regional weather and climate. Lee cyclones often attain maximum strength over northern Italy, and the associated strong southerly flow ahead of the cyclone can induce storm surges in the Adriatic sea (which occasionally threaten Venice) and advect moist Mediterranean air toward the Alps (which substantially contributes to the annual precipitation totals in northern Italy and the eastern Alps). Further downstream, lee cyclones propagate toward the east, following one of two major tracks (Whittaker and Horn 1984): One leads across the eastern Alps into eastern Europe, the other follows the northern border of the Mediterranean Sea. As a result, Alpine lee cyclones are crucial to the precipitation in much of the eastern Mediterranean region. Since lee cyclogenesis is sensitive to the meridional location of the midlatitude storm track (as well as to other factors), Mediterranean precipitation in the event of global climate change is thus difficult to assess.

2.3.2.3 Regional Wind Systems Several of the regional Alpine wind systems possess some characteristics of the archetypal flow configurations (figure 2.18) and occur transiently during episodes of frontal passage (figure 2.19). Key aspects of North and South Föhn (cf. Seibert 1990) are evident in the stream crossing the mountains. The densely packed isentropes in the vertical sections of figure 2.18a are indicative of the Föhn region, which is associated with low pressure and increased wind velocities. A lee-side warming results from latent heat release along the parcel's trajectories, and—presumably more importantly—from adiabatic warming through net descent if the parcels at the lee slopes stem from some higher level upstream. Downslope wind storms develop when the gravity wave response undergoes a nonlinear amplification involving the transition to supercritical-like flow (Smith 1985; Durran 1986), gravity-wave breaking (Clark and Peltier 1977), or both. The latter process is associated with the overturning of density surface, yields a configuration with dense fluid above lighter fluids, and induces clear air turbulence through the breakdown of this unstable configuration. The elongated shape of the Alps significantly facilitates the development of Föhn-like flows and makes them more two-dimensional both in cases of northerly (North Föhn) and southerly (South Föhn) flow. The upstream deceleration can then be interpreted as the blocking of the incident upstream flow (Pierrehumbert and Wyman 1985). The deep inner-alpine valleys experience particularly strong Föhn flows. To some extent this results from gap and channelling effects (cf. Wippermann 1984), but it is presumably also related to the reduced horizontal scale on the valley transects, which makes the flow more susceptible to gravity-wave effects as opposed to the effects of Earth rotation.

Mistral, Bora, and the Bise (for a description of these flows, see respectively Pettre 1982; Smith 1987; and Wanner and Furger 1990) can be associated with streams of air deflected horizontally by the Alpine massif (panels (c) and (d) of figure 2.18). In these cases, some of the stratification is often concentrated in a inversion layer that separates the cold air below (which flows around the Alps) from the potentially warmer air aloft (which flows over the Alps). Mistral and Bora usually occur when the cold air behind a cold-frontal passage is horizontally deflected around the western and eastern portions of the Alps, respectively (figure 2.19). The Bise results from the channeling of north-easterly flow between the Alps and the Jura mountains. The specifics of the local topography as well as other factors modify each of these wind systems. For instance, the Bora is on the one hand associated with a broad excursion of cold air around the Alps and yet flows itself over the northern portion of the Dinaric Alps and thereby induces high winds. The predominant flow configurations determine the major thermal anomalies of the Alpine climatology (see section 2.2.1). In the major Föhn valleys to the north of the Alps, South Föhn occurs on as many as 15 percent of the days in the month, with maximum frequency during the spring and autumn. During Föhns, the local surface air temperatures may increase by more than

10 K, and the combined effect of this Föhn warming on the longer-term local climatological mean is pronounced. The most significant factor on the regional scale, however, is related to the shielding of the lee against upstream influence. If the flow is around the Alps, it leaves a pool of almost stagnant air in its wake that greatly reduces the importance of advective effects. The prolonged residence time of air parcels in the Alpine wake then allows local boundary layer and radiative processes to determine the lower atmosphere's thermal structure. These processes, which are also affected by the orographically controlled distribution of cloud cover, lead to a pronounced climatological temperature contrast across the Alpine main crest and also make the often stagnant air mass in the Po valley susceptible to air pollution.

2.3.2.4 Thermally Driven Mountain Circulations In addition to influencing the structure and development of preexisting flow systems that impinge upon the Alpine topography, the terrain is also associated with and responsible for the in situ excitation of daily periodic circulations on scales varying from individual slopes and valleys to the massif itself.

Figure 2.21 sketches the daytime and nighttime patterns of the slope and valley circulations as typically observed in the Alpine region during clear-sky conditions (see, e.g., Urfer-Henneberger 1970; Oke 1987). Over individual slopes, the response to diurnal radiative variation takes the form of upslope winds during the day and downslope winds at night (see, for example, Defant 1949; Egger 1990). These slope wind circulations are accompanied by and coupled with low-level airflow up and down the valleys, referred to respectively as valley and mountain winds (see, e.g., Wagner 1938). The driving mechanism for these flows is directly related to the cycle of daytime solar heating and nighttime radiative cooling of near-surface air layers.

Again on the scale of the main Alpine ridge itself, there is some indication of a diurnal circulation between the ridge and the adjacent foreland regions (Kleinschmidt 1922; Burger and Eckhart 1937), with low-level inflow (drainage) during the day (night) and a weak return flow above crest height. A larger diurnal temperature cycle in the inner Alpine region as compared to that over the flatland (cf. section 2.2.1) and periodic, Alpine-confined pressure anomalies (Scherhag 1966; Hafner et al. 1987) during fair weather support the existence of such a circulation. These circulation systems have been attributed to two effects, either singly or in combination: The terrain provides an elevated heat source (sink) effect during the day (night) (see, for example, Flohn 1953), and the more rapid heating (cooling) of the inner-Alpine region associated with the lesser volume of air per unit horizontal area in the major Alpine valleys compared with the surrounding plain (the so-called volume effect; see, e.g., Steinacker 1984a; Whiteman 1990) enhances thermal response in the region.

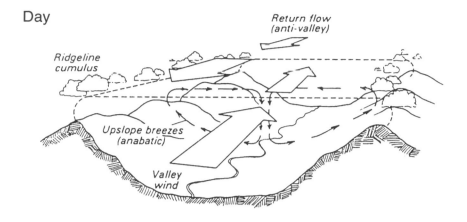

Day

Return flow
(anti-valley)

Ridgeline
cumulus

Upslope breezes
(anabatic)

Valley
wind

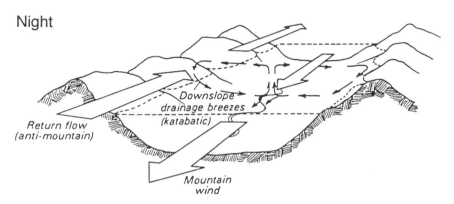

Night

Downslope
drainage breezes
(katabatic)

Return flow
(anti-mountain)

Mountain
wind

Figure 2.21 Schematics of the thermally driven circulations in a mountain valley during the day and at night. The view is up valley. (From Oke 1987.)

The mountain-foreland circulations can trigger or contribute to various flow phenomena characteristic of Alpine climate. For example, during the warm season, they contribute to establishing the preferred regions for convective activity and cloud formation (cf. Barry 1992; Banta 1990). Likewise in the cold season, the nighttime cold air drainage contributes to the buildup and maintenance of the persistent inversion layers and fog frequently experienced in the Swiss middle-land (see, for example, Wanner 1979).

2.3.2.5 Orographic Precipitation and Convection Seasonal or yearly precipitation totals generally tend to increase with height. Section 2.2.3 reviewed some of the corresponding literature for the Alpine region, and Smith (1979) and Kahlig (1986) offer comparative reviews covering a range of topographic obstacles. The precipitation-height relationship is in general far from linear, and hills as small as fifty meters can enhance precipitation by as much as 100 percent (Bergeron 1968; see also Browning 1980). The seeder-feeder mechanism (Browning 1985) is generally accepted to be of prime importance in this context. At low levels, topographic lifting leads to condensation and the formation of small, usually nonprecipitable water droplets. Appreciable precipitation can result when falling precipitation from aloft, induced either by orographic lifting or a preexisting synoptic-scale disturbance, collects these droplets. With the help of the seeder-feeder mechanism, the orographic enhancement of precipitation pertains to a deep layer (on the order of several kilometers) that can far exceed the mountain height.

The discussion of airflow regimes in section 2.3.2.1 suggests a significant sensitivity of the rainfall distribution and intensity to the predominant flow regime. Smith (1989) has documented such a sensitivity for the islands of Hawaii, but we are unaware of a corresponding systematic study for the Alpine region. If the flow is over the Alps, upslope precipitation occurs, and the low-level moisture is essentially extracted from the ambient flow. If the flow is instead around, low-level moisture is deflected laterally. Both flow regimes result in a shielding of the lee from precipitation, but the situation upstream is highly dependent upon the flow regime. Some of the case-to-case variability in upstream precipitation is likely a result of this effect. Precipitation signals for frontal passages along the Gotthard section (Phillips 1984) include both cases with light (for example, March 4, 1982, with precipitation totals of a few millimeters) and very heavy precipitation (such as April 29–30, 1982, with totals of about 100 millimeters), and the corresponding flows have been classified as around and over, respectively (Chen and Smith 1987).

An important aspect of Alpine precipitation is the comparatively frequent occurrence of widespread heavy and long-lasting precipitation and flooding events to the south of the Alps. Such events are most frequent during autumn, when there is an ample supply of warm, moist air over the Mediterranean Sea. The typical synoptic setting is associated with moist and warm southerly flow ahead of a cold front, deep trough, or lee cyclone. Since the incident air mass is weakly stratified, this southerly airstream is able to flow over the main alpine crest (cf. figure 2.19a). To the north, Föhn-like effects result, whereas lifting and adiabatic cooling to the south result in condensation and heavy precipitation. If such a synoptic situation persists for several days, devastating flooding can result. The improved understanding and forecasting of heavy precipitation in the Alpine region is currently one of the primary objectives of a large-scale international field campaign (cf. Binder and Schär 1996), and an assessment of the associated flooding potential is a vital aspect of climate change research in the Alpine region. Although some promising numerical results are available at intermediate computational resolution (e.g., Buzzi et al. 1995; Binder and Rossa 1995), the detailed simulation of flooding episodes might require very high horizontal and vertical resolution, because the precipitation processes might rely on the interaction of the topographic gravity-wave signal with the embedded convection.

The discussion above pertains mostly to stably stratified weather regimes. Even these regimes often have embedded convection, related either to frontal activity or to the generation of instability through ascent (see Smith 1979). The horizontal scale of convective updraughts is only a few kilometers and eschews the laminar description of the flow as given in section 2.3.2.1. Moist convection is particularly important in summer, when it provides the dominant contribution to the rainfall totals. The Alps can influence the flow in such situations either by inducing low-level convergence of moist air or by conditioning the local profiles of temperature, humidity, and wind. The latter effects include processes associated with the Alpine thermal anomaly as well as wind channeling effects along the Alpine foothills (cf. Binder, Davies, and Horn 1989; Houze et al. 1993). Though the Alps' thermal anomaly can itself be instrumental in inducing convection (e.g., air mass thunderstorms), most severe weather occurs when an active triggering takes place, usually in conjunction with an approaching synoptic system. Observational evidence for this type of triggering is available for convective systems both to the south (Buzzi and Alberoni 1992; Cacciamani, et al. 1995) and north of the Alps (Schiesser, Houze, and Huntrieser 1995; Huntrieser et al. 1997). The associated synoptic systems are important in both providing moisture supply and inducing ascent.

Precipitation in the Alpine region is a particularly important climate parameter, and not only for its relevance to ecological and economical systems. Because precipitation results from a chain of complex and highly nonlinear processes, it is very sensitive to external parameters and to the large-scale flow. Global warming could affect substantially the frequency and distribution of precipitation. Even a modest warming will significantly increase the air's potential for transporting water vapor. Current GCM numerical simulations of the $2 \times CO_2$ climate indeed suggest the global concentration of water vapor has increased on average by about 10 to 30 percent (Houghton, Jenkins, and Ephraums 1990), and regional climate simulations indicate that this could significantly increase precipitation, particularly in mountainous and coastal regions and during the synoptically active seasons (Schär et al. 1996; Frei et al. 1997).

2.3.3 Feedbacks to the Larger-Scale Flow

Several of the mesoscale processes referred to in the previous sections can feed back through nonlinear processes to the larger-scale flow and thereby affect the atmosphere's synoptic and finally planetary-scale circulation. Little is known about the Alps' relative contribution, but there can be no doubt that the earth's topography as a whole is key in defining the climatic zones' geographic distribution. Charney and Eliassen (1949) mooted that the stationary planetary-scale waves are "anchored" to the major topographic obstacles, but the extent to which topography controls this process has only recently become evident from general circulation numerical experiments. Figure 2.22 reproduces a spectacular example from such an experiment by Broccoli and Manabe (1992). The figure shows the results of a pair of global climate simulations. In both experiments, the geographical distribution of land and sea is prescribed, but the topography has been removed in the simulation shown in the right-hand panels. The bottom panels depict the simulated geographical distribution of wet and dry climatic zones on the Northern Hemisphere spring. The local effects of topography comprise precipitation anomalies (e.g., in the Alpine region) and shadow effects (which may be inaccurately represented because of the employed model's low resolution). In addition, there are remarkable effects far removed from mountains. For instance, large parts of Siberia are classified as dry in the topography run (in agreement with the observed climatology) but would experience substantially larger rainfall amounts in the absence of topography. The top panels in figure 2.22 imply that the control topography exerts is indeed through planetary-scale standing wave patterns and their embedded storm tracks.

The results of Broccoli and Manabe, which are supported by other numerical experiments, entail substantial up scale contributions, meaning that the response's horizontal scale significantly exceeds that of the topography. Such effects rely on the generation of flow anomalies through irreversible processes that can in turn quasi permanently alter the downstream conditions. The overall scale of the response is then governed by advective and dissipative processes downstream rather than by the scale of the topography itself. Dynamical instabilities, in the atmosphere primarily due to barotropic and baroclinic processes, can in addition amplify and affect the structure of orographic wakes.

There are several orographic mesoscale processes of the above category, including the dissipation of internal gravity waves at upper levels (gravity-wave drag), topographically induced condensation (orographic precipitation), processes associated with flow splitting and surface friction (formation of shear lines and vortices to the lee), and the formation and subsequent decay of an orographic cyclone (lee cyclogenesis). General circulation models cannot resolve several of these processes, which must be parameterized instead. The most important example is gravity-wave drag, considered an absolutely essential contributor to the planetary-scale circulation (Davies 1986; Palmer, Shutts, and Swinbank 1986; McFarlane 1987) as well as one of the key factors in forcing the planetary-scale quasi-stationary waves (figure 2.22, top diagrams). These waves not only control the distribution of climatic zones but are in addition highly sensitive and responsible for a considerable portion of the interseasonal and interannual variability in the Northern Hemisphere (e.g., Wallace, Zhang, and Lau 1993; Lau, Sheu, and Kang 1994). Currently the limited knowledge about these waves' behavior in a changed climate is of some concern (cf. Held 1993).

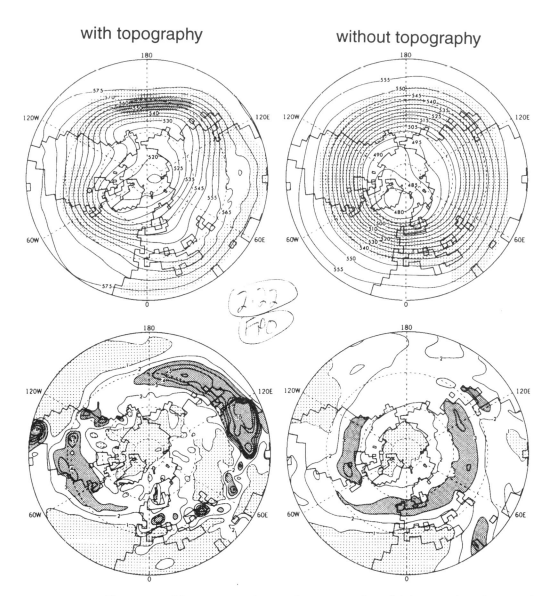

Figure 2.22 Mean spring circulation and precipitation from global numerical simulations with topography (left) and without topography (right). Top panels show the 500 hPa geopotential height in decameters. Light and dense stippling indicates winds greater than 12 ms^{-1} and greater than 24 ms^{-1}, respectively. Bottom panels show precipitation rates [mm d^{-1}]. Contours are given at 1, 2, 3, 4, 5, 6, 8, 10, 15, 20 and 30 mm d^{-1}. Light stippling indicates dry regions with precipitation less than 1 mm d^{-1}; dense stippling indicates wet regions with precipitation greater than 3 mm d^{-1}. (From Broccoli and Manabe 1992.)

2.4   FURTHER REMARKS

Details of the present-day climate were summarized using the available observational data and discussed in terms of the underlying dynamical processes. This approach highlighted the Alps' influence on the region's distinctive weather phenomena and climate. It also served to pinpoint both the incompleteness of our climatological database and the shortfalls in our understanding. From the observational standpoint, there is a pressing need to develop from the data already existing more comprehensive, accurate, homogenized, Alpine-wide data sets of the basic climatic elements. Likewise, the spatial structure and temporal development of many key Alpine-related flow systems have yet to be described adequately, and rectifying this will require specifically designed field experiments. From the standpoint of understanding it was emphasized that the geometry and height of the Alps rendered the dynamics of the orographically induced phenomena complex, nonlinear, and often multiscaled. This makes their study particularly challenging, but nevertheless recent theoretical developments and the availability of powerful computers has opened the way for significant progress.

The understanding of current Alpine climate forms a sound foundation for studies both of the Alpine region's paleoclimatic history and its fate in the event of a global climate change. Subsequent chapters will consider these aspects.

GLOSSARY

This section is partly based on the definitions given in the glossary produced by the United Kingdom Meteorological Office (Meteorological Office 1991).

adiabatic: An adiabatic (thermodynamic) process is one in which heat does not enter or leave the system. The adiabatic transport of an air parcel can nevertheless lead to temperature changes as a result of expansion and compression. An example of a nonadiabatic (diabatic) atmospheric process is latent heating through condensation of water vapor.

advection: Transfer of mass (or of an air mass property) by horizontal or vertical winds.

albedo: The fraction of incoming solar radiation reflected by the Earth's surface or the cloud cover.

baroclinic: An atmosphere that entails quasi-horizontal temperature gradients is said to be baroclinic. A process that depends on this property is also said to be baroclinic.

Bise: A cold, dry northeasterly wind to the north of the Alps.

blocking: In a blocked midlatitude westerly flow, a quasi-stationary high-pressure system interrupts cyclones' usual eastward progression. In a blocked circulation, the upper-level westerly flow is split into two branches.

Bora: A cold northeasterly wind that blows down from the Dinaric Alps onto the eastern Adriatic coast.

climatological mean: The average distribution of some climatic element (for example, temperature) over a climatological period (say, thirty years).

convection: Vertical heat transfer by rising warm and sinking cold air. Convection in the atmosphere often occurs in the form of moist convection, which involves condensation and convective cloud formation in the ascending branches.

downscaling: To infer the regional-scale climate from larger-scale information. The term is often used in conjunction with climate scenarios, where it relates to the construction of regional or local climate scenarios from larger-scale information.

empirical orthogonal function (EOF): The variability of some field can always be expanded as a superposition of orthogonal base functions. Empirical orthogonal functions are a special type of orthogonal base functions objectively constructed so as to describe the observed variability with as few base functions as possible.

Föhn: A warm, dry wind to the lee of a mountain range. The term originated from the Alpine South Föhn but is used as a general term of this type of winds.

gradient: The slope of a function or field. In the case of a topographic field, the gradient always points in the direction of the steepest uphill slope. In general, the gradient is a vector.

gravity wave: A type of wave that depends for its existence on the restoring force buoyancy provides. In a stably stratified fluid such as the atmosphere, the vertical displacement of fluid parcels (e.g., by topographic lifting) generates internal gravity waves.

inviscid: Not affected by viscosity.

isentropes: Surfaces of constant potential temperature. The adiabatic motion of an air parcel always follows isentropic surfaces.

Kelvin [K]: A temperature scale that starts at absolute zero ($-273.15\,°C$). To convert temperatures from degrees Celsius to Kelvin, add 273.15.

latent heat: The quantity of heat absorbed or emitted during a change of state. In the atmosphere, latent heating associated with condensation (evaporation) and freezing (melting) of water is important for cloud formation and for the transport of energy.

lee cyclogenesis: The formation of a cyclone (lee cyclone) to the lee of a large-scale mountain. Lee cyclogenesis to the lee of the Alps is often in the region of the Gulf of Genoa.

low-pass filter: A mathematical operation (or electronic circuit) that filters a time series so as to remove its high-frequency components.

mesoscale: The horizontal scale appropriate to atmospheric systems of a size between that of individual cumulus clouds (approximately 2 kilometers) on the one hand and that of major depressions and anticyclones (approximately 2,000 kilometers) on the other.

Mistral: A northerly wind that blows along the lower Rhone valley and offshore along the Mediterranean coast.

North Föhn: A dry northerly wind to the south of the Alps.

orography: A term used in meteorology to signify the variation of the height of the ground above sea level. It is synonymous with topographic height.

planetary scale: The scale appropriate to atmospheric features with horizontal scales comparable to that of the planet (several thousand kilometers).

planetary waves: Meandering waves of the midlatitude westerly flow with horizontal wave lengths typically of about 5,000 kilometers.

potential temperature: The temperature an air parcel would possess if it were adiabatically compressed (expanded) to the standard pressure, 1,000 hPa. Unlike temperature, potential temperature increases with height in most of the atmosphere.

storm track: A region with a high frequency of cyclonic activity, located along the typical tracks of low-pressure systems. The major storm tracks of the Northern Hemisphere are the Atlantic and Pacific storm tracks.

stratification: In an unstratified fluid (neutral stratification) the density is constant. In a stratified fluid (stable stratification), the density decreases with height. Most layers of the atmosphere are stably stratified.

stratosphere: The region of the atmosphere above the troposphere and below the mesosphere. It extends from the tropopause (approximately eleven kilometers in midlatitudes) to the stratopause (approximately fifty kilometers).

synoptic scale: The scale appropriate to atmospheric features related to low- and high-pressure systems (approximately 1,000 to 4,000 kilometers).

teleconnection: A connection (relation) between climate variations or anomalies at locations separated by distances greatly in excess of the normal synoptic scale.

troposphere: The lower region of the atmosphere, extending to about sixteen kilometers near the equator, eleven kilometers in midlatitudes, and nine kilometers near the pole. The troposphere is capped by the tropopause. Most clouds are located within the troposphere.

trough: A trough (of low pressure) is a pressure feature in the synoptic weather chart. It is characterized by isobars shaped like a trough.

Universal Time (UT): Western European time zone, formerly Greenwich mean time (GMT).

REFERENCES

Abegg, B., and R. Froesch. 1994. Climate change and winter tourism. In *Mountain Environments in Changing Climates*, ed. M. Beniston, 328–40. London: Routledge.

Aebischer, U., and C. Schär. 1998. Low-Level Potential Vorticity and Cyclogenesis to the Lee of the Alps. *Journal of the Atmospheric Sciences* 55:186–207.

Aellen, M. 1994. Die Gletscher der Schweizer Alpen im Jahr 1992–93. *Die Alpen. Zeitschrift des Schweizer Alpen-Clubs* 70:207–27.

Agassiz, L. 1840. *Etudes sur les glaciers*. Neuchatel: Private papers.

Auer, I., R. Böhm, and H. Mohnl. 1993. Climatic change on Sonnblick—A multielemental approach to describe climatic change using a centennial data set. In *Proceedings of the Eighth Conference on Applied Climatology*, American Meteorological Society, 249–51.

Banta, R. M. 1990. The role of mountain flows in making clouds. In *Atmospheric Processes over Complex Terrain*, Meteorological Monographs No. 23, ed. B. Blumen, 229–83. Boston: American Meteorological Society.

Bardossy, A., and H. J. Caspary. 1990. Detection of climate change in Europe by analysing European atmospheric circulation patterns from 1881–1989. *Theoretical and Applied Climatology* 42:155–67.

Barry, R. G. 1992. *Mountain Weather and Climate*. 2d ed. London: Routledge.

Baumgartner, A., E. Reichel, and G. Weber. 1983. *Der Wasserhaushalt der Alpen*. München: Oldenburg.

Beniston, M., M. Rebetez, F. Giorgi, and M. R. Marinucci. 1994. An analysis of regional climate change in Switzerland. *Theoretical and Applied Climatology* 49:135–59.

Bergeron, T. 1968. *Studies of the Orogenic Effect in the Areal Fine Structure of Rainfall Distribution*. Report No. 6, Meteorological Institute Uppsala.

Binder, P., and A. Rossa. 1995. The piedmont flood: operational prediction by the Swiss model. *MAP Newsletter* 2:12–16.

Binder, P., and C. Schär, eds. 1996. *The Mesoscale Alpine Programme: Design Proposal*. 2nd ed. Swiss Meteorological Institute, MAP Programme Office; CH - 8044 Zurich.

Binder, P., H. C. Davies, and J. Horn. 1989. Free atmosphere kinematics above the northern Alpine foreland during ALPEX-SOP. *Contributions to Atmospheric Physics* 62:30–45.

Bleck, R., and C. Mattocks. 1984. A preliminary analysis of the role of potential vorticity in Alpine lee cyclogenesis. *Contributions to Atmospheric Physics* 57:357–68

Blumer, F. P., and R. Spiess. 1990. "Investigations on the altitudinal dependence of precipitation in the Swiss Alps." Proceedings of the 21st international meeting on Alpine Meteorology (ITAM), Sept. 17–21, 1990, Engelberg, Switzerland, 415–418.

Böhm, R. 1993. Air temperature fluctuations in Austria 1775–1991: A contribution to the greenhouse warming discussion." In *Proceedings of the Eighth Conference on Applied Climatology*, American Meteorological Society, J26–J30.

Böhm, R. 1992. *Lufttemperaturschwankungen in Österreich seit 1775*. Österreichische Beiträge zu Meteorologie und Geophysik, no. 5. Wien: Zentralanstalt für Meteorologie und Geodynamik.

Bonelli, P., and R. Pelosini. 1992. Intense precipitation on the italian side of the alps: analysis and forecasting. In *Proceedings of the 22nd International Conference on Alpine Meteorology*, Toulouse, France, a1–a9.

Bresch, D. 1995. Die mittlere Zirkulation der Nordhemisphäre und deren Variabilität. Diplomarbeit, Atmosphärenphysik ETH, 8093 Zurich.

Broccoli, A. J., and S. Manabe. 1992. The effects of orography on midlatitude Northern Hemispheric dry climates. *Journal of Climate* 5:1181–1201.

Browning, K. A. 1985. Conceptual models of precipitation systems. *ESA Journal* 9:157–80.

Browning, K. A. 1980. *Structure, Mechanism and Prediction of Orographically-Enhanced Rain in Britain*. Global Atmospheric Research Programme Publication Series No. 23.

Burger, A., and E. Eckhart. 1937. Über die tägliche Zirkulation der Atmosphäre im Bereich der Alpen. *Gerlands Beiträge zur Geophysik* 49:341–67.

Buzzi, A., and P. P. Alberoni. 1992. Analysis and numerical modelling of a frontal passage associated with thunderstorm development over the Po Valley and the Adriatic Sea. *Meteorology and Atmospheric Physics* 48:205–24.

Buzzi, A., and Tosi, E. 1989. Statistical behaviour of transient eddies near mountains and implications for theories of lee cyclogenesis. *Journal of the Atmospheric Sciences* 46:1233–49.

Buzzi, A., N. Tartaglione, C. Cacciamani, T. Paccagnella, and P. Patruno. 1995. Preliminary meteorological analysis of the Piedmont flood of November 1994. *MAP Newsletter* 2:2–6.

Cacciamani, C., F. Battaglia, P. Patruno, L. Pomi, A. Selvini, and S. Tibaldi. 1995. A climatological study of thunderstorm activity in the Po Valley. *Theoretical and Applied Climatology* 50:185–203.

Cacciamani, C., S. Nanni, and S. Tibaldi. 1995. Mesoclimatology of winter temperature and precipitation in the Po Valley of northern Italy. *International Journal of Climatology* 14:777–814.

Charney, J. G., and A. Eliassen. 1949. "A numerical method for predicting the perturbations of the middle latitude westerlies." *Tellus* 1:38–54.

Chen, W.-D., and R. B. Smith. 1987. Blocking and deflection of airflow by the Alps. *Monthly Weather Review* 115:2578–97.

Clark, T. L., and W. R. Peltier. 1977. On the evolution and stability of finite amplitude mountain waves. *Journal of the Atmospheric Sciences* 34:1715–30.

Davies, H. C. 1986. Observational studies and interpretation of the mountain pressure drag during ALPEX. In *Proceedings ECMWF Seminar/Workshop on Observation, Theory and Modelling of Orographic Effects*, Reading, Great Britain, September 9–13 1986, 113–36.

Defant, F. 1949. Zur Theorie der Hangwinde nebst Bemerkungen zur Theorie der Berg- und Talwinde. *Archives for Meteorology, Geophysics and Bioclimatology* A1:421–50.

Denhard, M., and C.-D. Schönwiese. 1992. Non-parametric trend statistics and rank correlations of long European sea level pressure time series. In *Proceedings 5th International Meeting on Statistical Climatology*, edited by Environment Canada, 575–8.

De Quervain, M., and R. Meister. 1987. Fifty years of snow profiles on the Weissfluhjoch and relations to the surrounding avalanche activity (1936/37–1985/86). In *Proceedings of the IAHS Symposium, Davos, 1986*. IAHS Publ. 162, 161–81.

Deser, C., and M. Blackmon. 1993. Surface climate variations over the North Atlantic during winter, 1900–1989. *Journal of Climate* 6:1743–53.

Direction de la Météorologie Nationale. 1988. *Précipitation en France*. Direction de la Météorologie National, Service central d'éxploitation.

Direction de la Météorologie Nationale. 1989. *Atlas Climatique de la France* (réedition 1989). Ministère des Transport, Direction de la Météorologie National.

Durran, D. R. 1990. Mountain waves and downslope winds. In *Atmospheric Processes over Complex Terrain*, ed. B. Blumen. Meteorological Monographs 23:59–81. Boston, MA: American Meteorological Society.

Durran, D. R. 1986. Another look at downslope windstorms. Part I: On the development of analogs to supercritical flow in an infinitely deep, continuously stratified fluid. *Journal of the Atmospheric Sciences* 43:2527–43.

Egger, J. 1990. Thermally forced flows: Theory. In *Atmospheric Processes over Complex Terrain*, ed. B. Blumen. Meteorological Monographs 23:43–58. Boston, MA: American Meteorological Society.

Egger, J., and K. P. Hoinka. 1992. Fronts and orography. *Meteorology and Atmospheric Physics* 48: 3–36.

Etling, D. 1989. On atmospheric vortex streets in the wake of large islands. *Meteorology and Atmospheric Physics* 41:157–64.

Exner, F. M. 1924. Monatliche Luftdruck- und Temperaturanomalien auf der Erde. *Sitzungsberichte der Kaiserlichen Akademie der Wissenschaften, Wien (Abt.IIa Math.-naturw.)* 133:307–408.

Exner, F. M. 1913. Übermonatliche Witterungsanomalien auf der nördlichen Erdhälfte im Winter. *Sitzungsberichte der Kaiserlichen Akademie der Wissenschaften, Wien (Abt.IIa Math.—naturw.)* 122: 1165–1241.

Fliri, F. 1991: Die Schneeverhältnisse in Nord- und Osttirol in der Periode 1895–1991. Mitteilungen der Österreichischen Geographischen Gesellschaft 133: 7–25.

Fliri, F. 1984. *Synoptische Klimatographie der Alpen zwischen Mont Blanc und Hohen Tauern (Schweiz-Tirol-Oberitalien)*. Wissenschaftliche Alpenvereinshefte, no. 29. Innsbruck, Austria.

Fliri, F. 1974. *Niederschlag und Lufttemperatur im Alpenraum*. Wissenschaftliche Alpenvereinshefte, no. 24. Innsbruck, Austria.

Flohn, H. 1953. Hochgebirge und allgemeine Zirkulation. II Die Gebirge als Wärmequellen. *Archives for Meteorology, Geophysics and Bioclimatology* A5:265–279.

Föhn, P. 1990. Schnee und Lawinen. In *Schnee, Eis und Wasser der Alpen in einer wärmeren Atmosphäre*, ed. D. Vischer, 33–48. Mitteilungen der Versuchsanstalt für Wasserbau, Hydrologie und Glaziologie, no. 108, ETH Zürich.

Föhn, P., and C. Plüss. 1989. *Frühwinterliche Schneearmut: Fiktion oder Realität?* Interner Bericht SLF, Davos, Switzerland, no. 655.

Fraedrich, K. 1994. An ENSO impact on Europe? A review. *Tellus* 46A:541–52.

Fraedrich, K. 1990. European Grosswetter during warm and cold extremes of the El Nino/Southern Oscillation. *International Journal of Climatology* 10:21–31.

Fraedrich, K., and K. Müller. 1992. Climate anomalies in Europe associated with ENSO extremes. *International Journal of Climatology* 12:25–31.

Fraedrich, K., C. Bantzer, and U. Burckardt. 1993. Winter climate anomalies in Europe and their associated circulation at 500hPa. *Climate Dynamics* 8:161–75.

Frei, C., and C. Schär. 1998. A precipitation climatology of the Alps from high-resolution rain-gauge observations. *International Journal of Climatology* 181, in press.

Frei, C., M. Widmann, D. Lüthi, H. C. Davies, and C. Schär. 1997. Global warming and the occurrence of heavy precipitation over southern Europe and the Alpine Region. *Proc. International Symposium on Cyclones and Hazardous Weather in the Mediterranean.* Palma de Mallorca, Spain. April 14–17, 1997, 593–597.

Geiger, H., J. Zeller, and G. Röthlisberger. 1991. *Starkniederschläge des Schweizerischen Alpen- und Alpenrandgebietes.* Publikation der Eidgenössischen Forschungsanstalt für Wald, Schnee und Landschaft, Birmensdorf, Switzerland.

Gerstengarbe, F-W., and P. C. Werner. 1993. *Katalog der Grosswetterlagen Europas nach Paul Hess und Helmuth Brezowski, 1881–1992.* Berichte des Deutschen Wetterdienstes, no. 113. Offenbach a.M. 249 p.

Glowienka-Hense, R. 1990. The North Atlantic oscillation in the Atlantic-European SLP. *Tellus* 42A:497–507.

Hafner, T. A., M. E. Reinhardt, E. L. Weisel, and H. P. Fimpel. 1987. Boundary layer aspects and elevated heat source effects of the Alps. *Meteorology and Atmospheric Physics* 36:61–73.

Hagen, E., and G. Schmager. 1991. On mid-latitude airpressure variations and related SSTA in the tropical/subtropical North Atlantic during 1957–1974. *TOGA Notes* 4:10–16.

Halpert, M. S., and C. F. Ropelewski. 1992. "Sea surface temperature patterns associated with the Southern Oscillation." *Journal of Climate* 5:577–593.

Hamilton, K. 1988. A detailed examination of the extratropical response to tropical El Nino/Southern Oscillation events. *International Journal of Climatology* 8:67–86.

Heinemann, H.-J. 1994. Homogenisierung der Säkulareihe der Jahresmittel-temperturen von Bremen. *Meteorologische Zeitschrift*, N.F. 3:35–8.

Held, I. M. 1993. Large-scale dynamics and global warming. *Bulletin of the American Meteorological Society* 74:228–41.

Hess, P., and H. Brezowsky. 1952. *Katalog der Grosswetterlagen Europas.* Berichte des Deutschen Wetterdienstes in der US-Zone, no. 33.

Hoskins, B. J., J. J. Hsu, I. N. James, M. Masutani, P. D. Sardeshmukh, and P. G. White. 1989. *Diagnostics of the Global Atmospheric Circulation Based on ECMWF Analysis, 1979–1989.* WCRP-29 (WMO TD-326). World Meteorological Organization, Geneva, Switzerland.

Houghton, J. T., G. J. Jenkins, and J. J. Ephraums, eds. 1990. *Climatic Change: The IPCC Scientific Assessment.* World Meteorological Organization/United Nations Environment Programme. Cambridge: Cambridge University Press.

Houghton, J. T., L. G. Meira Filho, B. A. Callander, N. Harris, A. Kattenberg, and K. Maskell, eds. 1996. *Climate Change 1995: The Science of Climate Change.* Contribution of WGI to the Second Assessment Report of the Intergovernmental Panel on Climate Change. World Meteorological Organization/United Nations Environment Programme. Cambridge: Cambridge University Press.

Houze, R. A., Jr., W. Schmid, R. G. Fovell, and H. H. Schiesser. 1993. Hailstorms in Switzerland: Left movers, right movers, and false hooks. *Monthly Weather Review* 121:3345–70.

Huntrieser, H., H.-H. Schiesser, W. Schmid, and A. Waldvogel. 1997. Comparison of traditional and newly developed thunderstorm indices for Switzerland. *Weather and Forecasting*, 12, 108–125.

Hurrell, J. W., 1995. Decadal trends in North Atlantic Oscillation: Regional temperature and precipitation. *Science* 269: 676–679.

Kahlig, P. 1986. Orography and precipitation. In *Proceedings 19. Internationale Tagung für Alpine Meteorologie (ITAM)*, Rauris, Austria, 1–15.

Kiladis, G. N., and H. F. Diaz. 1989. Global climatic anomalies associated with extremes in the Southern Oscillation. *Journal of Climate* 2:1069–90.

Kirchhofer, W. 1974. *Classification of European 500 mb patterns*. Arbeitsbericht SMA, no. 43. Zürich.

Kleinschmidt, E. 1922. Der tägliche Gang des Windes in der freien Atmosphäre und auf Berggipfeln. *Beiträge zur Physik der freien Atmosphäre* 10:1–15.

Kuhn, M. 1990. Energieaustausch Atmosphäre—Schnee und Eis. In *Schnee, Eis und Wasser der Alpen in einer wärmeren Atmosphäre*, ed. D. Vischer, p. 21–32. Mitteilungen der Versuchsanstalt für Wasserbau, Hydrologie und Glaziologie, no. 108, ETH Zürich.

Kushnir, Y. 1994. Inter-decadal variations in North Atlantic sea surface temperature and associated atmospheric conditions. *Journal of Climate* 7:141–57.

Lamb, H. H. 1972. *British Isles Weather Types and a Register of the Daily Sequence of Circulation Patterns 1861–1971*. Geophysical Memoirs no. 116. Meterological Office, London.

Lamb, P. J., and R. A. Peppler. 1987. North Atlantic Oscillation: Concept and an application. *Bulletin of the American Meteorological Society* 68:1218–25.

Landeshydrologie und Geologie. 1992. *Hydrologischer Atlas der Schweiz*. Bern, Switzerland: EDMZ.

Lang, H. 1985. Höhenabhängigkeit der Niederschläge. In *Der Niederschlag der Schweiz*, ed. B. Sevruk. *Beiträge zur Geologie der Schweiz* 31: 149–57.

Lang, H. 1981. Is evaporation an important component in high Alpine hydrology? *Nordic Hydrology* 12:217–24.

Lang, H., and M. Rohrer. 1987. Temporal and spatial variations of the snow cover in the Swiss Alps. In *Large-Scale Effects of Seasonal Snow Cover*, ed. B. E. Goodison, R. G. Bary, and J. Dozier. IAHS publication no. 166. Oxfordshire, U.K.: IAHS Press.

Lau, K.-M., P.-J. Sheu, and I.-S. Kang. 1994. Multiscale low-frequency modes in the global atmosphere. *Journal of the Atmospheric Sciences* 51:1169–93.

Lauscher, F. 1976a. Weltweite Typen der Höhenabhängigkeit des Niederschlags. *Wetter und Leben* 28:80–90.

Lauscher, F. 1976b. Methoden zur Weltklimatologie der Hydrometeore. Der Anteil des festen Niederschlags am Gesamtniederschlag. *Archives for Meteorology, Geophysics and Bioclimatology* B24:129–76.

Lauscher, F. 1958. Studien zur Wetterlagen-Klimatologie der Ostalpenländer. *Wetter und Leben* 10:79–83.

Martinec, J. 1987. Importance and effects of seasonal snow cover. In *Large-Scale Effects of Seasonal Snow Cover*, ed. B. E. Goodison, R. G. Barry, and J. Dozier. IAHS publication no. 166. Oxfordshire, U.K.: IAHS Press.

McFarlane, N. A. 1987. The effect of orographically excited gravity wave drag on the general circulation of the lower stratosphere and troposphere. *Journal of the Atmospheric Sciences* 44:1775–1800.

Meteorological Office. 1991. *Meteorological Glossary*. HMSO Publications Center, London.

Moses, T., G. N. Kiladis, H. F. Diaz, and R. G. Barry. 1987. Characteristics and frequency of reversals in mean sea level pressure in the North Atlantic sector and their relationship to long-term temperature trends. *International Journal of Climatology* 7:13–30.

Müller, H. 1984. Review paper on the radiation budget in the Alps. *International Journal of Climatology* 5:445–62.

Müller-Westermeier, G. 1992. Untersuchung einiger langer deutscher Temperaturreihen. *Meteorologische Zeitschrift*, N.F. 1:155–71.

Murray, R. 1993. Bias in southerly synoptic types in the decade 1981–1990 over the British Isles. *Weather* 48:152–4.

Nobilis, F., T. Haiden, and M. Kerschbaum. 1991. Statistical considerations concerning probable maximum precipitation (PMP) in the Alpine country of Austria. *Theoretical and Applied Climatology* 44:89–94.

Ohmura, A., H. Gilgen, and M. Wild. 1989. *Global Energy Balance Archive (GEBA) Report 1: Introduction*. Zürcher Geografische Schriften no. 34. Zürich.

Ohmura, A., and H. Lang. 1989. Secular variation of global radiation in Europe. In *IRS '88: Current Problems in Atmospheric Radiation*, ed. J. Lenoble and J.-F. Geleyn, 298–301. Hampton, VA: DEEPAK Publishing Company.

Ohmura, A., G. Müller, K. Schroff, and T. Konzelmann. 1990. *Radiation Annual Report ETH No. 1, 1987 and 1989*. Zürcher Geografische Schriften no. 52. Zürich.

Oke, T. R. 1987. *Boundary Layer Climates*. 2d ed. London: Methuen.

Palmer, T. N., and D. L. T. Anderson. 1994. The prospects for seasonal forecasting. *Quarterly Journal of the Royal Meteorological Society* 120:755–94.

Palmer, T. N., and Z. Sun. 1985. A modelling and observational study of the relationship between sea surface temperature anomalies in the north-west Atlantic and the atmospheric circulation. *Quarterly Journal of the Royal Meteorological Society* 111:947–75.

Palmer, T. N., G. J. Shutts, and R. Swinbank. 1986. Alleviation of a westerly bias in general circulation models through an orographic gravity wave drag parameterization. *Quarterly Journal of the Royal Meteorological Society* 112:1001–39.

Pasquale, V., G. Flocchini, and G. Russo. 1992. Surface temperature field trend along an European traverse. Paper presented at International Conference on Alpine Meteorology, CIMA 1992, Toulouse, France, Sept. 7–11, 1992.

Patzelt, G., and M. Aellen. 1990. Gletscher. In *Schnee, Eis und Wasser der Alpen in einer wärmeren Atmosphäre*, ed. D. Vischer, 49–69. Mitteilungen der Versuchsanstalt für Wasserbau, Hydrologie und Glaziologie no. 108, ETH Zürich.

Pettre, P. 1982. On the problem of violent winds. *Journal of the Atmospheric Sciences* 39:542–54.

Pfister, C. 1992. Monthly temperature and precipitation patterns in Central Europe from 1525 to the present. In *Climate Since 1500 A.D.*, ed. R. S. Bradley and P. D. Jones, 118–43. London: Routledge.

Pfister, C. 1984. *Das Klima der Schweiz von 1525–1860 und seine Bedeutung in der Geschichte von Bevölkerung und Landwirtschaft*. Band I. Bern, Switzerland: Verlag Paul Haupt.

Phillips, P. D. 1984. *The ALPEX Gotthard Data Set.* LAPETH-22, Institute of Atmospheric Physics, ETH Zürich.

Pierrehumbert, R. T., and B. Wyman. 1985. Upstream effects of mesoscale mountains. *Journal of the Atmospheric Sciences* 42:977–1003.

Ratcliffe, R. A. S., and R. Murray. 1970. New lag associations between North Atlantic sea temperature and European pressure applied to long-range weather forecasting. *Quarterly Journal of the Royal Meteorological Society* 96:226–46.

Reiter, E. R. 1975. *Handbook for Forecasters in the Mediterranean. Part I: General Description of the Meteorological Processes.* Enviromental Prediction Research Facility. Monterey, CA: Naval Postgraduate School.

Rex, D. F. 1950. The effect of Atlantic blocking upon regional climate. *Tellus* 3:101–12.

Richner, H., and P. D. Phillips. 1984. A comparison of temperatures from mountaintops and the free atmosphere—Their diurnal variation and mean difference. *Monthly Weather Review* 112:1328–40.

Rogers, J. C. 1990. Patterns of low-frequency monthly sea level pressure variability (1899–1986) and associated wave cyclone frequencies. *Journal of Climate* 3:1364–79.

Rogers, J. C. 1984. The association between the North Atlantic Ocean and the Southern Oscillation in the Northern Hemisphere. *Monthly Weather Review* 112:1999–2015.

Ropelewski, C. F. and M. S. Halpert. 1987. Global and regional precipitation patterns associated with the El Nino/Southern Oscilation. *Monthly Weather Review* 115:1606–26.

Schär, C., and D. R. Durran. 1997. Vortex formation and vortex shedding in continously stratified flow past isolated topography. *Journal of the Atmospheric Sciences* 54: 534–554.

Schär, C., and R. B. Smith. 1993. Shallow-water flow past isolated topography. Part II: Transition to vortex shedding. *Journal of the Atmospheric Sciences* 50:1401–12.

Schär, C., C. Frei, D. Lüthi, and H. C. Davies. 1996. Surrogate climate change scenarios for regional climate models. *Geophysical Research Letters* 23: 669–72.

Scherhag, R. 1966. Die tägliche Druckvariation über Mitteleuropa im Sommer. *Tellus* 18:806–13.

Schiesser, H. H., R. A. Houze, and H. Huntrieser. 1995. The mesoscale structure of severe precipitation systems in Switzerland. *Monthly Weather Review* 123: 2070–97.

Schirmer, H., and V. Vent-Schmidt. 1979. *Das Klima der Bundesrepublik Deutschland; Lieferung 1: Mittlere Niederschlagshöhen für Monate und Jahr, 1931–1960.* Offenbach am Main: Deutscher Wetterdienst.

Schönwiese, C.-D., J. Rapp, T. Fuchs, and M. Denhard. 1994. Observed climate trends in Europe 1891–1990. *Meteorologische Zeitschrift,* N.F. 3:22–28.

Schönwiese, C.-D., J. Rapp, T. Fuchs, and M. Denhard. 1993. *Klimatrend-Atlas Europa 1891–1990.* Bericht no. 20, 4. Auflage. Zentrum für Umweltforschung, Universität Frankfurt am Main.

Schüepp, M. 1968. *Kalender der Wetter- und Wittterungslagen im zentralen Alpengebiet.* Veröffentlichungen der Schweizerischen Meteorologischen Zentralanstaltanstalt, no. 11.

Schüepp, M., and H. Schirmer. 1977. Climates of central Europe. In *Climates of Central and Southern Europe,* ed. C. C. Wallén, 3–73. World Survey of Climatology series, vol. 6. Amsterdam: Elsevier Scientific Publishing Company.

Schüepp, M., M. Bouët, M. Bider, and C. Urfer. 1978. *Regionale Klimabeschreibungen (1. Teil).* Beiheft Annalen SMA, Zürich.

Schüepp, M., G. Gensler, and M. Bouët. 1980. *Schneedecke und Neuschnee*. Beiheft Annalen Schweizerische Meteorologische Anstalt, Zürich.

Seibert, P. 1990. South Föhn studies since the ALPEX Experiment. *Meteorology and Atmospheric Physics* 43:91–104.

Sevruk, B. 1985. Correction of precipitation measurements. Paper presented at ETH/IAHS/WMO workshop, Zürich, April 1–3, 1985, Zürich, Switzerland.

Smith, R. B. 1989. Hydrostatic flow over mountains. *Advances in Geophysics* 31:1–41.

Smith, R. B. 1987. Aerial observation of the Yugoslavian Bora. *Journal of the Atmospheric Sciences* 44:269–97.

Smith, R. B. 1986. Mesoscale mountain meteorology in the Alps. *Proceedings of the Conference on Results of the Alpine Experiment*, Venice, Italy, 1985, 407–23. WMO/ICSU Geneva.

Smith, R. B. 1985. On severe downslope winds. *Journal of the Atmospheric Sciences* 42:2597–2603.

Smith, R. B. 1979. The influence of mountains on the atmosphere. *Advances in Geophysics* 21:87–230.

Smith, R. B., and S. Grønås. 1993. Stagnation points and bifurcation in 3-D mountain airflow. *Tellus* 45A:28–43.

Smolarkiewicz, P. K., and R. Rotunno. 1989. Low Froude number flow past three dimensional obstacles. Part I: Baroclinically generated lee vortices. *Journal of the Atmospheric Sciences* 46:1154–64.

Sneyers, R. 1990. On the statistical analysis of series of observations. WMO publication no. 415 (Technical note no. 143), Geneva.

Sneyers, R., R. Böhm, and S. Vannitsem. 1992. Temperature changes in the Austrian Alps (period 1775–1990): Comparison with concomitant changes in neighbouring stations. Paper presented at International Conference on Alpine Meteorology, CIMA 1992, Toulouse, France, Sept. 7–11, 1992.

Steinacker, R. 1984a. Area-height distribution of a valley and its relation to the valley wind. *Contributions to Atmospheric Physics* 57:64–71.

Steinacker, R. 1984b. Airmass and frontal movement around the Alps. *Rivista di Meteorologia Aeronautica* 43:85–93.

Steinhauser, F. 1953. Niederschlagskarte von Österreich für das Normaljahr 1901–1950. Beiträge zur Hydrographie Österreichs, no. 27. Wien.

Stone, P. 1992. *The State of the World's Mountains: A Global Report*. London: Zed Books Ltd.

Tafferner, A. 1990. Lee cyclogenesis resulting from the combined outbreak of cold air and potential vorticity against the Alps. *Meteorology and Atmospheric Physics* 43:31–48.

Thorpe, A. J., H. Volkert, and D. Heimann. 1993. Potential vorticity of flow along the Alps. *Journal of the Atmospheric Sciences* 50:1573–90.

Tibaldi, S., A. Buzzi, and A. Speranza. 1990. Orographic cyclogenesis. In *Extratropical Cyclones* the Eric Palmen memorial volume, ed. C. W. Newton and E. O. Holopainen, 107–27. Boston: American Meteorological Society.

Touring Club Italiano. 1989. *Atlante tematico d'Italia, Folio 12: Precipitationi*. Touring Club Italiano, Consiglio Nazionale delle Ricerche, Milano, Italy.

Trüb, J., and H. C. Davies. 1995. Flow over a mesoscale ridge: Pathways to regime transition. *Tellus* 47A:502–24.

Tyndall, J. 1863. On radiation through the earth's atmosphere. *Philosophical Magazine* 4:200.

Urfer-Henneberger, C. 1970. Neuere Beobachtungen über die Entwicklung des Schönwetter-windsystems in einem V-förmigen Alpental (Dischma bei Davos). *Archives for Meteorology, Geophysics and Bioclimatology* B18:21–42.

Uttinger, H. 1951. Zur Höhenabhängigkeit der Niederschläge in den Alpen. *Archives for Meteorology, Geophysics and Bioclimatology* B2:360–82.

Uttinger, H. 1949. *Die Niederschlagsmengen in der Schweiz, 1901–1940: Niederschlagskarte der Schweiz*. Zürich: Schweizerische Meteorologische Zentral Anstalt.

Van Loon, H., and R. A. Madden. 1981. The Southern Oscillation, Part I: Global associations with pressure and temperature in northern winter. *Monthly Weather Review* 113:2063–74.

von Ficker, H. 1920. Der Einfluss der Alpen auf Fallgebiete des Luftdrucks und die Entwicklung von Depressionen über dem Mittelmeer. *Meteorologische Zeitschrift* 37:350–63.

von Rudloff, H. 1986. Die Schwankungen und Pendelungen des Klimas in den Hochalpen seit dem Beginn der Instrumentenbeobachtung (1781). Paper presented at International Conference on Alpine Meteorology, ITAM 1986, Rauris, Austria, 321.

von Rudloff, H. 1971. Die jüngsten Änderungen im Ablauf der Haupt-Klima-Elemente in den Hochalpen. In *Annalen der Meteorologie* no. 5, 171–80. ITAM 1970, Oberstdorf.

von Rudloff, H. 1962. Die Klimaschwankungen in den Hochalpen seit Beginn der Instrumenten-Beobachtungen. *Archives for Meteorology, Geophysics and Bioclimatology* B13:303–51.

Wagner, A. 1938. Theorie und Beobachtung der periodischen Gebirgswinde. *Gerlands Beiträge zur Geophysik* 52:408–49.

Wagner, A. 1932. Der tägliche Luftdruck und Temperaturgang in der freien Atmosphäre und in Gebirgstälern. *Gerlands Beiträge zur Geophysik* 37:315–44.

Wallace, J. M., Y. Zhang, and K.-H. Lau. 1993. Structure and seasonality of interannual and interdecadal variability of the geopotential height and temperature fields in the Northern Hemisphere troposphere. *Journal of Climate* 6:2063–82.

Wanner, H. 1991. Ein Nussgipfel im Westwind: Zur Dynamik von Wetter und Klima im Alpenraum. In *Die Alpen—ein sicherer Lebensraum*, ed. J. P. Müller and B. Gilgen, 50–69. Publication of the Swiss Academy of Natural Sciences. Disentis, Switzerland: Desertina Verlag.

Wanner, H. 1980. Grundzüge der Zirkulation der mittleren Breiten und ihre Bedeutung für die Wetterlagenanalyse im Alpenraum. In *Das Klima—Analysen und Modelle, Geschichte und Zukunft*, ed. H. Oeschger, B. Messerli, and M. Svilar, 117–24. Berlin: Springer.

Wanner, H. 1979. *Zur Bildung, Verteilung und Vorhersage winterlicher Nebel im Querschnitt Jura-Alpen*. Geographica Bernensia, G7. Bern.

Wanner, H., and M. Furger. 1990. The Bise—Climatology of a regional wind north of the Alps. *Meteorology and Atmospheric Physics* 43:105–16.

Weber, R. O, P. Talkner, and G. Stefanicki. 1994. Asymmetric diurnal temperature change in the Alpine region. *Geophysical Research Letters* 21:673–6.

Whiteman, C. D. 1990. Observations of thermally developed wind systems in mountainous terrain. In *Atmospheric Processes over Complex Terrain*, ed. B. Blumen. Meteorological Monographs 23:5–42. Boston: American Meteorological Society.

Whittaker, L. M., and L. H. Horn. 1984. Northern Hemisphere extratropical cyclone activity for four mid-season months. *Journal of Climatology* 4:297–310.

Widmann, M., and C. Schär. 1997. A principal component and long-term trend analysis of daily precipitation in Switzerland. *International Journal of Climatology*, 17: 1333–1356.

Wilby, R. 1993. Evidence of ENSO in the synoptic climate of the British Isles since 1880. *Weather* 48:234–9.

Wild, M., A. Ohmura, H. Gilgen, and E. Roeckner. 1995. Regional climate simulation with a high resolution GCM: Surface radiative fluxes. *Climate Dynamics* 11:469–86.

Wippermann, F. 1984. Air flow over and in broad valleys: Channeling and counter-current. *Contributions to Atmospheric Physics* 57:92–105.

Witmer, U., P. Filliger, S. Kuny, and P. Küng. 1986. *Erfassung, Bearbeitung und Kartieren von Schneedaten in der Schweiz.* Geographica Bernensia G25. Bern.

World Meteorological Organization (WMO). 1970. *Climatic Atlas of Europe I: Maps of Mean Temperature and Precipitation.* World Meteorological Organization, Geneva, Switzerland.

3   Alpine Paleoclimatology

Guy S. Lister, David M. Livingstone, Brigitta
Ammann, Daniel Ariztegui, Wilfried Haeberli, André
F. Lotter, Christian Ohlendorf, Christian Pfister, Jakob
Schwander, Fritz Schweingruber, Bernard Stauffer, and
Michael Sturm

3.1   UNFOLDING CLIMATE HISTORY IN THE ALPINE REGION

An often stated concern today is that global warming may result in higher
surface air temperatures than have ever been experienced, with unpredictable
effects on the biosphere and on social and economic infrastructures in many
regions of the globe. Many of us have a subjective view of the past in this
context that is limited to the last century or so. One reason this historical
perspective is limited is that networks of meteorological stations delivering
high-quality instrumental data have been in existence for only the last 150
years at most. In addition, comparative media reports tend to focus on a time
span within modern human experience. This period, lying within the subjective time horizons of most of the inhabitants of the Alpine nations, therefore makes a suitable beginning to the story of Alpine paleoclimatology.

3.1.1   The Short-Term Perspective: Trends in Air Temperature over the Last Hundred Years

Instrumental data from individual meteorological stations can provide
detailed knowledge of temporal variations in key meteorological variables,
but unfortunately only over a relatively short period. A good example of
this is air temperature, for which many reliable series of instrumental measurements exist, some reaching back one, two, or even three centuries. Probably the best known of these, Manley's (1974) composite central England
data series, stretches back as far as the middle of the seventeenth century; in
the Alpine region, the Basle series (Bider, Schüepp, and von Rudloff 1959)
begins about a century later.

    Lack of information on the degree of spatial homogeneity, however, limits
the usefulness of data from individual stations. Distinguishing regional and
global variations from those occurring purely at the local level requires data
from national and international networks of stations. Extensive analyses
of such data by Jones, Wigley and Wright (1986), Jones et al. (1986), and
Hansen and Lebedeff (1987) have revealed that, over the last 100 years or so,
surface air temperatures in the Northern Hemisphere have been increasing at

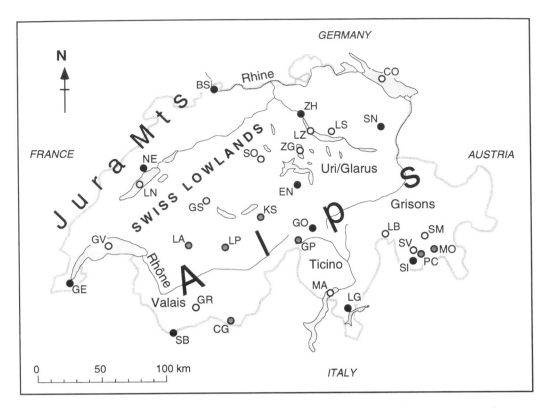

Figure 3.1 The Swiss Alpine region: locations of meteorological stations and sampling sites mentioned in text. Meteorological stations (black): Zurich (ZH), Basle (BS), Geneva (GE), Neucâtel (NE), Lugano (LG), Engelberg (EN), Säntis (SN), St. Bernard's Pass (SB), St. Gotthard's Pass (GO), and Sils-Maria (SI). Lakes (white): Lake Zurich (LZ), and Lake Geneva (GV), Lake Neuchâtel (LN), Lake Zug (ZG), Lützelsee (LS), Lej da San Murezzan (SM), Lej da Silvaplauna (SV), Gerzensee (GS), Soppensee (SO), Lago Basso (LB), Gouillé Rion (GR), Lago di Maggiore (MA), and Lake Constance (CO). Other sampling sites (grey): Colle Gnifetti (CG), Lötschenpass (LP), Piz Corvatsch (PC), Kleine Scheidegg (KS), Grimsel Pass (GP), Laufenen (LA), and Morteratsch Glacier (MO).

a mean rate of about 0.005–0.007 degrees Kelvin per year ($K \cdot yr^{-1}$). Because of the spatially and temporally heterogeneous nature of climatic phenomena, regional manifestations of global climate change are likely to differ markedly from average annual global or hemispherical trends (Jones and Kelly 1983; Jones and Briffa 1992). The detailed spatial and temporal knowledge of trends in meteorological variables that can be obtained by analysing instrumental data from networks of stations is therefore important for understanding the physical mechanisms underlying climate change, assessing the impact of climate change on society, and determining the representativeness of climatic reconstructions based on proxy data from a limited geographical region or time of year (Jones and Kelly 1983).

Swiss meteorological records from a variety of stations (figure 3.1) show that surface air temperatures in the Alpine region have been increasing per-

Table 3.1 Linear regressions of annual mean surface air temperature anomaly against time

Station	Altitude [m a.s.l.]	Population [10^3]	Period	n	$F_{1,n-2}$	r^2	$\delta T/\delta t \pm 95\%$ C.I. [$K \cdot yr^{-1}$]
Cities							
Zurich	556	343	1880–1987	108	49	0.31	0.012 ± 0.004
Basle	317	170	1880–1987	108	40	0.27	0.011 ± 0.003
Geneva	405	165	1880–1964	85	33	0.29	0.014 ± 0.005
Towns and Villages							
Neuchâtel	487	33	1880–1987	108	17	0.14	0.007 ± 0.004
Lugano	276	26	1880–1987	108	31	0.23	0.008 ± 0.003
Engelberg	1,018	3	1880–1987	108	21	0.16	0.008 ± 0.003
Mountain stations							
Mt. Säntis	2,500	0	1880–1987	108	20	0.16	0.009 ± 0.004
St. Bernard's Pass	2,479	0	1880–1987	106	28	0.21	0.010 ± 0.004
St. Gotthard's Pass	2,090	0	1880–1960	81	14	0.15	0.011 ± 0.006
Other measurements							
Central England	–	–	1880–1987	108	10	0.09	0.005 ± 0.003
Global I	–	–	1880–1984	105	74	0.42	0.005 ± 0.001
Global II	–	–	1880–1987	108	153	0.59	0.006 ± 0.001

Note: Data for central England from Manley 1974 and Parker, Legg, and Folland 1992; those for global mean data from Jones et al. 1986 (Global I) and Hansen and Lebedeff 1987, 1988 (Global II; cf. figure 3.2). Population figures are as of 1989. The period 1880–1987 was chosen whenever possible to correspond with the data of Hansen and Lebedeff (1987, 1988). The number of annual mean temperatures included in each regression is denoted by n, the ANOVA F value by $F_{1,n-2}$ ($p < 0.01$ in all cases), the coefficient of determination (fraction of variance explained by the regression) by r^2, the rate of temperature increase by $\delta T/\delta t$ and the 95 percent confidence interval for $\delta T/\delta t$ by 95 percent C.I. Figure 3.1 gives the locations of the Swiss stations.

sistently at an average rate of about 0.01 $K \cdot yr^{-1}$ since the end of the nineteenth century (table 3.1). Both the short-term structure and the longer-term rate of increase are similar not only in cities (Zurich, Geneva, Basle), where local warming effects due to urban expansion might be expected (Dronia 1967; Kukla, Gavin, and Karl 1986; Wang, Zeng, and Karl 1990), but also in towns (Neuchâtel, Lugano) and villages (Engelberg), as well as on mountain peaks (Mt. Säntis) and high alpine passes (St. Bernard's Pass, St. Gotthard's Pass) far from centers of population, which implies that the increase is truly a regional phenomenon.

The magnitude of the long-term warming trend over the last 100 years varies from 0.007 $K \cdot yr^{-1}$ to 0.014 $K \cdot yr^{-1}$ (table 3.1). Data from rural mountain sites (altitudes above 2,000 meters above sea level (m a.s.l.), which are most likely to reflect global or regional climatic trends because they are subject to the least local anthropogenic influence, depict a warming trend of about 0.01 $K \cdot yr^{-1}$. Warming trends in the cities (population exceeding 100,000, altitudes below 600 m a.s.l.) exceed those at the mountain sites by about 20 percent, which is to be expected in view of the urban heat island phenomenon. In towns and villages (populations less than 100,000, altitude below 1,500 m a.s.l.), however, warming trends tend to be about 25 percent lower than at the mountain sites and as much as 40 percent lower than in the cities, despite the fact that urban warming is known to affect even small towns with populations under 10,000 (Karl, Diaz, and Kukla 1988).

Long-term warming trends in the Swiss Alps and lowlands are large in comparison to global trends (Beniston et al. 1994). Even when the high warming rates measured at the city stations are disregarded, long-term regional warming rates in Switzerland are still about 50 to 100 percent higher than the global rate of increase of surface air temperature over the same time period found by Jones, Wigley and Wright (1986), Jones et al. (1986), Jones et al. (1988), and Hansen and Lebedeff (1987, 1988); they are also about 50 percent higher than the warming rate shown by Manley's (1974) central England surface air temperature data series (extended by Parker, Legg, and Folland 1992). Assuming the same forcing scenario applies, possible future global warming trends are therefore likely to manifest themselves particularly strongly in Switzerland in both Alpine and lowland areas.

Several studies have shown the Northern Hemisphere's reported warming during the last fifty years to be principally the result of an increase in nighttime temperatures, whereas daytime temperatures exhibit hardly any increase at all, implying a long-term decrease in the daily temperature range (Karl, Kukla, and Gavin 1984, 1986; Karl et al. 1991; Plantico et al. 1990; Bücher and Dessens 1991; Kukla and Karl 1993). Weber, Talkner, and Stefanicki (1994) have shown this also to be the case in Swiss lowland areas, but not at mountaintop stations in the Alps, suggesting that an increase in cloud cover or anthropogenic aerosol concentrations in the very low troposphere is decreasing (short-wave) daytime insolation and increasing (long-wave) nighttime back radiation in lowland areas. This may explain why warming trends at mountain sites tend to exceed those in lowland towns and villages. In any event, (Alpine) ecosystems situated above the lowermost part of the troposphere can be expected to respond to climate change differently than (lowland) ecosystems situated within it.

3.1.2 The Long-Term Perspective: Recognition of Climate Fluctuations in the Past

The recent trends mentioned above have been recognized on the basis of instrumental data, which are of course the most detailed and most precise indicators of past and present climate at our disposal. However, the use of instrumental data for the detection of climate trends has one major drawback: the relative brevity of the available time series. The fact that the distinction between trends and fluctuations is essentially subjective, depending on the length of the data time series at hand, means that natural low-frequency fluctuations in meteorological variables might be interpreted as trends instead of variations around a long-term mean if the time series is shorter than about half the period of the fluctuation. The length of the time series upon which the computation of, for example, global mean air temperatures is based (100–150 years) means that the change since about the middle of last century can be interpreted as a trend only with respect to this rather limited period of time. The question therefore arises of whether the regional and global warming that has occurred during the last hundred years is an anthropogenic effect (due, for instance, to GHGs) or merely part of a natural cycle unrelated to human activity. A look at some of the individual air temperature data series that stretch back farther than the middle of last century (e.g., Basle, Geneva, and central England in figure 3.2, or the Austrian data shown in figure 2.4) does suggest a warming trend beginning about the end of last century but also makes it clear that fluctuations on a timescale of decades are common. (Three such fluctuations, with amplitudes of about 0.5 K, are apparent in the Basle and Geneva air temperature series between 1750 and 1900; see figure 3.2). To put the changes over the last hundred years or so in a longer-term perspective, we must extend the data window even farther back in time. We can do this only by resorting to various types of so-called proxy data, extracted either from documentary records, which can reach back about 500 years, or from natural archives, including perennial ice, tree rings and lake sediments. For the Alpine region, the latter can extend the data window continuously back to about 12,000–15,000 years before the present, the morphological changes wrought during the last glacial episode of the Ice Age having largely destroyed earlier evidence.

Although Homo sapiens has experienced at least three glacial episodes, some 600 generations have passed since the last Glacial and no "tribal memories" of Ice Age conditions remain. Only after learning how to decipher the information left behind in nature's archives, a process that did not begin until last century, did mankind begin indirectly to rediscover Alpine glacial history. This process led to the ability to put the above-mentioned short-term modern human experience, especially with regard to climate studies, into a much longer-term perspective. Basic to any understanding of long-term climate development in the alpine region is the Glacial Theory.

3.1.2.1 The Glacial Theory The Glacial Theory views the Ice Age as massive expansions of continental ice occurring during cold climatic conditions. The debris Ice Age glaciers left behind after their retreat supplied initial evidence supporting this theory, the same debris principally responsible for the multiplicity of valued landforms and lakes in the Alpine region today. Subsequently, the recognition and investigation of new sources of information by a variety of disciplines allowed the accumulation of increasingly detailed evidence further supporting the theory. The perception of how climatic developments have affected the Alpine region has continued to evolve since the first recognition, just over a century ago, of the occurrence of the Ice Age Glacial and its climatic implications. During recent decades,

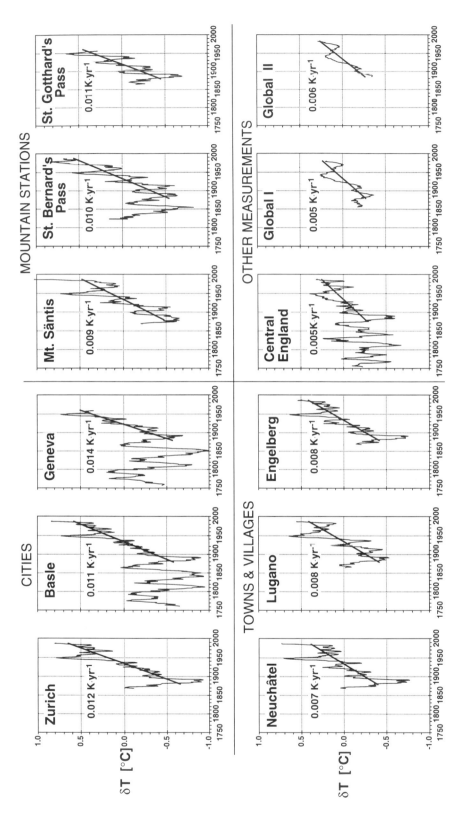

Figure 3.2 Annual mean surface air temperature anomalies (δT, eleven-year running means) computed for nine stations in Switzerland, for the central England data series of Manley (1974) as extended by Parker, Legg, and Folland (1992) and for the global mean data of Jones et al. (1986) (Global I) and Hansen and Lebedeff (1987, 1988) (Global II). Data from the Swiss stations (courtesy of the Swiss Meteorological Office) homogenised by Schüepp (1961) up to 1960. See table 3.1 for the altitudes of the Swiss stations and the population of the city, town, or village in which they are located; see figure 3.1 for the station locations. The straight lines (gradients given) represent linear regressions of the corresponding temperature anomaly against time (see table 3.1 for details).

because of improved access to and evaluation of natural evidence from a global variety of sources, this process has been especially intense.

The Glacial Theory required most of the nineteenth century to evolve, much of this evolution being centred on Switzerland (Flint 1971). In the late eighteenth century, so-called erratics, large boulders of locally foreign rock (for example, granite from the Alps found atop limestones in the Jura Mountains, separated from the Alps by the Swiss lowlands) aroused curiosity after observers realized that these erratics had been transported long distances from their original locations. Debates began about possible transport agents, involving especially catastrophic flooding, transgressive/regressive marine phases and icebergs (the term "drift" for what was later identified as glacial till is still used today). James Hutton (1795) and John Playfair (1802), however, suggested that transport had been by large paleo-glaciers.

Then, in 1815, an Alpine villager gave the Swiss civil engineer Ignaz Venetz Sitten the idea that rock debris may have been transported by glaciers, and Venetz-Sitten subsequently presented an argument for prehistorically extended glaciers at a meeting of the Swiss Natural History Society in Lucerne. Meanwhile, others were also beginning to realize that the same mechanism may have been responsible for depositing debris in northwestern Europe. In the early 1830s this idea received further support from Venetz-Sitten's colleague Jean de Charpentier and also from the zoologist Louis Agassiz and from the botanist Karl Schimper, after witnessing convincing evidence from the Diablerets Glacier and from moraines in the Rhône Valley (Schimper was apparently the first to use the expression *Eiszeit*, or *Ice Age*). At about the same time in Germany, Bernhardi (1832) interpreted the Ice Age in Europe as having been caused by a southward spread of ice from the polar regions. However, this was to prove erroneous for the Alpine region. By 1837 Agassiz had proposed a "great ice period" for the Alps as a result of climate change (Agassiz 1840), but it was another ten years before he recognized that the glaciers were of Alpine rather than of polar origin. Partly as a result of Agassiz' influential work, evidence was also being collected in the British Isles and North America. Agassiz' appointment as professor at Harvard in 1846 helped strengthen the Glacial Theory, but resistance from the proponents of the competing "iceberg theory" on both sides of the Atlantic continued for decades. Curiously, Danish scientific exploration of Greenland then provided a convincing analogy for the huge European ice masses. Ice-coring techniques have since recovered Glacial-aged ice containing detailed proxy data from the Greenland ice mass. Figure 3.3 outlines the ice limits for northern Europe and the Alps during the last Glacial maximum at about 18,000 ^{14}C years BP. These limits had been more or less established by the 1880s (e.g., Penck 1882), although more detailed work has continued up to the present.

The Ice Age was initially treated as a single event. However, during the 1840s and 1850s, discoveries of multiple layers of drift with intervening interglacial horizons in Switzerland, the United Kingdom and the United

Figure 3.3 Ice extent in Europe at the time of maximum glaciation, about 18,000 ^{14}C yr BP (after Flint 1957, in West 1977), equivalent to about 21,000 calendar yr BP (Bartlein et al. 1995).

States increased the number of glacial stages identified to two, three, and then four (e.g., Penck 1882). Fossil flora from interglacial sediments, first recognised by Oswald Heer (1858) in Switzerland, indicated that the climate during interglacial periods was close to that existing today, or, expressed slightly differently, that our present climate is that of an interglacial period. Later, Heer (1865) made possibly the first attempt ever at a scientific paleoclimate reconstruction, in this case for Switzerland, based on floral evidence. The four Alpine glacial stages were named (youngest to oldest) "Würm," "Riss," "Mindel," and "Günz" after four Bavarian rivers, with the interglacial warm stages named using both previous and subsequent glacial stage names (e.g., "Riss/Würm"). We now know that each glacial/interglacial cycle lasted approximately 100,000 years, but at the time of this early work, the durations could not be accurately determined; even the dating of the last Glacial maximum (ca. 18,000 ^{14}C yr BP, now equivalent to 21,000 calendar yr BP; Bartlein et al., 1995) had to wait for the development of suitable methods, such as radiocarbon dating and varve counting.

Although the Ice Age was believed to be the manifestation of a colder climate, it was some time before a more analytical treatment of climate mechanisms was embarked upon. Penck (1882) was apparently the first to state clearly that the dimensions of glaciers and the altitudes at which they occurred were systematically related to air temperature and precipitation. Secular climate change had thus started to become a legitimate research target.

3.1.2.2 Nonglacial Continental Evidence For large continental areas (periglacial and nonglacial) beyond the maximum ice extent, glacial evidence for past climatic conditions was lacking. Nevertheless, other indicators of a colder climate had been recognized, including physical features (frost shattering and wedging of rock, freeze and thaw structures in alluvium, postglacial isostatic uplift of coastal features) and widespread faunal evidence (mammoth and other vertebrate remains, fossil shells indicative of colder conditions). In addition, large stratigraphic deposits of silt, originally thought to be flood or lacustrine sediments, were identified in the 1870s and 1880s as the wind-blown fine fractions lifted from glacial outwash debris. This "loess," huge deposits of which are found in China and lesser deposits in North America, blanketed some two million square kilometres of Europe. Peat bogs also provided stratigraphic evidence as gradually accumulating floral remains provided evidence for the ecological impacts of climate change. The Scandinavian botanist Axel Blytt first used this method of reconstruction in 1876, although another forty years passed before statistical methods were introduced into pollen profile evaluations (cf. section 3.2.5).

Loess formation indicates dry, windy climatic conditions, yet in other regions evidence was found for conditions wetter than those prevailing today. In the latter part of the nineteenth century, the existence of so-called pluvial climates was suggested to account for the higher lake levels in parts of North Africa, America, and Asia during the Ice Age. Today, lake basins in these areas are commonly either dry or hydrologically closed (water loss occurring by evaporation only). It was thus becoming increasingly evident that global climate change meant different things in different places.

3.1.2.3 The Marine Record Glacial and periglacial stratigraphic records on land suffer from the disadvantage of rarely being continuous back through the glacial/interglacial episodes. The glacial processes of erosion, mixing, and deposition alternated cyclically between extremes, so records originating prior to the last glacial retreat have in most cases been damaged or destroyed. In contrast, marine sediment sequences have in places accumulated continuously throughout the whole of the Quaternary period (about the last 1.6 million years). These sequences have a huge stratigraphic potential, since they comprise inputs of both detrital material from land (changes in erosion rates) and temporally and spatially variable accumulations of various marine indicators (e.g., patterns of alternating "cold" and "warm" microfaunal relicts).

Although scientific investigations of deep-sea surface sediments were launched more than a century ago with the Royal Society of London's global expedition of 1872–76, research into short sequences of Quaternary sediments started only in the 1930s, after deep-sea core-recovery methods became practical. The investigation of long sediment sequences began to flourish in the 1960s with the availability of improved drilling and seismic-survey technologies and the development of the theories of sea floor spreading and plate tectonics (see, for example, Kennet 1982). The resulting findings have rewritten Quaternary Ice Age history and have helped initiate the investigation of global climate processes, including the modeling of the climate-ocean system. In addition to the measurement of classical strati-graphic data, which yielded information on regional sediment patterns and their dependence on ocean currents, on biological productivity, and on the terrestrial detritus delivered by rivers, ice and wind, oxygen isotope measurements provided indices for global climate conditions. Marine oxygen isotope stratigraphy has now proven one of the most powerful tools for quantitatively establishing the Quaternary history of climatically controlled ice advances and retreats.

At the Eidgenössische Technische Hochschule (ETH) in Zurich in 1946, H. Urey explained that a temperature-dependent partitioning, or fractiona-tion, of the stable isotopes of oxygen (^{16}O and ^{18}O) occurs between liquid water and the vapor derived from it, because these isotopes have differing vapor pressures and therefore evaporate and condense at different rates (Urey 1947). Urey had provided one of the most important tools, an isotopic thermometer, which, together with the parallel development of radiocarbon dating by W. F. Libby (Arnold and Libby 1951), enabled Ice Age history to be rewritten. Evaporation from the ocean surface preferentially removes ^{16}O, the lighter isotope, and causes the residual seawater to become enriched in the heavier isotope, ^{18}O (see box 3.1). Foraminifera, marine algae living in both surface and bottom waters, incorporate the oxygen isotope ratios of the seawater into their carbonate skeletons. The skeletal material accumulates incrementally on the sea floor; coring then provides sequences of this mate-rial from which the seawater's past isotopic evolution can be assessed. Emi-liani pioneered this method in 1954, producing the first of numerous Quaternary isotopic records of increasingly high resolution (e.g., figure 3.4). These records indicate primarily the degree of partitioning of water between the oceans and the continental ice masses, so that during the last Glacial maximum, when the sea level was about 130 meters lower than it is today, seawater's mean oxygen isotopic value was about 1 percent more positive than it is today (figure 3.4b). One of this method's greatest strengths is that the isotopic shifts happen essentially simultaneously in all oceans, allowing stratigraphies from different locations to be matched and links with regional indicators to be established.

The marine oxygen isotope record for the Quaternary (figure 3.4a) shows that the glacial/interglacial episodes have been fairly constant with respect to both frequency and amplitude during nearly the last million years (involving probably eleven glacial episodes). Each full cycle lasted about 100,000 years, with a slow buildup of ice on the continents, resulting in a fall in sea-level,

Box 3.1 Stable Isotopes

Isotopes are defined as atoms whose nuclei contain the same number of protons but a different number of neutrons. They thus have the same atomic number but different mass numbers. Most stable (nonradioactive) elements have at least two isotopes, with one usually being strongly predominant. For example, the two most important stable isotopes of oxygen, ^{16}O and ^{18}O, have natural mean abundances of 99.76 percent and 0.19 percent, respectively. Although almost identical, the isotopes of an element exhibit subtle behavioral differences during natural processes. The partial separation and partitioning that can thus occur during physical or chemical processes is referred to as "isotopic fractionation." For instance, because of differences in vapor pressure of the isotopic species in water, isotopic fractionation always concentrates the heavy isotopes of hydrogen and oxygen in the liquid phase during evaporation and precipitation processes. Isotopic fractionation can be due to kinetic effects, each isotopic species having its own individual rates in chemical reactions and physical processes (e.g., evaporation), or temperature-dependent equilibrium effects due to the differing thermodynamic properties of the isotopically substituted species in a molecule (such as water) or crystal (for example, carbonate precipitated from water). Comprehensive discussions of stable isotopes and their fractionation in nature are given in Faure (1977), Pearson and Coplen (1978), and Hoefs (1980) discuss stable isotopes and fractionation comprehensively.

For a particular element in a substance (such as hydrogen or oxygen in water), the stable isotope ratios are conventionally expressed in the delta (δ) notation as relative deviation (per mil, i.e., ‰) from the isotopic ratio of a standard substance: $\delta‰ = (R_{sample} - R_{standard}) \cdot 10^3 / R_{standard}$; where R = ratio of the heavier isotope to the lighter isotope of the given element. Calculated results are expressed within a range of positive through negative numbers. The isotopic-ratio values are said to be "getting lighter" or "decreasing" or "becoming more negative" if the lighter isotope is becoming relatively more abundant. Several different standard substances are in use; for example, there are water standards for water samples (e.g., SMOW = Standard Mean Ocean Water: Craig 1961) and carbonate standards for carbonate samples (e.g., PDB = Peedee Formation belemnite: Craig 1957), but the isotopic ratio given with respect to a particular standard can be converted to its equivalent with respect to another standard (e.g., the oxygen isotope value for a carbonate can be converted to that of the parent water, assuming the water temperature during carbonate formation can be established).

Globally, the isotopes of both oxygen and hydrogen in meteoric waters each have a constant ratio, but regionally, climatic and other environmental conditions control the ratios of each in both time and space. In meteoric waters, the ultimate source of which is evaporation from the ocean surface, the characteristic patterns for stable-isotope ratios are due almost entirely to fractionation during evaporation and condensation processes and can be related to geographical factors such as latitude, altitude and continentality, or to distance from the ocean source (Epstein and Mayeda 1953; Dansgaard 1964; Craig and Gordon 1965; Yurtsever 1975; Gat 1980). Air temperature, strongly related to latitude and altitude, is the most direct control during precipitation (cf. figure 3.30). A given continental locality under a stable climatic regime is therefore subject to precipitation with a limited isotopic range, characteristic for its set of geographic variables.

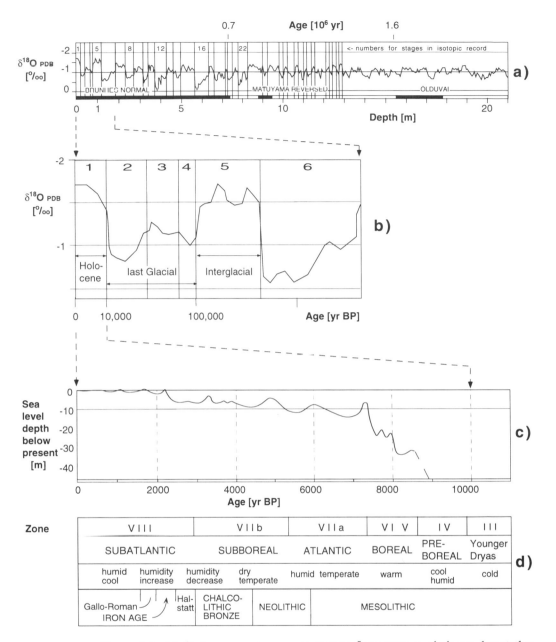

Figure 3.4 (a) Quaternary marine oxygen isotope fluctuations in planktonic foraminifera (deviations from Emiliani B1 standard) shown against the paleomagnetic record in piston core V28-239. Oxygen isotope stages 1 to 23 are numbered after Emiliani (1955, 1966) and Shackleton and Opdyke (1973) (after Kennet 1982); (b) An enlarged section of (a) showing details of the last 200,000 years; (c) Curve indicating Holocene sea level changes along the French Atlantic coast; and (d) Pollen zones, climate, and human cultural development (after Ters 1987). The beginning of the Holocene, shown here as having the age "10,000 yr BP" on the basis of radio-carbon age determinations, is now believed to have a calendar age of at least 11,250 cal. yr BP (Bartlein et al. 1995).

followed by a comparatively fast collapse associated with the onset of a warm period similar to that of our present Holocene epoch. During the first half of the Quaternary, however, the periods were shorter and the amplitudes smaller (figure 3.4a).

Considerable effort has gone into establishing the causal mechanisms of the Quaternary climatic fluctuations. In 1938 Milankovitch proposed that the earth's orbital variations were forcing the glacial cycles. He suggested that small cyclical changes in the earth's orbital parameters (*obliquity*, or tilt of rotation axis; *precession*, or circular wobble of axis; and *eccentricity*, or off-centeredness of path around sun: Milankovitch 1938, 1941) controlled changes in the amount of solar radiation reaching the earth's surface (insolation). Hays, Imbrie, and Shackleton (1976) obtained strong supporting evidence through their studies of marine sediment sequences: They found that the dominant spectral peaks at 100,000, 40,000, and 22,000 years for Quaternary climate fluctuations closely matched those of the earth's orbital parameters. However, the Milankovitch theory does not supply answers to all relevant questions: For example, why is the 100,000-year climate cycle so dominant? Other proposed mechanisms include changes in solar activity, the injection of volcanic aerosols into the stratosphere, and most recently, variations in the inclination of the earth's orbit with respect to the ecliptic, taking the planet through clouds of extraterrestrial dust (Muller and McDonald 1995).

In the 1970s, the CLIMAP project (Climate Mapping, Analysis and Prediction), conducted by a multi-institutional consortium, began to evaluate proxy data arriving from the Deep-Sea Drilling Program (CLIMAP Project Members 1976, 1981). Based on the Milankovitch theory, sea surface temperatures and magnitudes of the continental ice reservoirs and general circulation models were calculated, and global conditions during the last glacial maximum (18,000 yr BP) were estimated.

3.1.2.4 Polar Ice Cores The marine record has two shortcomings for high-resolution paleoclimate studies: (1) limited temporal resolution due to very low sedimentation rates (on the order of mm/century) and to bioturbation (disturbance of the uppermost deposits by bottom fauna), and (2) the spatially integrated nature of signals from tracers such as oxygen isotopes. However, paleoclimate information relevant to regional-scale, hemisphere-scale, or even global-scale phenomena can be obtained from another source of records with high temporal resolution: polar ice. Polar ice cores contain a multitude of tracers deposited directly from the atmosphere, including those in gases, moisture, aerosols and various dust-sized particulate matter. Since stable-isotope values in the ice are related primarily to the atmospheric temperature during periods of precipitation, the information they provide differs from that stable isotopes in marine sediment cores provide.

Of the world's perennial ice accumulations, polar ice provides some of the longest, most continuous high-resolution records containing multifaceted evidence for fluxes of atmospheric constituents on hemispheric to global scales. Chapter 2 documented the North Atlantic region's strong influence on Europe's climate in terms of ocean circulation patterns, sea surface temperatures, and atmospheric conditions. Thus the Greenland ice mass represents an ideally situated archive of high-resolution data reflecting the climatic influences affecting the Alpine region from that quarter during the last 250,000 years (two glacial episodes). The Greenland ice record can also be viewed as a linchpin connecting oceanic and continental archives, thus linking regional-scale alpine paleoclimate to paleoclimate on a wider scale.

Proxy data stored in perennial ice are linked much more directly to atmospheric conditions than such data from other natural archives. Mediating biological, chemical, or physical processes incorporate original atmospheric constituents directly into the ice without alteration (Schwander 1995). As annual increments of snow accumulate, the increasing overburden gradually compresses the lower layers and transforms it first to compact but open porous firn, and with increasing pressure, to ice. Through the firn-ice transition zone, the pore spaces close off, trapping air in bubbles. Layers below the firn-ice transition zone thus preserve samples of air as well as samples of precipitation (including stable isotopes), atmospheric dust, aerosols, and chemicals in continuous ice sequences. Ice core drilling, which began during the late 1950s for climatic studies in Greenland, can reach ice depths of 3,000 meters. Some fifty physical and chemical parameters are routinely measured in modern Greenland ice cores (Fuhrer et al. 1993; GRIP Members 1993), many related to atmospheric composition, precipitation rates and temperatures, and aerosols. Accurate dating of the ice stratigraphy is crucial, though. Methods include counting seasonally varying parameters, radionuclides, ice flow modeling, and stratigraphic markers like volcanic ash layers. Nevertheless, uncertainty in ice core dating increases with the ice's age and depth of the ice, starting at about 0 to 5 percent over the last 10,000 years and increasing to about 10 percent at 50,000 years BP. The temporal resolution at the latter age can, however, still be about one year, so the relative dating of successive events is extremely good.

In July 1992 the European Greenland Ice Core Drilling Project (GRIP) reached bedrock in the central part of the Greenland ice sheet (figure 3.5) at a depth of 3,028 meters (GRIP Members 1993). The ice core obtained represents the precipitation occurring over the past 250,000 years. In 10,000-year-old ice still has an annual layer thickness of about 50 millimeters, allowing very detailed reconstruction of climate changes. Table 3.2 lists the parameters and corresponding ice core properties that provide climate information.

Results from analysis of the GRIP ice core confirm that the transition from the last glaciation to the Holocene occurred in distinct, very dramatic climatic shifts. Twice (at 14,500 yr BP and 11,700 yr BP) the temperature in Greenland rose rapidly (perhaps more than 5°C) in less than two decades (figure 3.6), and the annual precipitation rate doubled during the same short

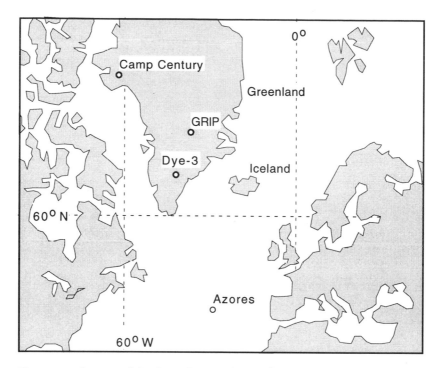

Figure 3.5 Location of the Camp Century, GRIP and Dye-3 coring sites in Greenland, seen from a North Atlantic perspective. Differences in the sea level pressure measured at stations on Iceland and the Azores are the basis of the North Atlantic Oscillation Index, which indicates the strengths of the Westerlies over the North Atlantic (see also chapter 2).

Table 3.2 Climatic parameters and corresponding ice core properties allowing extraction of climate information

Climatic parameter	Ice core data
Condensation temperature	Isotopes of H_2O
Precipitation rate	Isotopes, ions, dust, ^{10}Be, ^{36}Cl
Atmospheric circulation	Dust, ions, ^{10}Be, ^{36}Cl
Greenhouse effect	CO_2, CH_4, N_2O, ...
Cycles of C, S, O, ...	$\delta^{13}C$, $SO_4^=$, $\delta^{18}O$ of O_2
Atmospheric chemistry	CH_4, formaldehyde, H_2O_2
Solar activity	^{10}Be, ^{36}Cl
Volcanism	electrical conductivity, $SO_4^=$, Cl, ...
Sea-level changes	$\delta^{18}O$ of O_2
Biological activity	NH_4^+, methanosulfonic acid, CH_4

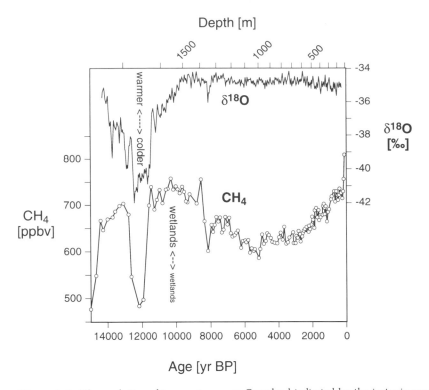

Figure 3.6 The evolution of temperature over Greenland indicated by the isotopic composition of preserved precipitation retrieved by ice coring, and the variation in atmospheric methane (CH_4) concentration in the Northern Hemisphere as revealed by the analysis of air bubbles occluded in ice cores from central Greenland. Methane is regarded as an indicator of global wetland extent. After the termination of the last glaciation, wetlands developed first at low latitudes, leading to a strong increase in methane concentration. The increase was interrupted during the cold, dry Younger Dryas around 12,000 cal. yr BP. Drier conditions prevailed after 8,000 cal. yr BP, leading to decreasing methane concentration, followed by a slow increase after 5,000 cal. yr BP as the northern wetlands started to grow. The anthropogenic methane increase began about 200 years ago. The record ends at 1900 AD. The present concentration is about 1750 p.p.b.v. The determination of the north-south methane concentration gradient by the analysis of Greenland, Antarctic, and mid-latitude ice cores will help assess the latitudinal distribution of wetlands in the past. This will improve knowledge of the global and regional water cycle, which is essential for investigating of regional climates.

span of time. Similar drastic climatic shifts during the Ice Age can in some cases also be correlated with large iceberg discharges recorded in deep-sea sediments. Further evidence for these rapid climate changes, which have been linked to major shifts in ocean circulation patterns, is now being obtained from an increasing number of continental archives in Europe. In addition, stable-isotope data from the GRIP ice core indicate that during the last interglacial, the Eemian epoch, global temperatures may have been on average 2°C warmer than they are today, so the Eemian epoch may yet provide an instructive analog for the warmer climate scenario predicted for the near future.

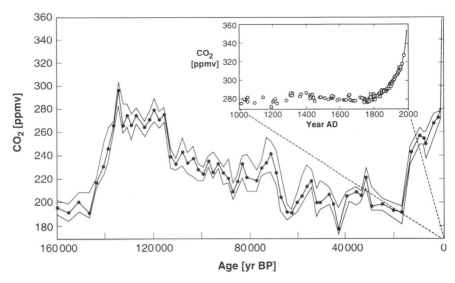

Figure 3.7 Composite carbon dioxide record from various Antarctic ice cores covering a full glacial cycle. The last thousand years are shown on an expanded scale. The ice core record overlaps with direct atmospheric measurements from Mauna Loa, shown as a solid line on the expanded scale (Keeling et al. 1995). Ice core data: Siple (Neftel et al. 1985), South Pole (Friedli et al. 1986; Siegenthaler et al. 1988), D47/D57 (Barnola et al. 1995), Byrd (Neftel et al. 1988), Vostok (Barnola et al. 1987).

The Greenland ice cores also contain environmental indicators derived from source areas ranging in extent from regional to hemispheric, as is shown for example in the methane concentration time series for the pre-industrial era (figure 3.6; Blunier et al. 1993; Blunier et al. 1995; Chappellaz et al. 1993). This methane was produced mainly in tropical and midlatitude wetlands, yet its concentration varied parallel to the local temperature in Greenland indicated by the stable-isotope record (figure 3.6). The rapid climate shifts were therefore not limited to the North Atlantic region.

Methane, like carbon dioxide, is a GHG. One of the more pressing questions about how climate may change over the next few decades concerns the role played by these atmospheric GHGs in determining climate. The key to a detailed understanding of this role is a record of past GHG concentrations. Very accurate atmospheric measurements of carbon dioxide concentrations have been made on Mauna Loa, Hawaii, since 1958, but no earlier instrumental records exist. The bubbles of air trapped in ice cores thus provide a unique opportunity to analyze the atmosphere's composition previous to the beginning of the Mauna Loa record (figure 3.7). The ice core record of carbon dioxide and methane overlaps smoothly with direct atmospheric measurements, thus tending to confirm the reliability of the records of past atmospheric composition. A complete record of these two GHGs over the last 150,000 years is now available. Concentrations of the two gases increased parallel to the global warming that occurred at the end of the last glaciation. The increased concentrations in part resulted from the temperature increase (mainly in the case of methane) and in part caused or at least amplified the increase (mainly in the case of carbon dioxide). Methane is also regarded as an indicator of global wetland extent, which is in turn a function of the climatically controlled water cycle. Concentrations of methane trapped in air bubbles in the ice thus indicate the changing extent of wetland areas.

Climate changes of the same magnitude as the warming we might expect from an increased greenhouse effect within the next century occurred last at the end of the Younger Dryas, near the termination of the last glacial stage (Eicher and Siegenthaler 1976; Oeschger 1987). Since then, a stability has characterized the climate of the Holocene epoch that has had no equivalent during the preceding 100,000 years. Why has this climate stability persisted over just the last 10,000 years, and what developments are programmed naturally for the near future? There is some evidence that the climate of the last interglacial (Eemian) era was less stable than that of the Holocene, at least in the North Atlantic region, so continuing studies of detailed ice core records from Greenland, and also from Antarctica, will help shed light on this critical question.

Polar ice core records reflect climate on a regional to global scale. Comparison of these records with records of a more regional and local character, like those from lake sediments, peat deposits, high-altitude glaciers, or tree rings (see section 3.2), can provide information on how a certain region has reacted to past global changes. This is particularly true for the climatically sensitive Alpine region, which is strongly influenced by the climatic developments over the North Atlantic, so ice core records from Greenland and the multiproxy records archived in the Alpine region should provide complementary time series for different ends of that large climatic system, helping reveal its changing character through time.

3.1.3 Overview

Throughout the earth's history, its climate has continually adjusted and readjusted in response to changes in the total net solar energy received and in this energy's dynamic distribution among atmosphere, hydrosphere, cryosphere, and biosphere. Thus, although the changes in global and regional climate that have undoubtedly been occurring over the last hundred years or so (e.g., figure 3.2) may appear alarming from the point of view of a complex human society that has come to rely tacitly on climatic stability for its smooth functioning, shifts and alterations in global and regional climate are normal, viewed from a long-term, less anthropocentric perspective. During the last 1.6 million years of the Quaternary period, major rhythmic alternations in the earth's climate between glacial and interglacial stages have left behind massive signatures on the continents and in the oceans. These alternations appear to be linked to the Milankovitch cycles, that is, periodic changes in the earth's orbital parameters that modify the amount of solar energy reaching the earth at any given latitude and time and that have peri-

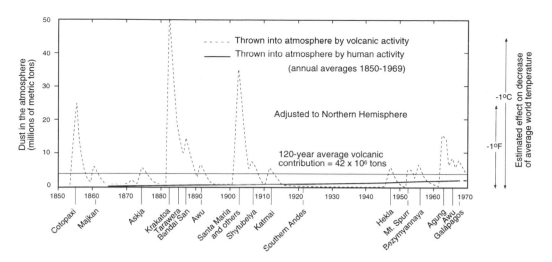

Figure 3.8 Comparison of dust in the earth's atmosphere due to volcanic eruptions since 1850 with that due to human activity. Possible effect on world temperature shown on right (after Mitchell 1970).

ods of approximately 100,000 years, 40,000 years, and 22,000 years. However, spectral analysis shows that the Milankovitch theory alone does not adequately explain climatic records. Other factors, including physical changes in the sun itself over various time scales (for example, eleven-yr sunspot cycles, total solar irradiance), global albedo, and a range of feedback processes occurring on earth (such as partitioning shifts in the carbon cycle and the greenhouse effect) are also involved. During the Holocene epoch (about the last 10,000 years), we have been in a relatively stable warm interglacial period, to which soils, vegetation, the animal kingdom and civilization have adapted. Nevertheless, low-amplitude climatic perturbations have continued to the present day. Less-regular regional phenomena acting on different timescales may also be of importance. Examples of these are the ENSO, periods of prolonged drought such as those experienced in the Sahel, and the climatic effects of volcanic eruptions (figure 3.8).

Because the climate signatures contained in both marine sediment records and polar ice records result from spatial integration over wide areas, they cannot provide detail for individual continental regions. Although the stable-isotope record from the GRIP ice core shows the climate during the Holocene to have been unusually stable on average, other continental paleo-evidence implies the occurrence of significant regional climate shifts during this epoch. Records of lake levels in tropical Africa, Central America, western Asia, and most recently, the montane Mediterranean zone of Morocco indicate substantial changes in humidity during the Holocene that the Milankovitch theory cannot explain (Kutzbach and Street-Perrot 1985) but that appear to be related to sea surface temperatures in the North Atlantic (Lamb et al. 1995; figure 3.9). Another example of a significant change in regional

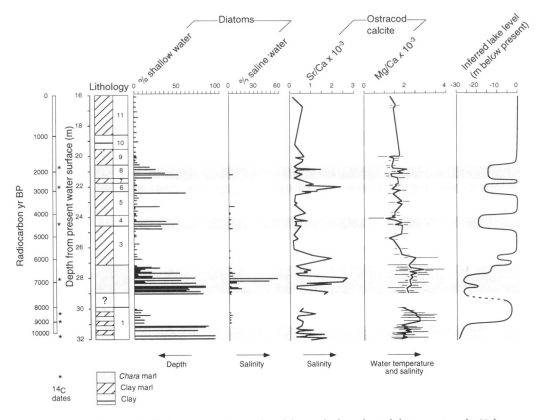

Figure 3.9 Shallow-water phases inferred from paleolimnological data spanning the Holocene in core C86 from Lake Tigalmamine, Middle Atlas Mountains, Morocco (after Lamb et al. 1995). Relative counts of shallow-water and salinity tolerant diatom species (lower levels) and plantonic species in intervening deep-water intervals supports interpretation of the lithographic units in terms of lake level. In addition, trace element measurements for Sr/Ca and Mg/Ca ratios in ostracod calcite, which respond to salinity and water temperature changes during evaporative drawdown, provide further evidence.

climate during the Holocene, one of direct relevance to the Alpine region, is the sharp long-term increase in summer precipitation runoff that isotopic and stratigraphic evidence (increased erosion and lacustrine productivity) from Swiss peri-Alpine lakes shows to have occurred at about 3,500–4,000 yr BP (Lister 1985). Global or hemispheric averages may therefore provide an overview of the total climate system, but the establishment of climate patterns, which involves distinguishing between different regions' or localities' responses to past climatic changes, logically requires regional and local proxy data with as high a spatial resolution as possible.

Having demonstrated the need to put recent short-term climate trends in a long-term perspective, we now consider in more detail the approaches and techniques employed in achieving this, with special reference to the Alpine region. The purpose of including these technical details is to give an impression of the difficulties and complexities involved in climate reconstruction.

3.2 CURRENT APPROACHES AND SIGNPOSTS TO THE FUTURE

3.2.1 Proxy Data

In 1977, COHMAP (the Cooperative Holocene Mapping Project) targeted the task of reconstructing continental paleoclimates. The COHMAP approach used both marine and continental paleodata and the CLIMAP results mentioned in section 3.1.2 to test the accuracy of model simulations intended to help elucidate climate forcing's spatial and temporal mechanisms. COHMAP's output to date includes a series of global maps of a range of meteorological variables at midwinter and midsummer for 18,000, 12,000, 9,000, 6,000 and 0 yr BP (COHMAP Members 1988; Kutzbach et al. 1993) and summaries of regional scenarios at the same points in time (Wright et al. 1993). As we saw in section 3.1, Europe had a comparatively long history of paleodata investigation prior to the COHMAP work, with respect particularly to relict glacial deposits but also to pollen records. More recently, the analysis of evidence from sources such as written historical reports, ice cores, tree rings and lake and swamp deposits, together with advances in radiometric dating, computing, and analytical techniques, has opened up new avenues of research and led to the reevaluation of earlier data sets.

Natural archives provide the longest data records, but for a variety of reasons, including the proxy nature of these data, differences in the evaluation methods employed, and natural patchiness of many data sets, they are often inherently problematic for obtaining quantitative reconstructions of climatic aspects. Nevertheless, accurate paleoclimate reconstructions based on natural archives are possible, assuming adequate calibration of individual proxy data, calibration cross-checks based on multiproxy data, and perhaps most critically, collaboration with climatology and atmospheric physics to maintain internal consistency and physical correctness.

Having examined a detailed short-term instrumental Alpine record (section 3.1.1) and considered some aspects of the larger spatial and temporal contexts (section 3.1.2), we now focus on some of the concrete approaches employed in reconstructing paleoclimates and look at some of the methodological issues, the processes involved, and the challenges we face in attempting to obtain accurate high-resolution paleoclimate reconstructions from the available proxy data.

Modern climate is usually defined in terms of directly measurable meteorological variables, such as air temperature, air pressure, and precipitation. Since no direct measurements of such meteorological variables exist prior to the very recent historical past, paleoclimates are defined in terms of estimates of these meteorological variables obtained from so-called proxy data. To be of value, proxy data must obviously exhibit a strong correlation with at least one of the meteorological variables of interest. The expressions "indicator" and "tracer" are generally used synonymously to denote the substance carrying the climate signature, that is, the substance yielding the proxy data. Concentrations of atmospheric gases, water temperatures, and aerosols in ice cores can be considered as direct proxies (primary indicators), whereas natural vegetation patterns derived from pollen studies might be considered a first-order indirect proxy, and freshwater plankton fossils, whose characteristics are most immediately determined by internal lake conditions, a second-order indirect proxy. The more indirect a proxy is in its representation of a climatic variable, the greater is the need to quantify the environmental steps through which that tracer gained its climatic information by investigating parallel tracer evidence. Furthermore, any one tracer may represent only one aspect of climate (in itself, a tracer indicating mean annual air temperature, for example, says little about humidity or wind characteristics). A multiproxy approach to reconstructions based on natural-archive evidence is therefore usually desirable.

3.2.1.1 Time Series and Synoptic (Time Slice) Approaches A time series approach involves the investigation of events of the same type occurring through time at one particular site. This approach is useful for establishing the rates of change (noncyclic) or the amplitudes and periods of fluctuations (cyclic) of a particular climatic variable at the given site.

A synoptic approach involves the display of spatial patterns (range, shapes, gradients) over an area for a limited period of time during which the temporal mean can be considered constant. The production of a synoptic map of a particular climatic variable at some given time in the past requires extremely accurate dating; in addition, a minimum density of suitably positioned network sites is needed to ensure adequate spatial resolution.

3.2.1.2 Chronology Chronostratigraphic control within paleodata time series is of critical importance for accurately correlating data from different sites or archive types, particularly in the case of time slice reconstructions. Data series with annual resolution (e.g., from lacustrine and marine varves, perennial ice, or tree rings) that extend continuously back from the present are clearly the most useful, but commonly series or parts thereof have "floating" chronologies, that is, are not anchored to an absolute date. Such series and those from other archives not annually resolvable therefore require the application of absolute dating methods. The most commonly used of these are radioisotope methods and event stratigraphy using marker horizons (e.g., ^{14}C, ^{137}Cs, ^{210}Pb, and U/Th). Sedimentological criteria can then be employed to assign ages to intervening levels.

3.2.1.3 Calibration Initial calibration of proxy data is commonly carried out against current climatic conditions and, if possible, over a range of environment types. However, such a calibration is not always entirely satisfactory, since the full range of meteorological conditions pertaining under any given climatic regime is rarely immediately obtainable. Further calibration of proxy data is possible by going back in time and calibrating first

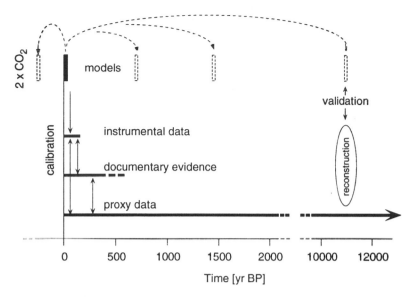

Figure 3.10 Schematic diagram illustrating typical time spans for which various types of observed data are available and indicating the potential for intercalibration of data sets and calibration of climatic models. Climate models provide time slices rather than time series and are validated against reconstructions based on proxy data for the past (e.g., the immediate past or the glacial maximum at 18,000 ^{14}C yr BP) with the intention of predicting the climatic effects of changing atmospheric conditions (such as a future doubling of the CO_2 content).

against available instrumental data (up to a century or so) and then against documentary evidence (three to eight centuries: figure 3.10). Clearly these calibration possibilities are not available to the same extent for all regions (for example, weather documentation is usually more comprehensive for those countries with long naval traditions or technological histories). The only hard evidence available for climate reconstructions extending further into the past lies in the natural archives (figure 3.10). Information in those archives, however, lacks "witness testimony," so evaluation can be problematic. Calibrations based on evidence from the last few centuries may not always be applicable further back in time, particularly if anthropogenic activity has modified tracer response. Furthermore, a given tracer may respond similarly to different environmental variables or remain unaltered between certain environmental thresholds.

This section outlines the applicability of a range of observational-data approaches to climate reconstruction for both the contemporary period and the more distant past, that is, on time scales of decades to millennia. In particular, a selection of proxy-data approaches currently being used, the applications and limitations of each, and their combined potential for the reconstruction of climatic conditions that have affected the Alpine region is discussed. The various archive types, their spatial and temporal record characteristics, the proxy data in each, and the range of evaluation approaches may be considered complementary elements in a broad hierarchy. Here we look at four kinds of proxy-data archives and the types of proxy evidence accessible in each. The first concerns *documented historical reports of weather and associated phenomena*, the other three the natural archives of *tree ring series*, *perennial ice accumulations*, and *lake sediments*.

3.2.2 Documentary Records of Weather and Its Effects

Data assemblages from critically evaluated historical reports of weather and associated phenomena form a useful archive for extending instrumental records back several centuries, thus also improving the calibration possibilities for natural-archive evidence. These kinds of data have been used to reconstruct time series reaching back several centuries for a range of climatic parameters and also for the synoptic mapping of pressure distribution fields associated with *Grosswetterlagen*. (See chapter 2 for a discussion of this concept).

Although systematic instrumental observations were first attempted during the late seventeenth century (e.g., Legrand and Le Goff 1992), official meteorological station networks have accumulated instrumentally measured data only over the last century or so. Before these networks, institutions, groups, or individuals in a range of official or private documents recorded information about weather. This documentary information is today classed as proxy data because it does not incorporate the standardized procedural objectivity associated with network instrumental data. Historically documented reports, often in annal or chronicle form and written in a variety of languages, commonly used various descriptive criteria or a mixture of description and simple instrumental measurements. Even though extreme weather events, because of their severe impacts on the community, were usually the most widely and comprehensively described situations, day-to-day weather conditions were also regularly recorded. Such material thus constitutes a pool of information that, suitably processed into a uniform format, can extend the climatic history available from network instrumental records by up to 500 years.

Recognition of the potential inherent in historical documentary data has recently led to a systematically coordinated and standardized methodology for normalizing and logging the meteorological information extracted from documents relating to Europe (Frenzel, Pfister, and Gläser 1994). The resulting Euro-Climhist database now comprises a range of standardized, high-resolution climate data contributed by partners from eighteen participating countries. This database makes it feasible to synthesize high-resolution spatial and temporal reconstructions of the weather conditions for Europe since the Middle Ages.

Historically documented information may include direct weather observation evidence or indirect testimony in the form of environmental impact evidence (Pfister 1992). Members of the urban élite recorded the bulk of information available. Chronicles from relatively remote Alpine valleys reported only extreme events, mostly floods, and so can be used only to reconstruct the magnitude and severity of natural disasters.

1. Direct documentary information, that is, a direct record of weather observations, has a number of strengths, including absolute dating control, high temporal resolution, high sensitivity to anomalies, and the ability to correlate with corresponding environmental impacts. Direct documentary information covers all seasons; this is particularly important for winter, which is often poorly represented by tracers related to seasonal biological activity. Shortcomings in this kind of information include record discontinuity, heterogeneity, and the selective perception biases different observers introduce. Nevertheless, normalized quantitative information can be derived from this source using relatively simple and robust mathematical techniques. Statistical comparisons can also be made if data derived from documentary sources overlaps with instrumental data from nearby meteorological stations.

2. Indirect documentary information, that is, observations of the impact of weather events on the environment, comprises mainly observations of events such as floods, low water table levels, the calendar dates of lake freeze-up and breakup and the duration of snow cover. Observations concerning the response of vegetation to weather events, including times of blossoming, ripening, and harvest, yields of grain and grape crops, and even the sugar content of grapes, also contribute seasonal climatic evidence. Some of these data types are continuous, quantitative, and homogeneous, allowing formal calibrations to be performed against instrumental records in much the same way as proxy data from natural archives are calibrated. Usually, if the written record was the work of one person, the extractable data series are limited to several decades at most, whereas plurisecular data series, such as records of grape harvest dates (Baulant and Le Roy Ladurie 1980), were accumulated within institutionalized frameworks, where continuous records generally spanned longer periods.

3.2.2.1 Methods Some fifteen years ago it was recognized that documentary evidence for various past climatic events was equivocal, compilations of such evidence often integrating both reliable and unreliable data. As a result, the earlier published overviews contain inaccurate or uncertain dating of particular events and spurious multiplication of events (Ingram, Underhill, and Farmer 1981; Pfister 1984; Alexandre 1987). This led to the development of a standardized methodology for evaluating sources and rejecting unreliable information.

Early this century, Brooks (1928) proposed a system of temperature indices based on documentary data to express the relative degree of winter severity. Lamb and Johnson (1966) later extended this procedure to calibrating summer precipitation trends. Le Roy Ladurie (1967) then calibrated grape harvest dates to produce a time series curve for estimated summer temperatures, which he compared to the known history of Alpine glacier fluctuations. Other types of serialized documentary data, such as the duration of snow cover, grape harvest yields, and tithe auction dates (Pfister 1984), and ice phenology and rye harvest dates (Vesajoki and Tornberg 1994; Tarand and Kuiv 1994), have subsequently also been calibrated in terms of meteorological variables. Pfister (1984, 1992) developed a scheme for obtaining weighted monthly temperature and precipitation indices based on calibrated proxies (snow and ice features, phenology) and historical descriptive data. This methodology, now proposed as an international standard within the Euro-Climhist project, has been used, for example, to synthesize a series of seasonal climatic conditions for the late Maunder Minimum (1675–1715; Pfister et al. 1994), during which Europe was affected by unusually low temperatures coinciding with a period of reduced sunspot activity.

The synoptic approach has produced useful spatial climatic syntheses (e.g., Lamb 1977) in which the lack of complete documentary coverage has proved less of a problem than originally anticipated. Test runs using contemporary meteorological data from sites distributed across Europe indicate that only a threshold number of sites is required for this purpose. The synoptic maps reconstructed from documentary evidence then provide a back check with respect to particular document entries; positioning single-site evidence within the reconstructed pattern can help identify inconsistencies. In addition, a mapped scenario can be used to interpolate enclosed areas for which no data are available, although this applies more to temperature (highly correlated over large areas) than to precipitation (more locally variable). Furthermore, reconstructed synoptic patterns also yield climatic information for

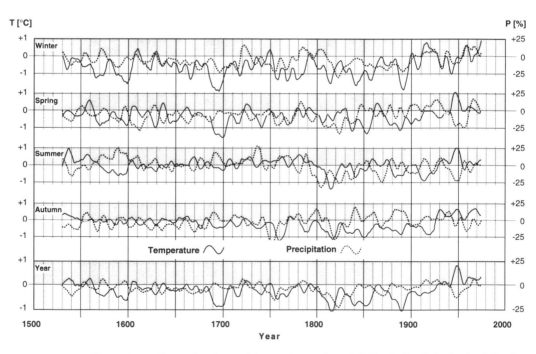

Figure 3.11 Seasonal and annual temperature and precipitation in the Swiss lowlands for the period 1560–1975, expressed as eleven-year moving averages of the departures from the respective mean values for 1901–60 (Pfister 1984, 1992).

areas beyond the mapped area, since these patterns can be characterized in terms of more-extensive climate scenarios.

The more common time series approach has been used to produce detailed series for seasonal and annual trends in temperature and precipitation. Figure 3.11 documents such records for the Swiss lowlands over the last 400 years (although autumn data are weak, and the annual information possibly includes statistical artifacts). Figure 3.12 depicts the frequencies and magnitudes of flood events in five areas of Switzerland for the same period. Such time series are particularly valuable for the calibration of and correlation with other proxy data sets. The winter information is rarely obtainable from any biologically influenced tracer or archive (vegetation, tree rings, insects, etc.).

3.2.2.2 Overlap with Evidence from Natural Archives The evidence of climate and its environmental and social impact contained in historical documents can have seasonal and site-specific aspects in much the same way as natural-archive proxy data do. The information documentary and natural-archive data provide can thus be used in a complementary manner. Focusing on the period of overlap between the two can result in better calibrations,

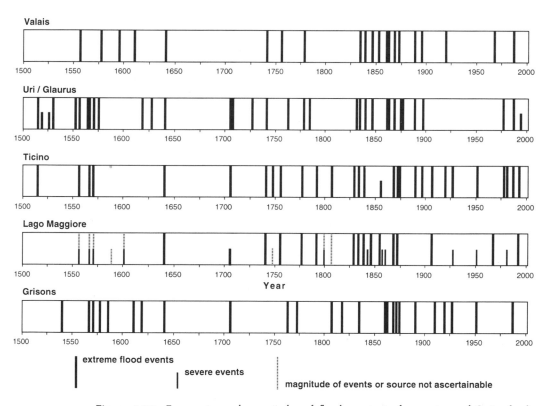

Figure 3.12 Frequencies and magnitudes of flood events in four regions of Switzerland (Valais, Uri/Glarus, Ticino and Grisons) and Lago Maggiore (Ticino), 1560–1975 (C. Pfister, in prep.). See Figure 3.1 for the locations of these regions.

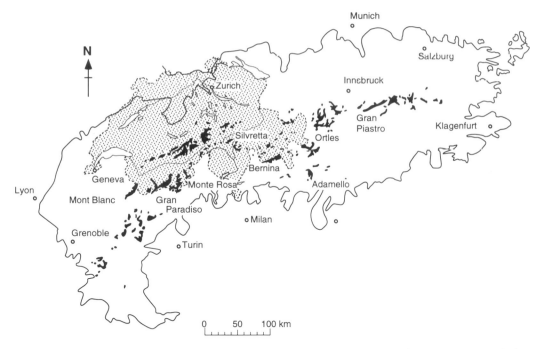

Figure 3.13 Extent of glaciation in the Alpine region today (black) in comparison with that during the last Ice Age (solid line). The hatched area represents Switzerlomd.

which, in conjunction with contemporary synoptic mapping, could provide a Rosetta stone allowing the interpretation of data from natural archives for the more distant past (Eddy and Oeschger 1991).

3.2.3 Alpine Glaciers and Permafrost

Glaciers and permafrost in cold mountain areas constitute the first of the natural archives addressed in this section. Because glaciers and permafrost react sensitively to changes in atmospheric conditions, the observation and reconstruction of their evolution through time furnishes valuable information about climate changes in the past. In the Alpine region, the information available relates mainly to the time since the last glacial maximum. Climate signals archived in the ice itself are also becoming increasingly available. Paleoglaciology, the science concerned with the analysis of this type of information, deals with the reconstruction and modeling of ice that has vanished as well as the various forms of old ice that still exist today.

Alpine glaciers are situated at altitudes cold and humid enough for perennial snow to have accumulated over the years (figure 3.13). Increasing overburden gradually compacts the snow into firn (multiyear snow) and ice, which eventually spills out from the main accumulation area and flows slowly down the valley to meet warmer conditions. The balance between the ice input above and melting losses below determines the position of the glacier tongue. In a steady-state condition, ice ablation exactly compensates for snow accumulation, and the glacier's mass balance is at equilibrium. During steady state, both the down-valley flow of ice within the glacier and the transport of rock debris to the accumulating terminal moraines continue. The glacier length remains stable unless shifting meteorological conditions affect mass input or loss, in which case a new equilibrium length is established. The length change that occurs directly relates to the glacier's original length and the change in mass balance at its surface. Present-day Alpine glaciers adjust to new equilibrium conditions within characteristic response times of several decades to about a century. Past glacier changes contain information on both precipitation and temperature. This combined information must be disentangled by using independent methods such as the analysis of permafrost distribution.

The term *permafrost* refers to subsurface material that remains frozen throughout the year. Permafrost's existence is closely related to air temperature; the reconstruction of past permafrost occurrences therefore provides information on past temperature conditions. In the Alps, permafrost is widespread above the 0°C isotherm of mean annual air temperature, which coincides roughly with the timberline. Depending on mean annual air temperature (altitude), solar radiation (slope aspect) and snow cover (which can be redistributed by wind and avalanches), permafrost depth may range from a few to hundreds of meters, corresponding to ages ranging from several years to hundreds of millennia. Perennially frozen debris cones and moraines, which can contain more ice than rocks, creep down mountain slopes like lava streams, building up landforms called rock glaciers. Changes in climate and surface conditions cause temperature changes at depth by heat conduction. This is a slow process, requiring decades to centuries in thin, warm permafrost and millennia in thick, cold permafrost.

The spatial relationship between glaciers and permafrost in mountain areas depends mainly on precipitation and hence on the degree to which the climate is continental. Under wet maritime conditions, the equilibrium line of glaciers (where snow accumulation and melting are exactly balanced) is at lower ("warmer") altitudes, glaciers are at melting temperature (temperate ice), and glacier tongues advance into permafrost-free and often forested areas. With decreasing precipitation, the equilibrium line retreats to higher ("colder") altitudes, where less melting can balance reduced snow accumulation, glacier ice is partly cold (polythermal) or even entirely cold, and glacier tongues are frozen to the underlying and surrounding permafrost. Observations of the distribution of glaciers in conjunction with that of permafrost thus provides information on paleoprecipitation. Ground ice in permafrost and, especially, cold firn and ice at high altitudes constitute interesting archives of paleoclimatic conditions.

The following deals first with paleoclimatic information from past glaciers and permafrost, followed by a summary of the results of research on ground ice and cold firn and ice at high altitudes.

3.2.3.1 Reconstruction and Modeling of Former Ice Conditions During the last glacial stage, glaciers were much larger than they are today, completely filling Alpine valleys and coalescing into large piedmont lobes after exiting to peri-Alpine areas. These large piedmont lobes were surrounded by permafrost and left behind eroded rock material that created typical geomorphological features as the ice melted back to the highest levels at the end of the last glacial stage. Paleoglaciological modeling, based on the relict physical evidence, has been used for reconstructions and to infer climatic conditions from the latest glacial stage in the Alpine region. Since the mid-1970s, a variety of modeling approaches have been used (Haeberli 1991) ranging from simple cryosphere schemes to complex flow models.

Comparisons of glacier and permafrost characteristics, ice flow and ice discharge calculations, and two-dimensional steady-state modeling indicate that progressive cooling and drying must have accompanied the buildup of ice to the point of its maximum advance. The piedmont glaciers extending out on to the peri-Alpine Swiss Plateau during the last glacial maximum at about 18,000 yr BP were polythermal (predominantly cold) and fringed by continuous permafrost (Haeberli and Schlüchter 1987). Glacier flow was weak, and mass exchange corresponded to "cold desert" climatic conditions. The mean annual air temperature calculated for that scenario is about 15°C colder and the mean annual precipitation about 80 percent less than at present. Rapid down-wasting and retreat of these ice bodies, largely completed by the mid-Bølling (about 12,500 ^{14}C yr BP: Lister 1985) and concurrent with sharp rises in air temperatures and precipitation rates, marked the closing stages of the last glacial stage. During the Younger Dryas readvance of Alpine glaciers prior to about 10,000 yr BP, mean annual air temperatures were about 4°C lower and annual precipitation about 30 percent lower than today (Kerschner 1985). The additional energy flux necessary to melt glaciers and permafrost completely during the time between 20,000 and 10,000 yr BP amounts to about 0.3 W·m^{-2}.

These reconstructions, however, still need to be tested against geophysical evidence (e.g., heat flow disturbance characteristics), ice core and glacial-deposit evidence, and proxy-data evidence from other natural archives (such as pollen records or lacustrine sediment cores) as well to improve the linkages between modeling, chronology, and environmental signal resolution (Ammann et al. 1994). A better knowledge of permafrost's spatial and temporal distribution during the latest glacial stage in conjunction with models relating glaciers to their bed conditions and groundwater and permafrost characteristics (Speck 1994) would contribute toward achieving these goals.

Since the last glacial stage, there have been a number of smaller Alpine glacier advances and retreats, for example during the Younger Dryas cold period at the very end of the last Ice Age or during the so-called Little Ice Age between about 1600 and 1850 AD. Such fluctuations in glacier length have left deposits visible today in various stages of preservation. Many of these have now been radiocarbon dated and form a Holocene time series of glacier length changes. Figure 3.14 presents an overview summarizing present knowledge of Alpine glacier length changes during the Holocene.

Secular mass balances essentially covering the twentieth century have been measured for six glaciers in the Alps by repeated precision mapping. The area-averaged annual mass loss over the entire period varies between about 0.2 and 0.6 meters water equivalent. Such values reflect an additional energy flux toward the earth's surface of a few $W \cdot m^{-2}$, corresponding roughly to the estimated magnitude of anthropogenic greenhouse forcing (Houghton, Callander, and Varney 1992; UNEP 1994). Average rates of glacier mass loss in the Alps during twentieth-century warming were probably about one order of magnitude higher than during the last glacial stage (table 3.3; figure 3.13). During the last decade (1980–90), they further increased by more than 50 percent with respect to the secular average. Glacier inventory data serve as a statistical basis for extrapolating such results and for simulating regional aspects of past and potential future climate change effects (Oerlemans 1993, 1994). A study of the Alps (Haeberli and Hoelzle 1995) indicates the total Alpine glacier volume to have been about 130 km^3 in the mid-1970s. The total loss in Alpine surface ice mass from 1850 to the mid-1970s can be estimated at about half the original value. An acceleration of this development, with annual mass losses corresponding to greater than 1 m water equivalent yr^{-1} or more as anticipated from IPCC scenario A (business as usual) for the coming century, could eliminate major parts of the presently existing Alpine ice volume within decades (cf. Maisch 1992).

Warming of Alpine permafrost (Vonder Mühll, Hoelzle, and Wagner 1994) since the late 1980s as observed in a borehole drilled in 1987 near Piz Corvatsch (Upper Engadine, eastern Swiss Alps) appears to have accelerated by a factor of about five to ten as compared with reconstructed secular permafrost warming (Haeberli 1994a). Moreover, the melting of frozen ground as inferred from rock glacier photogrammetry seems also to have accelerated markedly during 1980–90 in comparison with the previous decade. The ongoing degradation of Alpine permafrost involves energy fluxes smaller by about one order of magnitude than those involved in glacier melting (table 3.3). This is mainly a consequence of the retarding effects of heat conduction and latent heat exchange within the ground. Attempts are now being made to build a network for the systematic monitoring of Alpine permafrost (Haeberli et al. 1993).

3.2.3.2 Ice Cores and Other Glaciological Archives Ice cores from cold midlatitude glaciers are more complex to analyze and interpret than those from polar ice sheets, but they contain important climatic signals not available in any other natural archive (Wagenbach 1989). The small-scale glacier geometry at high elevations and the proximity of glaciers to continental source areas are the main features of midlatitude drilling sites relevant to ice core research. Proximity to continental sources is, on the one

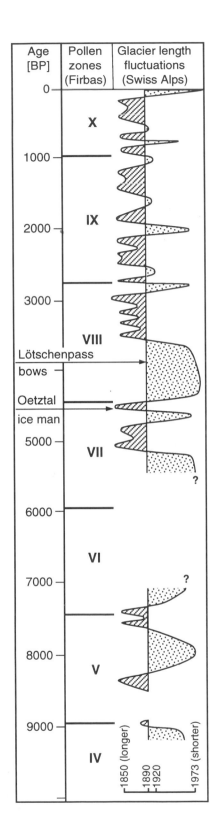

Table 3.3 Thermal characteristics, energy fluxes, and total amounts of energy involved in Alpine glacier and permafrost changes

Latent heat of fusion	3.338×10^5 J $\cdot$ kg^{-1}
Heat capacity of ice and rocks	2.100×10^3 J $\cdot$ kg^{-1} $\cdot$ K^{-1}
Characteristic long-term geothermal heat flow	0.1 W $\cdot$ m^{-2}
Estimated anthropogenic greenhouse forcing	2.5 W $\cdot$ m^{-2}
Ice Age glacier melting	0.3 W $\cdot$ m^{-2}
20th-century glacier melting	4.2 W $\cdot$ m^{-2}
1980–90 glacier melting	6.9 W $\cdot$ m^{-2}
Ice Age permafrost degradation	0.2 W $\cdot$ m^{-2}
20th-century permafrost degradation	0.3 W $\cdot$ m^{-2}
1980–90 permafrost warming	0.1 W $\cdot$ m^{-2}
1980–90 permafrost melting	0.5 W $\cdot$ m^{-2}
1980–90 permafrost degradation (warming + melting)	0.6 W $\cdot$ m^{-2}
Total energy in glacier melting since 1900	10^{10} J $\cdot$ m^{-2}
Total energy in permafrost degradation since 1900	5×10^8 J $\cdot$ m^{-2}

Note: from Haeberli 1994a.

hand, the most attractive reason for performing ice core studies on mountain glaciers, but on the other hand, a series of specific and sometimes unexpected problems emerge from this particular geographical situation, most connected with the seasonal and episodic changes of the atmospheric mixing height over the continent that controls the atmospheric impurity level at these sites (Wagenbach 1994a).

Exploratory core-drilling activities on cold, high-altitude Alpine glaciers have provided important experience upon which future projects can be based. Optimally placed drill sites should enable the investigation of the past few decades with an annual resolution in the meter range, of the past 200 years with an annual resolution in the decimeter range, and at least the past 1,000 years with less resolution (Haeberli and Stauffer 1994). Exploratory cores to some ten meters depth document present-day conditions and require a minimum of information on snow stratigraphy, borehole tempera-ture, and local depth to bedrock to be collected. Shallow cores to roughly half the ice depth, mainly involving the ice evolution during the twentieth century, can be dated quite safely by a combination of core analysis and two-dimensional modeling along the flow line to the borehole as calibrated by

Figure 3.14 Summary of current information on Alpine glacier length changes during the Holocene. The diagram indicates the radiocarbon ages of some archaeological artifacts (Oetztal ice man at Hauslabjoch and three bows from Lötschenpass) recently exposed by melting ice (Haeberli 1994a). See figure 3.1 for location of Lötschenpass.

measured profiles of bedrock depth, flow velocity, accumulation rates, snow stratigraphy and ten-meter temperatures. Deep cores reaching bedrock and involving Holocene ages must be treated with full three-dimensional flow considerations and correspondingly extended soundings and surveys around the drill site (Haeberli 1994b; Wagner 1994). Even then, the problem of time-dependent input to the model (for instance, ice geometry, accumulation rates, and so forth) remains unsolved.

Results from corresponding borehole and core analysis can provide important information with regard to the long term evolution of temperature, air mass circulation, snow accumulation, and atmospheric composition in mountain ranges. The $\delta^{18}O$ record measured in an ice core taken from Colle Gnifetti (Monte Rosa), for instance (illustrated in Figure 3.15), although deformed and only approximately dated, indicate that conditions in the Alpine region may have been warm during the early Holocene, but not warmer than during the twentieth century (Wagenbach 1994b). Information on the twentieth century, however, shows a change in $\delta^{18}O$ far larger than expected on the basis of observed air temperatures and not reflected in measured borehole temperatures (Haeberli and Funk 1991) but seemingly related to higher concentrations of insoluble dust (figure 3.15). The latter observation may suggest a change in the circulation pattern or in the fraction of summer precipitation contributing to the mean annual snow accumulation rather than a real increase in the local air temperature. Thorough inter-

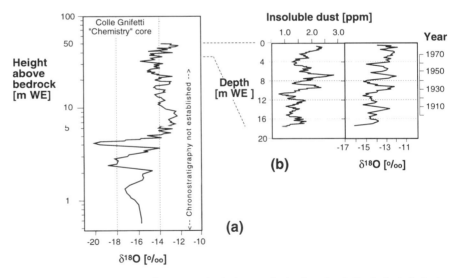

Figure 3.15 (a) Profile of $d^{18}O$ vs. ice depth down to bedrock in the Colle Gnifetti glacier ice core (slope with northern exposure; see figure 3.1 for location). Ice depth is displayed on a logarithmic scale to compensate partly for the thinning effect of the annual layers with increasing ice depth (low-pass filters are applied accordingly). After Wagenbach 1994b. A chronostratigraphy has not yet been established for the bulk of this record. (b) Low-pass filtered records of $\delta^{18}O$ and insoluble dust content for approximately the upper 17 meters of this core, which spans about the last 100 years.

pretation of such important records must be based on improved knowledge of the present-day situation with respect to local air chemistry, glacio-meteorology, and ice flow patterns (Wagenbach 1994a; cf. also Gäggeler, Baltensperger, and Schwikowski 1994). Colle Gnifetti and Col du Dôme (Mont Blanc) represent the main Alpine glacier reference sites (Funk 1994); continuation of work at these sites has high priority.

Scientific core drilling into ground ice of Alpine permafrost started at Murtèl/Corvatsch (Upper Engadine) in 1987. Flow considerations based on borehole deformation measurements and precision surveying of surface points indicate that the recovered cores of ice and frozen ground could date from the earlier part of the Holocene (Haeberli 1990). The potential of this newly opened archive requires further investigation.

Systematic investigation should also be undertaken of cold-ice patches. Even at low altitudes, wind-exposed ice crests and firn-ice divides are not temperate, but slightly cold and frozen to the underlying (permafrost) bed-rock. Such glaciological conditions (reduced heat flow through winter snow, no meltwater percolation, no basal sliding, low to zero basal shear stress at firn-ice divides) explain the perfect conservation of the Oetztal ice man (Haeberli 1994b), whose body had been buried under snow and ice in a small topographic bedrock depression on a saddle at Hauslabjoch (Austrian Alps; 3,200 m a.s.l.) about 4,500 conventional ^{14}C years ago (figure 3.14), remaining in place there until it melted free in 1991. At a lower altitude (2,700 m a.s.l.), but at a comparable site (Lötschenpass, Swiss Alps), three well-preserved wooden bows and a number of other archaeological objects were discovered as early as 1934 and 1944. Recent accelerator mass spectrometry (AMS) radiocarbon dating of the three bows gave dendrochronologically corrected ages of around 4,000 years (Bellwald 1992). Warming periods comparable to that of the twentieth century clearly have occurred before. The recent archaeological findings from melting ice in saddle configurations nevertheless confirm that the extent of glaciers and permafrost in the Alps may be less today than ever before during the Upper Holocene.

3.2.4   Tree Rings

Time series derived from tree rings became established as a paleoclimatic tool in the 1970s (Fritts 1976; Schweingruber 1988). Tree rings have both annual resolution and excellent radiocarbon dating possibilities, so they are ideal for correlating with time series obtained from meteorological stations, historical documents, and annually layered sediments or ice. Tree-ring properties depend largely on the climatic and hydrological conditions prevailing during the spring-to-autumn growth period, especially on the occurrence of exceptional droughts or cold periods, but also on mean summer temperatures in general. Tree-ring data for Switzerland are currently stored in a data bank at the Swiss Federal Institute for Forest, Snow, and Landscape Research in Birmensdorf and at the International Tree-Ring Data Bank in Boulder, Colorado.

Tree-ring studies can be divided into two categories: dendroecology and dendrochronology. Dendroecology, initiated by Hartig (1870), focuses on the relationships between growth conditions and the characteristics exhibited by the annual wood layers seen in cross section as tree rings. Dendrochronology concerns the chronological sequence of environmental events recorded in tree-ring series.

In temperate latitudes, trees add wood seasonally as a wide, light-colored zone of large, loosely packed cells in spring and as a narrower, dark-colored zone of small, densely packed cells in summer. Interannual differences in weather conditions affect each tree's annual growth characteristics, which in turn affect tree-ring widths and cell densities. Microclimatic and genetic factors also affect growth characteristics, so the number of specimens of a particular species in a stand of trees that react to a given climate condition is important. Suitable trees in Europe include the oak (Becker et al. 1985), pine, spruce, and larch. In addition to living trees, dead wood from various sources also provides valuable information. The dead wood employed includes old building timber and subfossil wood preserved in lakes, bogs (Eronen and Zetterberg 1992), or river sediments (Schweingruber, Schär, and Bräker 1984) or at the alpine and polar tree lines (Shiyatov 1992). Trees from tree lines are particularly useful for climatic work, because temperature is clearly the limiting growth factor there.

Trees in the Alpine region have an average age of about 100 years, so to construct long time series using wood with overlapping growth periods or to correlate overlapping series, cross dating is necessary. This method (Douglass 1941), critical to dendrochronology, assigns an absolute calendar year to each tree ring analyzed. The longest Alpine tree ring record to date, from spruce and fir trees at Lauenen in west Switzerland, spans the last 1,000 years (Schweingruber, Schär, and Bräker 1984).

Time series can be constructed of both tree ring widths and summer cell densities. Tree-ring width series form the basic dendrochronological framework for cross dating and correlation. For temperate regions, tree-ring width series are difficult to evaluate in terms of a continuous climate record, and it is likely that only extreme years (summer droughts, cold summers, and short-term temperature anomalies) can be climatologically interpreted (Kienast 1985). Tree ring cell density series, measured for summer cells at the widest part of each annual ring, indicate relative mean temperatures for the warm half of the year (April to September in the Northern Hemisphere), high cell densities being associated with high temperatures. A number of statistical procedures, initially developed by Polge (1966), must be applied to the raw data to relate cell density and temperature time series (e.g., long-term ring-width changes due to tree aging must be compensated for) and derive relationships between growth and climate (Cook and Kairiukstis 1990).

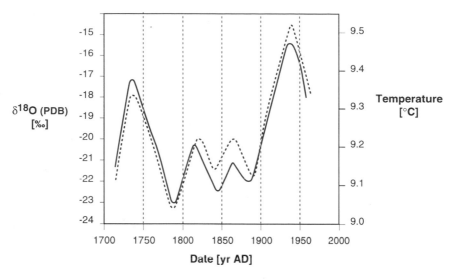

Figure 3.16 An oxygen isotope curve determined for a German oak (*Quercus petraea*: dotted line) compared with the annual mean air temperature curve for central England since the early 1700s (solid line). Curves have been smoothed with a thirty-year running mean. Air temperature curves from Basle and Geneva resemble that for central England but are shorter (after Libby 1987).

3.2.4.1 Stable Isotopes and Tree Rings Stable-isotope signatures for oxygen, hydrogen, and carbon in tree rings have also been measured and interpreted in terms of climatic and environmental factors. Such records have the benefits of excellent chronostratigraphic control, tree-ring sequences reaching back to 8,000 yr BP or earlier, and the availability of parallel data on growth characteristics. Oxygen and hydrogen isotope ratios have shown the most promising results to date. Libby (1987), for example, demonstrated a strong correlation between isotopic indices from German oak and instrumental records of mean annual air temperatures from Basle, Geneva, and central England spanning up to the last three centuries (figure 3.16). Using a multiproxy approach, including 2,000 years of tree-ring isotopic evidence from Japanese cedar, Greenland ice core oxygen isotope records, and marine core (fluctuations in organic carbon production), Libby (1987) has also demonstrated how solar activity, via sea surface and local air temperatures, controls the ultimate isotopic signatures preserved in tree rings.

Carbon isotope ratios in tree rings, affected by the $\delta^{13}C$ of atmospheric CO_2, the CO_2 concentration, and light and moisture stress (decreased stomatal conductance leading to elevated $\delta^{13}C$ values) show distinct seasonal variations. Weighted mean monthly $\delta^{13}C$ averages for short records have been correlated with solar radiation, measured soil moisture, and precipitation (Leavitt and Long 1991). Sauer, Siegenthaler, and Schweingruber (1995), in pilot studies using beech, pine, and spruce in Switzerland, concluded that short-term variations in precipitation and soil moisture can be reconstructed from carbon isotope signatures in tree rings for relatively dry sites. However, they also point out that poor understanding of relevant processes (e.g., a universal temperature coefficient has not yet been established) still hampers interpretation.

3.2.4.2 The Alpine Tree-Ring Record Both north and south of the Swiss Alps, normalized cell density time series from mountain fir, spruce, white pine, and larch correlate well with each other, with summer temperatures in Switzerland, and also with data from other parts of Europe (Schweingruber 1988; figure 3.17); and indeed correlations have also been established with records from as far away as Siberia and North America.

These results indicate that local growth factors such as light exposure, ground type, and precipitation play a subordinate role to mean summer temperature in controlling maximum cell densities, implying the existence of a potential for spatial reconstruction. Although specific maps for the Alpine region are not yet available, Schweingruber, Briffa, and Nogler (1991) have published 226 maps of mean summer temperatures in Europe from 1750 to 1975. For example, figure 3.18 compares the measured and reconstructed departures of summer temperature from the long-term mean pattern for Europe in 1963. Schweingruber, Briffa, and Nogler (1991) and Schweingruber and Briffa (1996) have also used conifer late-wood density series to reconstruct summer temperature distribution patterns for the Northern Hemisphere boreal zone since 1600. This ongoing work provides an example of how paleodata can be incrementally integrated upward from local through regional to global scales and provide selected levels of resolution for interfacing with other types of reconstructions and simulations.

3.2.5 Lakes

By their very nature, mountain regions tend to be well endowed with rivers and lakes. Because of the rapidity of their response and their relative isolation, lacustrine ecosystems are particularly suitable for studying climate change's ecological effects. This is important not only because of climate change's effect on aquatic ecosystems now and in the future but also because lake sediments represent one of the most abundant sources of high-resolution paleoecological proxy data for reconstructing past climates.

3.2.5.1 Trends in Lake Water Temperature and Ice Cover in the Alpine Region Climate change affects lacustrine ecosystems indirectly, mediated by the physical lake environment, of which water temperature and ice cover are two important aspects. Model studies suggest that one of climate change's effects is an increase in the temperature of rivers (Stefan and Sinokrot 1993) and lakes (Robertson and Ragotzkie 1990; Hondzo and Stefan 1993), and measurements support this scenario (Schindler et al. 1990). High-altitude or high-latitude lakes that freeze over annually are therefore likely to experience a long-term decrease in the annual duration of ice cover as the

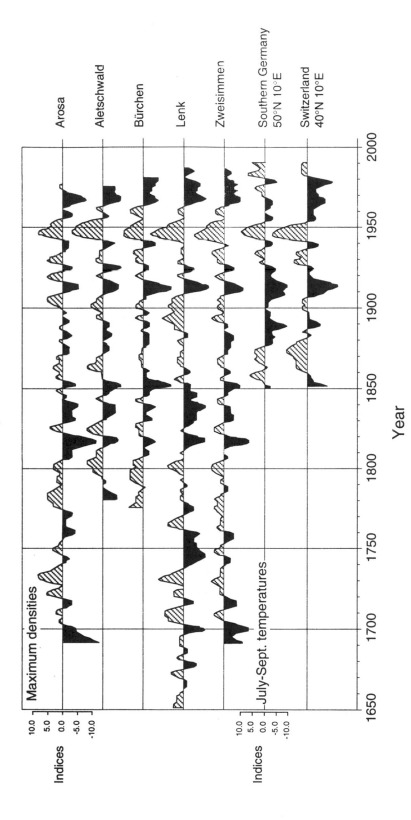

Figure 3.17 Time series showing the relationship between mean index curves of maximum cell density for spruce (*Picea abies*) from five sites close to the timberline in the Swiss Alps (the five upper curves) and the mean summer temperatures (July–September) for two stations, one at longitude 10°E, latitude 50°N in southern Germany, and the other at longitude 10°E, latitude 45°N in Switzerland (the two lower curves). Maximum cell density values for each curve have been standardized to eliminate the age trend. The temperature curves are derived from the Northern Hemisphere network data of Jones et al. (1986). All curves have been smoothed with a thirteen-year moving-average filter.

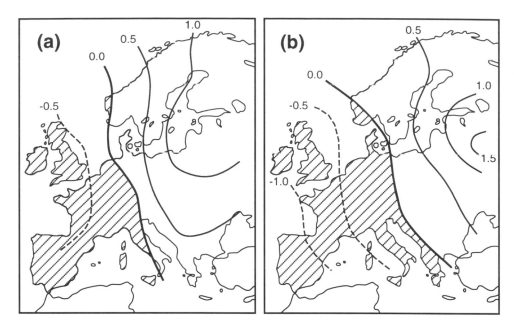

Figure 3.18 Isotherm patterns of the departures of annual summer temperatures in 1963 from the long-term mean. The meteorological observations (a) correspond closely with the indices obtained from tree ring cell density measurements (b).

climate warms. The impact on aquatic ecosystems of higher lake temperatures in summer and shorter ice seasons in winter is highly relevant to the employment of lake sediments as proxy data: Chironomid remains, for instance, are considered good indicators of lake water temperature (Walker et al. 1991; Walker, Mott, and Smol 1991), and diatom assemblages are sensitive to ice cover (Smol 1983, 1988). Equally important, however, is the potential value of the calendar dates of freeze-up and breakup of lake ice as proxy air temperature data in areas with a paucity of historical meteorological data (Gordon et al. 1985; Palecki and Barry 1986; Robertson, Ragotzkie, and Magnuson 1992; Assel and Robertson 1995).

As we saw in section 3.1.1, secular trends in air temperature over the last hundred years, both globally and regionally, are well documented; however, the response of lake water temperature and ice cover to recent climatic changes is not. Given such an obvious long-term increase in surface air temperature as that illustrated in figure 3.2, an increase in the temperature of surface waters over the same period of time might also be expected. Five main processes determine a lake's heat balance: (1) the absorption of direct and diffuse short-wave radiation from sun and atmosphere, respectively; (2) the absorption of long-wave radiation from the atmosphere; (3) the emission of long-wave radiation from the lake surface; (4) the exchange of latent heat between the lake surface and the atmosphere because of evaporation and condensation; and (5) the convective exchange of sensible heat between the lake surface and the atmosphere (e.g., Livingstone and Imboden 1989). The three radiative processes are the most important for lake heat budgets in absolute terms, whereas air temperature's direct causal influence is of only minor importance. Despite this, the essential similarity in the physical processes that determine surface air temperature and surface water temperature results in a high degree of correlation between the two variables. In shallow lakes, the mean lake temperature follows the surface air temperature closely. Figure 3.19a shows that the mean lake temperature of Lützelsee, a small, shallow lake (maximum depth 6.2 meters) situated about three kilometers from Lake Zurich in the Swiss lowlands, mirrors even short-term changes in air temperature faithfully. (Note the effect of the cold snap in June–July 1980.) In larger, deeper lakes such as Lake Zurich itself (maximum depth 136 meters), however, only the lake surface temperature reflects short-term variations in air temperature (figure 3.19b), and then only in spring and summer, when the surface mixed layer is comparatively thin and mixing with underlying colder water layers is minimal. The mean temperature of the uppermost 20 meters of the lake exhibits a longer-term integral response to meteorological forcing (figure 3.19c), which also involves a phase shift. The entire water body's overall response is even more inertial (figure 3.19d), the high-frequency temperature variations being progressively attenuated with increasing water depth (Livingstone 1993a). Air temperatures and water temperatures can therefore be viewed as indicators of essentially the same climatic trends, although with different response times and amplitudes (which in lakes are depth dependent).

Comparison of monthly surface air temperature anomalies (obtained by subtracting the seasonal variation) from the Zurich meteorological station with monthly Lake Zurich water temperature anomalies from (1) the lake surface, (2) the uppermost twenty meters of the lake and (3) the entire lake (figure 3.20) reveals that in the longer term, air and water temperatures exhibit a high degree of common response to climatic forcing even in the case of relatively large lakes. This similarity is especially evident in the case of the lake surface temperature anomaly (figure 3.20b), the temporal structure of which is not only qualitatively but also quantitatively comparable to that of the corresponding air temperature anomaly (figure 3.20a). The integrating effect of vertical mixing processes in the lake manifests itself in decreased amplitudes and a reduction in high-frequency fluctuations with increasing water depth (Livingstone 1993a). Pair-wise correlations between the annual mean values corresponding to the data series illustrated in figure 3.20 (forty-eight years of data) reveal that air temperature at Zurich correlates highly not only with lake surface water temperature (64 percent of variance explained) but also with mean lake temperature (30 percent of variance explained). Thus despite the (partly wind-dependent) integrating and smoothing effects involved, the interannual response of water temperatures in large lakes to climatic forcing is similar, although not identical, to that of air temperature.

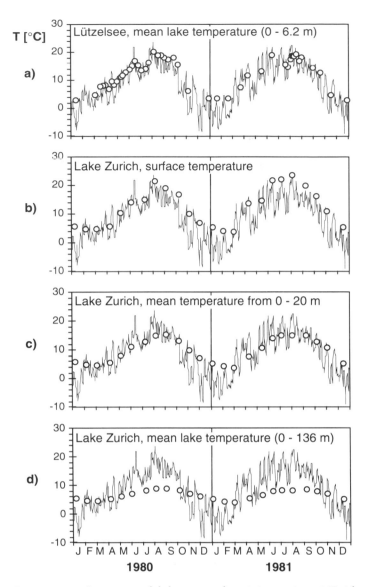

Figure 3.19 Comparison of daily mean surface air temperature at Zurich meteorological station in 1980 and 1981 (solid lines) with various lake water temperatures (open circles): (a) mean lake temperature of Lützelsee (adapted from Livingstone and Schanz 1994); (b) surface temperature of Lake Zurich; (c) mean temperature of the upper twenty meters of Lake Zurich; (d) mean lake temperature of Lake Zurich. The two lakes lie only three kilometers apart, but Lützelsee (maximum depth 6.2 meters) is much shallower than Lake Zurich (maximum depth 136 meters) and consequently responds faster and more completely to changes in meteorological factors. See figure 3.1 for locations of lakes.

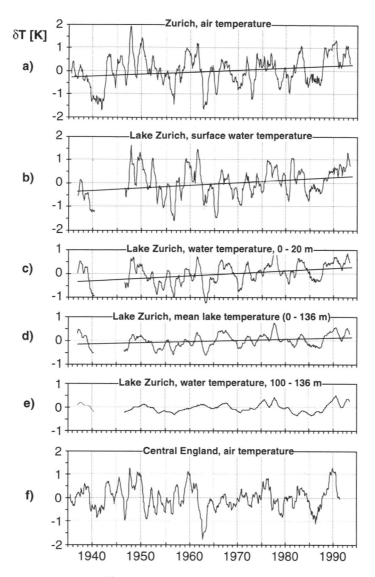

Figure 3.20 Monthly mean anomalies (δT, twelve-month running means) of (a) surface air temperature in Zurich; (b) surface water temperature of Lake Zurich; (c) mean water temperature from 0–20 meters in Lake Zurich; (d) mean lake temperature of Lake Zurich; (e) mean water temperature from 100 to 136 meters in Lake Zurich; and (f) surface air temperature in central England (data from Manley 1974 and Parker, Legg, and Folland 1992). The trend lines represent linear regressions based on the unsmoothed monthly anomalies. From Livingstone (in prep.). See figure 3.1 for locations of Zurich meteorological station and Lake Zurich.

Figure 3.20f illustrates the monthly mean anomalies of the central England surface air temperature data series of Manley (1974), extended by Parker, Legg, and Folland (1992), for the same time period as the Zurich data. The similarity in temporal structure between the Zurich and central England air temperature anomalies is immediately obvious; a linear relationship between the two explains 52 percent of the variance (cf. Pfister 1992). Consequently, the Lake Zurich water temperature anomalies exhibit a strong similarity to the central England air temperature anomalies. This implies that, to a first approximation, interannual variations in surface and mean lake water temperatures represent (at least in this case) a response to synoptic-scale climatic forcing. This does not necessarily apply to deep-water lake temperatures (e.g., figure 3.20e), however, which can respond in a highly idiosyncratic fashion to the constellation of meteorological events occurring in spring, depending, for instance, on local wind exposure or trophic state (Livingstone 1993a).

In addition to fluctuations on a time scale of several years, both the surface temperature and the temperature of the upper twenty meters of Lake Zurich (figure 3.20b,c) have apparently been undergoing a long-term increase at a rate of approximately $0.01 \text{ K} \cdot \text{yr}^{-1}$ since at least 1936, when observations began. This is essentially the same rate of increase observed for surface air temperature over the same time period. However, there is no statistically significant long-term trend in the deep-water temperature (below 100 meters) of Lake Zurich, or of other deep Swiss lakes such as Lake Geneva, Lake Zug, and Lake Neuchâtel (Livingstone 1993a), and so the rate of increase of the mean lake temperature is much less than that of the surface temperature. This implies that thermal stability in Lake Zurich, and presumably in other deep Swiss lakes, is undergoing a long-term increase as a result of the increase in surface temperatures. In the case of Lake Zurich, the long-term increase in thermal stability has reduced the period of potential lake circulation in winter by about ten days over the last fifty years and correspondingly prolonged the period of stratification (Livingstone, in prep.). Such long-term shifts in seasonal lake circulation patterns will change the environmental conditions to which phytoplankton, zooplankton, and other lake inhabitants are exposed. Some of these changes will likely be detrimental to aquatic ecosystems that depend on the presence of sufficient oxygen for their survival: Shorter circulation periods and longer stratification periods, for instance, can be expected to decrease deep-water oxygen concentrations, especially in eutrophic lakes, expanding the anoxic (oxygen-depleted) zone upward (Livingstone and Imboden 1996).

Long data sets from small, shallow lakes are rarely available; however, in view of the close correspondence between air temperature and the mean temperature of Lützelsee apparent in figure 3.19a, it can be assumed that shallow lakes are also being subjected to long-term temperature increases of about $0.01 \text{ K} \cdot \text{yr}^{-1}$. However, increasing temperatures in shallow lakes will increase oxygen consumption markedly (Livingstone and Schanz 1994) and will therefore also result in deteriorating environmental conditions of aerobic ecosystems.

The magnitude of the long-term trend in Lake Zurich's surface temperature corresponds to the results of model predictions made for other lakes. The modeling of lake temperatures in Wisconsin (Robertson and Ragotzkie 1990) and Minnesota (Hondzo and Stefan 1993), for instance, suggests that the increase in surface air temperature resulting from a doubling of atmospheric CO_2 will be paralleled by an increase in lake epilimnetic temperatures (essentially the temperatures of the upper five to fifteen meters) ranging from about 50 percent to 100 percent of the corresponding increase in surface air temperature. Hypolimnetic (deep-water) temperatures, on the other hand, are predicted to increase comparatively slightly in shallow lakes and even to decrease in deeper lakes. The earlier establishment of a stable water column at lower hypolimnetic temperatures because of higher net heat fluxes in spring would cause such a decrease. Climate change resulting from a doubling of atmospheric CO_2 is predicted to prolong the duration of summer stagnation by as much as sixty days in shallow lakes and forty days in deeper lakes (Hondzo and Stefan 1993). The observations from Lake Zurich tend to support these model predictions.

As mentioned above, the short-term temporal structure of the monthly surface air temperature and water temperature anomalies at Zurich is similar to that exhibited by Manley's (1974) central England data series (figure 3.20), but the latter lacks the linear trend apparent in the corresponding Swiss air and water data series from 1936–93. Thus, although temperature variations on a time scale of a few years appear to be a large-scale phenomenon, recent longer-term trends (i.e., over the last fifty years) may be less spatially homogeneous.

Because earlier data are not available, a detailed quantitative picture of interannual variations in lake temperature can be obtained only for the last few decades. However, the freeze-up and breakup dates of lakes that freeze over every year can provide some information on long-term changes in lake temperature regimes and, by implication, air temperature. For example, Palecki and Barry (1986) used lake freeze-up and breakup dates as indices of air temperatures in Finland, and Rannie (1983) used river freeze-up and breakup dates to obtain information on nineteenth century spring and autumn air temperatures in southern Manitoba. Robertson, Ragotzkie, and Magnuson (1992) and Assel and Robertson (1995) used records of lake freeze-up and breakup dates to determine regional winter temperature changes in the Lake Michigan area. In Japan, the 500-year-long Lake Suwa freeze-up data series (the first observation was recorded on December 18, 1397: Arakawa 1954) has been employed to estimate winter temperatures in central Japan (Gray 1974; Tanaka and Yoshino 1982), and, combined with tree ring data, to link the severity of winter in Japan to large-scale surface circulation patterns over the North Pacific from 1600 onward (Gordon et al. 1985).

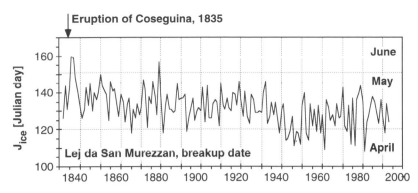

Figure 3.21 Breakup dates of ice cover (Jice, as Julian day) for Lej da San Murezzan, a lake in the Upper Engadine area of southeast Switzerland (see figure 3.1), from 1832 to the present (adapted from Livingstone 1997).

Breakup dates are available for various Swiss lakes, but usually only over periods of a few decades at most. One exception to this is Lej da San Murezzan (1,768 m a.s.l.; surface area 0.78 km^2; maximum depth 44 meters), an Alpine lake for which a continuous series of ice breakup dates from 1832 to the present is available (figure 3.21; Livingstone 1997). The long-term trend apparent in this series suggests that the breakup date has been moving forward over the last 150 years at an average rate of 7.6 days per century, from about May 18 at the beginning of the record to about May 5 now. Breakup dates of lakes in other parts of the world exhibit similar trends (e.g., 7.7 days per century for Lake Mendota, Wisconsin, and 5.6 days per century for Lake Baikal, Siberia). A trend toward earlier breakup of lake ice thus appears to be a global phenomenon, yet another indication of globally increasing surface air temperatures. Note the exceptionally late breakup dates for Lej da San Murezzan in 1836 (June 8) and 1837 (June 7) following the eruption of the volcano Coseguina (Nicaragua) in 1835. The violence of this eruption and its computed dust veil index (Lamb 1972) were comparable to those of the much better-known Tambora eruption nineteen years earlier. Cooling due to stratospheric aerosols of volcanic origin probably contributed to the unusual lateness of breakup in these two years.

The ice breakup dates for Lej da San Murezzan and other neighbouring lakes correlate strongly with mean April air temperatures all over Switzerland (Livingstone 1997). Although the highest proportion of variance explained (r^2) is associated with air temperatures at the station closest to the lake ($r^2 = 51$ percent at Bever, about nine kilometers from the lake), values of r^2 for stations situated all over Switzerland at various altitudes on both sides of the Alps are also high, ranging from 37–47 percent at eight stations located from 30 to 230 kilometers from the lake. The strong relationship between the thawing dates and mean April air temperatures in central England ($r^2 = 29$ percent for Manley's series) and the Netherlands ($r^2 = 27$ percent for the De Bilt series) further emphasize the large-scale character of the meteorological processes driving the thawing phenomenon. Pfister (1992) has already shown that Manley's (1974) data correlate fairly strongly with monthly Swiss temperature indices; nevertheless, it is perhaps surprising that data from central England, 1,200 kilometers distant, correlate so well with the breakup dates of an Alpine lake. Even the Northern Hemisphere April mean air temperature data of Jones et al. (1986) show a significant correlation with the lake ice breakup dates but explain only 4 percent of the variance. The thawing dates for Alpine lakes thus apparently depend to a large extent on regional climate. Lake ice breakup dates (where available) are therefore likely to constitute a useful set of proxy data for synoptic-scale meteorological processes.

In lakes subject to freezing over, the duration of ice cover is a major factor influencing spring oxygen concentrations (Livingstone 1993b). In contrast to its effect on lakes that do not freeze over, a warming trend resulting in a long-term decrease in the duration of ice cover will increase the spring oxygen content of lakes that do freeze over, which will tend to offset the expansion of the anoxic layer associated with the longer stratification period (Livingstone and Imboden 1996). Whether long-term warming will have a detrimental or beneficial effect on lake oxygen concentrations will therefore depend on the individual characteristics of the lake concerned.

3.2.5.2 Lake Sediment Characteristics Lake sediments provide excellent natural archives for a spectrum of atmospheric, terrestrial, and aquatic environmental indices, especially in deep-water basinal regions, where the characteristically slow, continuous settling of suspended sediment particles results in the gradual accumulation and preservation of natural tracer materials. Indeed, such depositional sequences provide continental records for past environmental conditions most closely analogous to those of the deep marine setting in terms of sedimentological criteria, continuity, and preservation. An added advantage of lake sediments over their marine counterparts is that the normally higher rates of sedimentation and limited bioturbation (bottom fauna reworking surface sediments) can result in continuous stratigraphic records spanning millennia that can have very high resolution (in some cases annual or even seasonal). These archives can contain unique sets of environmental tracers recording local, regional, and global signals, along with sufficient terrestrial organic matter to allow sequential radiocarbon dating.

Because few of us have the chance to familiarize ourselves with lacustrine sedimentary systems, the components of sediment, their sources, transport mechanisms, and the follow-on processes that contribute to the lacustrine sediment archive's final character will be outlined here.

Lacustrine sediment components originate from many different sources and result from a variety of processes. Water-borne inorganic materials transported into the lake by streams and rivers commonly include silt, sand, gravel, and cobble-sized material eroded from catchment rock and soils. Such material is termed "detrital" or "clastic." Specific mineral assemblages in a detrital sediment fraction usually allow the identification of their geological origins within the lake's catchment area. Fine detrital material ("eolian dust") and volcanic ash may also be transported aerially to a lake; such materials are therefore not necessarily representative of catchment geology. Volcanic ash, because of its characteristic mineral composition, spatial distribution, and eventlike deposition, can provide particularly good time marker horizons. Both water and wind transport to lakes a range of faunal and floral debris (e.g., pollen, seeds, insects) in addition to detrital rock components.

Organic particles may also be generated within a lake, particularly by phytoplankton and zooplankton in the biologically productive upper water levels (epilimnion and metalimnion), but also by organisms in the less productive deeper water levels (hypolimnion). Silt-sized calcite grains may precipitate as a by-product of plankton blooms, leading in some cases to thin, distinct summer deposits. At the bottom of the lake, diagenetic processes may result in minerals' being either precipitated out or dissolving in the sediments as chemical gradients build up over time in pore waters.

Particles of any kind that have been transported into a lake are termed "allochthonous," whereas those formed within the lake are referred to as "autochthonous." The proportion of allochthonous to autochthonous material depends mainly on the balance between erosion rates in the catchment area and biological productivity in the lake. For example, in a proglacial lake (a lake in which sedimentation is dominated by glacial silt), the combination of sediment load, low temperatures and lack of nutrients severely restricts biotic processes. Alternatively, a lowland lake surrounded by well-developed soils and vegetation may have low detrital inputs and be very productive, as evidenced for example by lakes' greenish color during algal blooms in spring. If a lake becomes too productive, the oxidation of settling organic material in the hypolimnion can deplete dissolved oxygen to the point where life forms depending on aerobic respiration can no longer exist and anaerobic microbiological and chemical processes take over. Anaerobic bottom-water conditions can result in the formation of black sediments (due to the presence of iron sulphide) and also tend to result in better preservation of organic matter.

River flow has sufficient energy to transport coarse material, but upon reaching a lake, the energy drops suddenly as the flow disperses. The coarsest sediment components come to rest in shallow water, usually close to the shore on deltas or along beaches. Offshore shelf areas subject to wave action (shallower than about five meters) are commonly sandy, drift currents having carried the finer silts further offshore. Down the lake-ward slopes to deeper waters, the sand gives way predominantly to silts and clays.

Because the deep waters generally have a very low energy regime and the particles in offshore suspension are finely grained (silt and clay), basinal regions mostly experience a continuous drizzle of particles that settle permanently, in contrast to the more discontinuous sediment deposition and resuspension that occur in shallow coastal waters. Grain size characteristics (range, size distribution, and degree of sorting) and depositional bedding criteria (thickness, internal structure, size grading, and bed contact) can be used to interpret modes of emplacement (rate, process, and duration), which, in turn, can be linked to a wider range of environmental controls (snow melt, storms, and long-term hydrological and hydrographic changes).

In contrast, grain size and sorting of particles formed in the water column itself depend more on particle growth rate and sinking rate. Exceptions to this general textural pattern across the bottom of a lake can result from underflow, whereby river water with a high density due to its sediment load or low temperature may flow into the lake along the lake bottom, transporting coarser detritus to deeper areas, and also from downslope redeposition of deposits by slumps, slides, and turbidity currents. Figure 3.22 illustrates some of the more important processes contributing to lacustrine sediment records.

Deep-water deposits receiving sediments settling from suspension commonly have a laminated depositional structure (sediments are said to be "laminated" if the individual layers are less than one centimeter thick and "bedded" if thicker). Typical laminated structures result from depositional pulses controlled by catchment drainage characteristics and seasonally dependent variations in the lake's physics and biology. In some lakes, the dominance of one particular seasonally periodic process, or of a limited number of such processes acting in concert, results in the deposition of annual laminations known as "varves," which can allow very accurate relative dating (e.g., Lotter 1991). Since the confirmed presence of varves is the exception rather than the rule, other dating methods, including radioisotope analysis and event stratigraphy (for example, volcanic-ash marker horizons) must also be employed, ages for intervening levels being obtained by interpolation, taking into account sedimentological criteria. High resolution is critical if sediment records are to be correlated either with other sediment records or with data series from other types of archives.

Because of their fine-grained, soft mud nature, basinal lacustrine sediments can usually, be penetrated to depths of up to ten meters or more with relatively simple and manageable coring devices (such as gravity corers and piston corers). Extruded cores are halved lengthwise to expose cross sections of the sediments for detailed description and logging and for the extraction and analysis of tracer materials. Seismic reflection profiling of the lake's sub-bottom structure usually precedes the coring process to determine the densities (grain size, compaction) and spatial extents of the various sediment layers and to decide on optimal coring sites. An approach combining seismic surveying and sediment coring thus allows the three-dimensional reconstruction of sedimentation history, a basic frame of reference needed for any core-based environmental investigations. In Swiss lakes, sediment cores shorter than ten meters usually contain continuous records spanning the last 10,000–15,000 years.

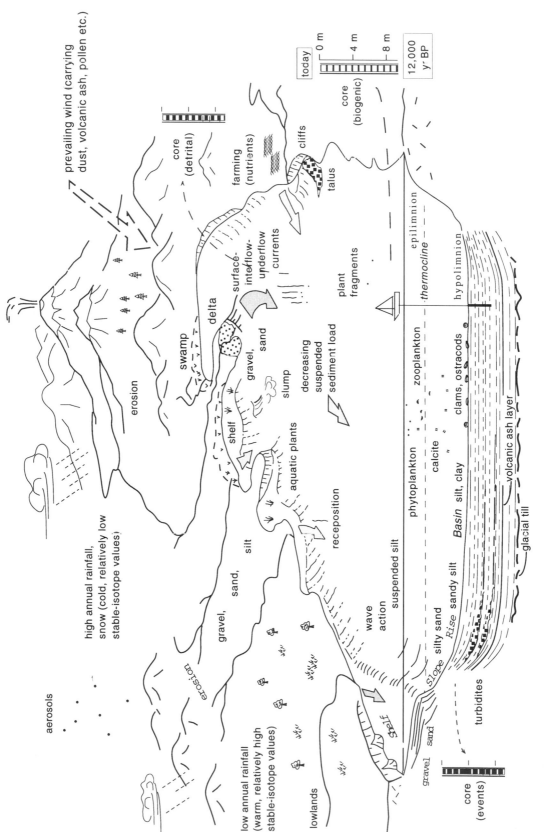

Figure 3.22 Schematic depiction of sediment sources and processes contributing to the final environmental record preserved in lacustrine sediment archives. Cores extracted from different parts of the lake, together with seismic reflection profiles showing sub-bottom depositional structures, allow a reconstruction of system responses to climate impact.

Environmental proxy data accumulating in lakes can be highly varied and can include evidence relatable to climate (e.g., stable isotope signatures, pollen, insect remains); atmospheric fallout (natural and artificial radionuclides and aerosols); productivity and mixing in lakes (aquatic floral and faunal remains, mineral precipitates); and frequencies and magnitudes of natural catastrophic events such as floods, volcanic eruptions and earthquakes (sedimentological criteria); as well as the downstream effects of deforestation, agriculture, and pollution.

Difficulties associated with obtaining good proxy-data sets from lacustrine sediment archives include the establishment of continuous time scales, gaps and disturbances in the sedimentary record due to sediment reworking (mass slumps and slides; turbidity currents induced by earthquakes, internal waves and floods; resuspension by bottom currents), bioturbation, and postdepositional chemical alterations. Separating the different levels of signals (local, regional, global) is a significant task requiring, as a first step, the calibration of tracer responses to modern conditions.

Lake systems, physically well defined, are found throughout a range of different climatic regions. Each lake accumulates environmental evidence, in the form of proxy data carried by tracers, unique to its setting under a given climatic regime. Any shift in climatic conditions will have an impact on the lake system and its archived tracer record, mediated, for example, by changes in hydrology, ecology, and catchment erosion stability. The Alpine region includes a large range of subenvironments and so reacts sensitively to climate change. The wealth of lacustrine systems in the Alpine region therefore makes the region particularly suitable for studying the impacts of climate change.

Processes governing the input of particles into a lake or the formation of particles within a lake depend in complex ways on the regional climate, the local vegetation, and the physical, chemical, and biological conditions prevailing within the lake. Comprehensive sediment analyses, including chemistry, mineralogy, biology, and paleomagnetic investigations, are worthwhile if continuous high-resolution archives are present.

3.2.5.3 Lacustrine Varves and Glaciers

If present, varves provide the highest level of resolution for the chronologies critical to lacustrine tracer studies (e.g., Peglar 1993). If the varves extend up to the modern sediment surface, they also facilitate the absolute dating of sediment levels. Investigation of varve sequences from Swiss Alpine lakes has shown that varve thicknesses and grain size can be linked, for example, to glacier lengths, mean summer temperatures, and rainfall (e.g., Lister et al. 1984; Leemann 1993).

De Geer (1913) introduced the term "varve" to the scientific literature, the Swedish word *varv* meaning either a cycle or a periodic repetition of layers. De Geer used this term to define a regular sedimentary sequence of dark and light couplets of laminae, each representing deposition during one year.

De Geer's definition of a varve, originally applied just to annually distinguishable detrital silt deposits in glacial or periglacial lakes, was subsequently used to construct high-resolution chronostratigraphies and to teleconnect deposits in Scandinavia and North America (Sauramo 1923; Antevs 1951; Schove 1969). Today, the term "varve" is used in a much wider sense and is applied to literally any annually laminated deposit, whether glacial or nonglacial, lacustrine or marine in origin (O'Sullivan 1983).

Two kinds of lacustrine varves illustrate the range of processes that lead to the deposition and preservation of annual laminae in freshwater lakes.

Biological/chemical varves are formed when particles produced in the lake water by biological productivity and/or chemical precipitation settle out differentially during the year (Nipkow 1920; Kelts and Hsü 1978; Sturm 1984; Lotter 1989). For example, in hard-water (carbonate-rich) lakes, seasonal phytoplankton blooms cause both a sharp decrease in PO_4-P and an increase in pH, resulting in the precipitation of silt-sized carbonate crystals (PO_4-P inhibits nucleation: see Kunz and Stumm 1984). Initially precipitated carbonate crystals reach about 40–60 µm in diameter and are deposited as a light-colored layer in spring and early summer. Later in the year, the deposition of smaller, more slowly formed carbonate crystals (less than 5 µm) together with the remains of the second phytoplankton bloom (pennate diatoms) forms the dark-colored, organic rich layer characteristic of the autumn-winter period in an individual varve (figure 3.23a).

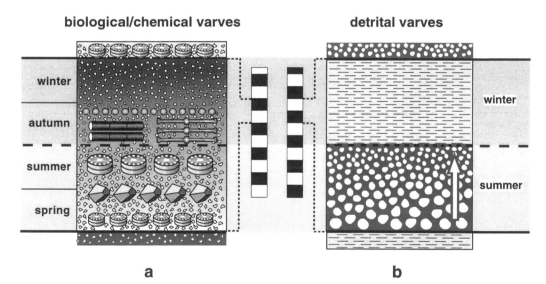

biological/chemical varves   **detrital varves**

Figure 3.23 Two characteristic varve types: (a) In biological/chemical varves, a light-colored layer of centric diatoms and large calcite crystals (deposited during spring and summer), alternating with a dark layer of organic material with pennate diatoms and very small crystals of calcium carbonate (deposited during autumn and winter), represents each year; (b) In proglacial lakes, detrital varves comprise a thick, coarse, light-colored summer layer alternating with a thinner, more finely grained dark layer deposited during autumn and winter. In distal glacial lakes, these varves can display a dark summer and a light winter layer.

Biological/chemical varves are restricted to lakes with very low detrital input. Such lakes commonly have small inflows, water renewal times exceeding one year, and low total sedimentation rates (less than $1 \ kg \cdot m^{-2} \cdot yr^{-1}$). They usually display strong thermal and sometimes chemical stratification, which may lead to meromictic water conditions during years of poor mixing. Many such lakes are classified as mesotrophic (lakes of medium productivity) or eutrophic (lakes of high productivity). Bottom waters are usually anaerobic, so benthic fauna, which can destroy laminae through bioturbation, are absent.

The composition of biological/chemical varves relates both to catchment conditions and lacustrine productivity. For example, evidence from biological/chemical varves in Soppensee in the Swiss lowlands provides proxy data for the rates and extents of forest expansion and for the change in lacustrine productivity that occurred in response to climatic warming at the end of the last glacial stage (Lotter, Ammann, and Sturm 1992). Another abrupt environmental response at the end of the Younger Dryas cold period indicates that the climate ameliorated within decades (Sturm and Lotter 1995), as also seen in the Greenland ice core data.

Detrital varves in lakes comprise essentially detrital particles that have been eroded and transported from within the catchment area in annual pulses of meltwater (such as that from glaciers) or precipitation and deposited in the low-energy lacustrine environment (figure 3.23b). Glacial varve couplets usually consist of a thicker, coarser, light-colored summer layer and a thinner, more finely grained dark layer deposited during fall and winter (Leonard 1986; Renberg 1986; Glen and Kelts 1991; Leemann and Niessen 1994a, 1994b). In contrast, varves formed in lakes situated far from a glacier that nonetheless receive glacial meltwater can consist of a dark summer layer and a light winter layer (Sturm and Matter 1978). Since detrital-varve lakes have, by definition, very low nondetrital components, such lakes are generally oligotrophic (i.e., productivity is low). They are usually thermally stratified during summer, whereas in winter, vertical mixing occurs throughout the whole lake. Water renewal times are less than a year, sedimentation rates are high (significantly greater than $1 \ kg \cdot m^{-2} \cdot yr^{-1}$), and the lakes are aerated down to the sediment-water interface (Sturm 1979). Benthic organisms have minimal bioturbating effects because of the low nutrient concentrations and high sedimentation rates. Detrital-varve deposits have low organic content, usually confined to sparse floral remains.

Detrital-varve thicknesses and grain size characteristics in Alpine lakes have been correlated with both mean annual air temperatures and cumulated snowfall (e.g., Lister 1984; Leonard 1985; Leemann and Niessen 1994b; Desloges 1994). Detrital varves in proglacial Lej da Silvaplauna (see figure 3.1 for location) exhibit temporal variations in thickness and grain size that correlate with two different controls: (a) in the short term (historical), with shifts in mean summer-temperatures, and (b), in the long term (Holocene), with glacier size changes in response to mean annual temperature, precipita-

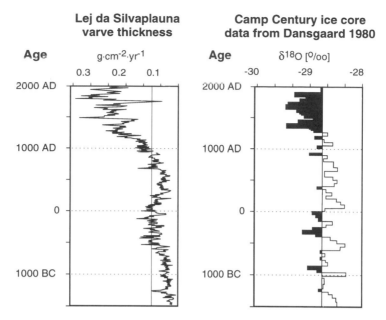

Figure 3.24 Comparison of detrital varve thicknesses (smoothed with a ten-year running mean) in Lej da Silvaplauna, a peri-Alpine proglacial lake (see figure 3.1 for location), with the Greenland ice core oxygen isotope record (redrawn from Dansgaard 1980). The long-term changes in varve thickness are related to glacier size, with the thicker varves here correlating with both a longer Alpine glacier and colder conditions in Greenland (Leemann and Niessen 1994b).

tion shifts, or both. The varve-thickness trends in this high-Alpine record also parallel trends in the oxygen isotope record obtained from the Greenland ice core (figure 3.24; Leemann and Niessen 1994b), additional evidence that regional climatic trends over the Alps during that time were strongly related to developments over the North Atlantic.

Lang (1885) was probably the first to compare meteorological data with data from glaciers. He suggested that glaciers retreat and advance mainly because of changes in precipitation and to a lesser extent, because of temperature changes. Glaciers are now known to react very sensitively to changes in one or more climate variables (e.g., Patzelt and Aellen 1990; Tangborn 1980; see also section 3.2.3). Glacier history does reflect climate history but is discontinuous, because generally only glacier high stands leave clear evidence (Leonard 1986). A proglacial lacustrine record is, however, continuous back to when the glacier last occupied a particular lake basin. De Geer (1912) used such varve records to infer changes in the rate of glacier retreat in southern Sweden during the last glacial stage.

Proglacial lacustrine varve records enable detailed reconstructions of up-valley glacial history (Leonard 1986; Leemann and Niessen 1994b; Karlén 1981), and linkages between sedimentological, geochemical, or geophysical characteristics of varved sediments and meteorological or hydrological parameters have been proposed (e.g., Leemann and Niessen 1994a; Leonard

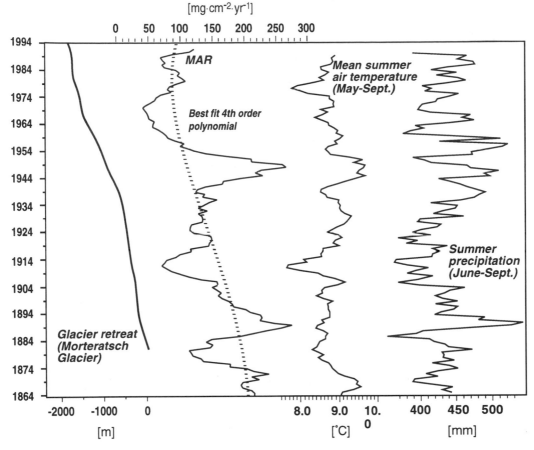

Figure 3.25 Variations in Lej da Silvaplauna varve thickness (expressed in terms of mass accumulation rate, MAR) compared with the history of the retreat of the nearby Morteratsch glacier, with mean summer air temperatures, and with cumulative summer precipitation (from the Sils-Maria meteorological station, located about one kilometers south of the lake). The curves for MAR values, temperature, and precipitation are smoothed with a five-point running mean. Note that the short-term varve thickness responses to summer temperature trends (i.e., thicker/warmer) are superimposed on the long-term varve thickness changes due to glacier size, which have the opposite relationship with temperature. See figure 3.1 for locations of Lej da Silvaplauna, Morteratsch glacier and Sils-Maria meteorological station.

1985; Leonard 1986; Desloges 1994). Recently obtained varve data from proglacial Alpine lakes show that temporal variations in varve thickness or, equivalently, in mass accumulation rate (MAR) can be viewed as integrated signals for both long- and short-term environmental changes (Leemann and Niessen 1994b; Leonard 1985) that can be traced separately. Low-frequency (decadal to century) MAR variations reflect changes in glacier size (figure 3.25), high MAR values being associated with large glaciers and low summer temperatures. In contrast, high-frequency (annual) MAR variations reflect interannual changes in runoff. In partly glaciated catchments, runoff is controlled by both rainfall and glacial melting. Consequently, in peak runoff years, sediment load transport is high. The high MAR values associated with this are related to either exceptionally high rainfall and/or exceptionally warm mean summer temperatures enhancing glacial melting. The combined data (figure 3.25) indicate that mean summer temperature exerts the stronger control on MAR, accounting for most of the variability in the MAR curve. The correlation between the MAR data and summer air temperatures are particularly good for the periods 1906–22 and 1939–57. However, it is also clear from Figure 3.25 that the MAR curve cannot be fully explained by summer air temperatures alone. For example, the peak in the MAR curve from 1885–97 is most probably due to exceptional summer rainfall.

Weather extremes—for example, the unusually cold summers of 1909–16, usually warm summers of 1942–53, or high rainfall around 1890—leave MAR signatures that correlate well with meteorological records (cf. Saarnisto 1983), thus providing a tool for extending the extreme-event record obtained from documented reports (see section 3.2.2).

In summary, annually laminated lacustrine sediments provide a unique tool for the study of past environmental change. With respect to temporal resolution, their potential is exceptional. The ability to distinguish single years, or even seasons, makes it possible to reconstruct high-resolution patterns for a range of natural indices. Past environmental events of short duration and high intensity can be traced, and biotic responses to environmental disturbances studied, at high resolution over long periods. Varve sequences also provide sensitive monitors of the anthropogenic use of both catchment areas and lake waters.

3.2.5.4 Biostratigraphic Tracers

Paleoecological and paleolimnological studies of lacustrine deposits may provide the long time series needed not only to reconstruct past environmental conditions, but also to assess biotic and abiotic systems' natural variability. Furthermore, they can yield information about these systems' reaction to different perturbations and, if a good time control is available, it is possible to estimate the phases and amplitudes of disturbances. Among subfossil biotic remains commonly studied in sediments are pollen, diatoms, coleoptera (beetles) and chironomids (midges).

Pollen grains are microspores (containing the male gametophytes) higher terrestrial and aquatic vegetation produces. Their cell walls consist of sporopollenins; these highly polymerized carotenes and carotene esters resist decomposition in reducing environments such as lake and mire deposits. Various media, such as insects, wind, and water, disperse pollen and spores, usually with a production and dispersal bias toward wind-pollinated taxa. Lakes trap pollen; hence a particular lake's pollen catchment depends on its open-water surface area (Jacobson and Bradshaw 1981; Prentice 1988; Sugita 1994). Pollen and spores can be identified down to family, genus, and in some instances species level on the basis of their rich morphology and aperture types (Fegri and Iversen 1989; Moore, Webb, and Collinson 1991). The concepts underlying the reconstruction of past climates from biostratigraphies, many originally developed for pollen records (e.g., Birks 1981; Bartlein, Prentice, and Webb 1986; Guiot 1987), are discussed below.

Diatoms are algae with siliceous cell walls. Lake sediments usually preserve the skeletal silica well. Diatoms have played an important role in the development of quantitative methods for environmental reconstruction. Early work in this direction concentrated on the relationship between diatom assemblages and the degree of acidity or alkalinity (pH) of lake water, which is related to bedrock geology and to the input of acid pollutants from the atmosphere (acid rain). First attempts to derive past lake pH from fossil diatom assemblages were based on regression analysis (see, for example, Battarbee, Smol, and Meriläinen 1986; Renberg and Hellberg 1982). In the late 1980s, much effort was put into developing robust mathematical methods for calibrating and for estimating errors reliably (ter Braak 1987; ter Braak and Looman 1987; Charles and Smol 1988; Birks et al. 1990). Using these techniques, it became possible to reconstruct, from fossil diatom assemblages, not only past pH values but also dissolved organic carbon (DOC) concentrations and salinity. Hall and Smol (1992), Agbeti (1992), and Anderson, Rippey and Gibson (1992) have reconstructed past concentrations of phosphorus, a major nutrient for algal growth, using diatom-based calibration functions.

Duration of ice cover on a lake may strongly favor benthic species among planktonic diatom taxa (Smol 1983, 1988). Diatom assemblage characteristics related to salinity changes may reflect past fluctuations in the water levels of closed-basin lakes, governed principally by the difference between evaporation and precipitation. Fritz et al. (1991) presented calibration functions for diatoms that allow quantitative reconstruction of past salinities in North American lakes. Direct relationships between the occurrence and abundance of diatom species and summer water temperatures have recently been described (Pienitz, Smol, and Birks 1995; Vyverman and Sabbe 1995); studies similar to these are presently being conducted in the Swiss Alps.

Coleoptera are insects that in many cases are individually adapted to a narrow environmental niche (see, e.g., Coope 1986). Wind and water carry insect remains from surrounding terrestrial ecosystems, or these remains originate in the lake itself. Their chitinous exoskeletons resist decomposition in lake sediments, where oxygen is usually absent. Classical studies have used coleopteran remains to infer average July air temperatures (e.g., Elias and Wilkinson 1985; see figure 3.26). Such studies employ only those taxa independent of vegetation to avoid circular argumentation with respect to climate reconstruction. Atkinson et al. (1986) developed the "mutual climatic range method," a numerically based calibration technique for coleoptera data that expresses species composition changes in terms of mean January and July air temperatures.

Lacustrine sediments also commonly preserve head capsules of chironomid larvae, which live in lakes and streams. These larvae exist in deep lake waters in relation to the presence of oxygen in the hypolimnion, which in turn is

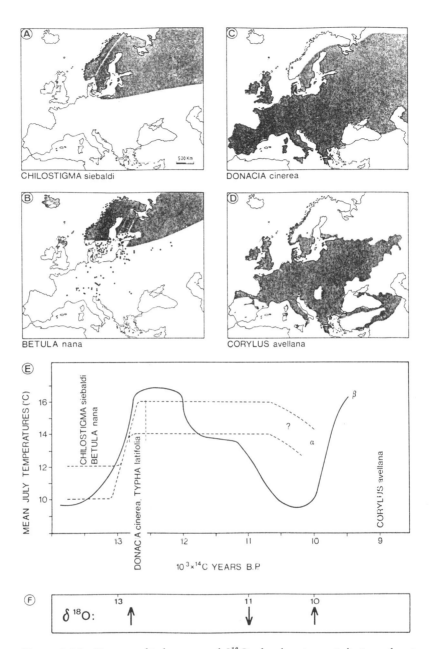

Figure 3.26 Biogeographical ranges and $\delta^{18}O$ of carbonates as indicators of past summer temperatures. (A-D): present-day ranges of species of (A) a caddis fly, (B) the dwarf birch, (C) a beetle, and (D) the hazel. (E): derived mean July temperatures between 14,000 ^{14}C yr BP and 9,500 ^{14}C yr BP for (a) the Swiss site Lobsigensee (Elias and Wilkinson 1983) and (b) central England (Coope 1977). (F): Three main shifts in the $\delta^{18}O$ values of precipitated carbonates indicating rapid warming around 12,800 ^{14}C yr BP and 10,000 ^{14}C yr BP and a cooling at 11,000 ^{14}C BP (beginning of the Younger Dryas: Eicher and Siegenthaler 1983). The two cold-adapted species, (A) and (B), co-occur before 12,800 ^{14}C yr BP. The beetle (C) and the aquatic plant *Typha latifolia* (cat's tail) record the rapid warming around this time. The late arrival of (D) may be due either to migrational lag or to climate (e.g., sufficiently warm summers, but very cold winters or cold springs).

related to the trophic state (productivity level) of the lake (Hofmann 1986). North American studies indicate that the presence and abundance of chironomids is also related to summer water temperatures (Walker, Mott, and Smol 1991).

Floral and faunal remains such as the above, preserved in lake sediments and peat deposits, provide useful evidence in establishing past climatic and other environmental conditions. However, to establish these conditions adequately, a number of assumptions must be made and requirements met. Each taxon's geographic range reflects its specific climatic requirements (e.g., Scandinavian species or Mediterranean species). The principle of uniformitarianism—"the present is the key to the past"—assumes that these climatic requirements have changed minimally with time, at least over the period concerned. Moreover, there must be enough floral and faunal remains in a suitable condition to permit definite identification. The validity of these assumptions, however, has some restrictions. Climate does not exclusively control biogeographical limits, for instance. Ecological factors, such as soil development or the migration of taxa, that occurred after the last glacial stage may also be relevant. Furthermore, many species develop different ecotypes. Although these ecotypes all belong to the same species morphologically, their physiology varies somewhat, allowing the species to adapt to a variety of specific environmental conditions (such as temperature, humidity, and nutrients); and, in contrast to morphology, physiology unfortunately does not usually fossilize. In addition, incorporation into an archive (taphonomy) may transform the signal of interest and may also add supplementary information, or "noise," to the signal of interest. The best ecological (and therefore climatological) information requires fossil remains to be identified down to the species level (e.g., identification of pollen of the Swiss stone pine, *Pinus cembra*, provides more ecological information than just "pine pollen," because many species of pine occur between Scandinavia and the Mediterranean).

Traditional paleoecological studies are often descriptive and narrative (Birks 1992). Processes responsible for observed patterns in the proxy records must be inferred from the fossil record and the sediment in which these fossils are found. The studies often describe changes in past environmental conditions (climate, vegetation, and so forth) only qualitatively (e.g., warm/cold, wet/dry). To test hypotheses concerning past environmental changes and to validate biological models and climate reconstructions, it is necessary to calibrate biotic proxy data against the relevant environmental control variables.

Two basic approaches to reconstructing past climates use the relationship between climate and the biogeographical range of a species as observed today (Birks 1981):

1. The *indicator-species approach* uses the presence or absence of certain species to estimate one or several climatic parameters (most often the mean January and July temperatures: see, for example, Iversen 1944 or Grichuk

1969). This approach was later developed into the methods of response surfaces of taxa to climatic gradients (see Bartlein, Prentice, and Webb 1986) and to the "mutual range method," based on overlapping the climate space of several species (see Atkinson, Briffa, and Coope 1987). Nevertheless, mean January and July temperatures may be less important to organisms than extreme temperatures. Moreover, the effects of competition between species may at times override climatic constraints.

2. The *assemblage approach* takes into account the quantitative composition of both surface samples and fossil samples (multivariate approach).

Past environments are quantitatively reconstructed in two steps. First, calibration functions that require a training set of modern surface sediment samples containing the organism or parameter of interest and associated environmental data must be established by regression. Second, the calibration of subfossil proxy data in a sediment core can be attempted by applying these modern calibration functions to the subfossil data. Several basic assumptions, however, underlie any quantitative paleoenvironmental reconstruction (Birks 1995):

• The taxa in the modern surface sediment samples are systematically related to the physical environment in which they live.

• The environmental variable to be reconstructed (for example, summer temperature) is, or is linearly related to, an ecologically important variable in the system.

• The same taxa are found in the modern surface sediment and in the fossil data, and their ecological responses have not changed over the intervening time.

• The mathematical methods used model adequately the responses of the biota of interest to the environmental variable.

• Other environmental variables have negligible influence, or their joint distribution in the past is the same as today.

Furthermore, all reconstructions must be evaluated critically, both statistically and ecologically. Using this approach, Imbrie and Kipp (1971) reconstructed past surface water temperatures for the Atlantic on the basis of modern surface water temperatures and foraminifera from marine surface sediments. From chironomid evidence, Walker, Mott, and Smol (1991) reconstructed past summer water temperatures in North American lakes; a similar study is now in progress in the Alpine region. Coleoptera have been studied at numerous North American, British, and Scandinavian sites (Elias 1994; Coope 1977, 1987; Atkinson, Briffa, and Coope 1987; Lemdahl 1991), but only a few such sites exist in the Alpine region (Elias and Wilkinson 1983; Ponel and Coope 1990; Ponel, de Beaulieu, and Tobolski 1992; Gaillard and Lemdahl 1994). The method has been applied in various ways at a large number of pollen sites all across Europe (e.g., Guiot et al. 1989; Guiot et al. 1992). Nevertheless, several inherent problems call for caution. Man has heavily influenced today's vegetation in Europe, for example, and therefore the surface samples and their relationship to today's climate may be irrelevant to periods before prehistoric human impact. Furthermore, modern analogues to past plant communities are sometimes lacking (especially for the 16,000–10,000 ^{14}C yr BP period). Pollen values are expressed as percentages of the total pollen count, a calculation that creates internal dependences. A time lag may exist between a perturbation, such as climatic change, and the vegetation's response to the perturbation, especially in the case of forest vegetation comprising mainly slowly migrating species. Species responding with shorter time lags, such as aquatic plants and insects, are therefore more useful. Because of the Alps' complex relief, the quantitative reconstruction of past climates requires much higher spatial resolution than is currently available on the broad European scale.

Until now, fossilized biotic remains have been employed to make only crude quantitative reconstructions of past climate in the Alpine region. Spatial and temporal resolution still need to be increased, and ecological constraints other than climate need to be taken into account. Multiproxy approaches are therefore needed to test specific hypotheses. Once these things happen, progress can be expected in reconstructing seasonal air temperatures. Information on changes in precipitation (or precipitation minus evaporation, that is, effective moisture) will, however, rely more on studies of stable isotopes, peat growth in mires, and lake level changes than on fossil pollen and insects.

The steepness of the environmental gradients in mountain areas creates one of the most fascinating ecotones: the timberline. Temperature is considered mainly to control the location of the upper tree line in many mountain systems and in this context can be expressed in various ways: in degree-days; as the location of the mean July isotherm (e.g., the 10°C isotherm for some sub-Alpine tree species, such as spruce); as a "temperature sum"; or in terms of the duration of the growth period. Usually we can distinguish between an upper limit of closed forest (the forest limit) and an upper limit of single trees (the tree limit or tree line); these two limits bound the ecotone or ecocline of the timberline (see also chapter 5).

Many attempts have been made to reconstruct Holocene temperatures by tracing timberline fluctuations. Because valley winds easily blow pollen up to Alpine areas, pollen percentages alone provide an unreliable measure of the migration of trees over the centuries. Better estimates of the presence or absence of trees can be derived either from pollen influx (i.e., number of pollen grains per surface area and year) or from macrofossils such as conifer needles, seeds, or bud scales, which are less easily transported because of their weight. Unfortunately, the number of sites in the Alps and in other European mountain areas where plant macrofossils are analyzed is still very low; however, Ponel, de Beaulieu, and Tobolski (1992) have given a fine example. As a first approximation to plant macrofossils, the counts of stomata guard cells of conifer needles preserved in pollen slides has proved useful (Trautmann 1953; Ammann and Wick 1993). As illustrated in figure 3.27, Wick and Tinner (1996) have reconstructed timberline fluctuations using high-resolution macrofossil analysis and AMS radiocarbon dating at two Alpine sites located at sensitive altitudes: Lago Basso (Splügen Pass, 2,250 m a.s.l.) and Gouillé Rion (Val d'Hérémence, 2,343 m a.s.l.). Their results agree well with the classical findings of Zoller (1960, 1977a) and Patzelt (1977), but their time control is much better. In some few cases, mega-fossils—wood—can be used to trace upper tree lines. Thinon (1992) and Tinner, Ammann, and Germann (1996) used charcoal in soils, and Bauerochse (1996) found *Pinus cembra* wood at the unusual altitude of 2,365 m a.s.l. that he dated to 6,860–6,140 radiocarbon years BP. Records in ancient forest soils above today's tree line (Tinner, Ammann, and Germann 1996) are promising because of the soil record's spatial extent compared to that of bog and lake sediment records.

3.2.5.5 Lacustrine Organic Geochemistry The relatively new and expanding field of lacustrine organic geochemistry has already been used to extract environmental information from Alpine lakes relatively barren of other tracer materials.

A combination of factors, including biomass production, a range of biochemical degradation effects, and chemical and physical transport processes, controls the accumulation and preservation characteristics of organic matter in lakes. The organic matter preserved in lake sediments thus records several aspects of a lake's environmental history (Ogura 1990). That organic matter comprises a very complex and variable mixture of compounds, such as lipids, carbohydrates, proteins, and other biochemicals derived from the tissues of organisms that lived in the lake or its catchment area. The relative proportions of organic matter derived from these two sources are largely a function of lake and catchment morphologies and productivities, and so can indicate the environmental conditions at the time of deposition. Nutrient-dependent primary (planktonic) productivity in lakes determines their trophic status: oligotrophic (low productivity), mesotrophic (medium productivity), or eutrophic (high productivity). Many studies of paleolimnological records taken from lakes with various geomorphological characteristics and exposed to different climatic regimes have shown that not only local, but also regional and more global climatic conditions directly influence both the accumulation and preservation of lacustrine organic matter (Gasse et al. 1991; Meyers, Takemura, and Horie 1993; Meyers and Ishiwatari 1993). This section outlines the application of lacustrine organic geochemistry to paleoclimatic problems illustrated by the results of a recent study conducted on Lej da San Murezzan (see figure 3.1 for location).

Organic matter deposited in different lacustrine environments displays characteristic molecular signals that reflect both the specific assemblages of contributing organisms and prevailing lacustrine conditions. Molecular stratigraphy, the study of these molecular signals, provides a useful approach for

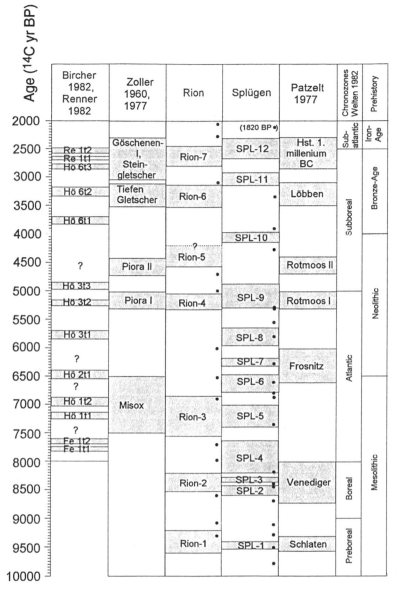

Figure 3.27 Climate oscillations in the central and eastern Alps (Switzerland, Austria, and northern Italy). Cold periods are associated with high tree-ring cell densities (Bircher 1982; Renner 1982), glacier advances, and minima in the ratio of arboreal to nonarboreal pollen (Zoller 1960, 1977b; Patzelt 1977). At Gouillé Rion and Lago Basso (see figure 3.1), a lowering of or openings in the timberline characterizes the cold periods (Wick and Tinner 1997). The concentrations of pollen grains and plant macrofossils reflect these changes and give rise to minima in the ratio of arboreal to nonarboreal pollen.

evaluating a wide range of environments (Brassel et al. 1986). Molecular stratigraphy is based on the concept of "biomarkers" (Eglinton and Calvin 1967) or "chemical fossils" (Tissot and Welte 1984), organic compounds having molecular structures that can be linked to specific biological source materials. Although microbes alter the organic matter both within the water column and during early burial, the stable stratigraphic record finally incorporates a range of residual molecular components. Biomarkers can help identify a lake's trophic status, thermal structure, and salinity, which in turn are climate dependent (Ariztegui, McKenzie, and Farrimond 1993; Ariztegui, Farrimond, and McKenzie 1996).

Several bulk-sediment parameters, including carbon and nitrogen isotope compositions, carbon-nitrogen ratios (a secondary source indicator) and "hydrogen index" values (Talbot and Johannessen 1992; Lallier-Vergès et al. 1993; Ariztegui and McKenzie 1995), are routinely determined to characterize the type and origins of sedimentary organic matter. "Hydrogen index" values, obtained by the Rock-Eval pyrolysis method, are expressed as the ratio of free hydrocarbons to total organic carbon in the sample. These analyses help to define aquatic productivity characteristics and to evaluate the degree of postdepositional biological and diagenetic alterations to the primary materials. The range of biochemical alterations to the original material and the complexity of the residual organic matter usually require a combination of these bulk-sediment techniques.

The same constituents, including proteins, carbohydrates, lignin (in plants), and lipids, essentially compose all organisms. The term "lipid" denotes all organism-produced substances that are practically insoluble in water but subsequently extractable with organic solvents, so that they tend to persist in the particulate organic fraction of the sediment record. Lipids include fat substances like animal fat and vegetable oils or waxes (e.g., surface coatings of leaves) and form a large part of the biomass of bacteria, phytoplankton, zooplankton, and higher plants (Tissot and Welte 1984). The lipid group includes several other groups of compounds, such as hydrocarbons, sterols, alcohols, and fatty acids. Relative variations in the compound proportions, both within and between these groups, can be related to environmental changes.

Because lipids are more readily analyzed than carbohydrates and proteins, they provide the most useful tracers for molecular stratigraphy. For example, fatty acids derived from lacustrine organisms characteristically have molecular chains of sixteen carbon atoms, that is, C-16 or less, whereas those from land plants are typically in the C-30 range. The sediments of most lakes show a bimodal distribution of fatty acids, the relative mode sizes depending on the balance of aquatic versus terrestrial input. In low-productivity lakes, the dominant mode is C-30; in high-productivity lakes, it is C-16.

Molecular stratigraphic indicators have provided the main evidence for reconstructing a history of primary productivity and trophic status spanning the last 13,000 years for Lej da San Murezzan, mentioned above in connection with lake ice cover. Researchers used biomarker studies of the lipid fraction (including sterols, fatty acids, n-alkanes, and alcohols) from organic matter preserved in the profundal sediments to quantify the relative amounts of autochthonous and allochthonous contributions. This information was used in making initial interpretations of the lake's trophic status in the past. Additional evidence from parallel paleoenvironmental tracers (sedimentological, isotopic, and organic geochemical data) for the lake then contributed to the final interpretation, which was subsequently correlated with independent climate indicators for the region (figure 3.28).

The fatty acid evidence indicates that relatively high productive phases characterized Lej da San Murezzan in the Early Holocene and relatively low productivity (oligotrophic lake conditions) during the late Holocene, prior to the anthropogenically induced eutrophication. Parallel evidence from distributions of dinosterol (a sterol exclusive to phytoplankton communities dominated by dinoflagellates) and pigment studies (temporal variations in total pigments indicate productivity), diadinoxanthin (a pigment indicative of diatoms), phytoplankton chlorophylls, carbon isotope ratios in the bulk organic matter, and the hydrogen index (HI) support this interpretation. The Holocene history of Lej da San Murezzan's trophic status can be correlated with the relative temperature variations for central Europe deduced from other proxy data (Züllig 1988).

The environmental and climatic information contained in lacustrine sedimentary organic matter has not yet been fully exploited. However, continuing research and improvements in analytical techniques are currently opening new prospects for the quantitative evaluations of organic material preserved in lake deposits. Despite the fact that organo-geochemical approaches are becoming more quantitative, considerable work is still required to obtain better assessments of the controls various environmental factors exert on the characteristic molecular structures found in fossil lacustrine organic matter (i.e., calibration studies). The range of definitive biomarkers still needs to be expanded, and new models that more rigorously correlate the organic geochemical indicators with quantitative determinations of autochthonous and allochthonous fluxes are needed for improved mass balance calculations. Eutrophic lakes should now be investigated with respect to the simultaneous organo-geochemical processes operating in the water column and at the sediment surface to determine quantitatively the process-effect-tracer sequence due to anthropogenic activities separately from that due to natural controls.

3.2.5.6 Stable Isotopes in Meteoric Waters and Lake Carbonates
The isotopic calibration of processes determining the characteristics of stable isotopes in natural drainage waters and lacustrine carbonates, also its application in a paleo-isotopic mapping concept for the Alpine area, should indicate the potential of stable isotope data in the reconstructions of regional meteorological patterns and in the validation of climate simulations. Box 3.1 presents a background introduction to stable isotopes within this context.

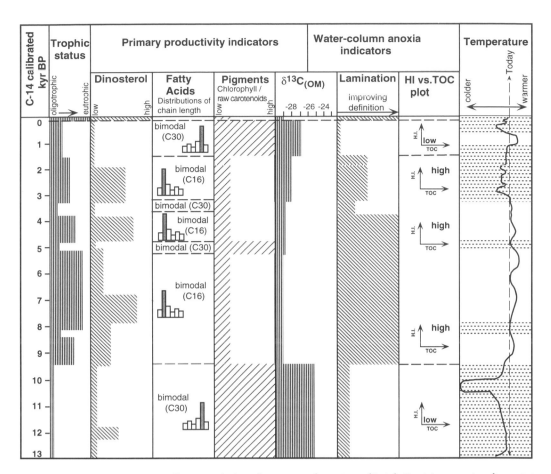

Figure 3.28 Changing algal productivity in the waters of Lej da San Murezzan (see figure 3.1 for lake location) since 13,000 ^{14}C yr BP, as interpreted from a range of tracer types present in a core taken from the bottom sediments. Oligotrophic status indicates periods of low productivity and eutrophic status those of high productivity. The trophic status is interpreted largely from a combination of different algal relicts, including sediment dinosterol content, dominance of the C-16 chain lengths fatty acid, and the variable chlorophyll pigment content versus the carotenoid pigment content relationship. In addition, the carbon isotope (^{13}C) ratios in the bulk organic matter have more negative values when the algal matter dominates in the bulk organic matter content and when more organic matter is preserved in the sediments because of anoxic bottom conditions. Since bioturbating bottom fauna are absent under anoxic bottom conditions, such conditions enhance the definition of the sediment lamination structure and raise values on the hydrogen index (HI) versus total organic carbon (TOC) plots. The far right panel correlates the mean annual air temperature curve for central Europe (compiled from a range of naturally archived tracer evidence: Züllig 1988) with evidence for temporal primary productivity changes in Lej da San Murezzan.

Stable-isotope ratios for oxygen and hydrogen in meteoric water (atmospheric precipitation, surface waters, and groundwaters) may be considered meteorological parameters that have characteristic patterns in time and space, much as air temperature or rainfall do. Stable-isotope ratios are, however, unique meteorological parameters in that they can, to an extent, be preserved in natural archives such as the accumulations of sediments at the bottoms of lakes. Oxygen isotope ratios in water are built into lacustrine carbonates (hydrogen isotope ratios possibly into lacustrine plankton tissue) and can be measured routinely using mass spectrometry. In the case of rainfall onto a catchment followed by stream and river drainage to a lake and then precipitation of carbonate particles from the lake water, isotopic information generally transfers according to the following schema:

1. Incoming atmospheric moisture has an isotopic composition that reflects its source and history of gain and loss.

2. Decreasing proportions of the heavier isotopic species of oxygen and hydrogen in precipitation accompany decreasing air temperature (season, altitude).

3. Evaporation enriches the heavier isotopic species in the residual water.

4. Evapotranspiration from vegetation does not involve isotopic fractionation but can alter the mean isotopic composition of runoff by selectively removing part of the total.

5. Carbonates forming in a lake assumes an isotopic signature determined by that of the water and a temperature-dependent fractionation effect (colder water temperatures producing heavier isotopic signatures in the carbonate).

Thus during the isotopes' journey through from the original meteoric moisture to their final resting place in the crystal structure of a carbonate, several isotopic fractionation processes, working in either the same or opposite directions, can manifest themselves (figure 3.29).

Oxygen isotope records archived in lacustrine sediments can thus be used to reconstruct a number of aspects of past climatic and hydrological conditions, including mean precipitation temperatures and sources, history of atmospheric moisture, shifts in runoff characteristics, and spatial representations of shifting climate patterns. Because the isotopic signals are recorded instantaneously, they provide a frame of reference for application together with other environmental tracers such as those discussed earlier in this chapter. In particular, they can be used to calibrate tracers that exhibit response time lags to climate shifts. Hydrogen-isotope (hydrogen ^{1}H; deuterium ^{2}H) ratios (δD) in meteoric waters, which may show evaporation-related deviations from covariance with oxygen isotope ratios (or from the gradient of the best fit line, $\delta D = 8 \cdot \delta^{18}O + 10$, termed the "global meteoric water line," the "deuterium excess" being defined as $d = \delta D - 8 \cdot \delta^{18}O$), can help identify the source and history of atmospheric moisture and subsequent evaporation characteristics. Hydrogen isotope ratios are not built into carbo-

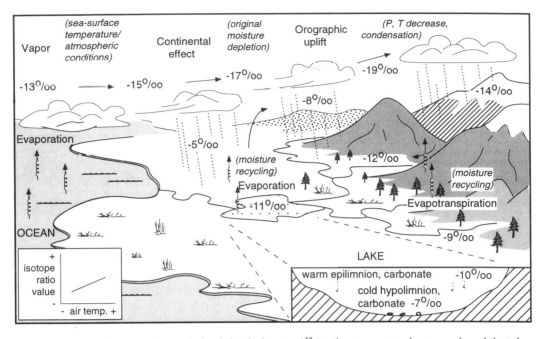

Figure 3.29 An idealized sketch showing $\delta^{18}O$ values in a regional water cycle and the information transfer to lacustrine archives. Atmospheric/sea-surface conditions at the oceanic source, especially temperature, determine initial isotopic ratios. The isotopic ratios in precipitation are also strongly related to the air temperature at which condensation occurs (latitude, season, altitude). With increasing distance from the oceanic source, successive rain outs deplete the heavier isotopes, leading to lighter isotopic ratios in the remaining vapor, the orographic effects of mountains particularly enhancing such depleton. After precipitation reaches the catchment surface, evaporation and evapotranspiration, especially during summer months, can recycle substantial amounts back to the atmosphere, modifying the isotopic signal in subsequent runoff. The reservoir effects of snow, ice, groundwater, and lakes may, in turn, delay runoff . The modified isotopic signal in lake water may then be fixed, with a water temperature–dependent fractionation, into lacustrine carbonates accumulating in the sediment archive (figure inset).

nates, though current efforts are attempting to extract useful δD signatures from zooplankton chitin, so this section will address primarily the oxygen isotope signals. To follow the processes by which oxygen isotope ratios are fixed in atmospheric precipitation, modified by catchment processes, and finally preserved in lacustrine sediment archives, it is useful and instructive to follow the aquatic pathway through each of these subenvironments.

Evaporative conditions at the sea surface of origin, initially determine stable-isotope ratios in atmospheric moisture, which are then modified by series of moisture losses and gains during transport strongly related to atmospheric temperature conditions (latitude, altitude, season). Measurements of isotopic values in precipitation have been gathered worldwide since 1961, with the database centered in Vienna at the International Atomic Energy Agency (IAEA). In a cooperative effort with the WMO, the IAEA now manages data from the Global Network of Isotopes in Precipitation (GNIP). Dansgaard (1964) originally constructed a model for the global stable-

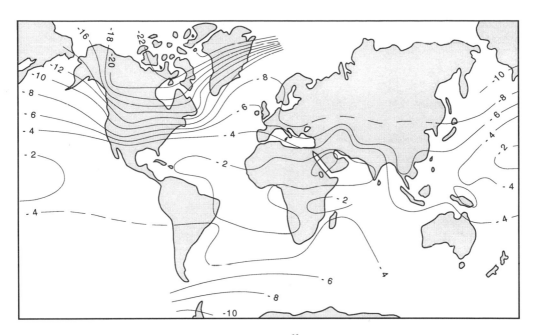

Figure 3.30 Global distribution of mean $\delta^{18}O$ in precipitation (averaged over two years or more). Considerable deviations from this general pattern exist over continental areas because of regional or local climatic and topographic features (after Gat 1981).

isotope data, and numerous publications (e.g., Yurtsever and Gat 1981; Siegenthaler and Matter 1983; Joussame and Jouzel 1987; International Atomic Energy Agency 1993; Schotterer, Oldfield, and Fröhlich 1996) have since addressed the relationships between stable isotopes and atmospheric conditions.

The mean global $\delta^{18}O$ pattern for modern atmospheric moisture displays latitudinal gradients with values generally decreasing away from the equatorial region, becoming steeper toward the polar regions, and significantly modified over the continents (figure 3.30). At any single location, the isotopic values typically can include large fluctuations within a single precipitation event, regular seasonal variations (lower values in winter), and little interannual variation. A different climatic state, though, is accompanied by its own isotopic scenario, as a paleodata-validated GCM simulation for the global isotopic pattern during the last glacial maximum (Jouzel et al. 1994) demonstrates. Considerably more detail at the regional level could still be added to the global isotopic pattern for modern precipitation, but as yet this has been done for only limited continental areas. In Europe, for example, preferential rain out of the heavier isotope (^{18}O) along principal meteorological trajectories has resulted in a NNW-to-SSE decreasing trend of 6 percent in the mean $\delta^{18}O$ pattern through Germany, the continental and orographic effects there having regionally overprinted the general global pattern (figure 3.31; Förstel and Hützen 1983; Rozanski 1995).

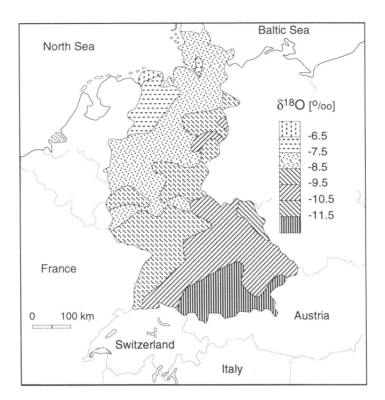

Figure 3.31 Regional distribution of mean $\delta^{18}O$ values in precipitation over the former Federal Republic of Germany (after Förstel and Hützen 1983). The decrease in ^{18}O content from north to south mainly results from the continental rain out effect. Orographic effects over Switzerland result in $\delta^{18}O$ values there having a larger range and considerably more spatial texture than is displayed for the Federal Republic of Germany.

In Switzerland the isotopic signatures in meteorological precipitation at network stations are known, although comparable mappings of precipitation values have not yet appeared. Since 1992, the Swiss National Isotopic Network, centered in Bern, has been obtaining regular measurements for 11 precipitation and 6 runoff stations, but also has isotopic time series reaching back some twenty-five years (Schotterer et al. 1995). Isotopic measurements for precipitation at Alpine sites during the last decade or so (Schotterer et al. 1993) have shown that shifts in isotopic-ratio values have closely paralleled those of air temperature during precipitation, both at the seasonal and interannual levels (figure 3.32). From site to site, though, absolute isotopic values depend also on local geographic/climatic boundary conditions, in effect producing a physico-climatic isotope topography, so that spatial reconstructions cannot rely on the isotopic gradient from one transect alone. Because the Alps significantly affect the local and regional climate and because both vertical and horizontal isotopic gradient components are present, stable-isotope patterns there must exhibit considerably more detail than for much of the rest of Europe. Because the Alps are likely to exhibit enhanced responses to global climate change (see section 3.1.1 and also chapter 2), isotopic patterns

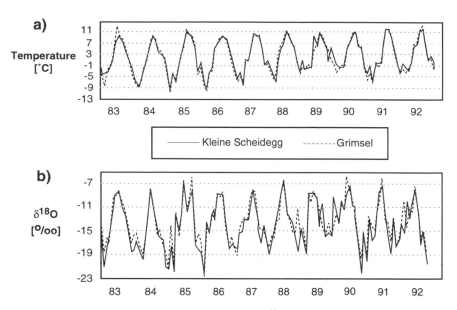

Figure 3.32 Mean monthly air temperature (a) and $\delta^{18}O$ in precipitation (b) from 1983 to 1992 at the Grimsel and Kleine Scheidegg stations, both located in central southern Switzerland (see figure 3.1) at 2,000 m a.s.l. (after Schotterer et al. 1993). The trends in $\delta^{18}O$ values in precipitation closely mimic those of air temperature at both the seasonal and interannual levels at both sites. The warming trend is primarily a winter feature (see also section 3.1.1).

in the Alpine region should provide one of the more sensitive indicators of that climate change.

In an initial isotopic investigation of the Swiss Alps, Siegenthaler and Oeschger (1980) quantified the dependence of $\delta^{18}O$ gradients for modern precipitation on both seasonal and altitudinal air temperature gradients but warned that these gradients might be invalid for paleoreconstructions. Yet more recent work involving the comparison of a GCM simulation of global isotopic gradients for the last glacial maximum with the present situation suggests that the relationship between $\delta^{18}O$ and air temperature differed little between these different climate regimes (Jouzel et al. 1994), though of course both the absolute isotopic-ratio values in precipitation at given sites and the modifying catchment effects differed significantly. For the Alpine region, past $\delta^{18}O$/air temperature gradients may be obtained from isotopic records in sediment cores taken from a series of Alpine lakes at different altitudes.

The original isotope ratios in water may be modified during drainage because of evaporation (resulting in fractionation, i.e., the preferential loss of lighter isotopes) and evapotranspiration (resulting not in fractionation but in the loss of part of the total). The extent of such modifications depends on the nature of catchment hydrology and vegetation and on prevailing meteorological conditions (Gat and Lister 1995). In addition, groundwater and snow reservoir effects result in flow delay, mixing, and seasonal release, producing

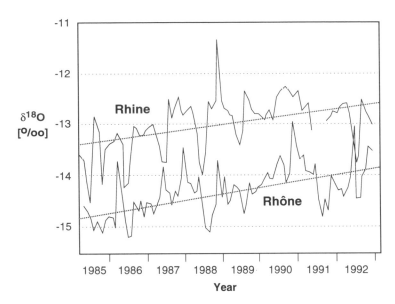

Figure 3.33 Temporal $\delta^{18}O$ trends in the Rhine and Rhône rivers (prior to their entering Lake Constance and Lake Geneva, respectively; see figure 3.1 for locations) from 1985 to 1993 (after Schotterer et al. 1993). The absolute values differ because of differences in mean catchment altitudes and local climate effects. The Rhône receives a higher proportion of glacier meltwater, and hydroelectric stations affect it more strongly than the Rhine. Linear-regression lines show that the isotopic trends for both were similar over the period shown, reflecting both shifting annual precipitation distributions and a winter warming trend (cf. figure 3.32).

characteristically variable integrations of isotopic signals with time. With increasing distance downstream, streams spatially integrate input from increasingly larger areas, averaging out much of the isotopic "noise" associated with individual precipitation events in the process. The larger Alpine rivers, however, retain the isotopic signals due to seasonal trends as well as those due to altitudinal and local climate effects (e.g., Schotterer et al. 1993; figure 3.33).

The reservoir holdup of drainage waters in a lake adds a further temporal component to the integration process whose magnitude depends on the relationship between runoff rate (catchment area) and lake volume (a lake may be little more than a wide place in a river or have a water residence time of years). Lake waters then tend to re-equilibrate isotopically with the overlying atmosphere, eventually stabilizing (at least at depth) at some value intermediate between that of the inflow, which can import high-altitude or winter isotopic signatures, and that of the atmosphere. The lake waters' final isotopic value therefore also depends on the lake surface area-to-volume relationship and the mixing characteristics (thermal and chemical structure) in the lake. Isotopic values in surface lake waters are generally more dynamic than those at depth, since they must continually adjust to short-term shifts (in temperature, direct precipitation, exchange and evaporation processes, and the like) near the air-water interface.

Several spatial and temporal effects thus influence mean stable-isotope values in lake waters. Yet under steady-state conditions, the isotopic values in a freshwater lake with a residence time of years should remain essentially constant, the event and seasonal effects being averaged out. Significant long-term isotopic changes in such a lake's history may relate to changes in climatic conditions. To define those changes in terms of climate, however, the relationships between the component processes contributing to the final isotopic signal (e.g., catchment effects, hydrological inconsistencies, changing lake water residence times), some or all of which may shift in concert with climate change, need to be established. For this purpose, parallel evidence (such as paleovegetation data and profundal lacustrine sedimentological criteria), from the same depositional sequence where possible, needs to be considered, and process modeling and multivariate statistical methods need to be applied.

Calcium carbonates are formed in many lakes, commonly either as silt-sized particles that precipitate in the surface waters during phytoplankton growth or as biogenic shell material (snails, clams, ostracods) in the bottom waters. For stable-isotope purposes, these lacustrine carbonates may be considered the continental equivalents of marine microfaunal carbonates (foraminifera), which have provided most of the now widely accepted marine stable-isotope evidence for past climatic change (e.g., Emiliani 1955; Shackleton and Opdyke 1973; Duplessy et al. 1986). During formation, carbonates incorporate modified oxygen isotope signatures from the water and are then successively buried in the accumulating sediments, forming continuous high-resolution records. Carbon isotope ratios are also derived from ambient waters during carbonate crystallization but are related in the first instance to lacustrine bio-organic budgets (Oana and Deevey 1960; Lister 1985; McKenzie 1985) and prevailing hydrological conditions (Talbot 1990).

Isotopic modification during incorporation into carbonate is due primarily to a water temperature–dependent fractionation, calculated to be $0.26‰ \cdot K^{-1}$ (Craig 1965; Stuiver 1970) for calcite precipitating at a rate allowing isotopic equilibrium. Thus to obtain an isotopic value for the lake water ($\delta^{18}O_{SMOW}$) from the isotopic value measured for the carbonate ($\delta^{18}O_{PDB}$) (necessary both for isotopic calibration with atmospheric precipitation and for normalizing interlake isotopic correlations), the lake water temperatures during carbonate precipitation must be established (Lister 1988a).

Carbonate silt commonly precipitates seasonally in the surface waters of midlatitude freshwater lakes in response to algal blooms (Kelts and Hsü 1978) and is thus subject primarily to the spring surface water temperature ranges, which may significantly affect the temperature-dependent isotopic fractionation within a season and interannually. Algal productivity rates can also affect calcite crystallization rates and thus the isotopic disequilibrium between the dissolved and solid carbonate species. Further complications may arise because of diagenetic alteration (e.g., in situ exchange, gains or losses to the original carbonate crystals by dissolution or overgrowth) that

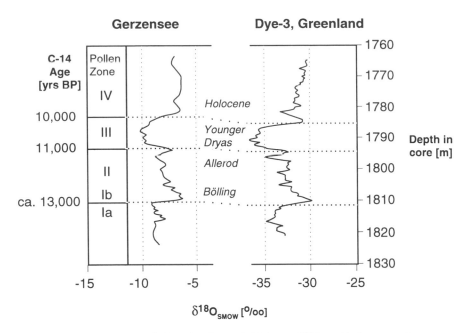

Figure 3.34 Comparison of a late glacial to early Holocene $\delta^{18}O$ profile from a marl sequence in Gerzensee, Switzerland, with that from a section of the Dye-3 ice core, southern Greenland (after Oeschger et al. 1984; see figures 3.1 and 3.5 for locations), with the European pollen zones added. In the later GISP-2 (Greenland Ice Sheet Project) ice core record from Greenland, negative $\delta^{18}O$ peaks corresponding to those shown here for Dye-3 close to the ^{14}C ages of ca. 10,000 yr BP, 11,000 yr BP, and 13,000 yr BP would correspond respectively to ca. 11,650 calendar yr BP, 12,890 cal. yr BP, and 14,670 cal. yr BP, on the basis of ice layer counts for GISP-2 (Stuiver et al. 1995). The strongly parallel nature of the two isotopic records gives evidence for substantial climate interdependence over the North Atlantic.

can be almost impossible to establish quantitatively for carbonate silts; and because of detrital carbonate silt admixtures (catchment rock silt washed in by rivers), leading to multicomponent isotopic signals. Yet significant isotopic data have been extracted from authigenic carbonate silt sequences from Swiss lakes. For example, Eicher and Siegenthaler (1976) measured an isotopic curve for the late glacial Holocene from carbonates in a core from Gerzensee in the Swiss lowlands. Together with newer isotopic measurements, that curve runs strongly parallel to (although some 20–25‰ more positive than) the oxygen isotope profile for the Dye-3 ice core from South Greenland (Oeschger et al. 1984; figure 3.34), and von Gravenstein et al. (1994) reported similar synchroneity of events on the basis of benthic isotopic records in Southern Germany, suggesting strongly linked climatic cause-and-effect scenarios over the North Atlantic at that time.

Benthic carbonates (usually biogenic shell material) may form in bottom waters if sufficient oxygen for the fauna is present, although faunal activity commonly destroys the chronostratigraphic control any varves initially present provide. In lakes of sufficient depth, bottom-water temperatures can remain essentially constant, so the temperature-dependent fractionation factor is less problematic, and absolute isotopic values for the lake waters can be derived from those measured in the benthic carbonates (Lister 1988a). In this respect benthic carbonates have an advantage over those formed in the surface waters (epilimnetic carbonates), where short-term temperature changes are greatest (cf. section 3.1.1). In addition, shell morphological characteristics obviate the diagenetic and detrital contamination problems. Isotopic profiles measured from benthic faunal carbonates in Lake Zurich for the late glacial Holocene period (Lister 1988a, 1988b) and Lake Neuchâtel (Schwalb, Lister, and Kelts 1994) in Switzerland have been used to indicate the magnitudes and timing of effects due to climate, glacial-ice melting, hydrological changes, and air temperature shifts. For example, figure 3.35 shows the isotopic curve for Lake Zurich water ($\delta^{18}O_{SMOW}$), calculated from the isotopic curves ($\delta^{18}O_{PDB}$) measured for benthic biogenic carbonates (ostracods and microscopic bivalves), together with stratigraphic evidence for contributing catchment events in that changing environment since about 14,500 ^{14}C yr BP.

How can this isotopic information be linked to characteristic regional climate patterns that exist at different times? In the past, oxygen isotope records from individual lakes were usually assumed to represent some associated but undefined region and were interpreted in terms of relative climatic shifts (e.g., positive isotopic shift implies warmer, negative isotopic shift implies cooler conditions) rather than in absolute air temperature values. Such individual records suffer drawbacks, though: (a) the absolute isotopic values are catchment specific; (b) catchment-specific changes, such as hydrological shifts, cannot be eliminated, and the effects at upstream lakes cannot be quantified; and (c) $\delta^{18}O$ values at one site say little about spatial isotopic patterns in precipitation or the governing climatic scenarios. Although trends in isotopic values at any single site may closely follow trends in local air temperature, site-to-site differences are not necessarily just temperature dependent (for example, local air mass histories). Figure 3.36 illustrates how two lakes in a mountainous region might provide different isotopic profiles for the same sequence of climatic events. In this illustration, isotopic profiles from additional lakes would be required to key into the regional pattern and establish the nature of its change through time.

Investigations of single-lake isotopic records, however, yield valuable insights useful for widening the approach. The current goal is the isotopic calibration of sets of lakes under present conditions, so that the elements can be quantitatively tracked around the following loop: regional climatic patterns → spatial isotopic patterns in atmospheric moisture → spatial isotopic patterns in hydrological network → isotopic records in lake sets → time-slice patterns linked to past climatic conditions. Paleo-isotopic mapping for the Alpine region is based on the following hypothesis: Switzerland lies in a boundary region where Alpine orography both strengthens and differentiates the meteorological effects resulting from the North Atlantic, Mediterranean, Polar and Continental climate systems, and it has a system of

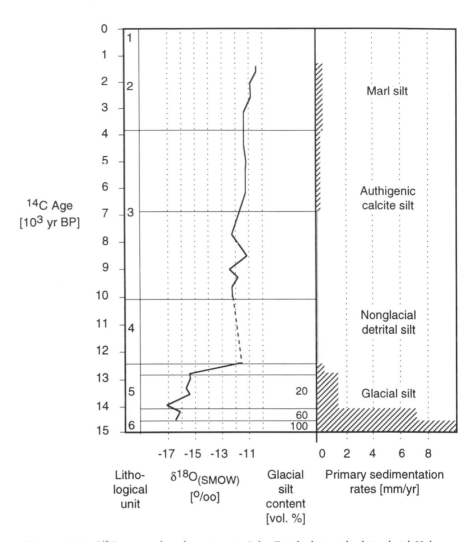

Figure 3.35 $\delta^{18}O_{SMOW}$ values for waters in Lake Zurich during the late glacial Holocene period calculated from $\delta^{18}O_{PDB}$ values measured for benthic biogenic carbonates. Primary sedimentation rates for deep-water silts reflect first the decreasing input of glacial silt during the final retreat stages of the Ice Age glacier (low $\delta^{18}O$ value of glacier meltwater), then the residual influx of detrital silt from the catchment, followed by the onset of increased carbonate silt production in the lake, and at about 3,500–4,000 ^{14}C yr BP, a sharp decline in the relative detrital silt influx (Lister 1988a). The gap in the $\delta^{18}O$ record (12,400–10,000 ^{14}C yr BP, approximately the period covered by the Gerzensee core (figure 3.34), is due to early postdepositional carbonate dissolution. See figure 3.1 for locations of lakes.

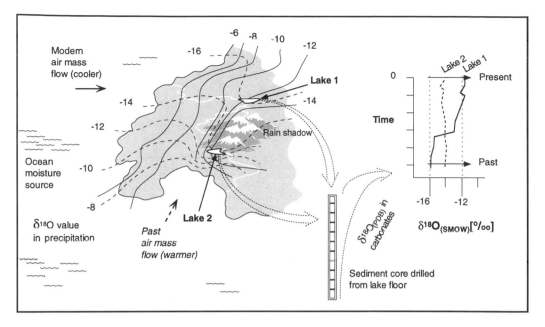

Figure 3.36 Sketch of hypothetical land-sea area with an exaggerated shift in prevailing wind direction and its associated $\delta^{18}O$ isoline pattern from some time in the past (dotted lines) to the present (solid lines). Sediment cores from lakes 1 and 2 would yield $\delta^{18}O$ time series like those shown on the right. $\delta^{18}O$ time series from a number of other lakes across this region would be required to reconstruct spatial isotopic patterns for selected time slices.

rivers and lakes diverging from major divides. The stable-isotope pattern for precipitation across the region should reflect the shifting balance of conditions through both time and space, and this pattern is in some form continuously transferred via the hydrological network to the lacustrine archives. Stable isotopes are the only meteorological components archived in this way, so the climatic information they harbor is unique. Quantitative definition of the information transfer and storage processes should thus enables reconstruction of the original isotopic patterns and thereby a key to the associated meteorological conditions. This paleo-isotopic mapping approach is currently being applied along a north-south Alpine transect area in Switzerland and is intended to function as an interface in the multidisciplinary approach between climate simulation outputs and reconstructions based on multiproxy data for the surface environment.

3.2.6 Summary and Conclusions

Understanding the cause-and-effect relationships for expected climate change is important to the community for three reasons: (1) to predict the character of the change; (2) to see if it can be prevented, or at least minimized, should it be perceived as threatening; and (3) to plan for the change if it cannot be avoided. Its cause, anthropogenic or natural, is a central issue. If the change were perceived, for example, to be a natural cooling (e.g., a drift toward another glacial stage), the management strategy might be changed to "Increase the atmospheric CO_2 content!" This emphasizes the state of the climate as such is not of importance to man so much as the impact at ground zero (hydrology, vegetation, erosion, frequency and severity of extreme events, ice melting, sea-level change). In many cases proxy data are actually more closely related to the impacts than the climate; indeed, climate reconstructions from proxy data are commonly routed through the impacts.

In this chapter we have presented a selection of the data types and methodologies currently used for Alpine climatic and impact reconstructions, their qualitative and quantitative advantages and limitations, and examples of the deliverable results. Table 3.4 provides an overview of the applicability of each and indicates the range of potential cross correlations between subsets. Large amounts of various kinds of proxy data from the Alpine region have already been published, much in the form of time series profiles with interpretations, but encompassing techniques for modeling proxy data are mostly still at the developmental stage. In particular, models must link observations with regional climatic scenarios correctly in space and time to realize the full potential of the proxy data. This goal can be attained only through an interdisciplinary approach involving regional climate simulations and reconstructions and should lead to innovative high-resolution modeling approaches in which both kinds of information guide the design of simulations.

There is currently no interface in climate models for the direct input of climatic information based on proxy data. Each discipline essentially has different operational data requirements, methodologies, and spatial and temporal scales of approach and outputs. Based on full-scale physical process simulations, model outputs are in the nature of time-slice snapshots for large areas and exhibit only limited close-up detail. Continental proxy-data reconstructions use collections of documented or natural evidence to synthesize a local or regional scenario, usually without addressing the role of large-scale physics. Archive evidence commonly spans long time periods and has good local resolution, but particularly in the case of natural archives, limited temporal fine tuning (seasonal or longer). Archive evidence may be discontinuous from site to site, and climate signals mostly need to be teased out of local bulk environmental signatures. Topographic representation in both approaches are poor to very poor (improvement may require nested regional models). Comparisons can only be made between the spatial outputs from both disciplines if they have similar scales and sufficient detail. In such cases, validation checks of model outputs may result in arbitrary parameter adjustments to improve model performance. Proxy data also require calibration, and the output requires validation. Calibration and validation for the very recent past are based directly on meteorological and hydrological network instrumental data; for periods farther back in time, documentary records, multiproxy approaches and various statistical techniques (regression analysis, multivariate statistics, geostatistics) are employed. The strength of proxy-

Table 3.4 Type, range, and applicability of data types and methods currently employed in Alpine climate reconstructions

Archive type	Typical record length (yr)	Typical resolution	Continuity	Meteorological variables							
				Mean annual temperature	Mean temp. of annual precip.	Summer temp. (July)	Winter temp. (Jan.)	Precipitation (total, distrib.)	Extreme events	Time series	Spatial patterns
Instrumental records	100	daily (monthly, annual)	excellent	measured	measured	measured	measured	measured		excellent	excellent
Documentary records	300–500	daily (monthly, annual)	good	reported or derived	reported or derived	reported or derived	reported or derived	reported or derived	excellent	good	limited
Polar ice (range of tracers)	100,000–250,000	annual (seasonal)	excellent		derived			derived	derived	excellent	excellent
Alpine glaciers (range of tracers)	1,000–15,000?	annual (seasonal)	poor to excellent	relative (mass balance)	derived	relative (mass balance)		relative (mass balance)	limited	good to limited	
Tree rings	1,500–8,000	annual	excellent	derived	derived	derived	derived	growing season	derived	excellent	summer temperatures
Pollen	10,000–125,000	seasonal to decadal	good to excellent	inferred and derived		inferred and derived	inferred and derived	inferred and derived	derived	good to excellent	good
Coleoptera	10,000–15,000	seasonal to decadal	poor to good	derived		inferred	inferred			poor to good	
Varves	100–10,000	annual (seasonal)	poor to excellent			relative (if proglacial)	relative (if freezes)	relative (if detrital)	observed	poor to excellent	
Chironomids, cladocera, and diatoms	10,000–15,000	seasonal to decadal	poor to excellent			relative (inferred)				poor to excellent	
Organo-geochemistry	10,000–15,000	seasonal to decadal	poor to excellent	derived		relative (inferred)		relative (inferred)		poor to excellent	
Stable isotopes	10,000–15,000	seasonal to decadal	poor to excellent	inferred	derived	derived	derived or inferred	qualitative		poor to excellent	poor to good

Row groupings: **Natural archives** (all rows); **Lake sediments** (Varves, Chironomids/cladocera/diatoms, Organo-geochemistry, Stable isotopes).

Note: "Derived" implies absolute values can be derived by calculation; "relative" implies that absolute endpoint values cannot be obtained (e.g., the temperature changed by 5 °C, but it is undetermined from which value to which value); "inferred" implies that only a qualitative interpretation can be gained (e.g., it became wetter).

data archives lies in the length of time they cover, which permits identification and statistical analysis of climatic fluctuations of much lower frequency than would otherwise be possible. The record of tree ring cell densities calibrated using summer temperatures (figure 3.17) provides an elegant example of this. Indeed, the regional warming trend during the last hundred years appears less of an omen for the future, at least in terms of human causes, when viewed on the longer time scale (see, e.g., figure 2.4), in which a warming trend appears in the context of a recovery from the mid-nineteenth century cool period.

It has at times been argued that, for an area as small as the Alpine region, climate reconstructions based on proxy data are unlikely to be relevant for correlation with the output of larger-scale models. However, (1) a set of climatic and environmental conditions defined for the Alpine region has strong implications for conditions farther afield (e.g., Alpine glacier responses during the Little Ice Age indicated a climatic condition affecting the entire continent), and (2) a reconstruction of climate in the Alpine region can be viewed as a single station in a network, much the same way as a single meteorological station is. The output from validated models may in turn be able to fill in the gaps in climate reconstructions and indicate potential critical sites for proxy-data investigations.

As yet, the goal of defining the regional spatial response patterns associated with the decadal and century-length climatic perturbations observed in archived time series has eluded climate research. An intermediate linking approach may include several steps:

1. defining the modern spatial pattern

2. establishing how the topography of that pattern changes shape and amplitude in response to larger-scale forcing

3. identifying a minimum number of definitive-response sites keyed to the pattern

4. moving the pattern through time using the evidence provided by the set of response site records as a guide

5. employing limited-area models for selected time slices.

The simultaneous response of local stations to regional climate forcing is well documented: For instance, irrespective of geographical location or altitude, both the short-term structure and the longer-term secular increase in surface air temperature in Switzerland have been similar over the last hundred years (figure 3.2); in addition, surface lake water temperatures show essentially the same signal as air temperatures (figure 3.20). Beniston et al. (1994) have linked decadal trends in Swiss meteorological parameters, including air temperature, via pressure field characteristics to the North Atlantic Oscillation Index (i.e., to the difference in sea level pressures between the Azores and Iceland, also a measure of the strengths of the Westerlies over the North Atlantic: see also chapter 2) Thus proxy indicator characteristics in a given Swiss lake that react to temperature (lacustrine thermal structure, primary productivity, deep-water oxygen content, remobilization of ionic species, and the like) can exhibit a measurable response to synoptic-scale events. For example, oxygen isotope values in meteoric waters closely follow temperature trends (figure 3.32), so the detailed similarities the isotopic records from Gerzensee and from the Dye 3 ice core display (figure 3.34) would appear to signify responses to synoptic scenarios that have overridden local controls.

Global climate research involves both diagnosis of the past state of the climate and estimation of its future evolution. Modelers aim to predict future climate scenarios accurately under the assumption of, for instance, a doubling of atmospheric carbon dioxide concentrations, but are still testing and refining their models. Proxy-data reconstructions focus on the past to obtain insights into the frequencies, amplitudes, extents, and impacts of climatic conditions both within a regime and between regimes, but the necessary tools are still being calibrated. That the current anthropogenic contribution to climate has no precedent is of concern to both approaches. The climate during the last interglacial (Eemian) has been suggested as one analogy for future climate because it was warmer than the present by up to 2°C, although this was clearly not due to anthropogenic causes. Indeed part of the problem, with respect to scientific precision, is that the expected change will not be large in comparison to natural changes in the past. Despite this, its impact on civilization is likely to be severe, largely because operational structures have been constructed so precisely in accordance with the most recent prevailing climate state. Coastal lowlands, delta flats, port cities, and other shoreline infrastructures, for example, have been developed under the tacit assumption of constant marine conditions. Resource and environmental management practices (for example, in farming, forestry, and water resources) have developed under prevailing meteorological conditions and therefore function optimally under precisely these, and not future conditions. With the trend toward higher (mesoscale) spatial resolution in applying climate models and the trend toward more spatially coherent paleoclimate reconstructions based on (regional) proxy data sets, the disciplines of modeling and paleoclimatology are now approaching one another. Moreover, these data sets will help define the regional effects of global climate shifts climate modeling predicts, an area of critical importance for future socioeconomic and ecological planning. Thus a strong interdisciplinary approach is desirable, involving the development of common goals in the fields of climate modeling and paleoclimatological reconstructions based on proxy data, and optimum progress is likely to depend on a complementary approach. Figure 3.37 summarizes the current conceptual approach to interdisciplinary climate reconstruction and simulation. The convergence of approaches combining observational evidence and modeling results should help further our understanding of the fundamental processes involved in determining Alpine climates. Ultimately, the results of this research should be useful not only to

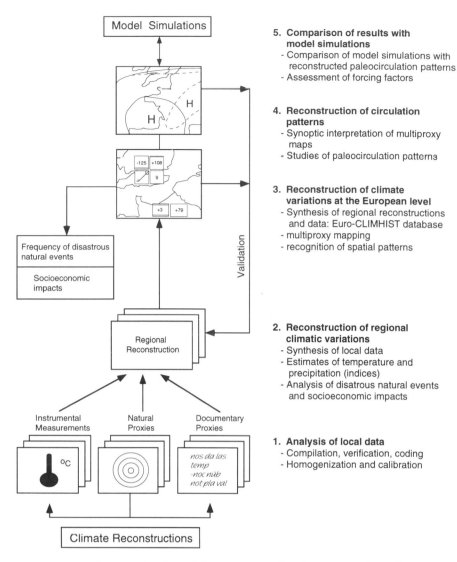

Figure 3.37 Conceptual outline of the current interdisciplinary approach to climate reconstruction and simulation (from Pfister et al. 1994).

other members of the climate research community but also to those policymakers charged with coordinating the responses of society to *any* likely future change in climate, regardless of whether such a change is anthropogenic or natural in origin.

REFERENCES

Agassiz, L. 1840. *Études sur les glaciers.* Neuchâtel: Privately published.

Agbeti, M. D. 1992. Relationship between diatom assemblages and trophic variables: A comparison of old and new approaches. *Canadian Journal of Fisheries and Aquatic Sciences* 49:1171–75.

Alexandre, P. 1987. *Le Climat en Europe au Moyen Age. Contribution à l'histoire des variations climatiques de 1000–1425, d'après les sources narratives de l'Europe occidentale.* Paris: Editions de l'Ecole des Hautes Etudes en Sciences Sociales.

Ammann, B., A. F. Lotter, U. Eicher, M. J. Gaillard, W. Haeberli, G. S. Lister, M. Maisch, F. Niessen, C. Schlüchter, and B. Wohlfarth. 1994. The Würmian Late-Glacial in lowland Switzerland. *Journal of Quaternary Science* 9(2):119–25.

Ammann, B., and L. Wick. 1993. Analysis of fossil stomata of conifers as indicators of the Alpine tree line fluctuations during the Holocene. In *Oscillations of the Alpine and Polar Tree Limits in the Holocene*, ed. B. Frenzel, 175–86. Stuttgart, Germany: Gustav Fischer Verlag.

Anderson, N. J., B. Rippey, and C. E. Gibson. 1992. A comparison of sedimentary and diatom-inferred phosphorus profiles: Implications for defining pre-disturbance conditions. *Hydrobiologia* 253:357–66.

Antevs, E. 1951. Glacial clays in Steep Rock Lake, Ontario, Canada. *Bulletin of the Geological Society of America* 62:1223–62.

Arakawa, H. 1954. Fujiwhara on five centuries of freezing dates of Lake Suwa in the central Japan. *Archiv für Meteorologie, Geophysik und Bioklimatologie, series B*, 6:152–66.

Ariztegui, D., P. Farrimond, and J. A. McKenzie. 1996. Compositional variations in sedimentary lacustrine organic matter and their implications for high Alpine Holocene environmental changes: Lake St. Moritz, Switzerland. *Organic Geochemistry* 24(4): 453–461.

Ariztegui, D., and J. A. McKenzie. 1995. Temperature-dependent carbon-isotope fractionation of organic matter: A potential paleoclimatic indicator in Holocene lacustrine sequences. In *Problems of Stable Isotopes in Tree-Rings, Lake Sediments and Peat-Bogs as Climatic Evidence for the Holocene*, ed. B. Frenzel, B. Stauffer, and M. M. Weiss. Stuttgart, Germany: Gustav Fischer Verlag, 17–28.

Ariztegui, D., J. A. McKenzie, and P. Farrimond. 1993. Fatty acid distributions in Holocene sediments from Lake St. Moritz, Switzerland: A record of fluctuating trophic conditions. *Advances in Organic Geochemistry* 10:470–2.

Arnold, J. R., and W. F. Libby. 1951. World-wide distribution of natural radiocarbon. *Physics Review* 81:64–9.

Assel, R. A., and D. M. Robertson. 1995. Changes in winter air temperatures near Lake Michigan, 1851–1993, as determined from regional lake-ice records. *Limnology and Oceanography* 40(1):165–76.

Atkinson, T. C., K. R. Briffa, and G. R. Coope. 1987. Seasonal temperatures in Britain during the past 22,000 years, reconstructed using beetle remains. *Nature* 325:587–92.

Atkinson, T. C., K. R. Briffa, G. R. Coope, M. J. Joachim, and D. W. Perry. 1986. Climatic calibration of coleopteran data. In *Handbook of Holocene Palaeoecology and Palaeohydrology*, ed. B. E. Berglund, 851–8. Chichester, U.K.: Wiley.

Barnola, J.-M., M. Anklin, J. Porcheron, D. Raynaud, J. Schwander, and B. Stauffer. 1995. CO_2 evolution during the last millenium as recorded by Antarctic and Greenland ice. *Tellus* 47B:264–72.

Barnola, J.-M., D. Raynaud, Y. S. Korotkevich, and C. Lorius. 1987. Vostok ice core provides 160,000 year record of atmospheric CO_2. *Nature* 329:408–14.

Bartlein, P. J., M. E. Edwards, S. L. Shafer, and E. D. Barker. 1995. Calibration of radiocarbon ages and the interpretation of palaeoenvironmental records. *Quaternary Research* 44:417–24.

Bartlein, P. J., I. C. Prentice, and T. Webb III. 1986. Climatic response surfaces based on pollen from some eastern North America taxa. *Journal of Biogeography* 13:35–57.

Battarbee, R. W., J. P. Smol, and J. Meriläinen. 1986. Diatoms as indicators of pH: An historical review. In *Diatoms and Lake Acidity*, eds. J. P. Smol, R. W. Battarbee, R. B. Davis, and J. Meriläinen, 5–14. Dordrecht, the Netherlands: W. Junk.

Bauerochse, A. 1996. Das Fimbertal. Klima- und Vegetationsgeschichtliche Untersuchungen in einem zentralalpinen Tal (Tirol/Graubünden). Ph.D. diss., Universität Hannover.

Baulant, E., and E. Le Roy Ladurie. 1980. Grape harvests from the fifteenth to the nineteenth century. *Journal of Interdisciplinary History* 10(4):839–49.

Becker, B., A. Billamboz, H. Egger, P. Gassmann, A. Orcel, C. Orcel, and U. Ruoff. 1985. Dendrochronologie in der Ur- und Frühgeschichte. Die absolute Datierung von Pfahlbausiedlungen nördlich der Alpen im Jahrringkalender Mitteleuropas. *Antiqua* 11:1–68.

Bellwald, W. 1992. Drei spätneolithisch/frühbronzezeitliche Pfeilbogen aus dem Gletschereis am Lötschenpass. *Archäologie der Schweiz* 15(4):166–71.

Beniston, M., M. Rebetez, F. Giorgi, and M. R. Marinucci. 1994. An analysis of regional climate change in Switzerland. *Theoretical and Applied Climatology* 49:135–59.

Bernhardi, A. 1832. Wie kamen die aus dem Norden stammenden Felsbruchstücke und Geschiebe, welche man in Norddeutschland und den benachbarten Ländern findet, an ihre gegenwärtigen Fundorte? *Jahrbuch für Mineralogie, Geognosie und Petrefaktenkunde* (Heidelberg) 3:257–67.

Bider, M., M. Schüepp, and H. von Rudloff. 1959. Die Reduktion der 200jährigen Basler Temperaturreihe. *Archiv für Meteorologie, Geophysik und Bioklimatologie*, series B, 9(3/4):360–412.

Bircher, W. 1982. Zur Gletscher- und Klimageschichte des Saastales. Glazialmorphologische und dendroklimatologische Untersuchungen. *Physische Geographie, Geographisches Institut, Universität Zürich* 9:1–233.

Birks, H. J. B. 1995. Quantitative palaeoenvironmental reconstructions. In *Statistical Modelling of Quaternary Science Data, Quaternary Research Association*, eds. D. Maddy and J. S. Brew, 161–254.

Birks, H. J. B. 1992. Some reflections on the application of numerical methods in Quaternary palaeoecology. *University of Joensuu, Publications of the Karelian Institute* 102:7–20.

Birks, H. J. B. 1981. The use of pollen analysis in the reconstruction of past climates: A review. In *Climate and History*, eds. T. M. L. Wigley, M. J. Ingram, and G. Farmer, 111–38. Cambridge: Cambridge University Press.

Birks, H. J. B., J. M. Line, S. Juggins, A. C. Stevenson, and C. J. F. ter Braak. 1990. Diatoms and pH reconstruction. *Philosophical Transactions of the Royal Society of London B* 327:263–78.

Blunier, T., J. Chappellaz, J. Schwander, J.-M. Barnola, T. Desperts, B. Stauffer, and D. Raynaud. 1993. Atmospheric methane: Record from a Greenland ice core over the last 1000 years. *Geophysical Research Letters* 20:2219–2222.

Blunier, T., J. Chappellaz, J. Schwander, B. Stauffer, and D. Raynaud. 1995. Variations in atmospheric methane concentration during the Holocene epoch. *Nature* 374:46–9.

Blytt, A. G. 1876. *Essays on the Immigration of the Norwegian Flora during Alternating Rainy and Dry Periods*. Christiania (Oslo), Norway: A. Cammermeyer.

Brooks, C. E. P. 1928. *Climate through the Ages*. London: Brill.

Bücher, A., and J. Dessens. 1991. Secular trend of surface temperature at an elevated observatory in the Pyrenees. *Journal of Climate* 4:859–68.

Charles, D. F., and J. P. Smol. 1988. New methods for using diatoms and chrysophytes to infer past pH of low-alkalinity lakes. *Limnology and Oceanography* 33:1451–62.

CLIMAP (Climate Mapping, Analysis, and Prediction) Project Members. 1981. Seasonal reconstructions of the earth's surface at the last glacial maximum. *Geological Society of America Map and Chart Series* MC-36.

CLIMAP Project Members. 1976. The surface of the Ice-Age earth. *Science* 191:1138–44.

COHMAP (Cooperative Holocene Mapping Project) Members. 1988. Climatic changes of the last 18,000 years: observations and model simulations. *Science* 241:1043–52.

Cook, E. R., and L. A. Kairiukstis, eds. 1990. *Methods of Dendrochronology: Applications in the Environmental Sciences.* Dordrecht, the Netherlands: Kluwer.

Coope, G. R. 1987. Fossil beetle assemblages as evidence for sudden and intense climatic changes in the British Isles during the last 45,000 years. In *Abrupt Climatic Change,* eds. W. H. Berger and L. D. Labeyrie 147–50. Dordrecht, the Netherlands: Reidel.

Coope, G. R. 1986. Coleoptera analysis. In *Handbook of Holocene Palaeoecology and Palaeolimnology,* ed. B. E. Berglund, 703–13. Chichester, U.K.: Wiley.

Coope, G. R. 1977. Fossil coleopteran assemblages as sensitive indicators of climatic change during the Devensian (last) cold stage. *Philosophical Transactions of the Royal Society of London B* 280:313–40.

Craig, H. 1965. The measurement of oxygen isotope paleotemperatures. In *Stable Isotopes in Oceanographic Studies and Paleotemperatures,* ed. E. Tongiorgi, 161–82. Pisa, Italy: Consiglio Nazionale delle Richerche, Laboratorio di Geologia Nucleare.

Craig, H. 1961. Standards for reporting concentrations of deuterium and oxygen-18 in natural waters. *Science* 133:1833–4.

Craig, H. 1957. Isotopic standards for carbon and oxygen and correction factors for mass-spectrometric analysis of carbon dioxide. *Geochimica et Cosmochimica Acta* 12:133–49.

Craig, H., and L. I. Gordon. 1965. Deuterium and oxygen 18 variations in the ocean and marine atmosphere. In *Stable Isotopes in Oceanographic Studies and Paleotemperatures,* ed. E. Tongiorgi, 9–130. Pisa, Italy: Consiglio Nazionale delle Richerche, Laboratorio di Geologia Nucleare.

Dansgaard, W. 1980. Palaeoclimatic studies on ice cores. In *Das Klima: Analysen und Modelle, Geschichte und Zukunft,* eds. H. Oeschger, B. Messerli, and M. Svilar, 237–45. Berlin: Springer-Verlag.

Dansgaard, W. 1964. Stable isotopes in precipitation. *Tellus* 16:436–68.

De Geer, G. 1912. A geochronology of the last 12,000 years. *Compte Rendu: Proceedings of the 11th International Geological Congress,* 1910, Stockholm, 241–53.

Desloges, J. R. 1994. Varve deposition and the sediment yield record at three small lakes of the southern Canadian cordillera. *Arctic and Alpine Research* 26:130–40.

Douglass, A. E. 1941. Crossdating in dendrochronology. *Journal for Forestry* 39:825–31.

Dronia, H. 1967. Der Stadteinfluss auf den weltweiten Temperaturtrend. *Meteorologische Abhandlungen, Institut für Meteorologie und Geophysik der freien Universität Berlin* 74(4):1–65.

Duplessy, J. C., M. Arnold, P. Maurice, E. Bard, E. Duprat, and J. Moyes. 1986. Direct dating of the oxygen-isotope record of the last glaciation by ^{14}C accelerator mass spectrometry. *Nature* 320:350–2.

Eddy, J. A., and H. Oeschger, eds. 1991. *The Pages Project: Proposed Implementation Plans for Research Activities.* Global Change Report no. 19. Stockholm: International Geosphere-Biosphere Programme.

Eglinton, G., and M. Calvin. 1967. Chemical fossils. *Scientific American* 216:32–43.

Eicher, U., and U. Siegenthaler. 1983. Stable isotopes in lake marl and mollusc shells from Lobsigensee (Swiss Plateau): Studies in the Late Quaternary of Lobsigensee 6. *Revue de Paléobiologie* 2:217–20.

Eicher, U., and U. Siegenthaler. 1976. Palynological and oxygen isotope investigations on Late-Glacial sediment cores from Swiss lakes. *Boreas* 5:109–117.

Elias, S. A. 1994. *Quaternary Insects and Their Environments*. Washington: Smithsonian Institution Press.

Elias, S. A., and B. Wilkinson. 1985. Fossil assemblages of coleoptera and trichoptera at Lobsigensee. *Dissertationes Botanicae* 87:157–62.

Elias, S. A., and B. Wilkinson. 1983. Lateglacial insect fossil assemblages from Lobsigensee (Swiss Plateau). Studies in the Late Quaternary of Lobsigensee 3. *Revue de Paléobiologie* 2:189–204.

Emiliani, C. 1966. Palaeotemperature analysis of Caribbean cores P6 304-8 and P6 304-9 and a generalised temperature curve for the past 425,000 years. *Journal of Geology* 74:109–26.

Emiliani, C. 1955. Pleistocene temperatures. *Journal of Geology* 63:538–78.

Emiliani, C. 1954. Depth habitats of some species of pelagic foraminifera as indicated by oxygen isotope ratios. *American Journal of Science* 252:149–58.

Epstein, S., and T. Mayeda. 1953. Variations in ^{18}O content of waters from natural sources. *Geochimica et Cosmochimica Acta* 27:213–24.

Eronen, M., and P. Zetterberg. 1992. Dendrochronology and climate history in the subarctic area of Fennoscandia. In *The Finnish Research Programme on Climate Change: Progress Report*, eds. M. Kanninen and P. Anttila, 13–8. Vapk, Helsinki, Finland: Academy of Finland.

Faegri, K., and J. Iversen. 1989. *Textbook of Pollen analysis*. Edited by K. Faegri, P. E. Kaland and K. Krzywinski. 4th ed. Chichester, U.K.: Wiley.

Faure, G. 1977. *Principles of Isotope Geology*. New York: Wiley.

Flint, R. F. 1971. *Glacial and Quaternary Geology*. New York: Wiley.

Flint, R. F. 1957. *Glacial and Quaternary Geology*. New York: Wiley.

Förstel, H., and H. Hützen. 1983. $^{18}O/^{16}O$ ratio of water in a local ecosystem as a basis of climate record. In *Palaeoclimate and Palaeowaters: A Collection of Environmental Isotope Studies*, 67–84. Vienna: International Atomic Energy Agency.

Frenzel, B., C. Pfister, and B. Gläser, eds. 1994. *Climatic Trends and Anomalies in Europe, 1675–1715*. Edited by B. Frenzel and B. Gläser. Stuttgart: Gustav Fischer Verlag.

Friedli, H., H. Lotscher, H. Oeschger, U. Siegenthaler, and B. Stauffer. 1986. Ice core record of the $^{13}C/^{12}C$ ratio of the atmospheric CO_2 in the past two centuries. *Nature* 324:237–8.

Fritz, S. C., S. Juggins, R. W. Battarbee, and D. R. Engstrom. 1991. Reconstruction of past changes in salinity and climate using a diatom-based transfer function. *Nature* 352:706–8.

Fuhrer, K., A. Neftel, M. Anklin, and V. Maggi. 1993. Continuous measurements of hydrogen peroxide, formaldehyde, calcium and ammonium concentrations along the new GRIP ice core from Summit, central Greenland. *Atmospheric Environment* 27A:1873–80.

Funk, M. 1994. Possible Alpine ice-core drilling sites: An overview. In *Greenhouse Gases, Isotopes and Trace Elements in Glaciers as Climatic Evidence of the Holocene*, eds. W. Haeberli and B. Stauffer, 40–4. Zurich: Versuchsanstalt für Wasserbau, Hydrologie und Glaziologie der Eidgenössischen Technischen Hochschule Zürich, Arbeitsheft 14.

Gaillard, M.-J., and G. Lemdahl. 1994. Late-glacial insect assemblages from Grand-Marais, south-western Switzerland: Climatic implications and comparison with pollen and plant macrofossils. *Dissertationes Botanicae* 234:287–308.

Gasse, F., M. Arnold, J. C. Fontes, M. Fort, E. Gibert, A. Huc, L. Bingyan, L. Yuanfang, L. Qing, F. Mélières, E. van Campo, W. Fubao, and Z. Qingsong. 1991. A 13,000-year climate record from western Tibet. *Nature* 353:742–5.

Gat, J. R. 1981. Lakes. In *Stable Isotope Hydrology: Deuterium and Oxygen-18 in the Water Cycle*, eds. J. R. Gat and R. Gonfiantini, 203–22. Vienna: International Atomic Energy Agency.

Gat, J. R. 1980. The isotopes of hydrogen and oxygen in precipitation. In *Handbook of Environmental Isotope Geochemistry*, eds. P. Fritz and J. C. Fontes, 21–47. Amsterdam: Elsevier.

Gat, J. R., and G. S. Lister. 1995. The "catchment effect" in the isotopic composition of lake waters: Its importance in palaeo-limnological interpretations. In *Problems of Stable Isotopes in Tree-Rings, Lake Sediments and Peat-Bogs as Climatic Evidence for the Holocene*, eds. B. Frenzel, B. Stauffer and M. M. Weiss, 1–16. Stuttgart: Gustav Fischer Verlag.

Glen, C. R., and K. Kelts. 1991. Sedimentary rhythms in lake deposits. In *Cycles and Events in Stratigraphy*, eds. G. Einsele, W. Ricken, and A. Seilacher, 188–221. Berlin: Springer-Verlag.

Gordon, G. A., J. M. Lough, H. C. Fritts, and P. M. Kelly. 1985. Comparison of sea level pressure reconstructions from western North American tree rings with a proxy record of winter severity in Japan. *Journal of Climate and Applied Meteorology* 24:1219–24.

Gray, B. M. 1974. Early Japanese winters. *Weather* 29:103–7.

GRIP (European Greenland Ice Core Drilling Project) Members. 1993. Climate instability during the last interglacial period recorded in the GRIP ice core. *Nature* 364:203–7.

Grischuk, V. P. 1969. An attempt to reconstruct certain elements of the climate of the Northern Hemisphere in the Atlantic period of the Holocene. In *Golotsen*, ed. M. I. Neistadt, 41–57. Moscow: Izd-vo Nauka.

Guiot, J. 1987. Late Quaternary climatic change in France estimated from multivariate pollen time series. *Quaternary Research* 28:100–18.

Guiot, J., A. Pons, J. L. De Beaulieu, and M. Reille. 1989. A 140,000-year continental climate reconstruction from two European pollen records. *Nature* 338:309–13.

Guiot, J., M. Reille, J. L. De Beaulieu, and A. Pons. 1992. Calibration of the climatic signal in a new pollen sequence from La Grande Pile. *Climate Dynamics* 6:259–64.

Haeberli, W. 1994a. Accelerated glacier and permafrost changes in the Alps. In *Mountain Environments in Changing Climates*, ed. M. Beniston, 91–107. London: Routledge.

Haeberli, W. 1994b. Glaciological basis of ice-core drilling on mid-latitude glaciers. In *Greenhouse Gases, Isotopes and Trace Elements in Glaciers as Climatic Evidence of the Holocene: Report of the ESF/EPC Workshop, Zurich, 27–28 October 1992*, eds. W. Haeberli and B. Stauffer, 3–9. Zurich: Versuchsanstalt für Wasserbau, Hydrologie und Glaziologie der Eidgenössischen Technischen Hochschule Zürich, Arbeitsheft 14.

Haeberli, W. 1991. Glazialmorphologische und palaeoglaziologische Modelle. In *Modelle in der Geomorphologie-Beispiele aus der Schweiz*, eds. M. Monbaron and W. Haeberli, Rapports et Recherches no. 3, 7–19. Fribourg, Switzerland: Geographical Institute of Fribourg.

Haeberli, W., ed. 1990. *Pilot Analyses of Permafrost Cores from the Active Rock Glacier Murtèl, Piz Covatsch, Eastern Swiss Alps: A Workshop Report*. Zurich: Versuchsanstalt für Wasserbau, Hydrologie und Glaziologie der Eidgenössischen Technischen Hochschule Zürich, Arbeitsheft 9.

Haeberli, W., and M. Funk. 1991. Borehole temperatures at the Colle Gnifetti core-drilling site (Monte Rosa, Swiss Alps). *Journal of Glaciology* 37(125):37–46.

Haeberli, W., and M. Hoelzle. 1995. Application of inventory data for estimating characteristics of and regional climate change effects on mountain glaciers: A pilot study with the European Alps. *Annals of Glaciology* 21:206–12.

Haeberli, W., M. Hoelzle, F. Keller, W. Schmid, D. Vonder Mühll, and S. Wagner. 1993. Monitoring the long-term evolution of mountain permafrost in the Swiss Alps. Proceedings of the VIth International Conference on Permafrost, Beijing, 214–19.

Haeberli, W., and C. Schlüchter. 1987. Geological evidence to constrain modelling of the Late Pleistocene Rhonegletscher (Switzerland). *The Physical Basis of Ice Sheet Modelling: Proceedings of an International Symposium Held during the XIXth General Assembly of the International Union of Geodesy and Geophysics, Vancouver, British Columbia, Canada,* 333–46. IAHS Publication no. 170.

Haeberli, W., and B. Stauffer, eds. 1994. *Greenhouse Gases, Isotopes and Trace Elements in Glaciers as Climatic Evidence of the Holocene.* Zurich: Versuchsanstalt für Wasserbau, Hydrologie und Glaziologie der Eidgenössischen Technischen Hochschule Zürich, Arbeitsheft 14.

Hall, R. I., and J. P. Smol. 1992. A weighted-averaging regression and calibration model for inferring total phosphorus concentration from diatoms in British Columbia (Canada) lakes. *Freshwater Biology* 27:417–434.

Hansen, J., and S. Lebedeff. 1988. Global surface air temperatures: Update through 1987. *Geophysical Research Letters* 15(4):323–6.

Hansen, J., and S. Lebedeff. 1987. Global trends of measured surface air temperature. *Journal of Geophysical Research* 20:13345–13372.

Hartig, R. 1870. Zur Lehre vom Dickenwachstum der Waldbäume. *Botanische Zeitung* 38:1–50.

Hays, J. D., J. Imbrie, and N. J. Shackelton. 1976. Variations in the earth's orbit: Pacemaker of the ice ages. *Science* 241:687–90.

Heer, O. 1865. *Die Urwelt der Schweiz.* Zurich: F. Schulthess.

Heer, O. 1858. *Die Schieferkohlen von Uznach und Dürnten.* Zurich: Orel Füssli.

Hoefs, J. 1980. *Stable Isotope Geochemistry.* 2nd ed. Berlin: Springer-Verlag.

Hofmann, W. 1986. Chironomid analysis. In *Handbook of Holocene Palaeoecology and Palaeohydrology,* ed. B. E. Berglund, 715–27. Chichester, U.K.: Wiley.

Hondzo, M., and H. G. Stefan. 1993. Regional water temperature characteristics of lakes subjected to climate change. *Climatic Change* 24:187–211.

Houghton, J. T., B. A. Callander, and S. K. Varney, eds. 1992. *Climate Change 1992: The Supplementary Report to the IPCC Scientific Assessment Report Prepared for IPCC by Working Group I.* Cambridge: Cambridge University Press.

Hutton, J. 1795. *Theory of the Earth.* Vol. 2. Edinburgh: William Creech.

IAEA (International Atomic Energy Agency). 1993. *Isotope Techniques in the Study of Past and Current Environmental Changes in the Hydrosphere and the Atmosphere.* Vienna: International Atomic Energy Agency.

Imbrie, J., and N. G. Kipp. 1971. A new micropalaeontological method for quantitative paleoclimatology: application to late Pleistocene Caribbean core V28–238. In *The Late Cenozoic Glacial Ages,* ed. K. K. Turekian, 77–181. New Haven, CT: Yale University Press.

Ingram, M. J., D. J. Underhill, and G. Farmer. 1981. The use of documentary sources for the study of past climates. In *Climate and History Studies in Past Climates and Their Impact on Man,* eds. T. M. L. Wigley, M. J. Ingram, and G. Farmer, 180–213. Cambridge: Cambridge University Press.

Iversen, J. 1944. *Viscum, Hedera* and *Ilex* as climatic indicators. *Geologiska Föreningen i Stockholm Förhandlingar* 66:463–483.

Jacobson, G. L., and R. H. W. Bradshaw. 1981. The selection of sites for paleovegetational studies. *Quaternary Research* 16:80–96.

Jones, P. D., and K. R. Briffa. 1992. Global surface air temperature variations during the twentieth century. Part I: Spatial, temporal and seasonal details. *The Holocene* 2(2):165–79.

Jones, P. D., and P. M. Kelly. 1983. The spatial and temporal characteristics of Northern Hemisphere surface air temperature variations. *Journal of Climatology* 3:243–52.

Jones, P. D., S. C. B. Raper, R. S. Bradley, H. F. Diaz, P. M. Kelly, and T. M. L. Wigley. 1986. Northern Hemisphere surface temperature variations: 1851–1984. *Journal of Climate and Applied Meteorology* 25:161–79.

Jones, P. D., T. M. L. Wigley, C. K. Folland, D. E. Parker, J. K. Angell, T. S. Lebedeff, and J. E. Hansen. 1988. Evidence for global warming in the past decade. *Nature* 332:790.

Jones, P. D., T. M. L. Wigley, and P. B. Wright. 1986. Global temperature variations between 1861 and 1984. *Nature* 322:430–4.

Joussaume, S., and J. Jouzel. 1987. Simulation of palaeoclimatic tracers using atmospheric general circulation models. In *Abrupt Climatic Change: Evidence and Implications*, eds. W. H. Berger and L. D. Labeyrie, 369–82. Dordrecht, the Netherlands: Reidel.

Jouzel, J., R. D. Koster, R. J. Suozzo, and G. G. Russell. 1994. Stable water isotope behaviour during the last glacial maximum: A general circulation model analysis. *Journal of Geophysical Research* 99(D12):25791–801.

Karl, T. R., H. F. Diaz, and G. Kukla. 1988. Urbanization: Its detection and effect in the United States climate record. *Journal of Climate* 1:1099–123.

Karl, T. R., G. Kukla, and J. Gavin. 1986. Relationship between decreased temperature range and precipitation trends in the United States and Canada, 1941–80. *Journal of Climate and Applied Meteorology* 25:1878–86.

Karl, T. R., G. Kukla, and J. Gavin. 1984. Decreasing diurnal temperature range in the United States and Canada from 1941 through 1980. *Journal of Climate and Applied Meteorology* 23(11):1489–504.

Karl, T. R., G. Kukla, V. N. Razuyayev, M. J. Changery, R. G. Quayle, R. R. Heim Jr., D. R. Easterling, and C. B. Fu. 1991. Global warming: Evidence for asymmetric diurnal temperature change. *Geophysical Research Letters* 18(12):2253–6.

Karlén, W. 1981. Lacustrine sediment studies: A technique to obtain a continuous record of Holocene glacier variations. *Geografiska Annaler* 63A:273–81.

Keeling, C. D., T. P. Whorf, W. Wahlen, and J. van der Plicht. 1995. Interannual extremes in the rate of rise of atmospheric carbon dioxide since 1980. *Nature* 375:666–670.

Kelts, K., and K. J. Hsü. 1978. Freshwater carbonate sedimentation. In *Lakes: Chemistry, Geology, Physics*, ed. A. Lerman, 295–323. New York: Springer Verlag.

Kennet, J. 1982. *Marine Geology*. London: Prentice Hall.

Kerschner, H. 1985. Quantitative paleoclimatic inferences from late-glacial snowline, timberline and rock glacier data, Tyrolean Alps, Austria. *Zeitschrift für Gletscherkunde und Galzialgeologie* 21:363–9.

Kienast, F. 1985. Dendroökologische Untersuchungen an Höhenprofilen aus verschiedenen Klimabereichen: Eine radiodensitometrische Studie ueber den Einfluss der Witterung, der Höhenlage und der Standortseigenschaften auf das Jahrringwachstum von Nadelbäumen. Ph.D. diss. Universität Zürich.

Kukla, G., J. Gavin, and T. R. Karl. 1986. Urban warming. *Journal of Climate and Applied Meteorology* 25:1265–70.

Kukla, G., and T. R. Karl. 1993. Night-time warming and the greenhouse effect. *Environmental Science and Technology* 27(8):1468–74.

Kunz, B., and W. Stumm. 1984. Kinetik der Bildung und des Wachstums von Calciumcarbonat. *Vom Wasser* 62:85–93.

Kutzbach, J. E., P. J. Guetter, P. J. Behling, and P. Selin. 1993. Simulated climatic changes: Results of the COHMAP climate-model experiments. In *Global Climates since the Last Glacial Maximum*, eds. H. E. Wright, J. E. Kutzbach, T. Webb, W. F. Ruddiman, F. A. Street-Perrott, and P. J. Bartlein, 24–93. Minneapolis: University of Minnesota Press.

Kutzbach, J. E., and F. A. Street-Perrot. 1985. Milankovitch forcing of fluctuations in the levels of tropical lakes from 18 to 0 kyr B.P. *Nature* 317:130–4.

Lallier-Vergès, E., A. Sifeddine, J. L. de Beaulieu, M. Reille, N. Tribovillard, P. Bertrand, T. Mongenot, N. Thouveny, J. R. Disnar, and B. Guillet. 1993. The lacustrine organic sedimentation as a response to the Late Würmian Holocene climatic variations, Lac du Bouchet (Haut-Loire, France). *Bulletin de la Société Géologique de France* 164/5:661–73.

Lamb, H. H. 1977. *Climate: Present, Past and Future*. Vol. 2: *Climatic History and the Future*. London: Methuen.

Lamb, H. H. 1972. *Climate: Present, Past and Future*. Vol. 1: *Fundamentals and Climate Now*. London: Methuen.

Lamb, H. H., F. Gasse, A. Benkaddour, N. El Hamouti, S. van der Kaars, W. T. Perkins, N. J. Pearce, and C. N. Roberts. 1995. Relation between century-scale Holocene arid intervals in tropical and temperate zones. *Nature* 373:134–7.

Lamb, H. H., and A. I. Johnson. 1966. Secular variations of the atmospheric circulation since 1750. H. M. Meteorological Office *Geophysical Memoir* Number 110.

Lang, C. 1885. Der seculäre Verlauf der Witterung als Ursache der Gletscherschwankungen in den Alpen. *Zeitschrift der Österreichischen Gesellschaft für Meteorologie* 20:443–57.

Leavitt, S. W., and A. Long. 1991. Seasonal stable-carbon isotope variability in tree rings: Possible palaeoenvironmental signals. *Chemical Geology* (Isotope Geoscience Section) 87:59–70.

Leemann, A. 1993. Rhythmite in alpinen Vorgletscherseen - Warvenstratigraphie und Aufzeichnung von Klimaveränderungen. Ph.D. diss. no. 10386, Swiss Federal Institute of Technology (ETH), Zurich.

Leemann, A., and F. Niessen. 1994a. Varve formation and the climatic record in an Alpine proglacial lake. *The Holocene* 4:1–8.

Leemann, A., and F. Niessen. 1994b. Holocene glacial activity and climatic variations in the Swiss Alps: Reconstructing a continuous record from proglacial lake sediments. *The Holocene* 4(3):259–68.

Legrand, J.-P., and M. Le Goff. 1992. *Les observations météorologiques de Louis Morin*. Vol. Monographie 6. Paris: Direction de la Météorologie Nationale.

Lemdahl, G. 1991. A rapid climatic change at the end of the Younger Dryas in south Sweden: Palaeoclimatic and palaeoenvironmental reconstructions based on fossil insect assemblages. *Palaeogeography-Palaeoclimatology-Palaeoecology* 83:313–31.

Leonard, E. M. 1986. Varve studies in Hector Lake, Alberta, Canada, and the relationship between glacial activity and sedimentation. *Quaternary Research* 25:199–214.

Leonard, E. M. 1985. Glaciological and climatic controls on lake sedimentation, Canadian Rocky Mountains. *Zeitschrift für Gletscherkunde und Glazialgeologie* 21:35–42.

Le Roy Ladurie, E. 1967. *L'histoire du climat depuis l'an mil*. Paris: Flammarion.

Libby, L. M. 1987. Evaluation of historic climate and prediction of near-future climate from stable-isotope variations in tree rings. In *Climate: History, Periodicity and Predictability*, eds. M. R.

Rampino, J. E. Sanders, W. S. Newman, and L. K. Königsson, 81–9. New York: Van Nostrand Reinhold Company.

Lister, G. S. 1988a. A 15,000-year record from Lake Zurich of deglaciation and climatic change in Switzerland. *Quaternary Research* 29:129–41.

Lister, G. S. 1988b. Stable isotopes from lacustrine ostracoda as tracers for continental palaeo-environments. In *Ostracoda in the Earth Sciences*, eds. P. De Deckker, J.-P. Colin, and J.-P. Peypouquet. Amsterdam: Elsevier,.

Lister, G. S. 1985. Late Pleistocene Alpine deglaciation and post-glacial climatic developments in Switzerland: The record of sediments in a peri-Alpine lake basin. *Mitteilungen aus dem Geologischen Institut der Eidgenössischen Technischen Hochschule und der Universität Zürich, Neue Folge* 249:1–115.

Lister, G. S. 1984. Deglaciation of the Lake Zurich area: A model based on the sedimentological record. *Contributions to Sedimentology* 13:177–85.

Lister, G. S., F. Giovanoli, G. Eberli, P. Finckh, W. Finger, Q. He, C. Heim, K. J. Hsü, K. Kelts, C. Peng, C. Sidler, and X. Zhao. 1984. Late Quaternary sediments in Lake Zürich, Switzerland. *Environmental Geology* 5(4):191–205.

Livingstone, D. M. 1997. Break-up dates of Alpine lakes as proxy data for local and regional mean surface air temperatures. *Climatic Change* 37:407–439.

Livingstone, D. M. (in prep.).

Livingstone, D. M. 1993a. Temporal structure in the deep-water temperature of four Swiss Lakes: A short-term climatic change indicator? *Verhandlungen der internationalen Vereinigung der Limnologie* 25:75–81.

Livingstone, D. M. 1993b. Lake oxygenation: Application of a one-box model with ice cover. *Internationale Revue der gesamten Hydrobiologie* 78(4):465–80.

Livingstone, D. M., and D. M. Imboden. 1996. The prediction of hypolimnetic oxygen profiles: A plea for a deductive approach. *Canadian Journal of Fisheries and Aquatic Sciences* 53(4):924–32.

Livingstone, D. M., and D. M. Imboden. 1989. Annual heat balance and equilibrium temperature of Lake Aegeri, Switzerland. *Aquatic Sciences* 51(4):351–69.

Livingstone, D. M., and F. Schanz. 1994. The effects of deep-water siphoning on a small, shallow lake: A long-term case study. *Archiv für Hydrobiologie* 132(1):15–44.

Lotter, A. F. 1991. Absolute dating of the Late-Glacial period in Switzerland using annually laminated sediments. *Quaternary Research* 35:321–30.

Lotter, A. F. 1989. Evidence of annual layering in Holocene sediments of Soppensee, Switzerland. *Aquatic Sciences* 51:19–30.

Lotter, A. F., B. Ammann, and M. Sturm. 1992. Rates of change and chronological problems during the late-glacial period. *Climate Dynamics* 6:233–239.

Maisch, M. 1992. Die Gletscher Graubündens - Rekonstruktion und Auswertung der Gletscher und deren Veränderungen seit dem Hochstand von 1850 im Gebiet der östlichen Schweizer Alpen (Bündnerland und angrenzende Regionen). Habilitation thesis, Universität Zürich.

Manley, G. 1974. Central England temperatures: Monthly means 1659 to 1973. *Quarterly Journal of the Royal Meteorological Society* 100:389–405.

McKenzie, J. A. 1985. Carbon isotopes and productivity in the lacustrine and marine environments. In *Chemical Processes in Lakes*, ed. W. Stumm, 99–118. New York: Wiley.

Meyers, P. A., and R. Ishiwatari. 1993. Lacustrine organic geochemistry: An overview of indicators of organic matter sources and diagenesis in lake sediments. *Organic Geochemistry* 20(7):867–900.

Meyers, P. A., K. Takemura, and S. Horie. 1993. Reinterpretation of Late Quaternary sediment chronology of Lake Biwa, Japan, from correlation with marine Glacial-Interglacial cycles. *Quaternary Research* 39:154–62.

Milankovitch, M. 1941. Kanon der Erdbestrahlung und seine Anwendung auf das Eiszeitproblem. *Academie Royale Serbe* (special edition) 133:1–633.

Milankovitch, M. 1938. Astronomische Mittel zur Erforschung der erdgeschichtlichen Klimate. *Handbuch der Geophysik* 9:593–698.

Mitchell, J. M. 1970. Pollution as a cause of global temperature fluctuation. In *Global Aspects of Pollution*. New York: Springer Verlag.

Moore, P. D., J. A. Webb, and M. E. Collinson. 1991. *Pollen Analysis*. 2d ed. London: Blackwell.

Muller, R. A., and G. J. McDonald. 1995. Glacial cycles and orbital inclination. *Nature* 377:107–8.

Neftel, A., E. Moor, H. Oeschger, and B. Stauffer. 1985. Evidence from polar ice cores for the increase in atmospheric CO_2 in the past two centuries. *Nature* 315:45–47.

Neftel, A., H. Oeschger, T. Staffelbach, and B. Stauffer. 1988. CO_2 record in the Byrd ice core 50000–5000 years BP. *Nature* 331:609–11.

Nipkow, H. F. 1920. Vorlaufige Mitteilungen über Untersuchungen des Schlammabsatzes im Zürichsee. *Schweizerische Zeitschrift für Hydrologie* 1:1–28.

Oana, S., and E. S. Deevey. 1960. Carbon-13 in lake waters and its possible bearing on paleolimnology. *American Jounal of Science* 258A:253–72.

Oerlemans, J. 1994. Quantifying global warming from retreat of glaciers. *Science* 264:243–5.

Oerlemans, J. 1993. Modelling of glacier mass balance. In *Ice in the Climate System*, ed. W. R. Peltier, 101–16 Berlin: Springer.

Oeschger, H. 1987. Die Ursachen der Eiszeiten und die Möglichkeit der Klimabeeinflussung durch den Menschen. *Mitteilungen der naturforschenden Gesellschaft Luzern* Sonderband Eiszeitforschung, 29:51–76.

Oeschger, H., J. Beer, U. Siegenthaler, B. Stauffer, W. Dansgaard, and C. C. Langway. 1984. Late glacial climate history from ice cores. In *Climate Processes and Climate Sensitivity*. Geophysical Monograph 29, eds. J. E. Hansen and T. Takahashi, 299–360. Washington, D.C.: American Geophysical Union.

Ogura, K. 1990. Lipid compounds in lake sediments. *Verhandlungen der internationalen Vereinigung der Limnologie* 24:274–8.

O'Sullivan, P. E. 1983. Annually-laminated lake sediments and the study of quarternary environmental changes: A review. *Quaternary Science Reviews* 1:245–313.

Palecki, M. A., and R. G. Barry. 1986. Freeze-up and break-up as an index of temperature changes during the transition seasons: A case study for Finland. *Journal of Climate and Applied Meteorology* 25:893–902.

Parker, D. E., T. P. Legg, and C. K. Folland. 1992. A new daily central England temperature series, 1772–1991. *International Journal of Climatology* 12:317–42.

Patzelt, G. 1977. Der zeitliche Ablauf und das Ausmass postglazialer Klimaschwankungen in den Alpen. In *Dendrochronologie und postglaziale Klimaschwankungen in Europa*, ed. B. Frenzel, 248–59. Wiesbaden, Germany: Steiner.

Patzelt, G., and M. Aellen. 1990. Gletscher. In *Schnee, Eis und Wasser in einer wärmeren Atmosphäre*, ed. D. Vischer, 49–69. Zurich: Mitteilungen der Versuchsanstalt für Wasserbau, Hydrologie und Glaziologie der Eidgenössischen Technischen Hochschule Zürich.

Pearson, F. J., and T. B. Coplen. 1978. Stable isotope studies of lakes. In *Lakes: Chemistry, Geology and Physics*, ed. A. Lerman, 325–34. New York: Springer-Verlag.

Peglar, S. M. 1993. The mid-Holocene Ulmus decline at Diss Mere, Norfolk, U.K.: A year-by-year pollen stratigraphy from annual laminations. *The Holocene* 3:1–3.

Penck, A. 1882. *Vergletscherung der deutschen Alpen*. Leipzig, Germany: J. A. Barth.

Pfister, C. (in prep.).

Pfister, C. 1992. Monthly temperature and precipitation in central Europe 1525–1979: Quantifying documentary evidence on weather and its effects. In *Climate since A.D. 1500*, eds. R. S. Bradley and P. D. Jones, 118–42. London: Routledge.

Pfister, C. 1984. *Das Klima der Schweiz von 1525–1860 und seine Bedeutung in der Geschichte von Bevölkerung und Landwirtschaft*. Vol. 1, *Academica Helvetica 6*. Bern, Switzerland: Verlag Paul Haupt.

Pfister, C., J. Kington, G. Kleinlogel, H. Schüle, and E. Siffert. 1994. High resolution spatio-temporal reconstructions of past climate from direct meteorological observations and proxy data: Methodological considerations and results. In *Climatic Trends and Anomalies in Europe 1675–1715*, eds. B. Frenzel, C. Pfister, and B. Gläser, 329–75. Stuttgart, Germany: Gustav Fischer Verlag.

Pienitz, R., J. P. Smol, and H. J. B. Birks. 1995. Assessment of freshwater diatoms as quantitative indicators of past climatic change in the Yukon and Northwest Territories, Canada. *Journal of Paleolimnology* 13:21–49.

Plantico, M. S., T. R. Karl, G. Kukla, and J. Gavin. 1990. Is recent climate change across the United States related to rising levels of anthropogenic greenhouse gases? *Journal of Geophysical Research* 95(D10):16617–37.

Playfair, J. 1802. *Illustrations of the Huttonian Theory of the Earth*. Edinburgh: William Creech.

Polge, H. 1966. Etablissement des courbes de variation de la densité du bois par l'exploration densitométrique de radiographie d'echantillons prélevés à la tarière sur des arbres vivants. *Annales des sciences forestières* 23:1–206.

Ponel, P., and G. R. Coope. 1990. Late-glacial and Early Flandrian coleoptera from La Taphanel, Massif Central, France: Climatic and ecological implications. *Journal of Quaternary Science* 5:235–50.

Ponel, P., J.-L. de Beaulieu, and K. Tobolski. 1992. Holocene palaeoenvironment at the timberline in the Taillefer Massif: Pollen analysis, study of plant and insect macrofossils. *The Holocene* 2:117–30.

Prentice, I. C. 1988. Records of vegetation in time and space: The principles of pollen analysis. In *Vegetation History*, eds. B. Huntley and T. Webb, 17–42. Dordrecht, the Netherlands: Kluwer.

Rannie, W. F. 1983. Breakup and freezeup of the Red River at Winnipeg, Manitoba, Canada in the 19th century and some climatic implications. *Climate Change* 5:283–96.

Renberg, I. 1986. Photographic demonstration of the annual nature of a varve type common in N. Swedish lake sediments. *Hydrobiologia* 140:93–5.

Renberg, I., and T. Hellberg. 1982. The pH history of lakes in southwestern Sweden, as calculated from the subfossil diatom flora of the sediments. *Ambio* 11:30–3.

Renner, F. 1982. Beiträge zur Gletschergeschichte des Gotthardgebietes und dendroklimatologische Analysen an fossilen Hölzern. *Physische Geographie, Geographisches Institut, Universität Zürich* 8:182.

Robertson, D. M., and R. A. Ragotzkie. 1990. Changes in the thermal structure of moderate to large sized lakes in response to changes in air temperature. *Aquatic Sciences* 52(4):360–80.

Robertson, D. M., R. A. Ragotzkie, and J. J. Magnuson. 1992. Lake ice records used to detect historical and future climatic changes. *Climatic Change* 21:407–27.

Rozanski, K. L. 1995. Climate control of stable isotopes in precipitation as a basis for palaeoclimate reconstructions. In *Problems of Stable Isotopes in Tree-Rings, Lake Sediments and Peat-Bogs as Climatic Evidence for the Holocene*, eds. B. Frenzel, B. Stauffer and M. M. Weiss, 171–86. Stuttgart, Germany: Gustav Fischer Verlag.

Saarnisto, M. 1983. Päijänteen Kinisselän Lustosedimentit 1900-Luvulla. In *Suomalaista sedimenttitutkimusta symposio Koilla, Joensuun korkeakoulu, Karjalan tutkimuslaitoksenjulkaisuja*, ed. H. Simola, 59–60.

Sauramo, M. 1923. Studies on the Quaternary varve sediments in southern Finland. *Bulletin de la Commission Geologique de Finlande* 60:1–164.

Saurer, M., U. Siegenthaler, and F. Schweingruber. 1995. The climate–carbon isotope relationship in tree rings and the significance of site conditions. *Tellus* 47B:1–11.

Schindler, D. W., K. G. Beatty, E. J. Fee, D. R. Cruikshank, E. R. DeBruyn, D. L. Findlay, G. A. Linsey, J. A. Shearer, M. P. Stainton, and M. A. Turner. 1990. Effects of climatic warming on lakes of the central boreal forest. *Science* 250:967–70.

Schotterer, U., K. Fröhlich, W. Stichler, and P. Trimborn. 1993. Temporal variation of ^{18}O and deuterium excess in precipitation, river and spring waters in Alpine regions of Switzerland. Proceedings of an International Symposium on Applications of Isotope Techniques in Studying Past and Current Environmental Changes in the Hydrosphere and the Atmosphere, Vienna, 53–64.

Schotterer, U., F. Oldfield, and K. Fröhlich. 1996. *GNIP: Global Network for Isotopes in Precipitation.* Bern, Switzerland: Läderach AG.

Schotterer, U., T. Stocker, J. Hunziker, P. Buttet, and J.-P. Tripet. 1995. Isotope im Wasserkreislauf: Ein Neues eidgenössisches Messnetz. *Gas, Wasser, Abwasser* 9:714–20.

Schove, D. J. 1969. A varve teleconnection project. In *Etudes sur le Quaternaire dans le Monde*, ed. M. Ters. Paris: International Quaternary Association, 927–35.

Schüepp, M. 1961. Klimatologie der Schweiz, Vol. C: Lufttemperatur, Part 2. *Beiheft zu den Annalen der Schweizerischen Meteorologischen Zentralanstalt (Jahrgang 1960)*:C15–C62.

Schwalb, A., G. Lister, and K. Kelts. 1994. Ostracode carbonate δ^{18}O and δ^{13}C signatures of hydrological and climatic changes affecting Lake Neuchatel, Switzerland, since the latest Pleistocene. *Journal of Palaeolimnology* 11:3–17.

Schwander, J. 1995. Eisbohrkerne: Eis als Archiv für Klima- und Umweltvorgänge. In *Gletscher im ständigen Wandel.* Zürich: Hochschulverlag AG, 65–79.

Schweingruber, F. H. 1988. *Tree Rings: Basics and Applications of Dendrochronology.* Dordrecht, the Netherlands: Kluwer.

Schweingruber, F. H., and K. R. Briffa. 1996. Tree ring density networks for climate reconstruction. In *Climatic Variations and Forcing Mechanisms*, eds. P. D. Jones, R. S. Bradley, and J. Jouzel, 43–65. Berlin: Springer-Verlag.

Schweingruber, F. H., K. R. Briffa, and P. Nogler. 1991. Yearly maps of summer temperatures in Western Europe from A.D. 1750 to 1975 and western North America from 1600 to 1982. *Vegetatio* 92:5–71.

Schweingruber, F. H., E. Schär, and O. U. Bräker. 1984. Jahrringe aus sieben Jahrhunderten. In *Saaner Jahrbuch*, 1–30. Gstaad: Müller.

Shackleton, N. J., and N. D. Opdyke. 1973. Oxygen isotope and palaeomagnetic stratigraphy of equatorial Pacific core V28–238: Oxygen isotope temperatures and ice volumes on a 10^5 and 10^6 year scale. *Quaternary Research* 3:39–55.

Shiyatov, S. G. 1992. The upper timberline dynamics during the last 1,100 years in the polar Ural Mountains. In *Oscillations of the Alpine and Polar Tree Limits in the Holocene*, eds. B. Frenzel, M. Eronen, K.-D. Vorren, and B. Gläser, 195–203. Stuttgart, Germany: Gustav Fischer Verlag.

Siegenthaler, U., H. Friedli, H. Lotscher, E. Moor, A. Neftel, H. Oeschger, and B. Stauffer. 1988. Stable-isotope ratio and concentration of CO_2 in air from polar ice core. *Annals of Glaciology* 10:151–56.

Siegenthaler, U., and H. A. Matter. 1983. Dependence of $\delta^{18}O$ and δD in precipitation on climate. In *Palaeoclimates and Palaeowaters: A Collection of Environmental Isotope Studies*, 37–51. Vienna: International Atomic Energy Agency. (STI/PUB/621).

Siegenthaler, U., and H. Oeschger. 1980. Correlation of $\delta^{18}O$ in precipitation with temperature and altitude. *Nature* 285:314–7.

Smol, J. P. 1988. Paleoclimate proxy data from freshwater arctic diatoms. *Verhandlungen der internationalen Vereinigung der Limnologie* 23:837–44.

Smol, J. P. 1983. Paleophycology of a high Arctic lake near Cape Herschel, Ellesmere Island. *Canadian Journal of Botany* 61:2195–204.

Speck, C. K. 1994. Änderung des Grundwasserregimes unter dem Einfluss von Gletschern und Permafrost. *Mitteilungen der Versuchsanstalt fuer Wasserbau, Hydrologie und Glaziologie der Eidgenössischen Technischen Hochschule Zuerich* 134:1–164.

Stefan, H. G., and B. A. Sinokrot. 1993. Projected global climate change impact on water temperatures in five north central U.S. streams. *Climate Change* 24:353–81.

Stuiver, M. 1970. Oxygen and carbon isotope ratios of fresh-water carbonates as climatic indicators. *Journal of Geophysical Research* 75(27):5247–57.

Stuiver, M., P. M. Grootes, and T. F. Braziunas. 1995. The GISP2 $\delta^{18}O$ climate record of the past 16,500 years and the role of the sun, ocean, and volcanoes. *Quaternary Research* 44:341–54.

Sturm, M. 1984. Suspended particles in lakes. *EAWAG News* 16/17:3–6.

Sturm, M. 1979. Origin and composition of clastic varves. In *Moraines and Varves*, ed. C. Schlüchter, 281–5. Rotterdam, the Netherlands: Balkema.

Sturm, M., and A. Lotter. 1995. Lake sediments as environmental archives. *Eidgenössische Anstalf für Wasserversorgung Abwasserreinigung und Gewässerschutz EAWAG News* 38E:6–9.

Sturm, M., and A. Matter. 1978. Turbidites and varves in Lake Brienz (Switzerland): Deposition of clastic detritus by density currents. *International Association of Sedimentologists* Special Publication 2:147–68.

Sugita, S. 1994. Pollen representation of vegetation in Quaternary sediments: Theory and method in patchy vegetation. *Journal of Ecology* 82:881–97.

Talbot, M. R. 1990. A review of the palaeohydrological interpretation of carbon and oxygen isotopic ratios in promary lacustrine carbonates. *Chemical Geology* (Isotope Geoscience Section) 80:261–79.

Talbot, M. R., and T. Johannessen. 1992. A high resolution paleoclimate record for the last 27,500 years in tropical West Africa from the carbon and nitrogen isotopic composition of lacustrine organic matter. *Earth and Planetary Science Letters* 110:23–27.

Tanaka, M., and M. M. Yoshino. 1982. Re-examination of the climatic change in central Japan based on freezing dates of Lake Suwa. *Weather* 37:252–9.

Tangborn, W. 1980. Two models for estimating climate-glacier relationships in the North Cascades, Washington, USA. *Journal of Glaciology* 25:3–21.

Tarand, A., and P. Kuiv. 1994. The beginning of the rye harvest: A proxy indicator of summer climate in the Baltic area. In *Climatic Trends and Anomalies in Europe 1675–1715*, eds. B. Frenzel, C. Pfister, and B. Gläser, 61–72. Stuttgart, Germany: Gustav Fischer Verlag.

ter Braak, C. J. F. 1987. Calibration. In *Data Analysis in Community and Landscape Ecology*, eds. R. H. G. Jongman, C. J. F. ter Braak, and O. F. R. van Tongeren, 78–90. Wageningen, Netherlands: Pudoc.

ter Braak, C. J. F., and C. W. N. Looman. 1987. Regression. In *Data Analysis in Community and Landscape Ecology*, eds. R. H. G. Jongman, C. J. F. ter Braak, and O. F. R. van Tongeren, 29–77. Wageningen, Netherlands: Pudoc.

Ters, M. 1987. Variations in Holocene sea level on the French Atlantic coast and their climatic significance. In *Climate History, Periodicity and Predictability*, eds. M. R. Raampino, J. E. Sanders, W. S. Newman, and L. K. Königsson, 204–37. New York: Van Nostrand Reinhold.

Thinon, M. 1992. L'analyse pédoanthracologique: Aspects méthodologiques et applications. Ph.D. diss. Université de Marseille.

Tinner, W., B. Ammann, and P. Germann. 1996. Treeline fluctuations recorded for 12,500 years by soil profiles, pollen and plant macrofossils in the central Swiss Alps. *Arctic and Alpine Research* 28:131–47.

Tissot, B. P., and D. H. Welte. 1984. *Petroleum Formation and Occurrence*. Amsterdam: Springer Verlag.

Trautmann, W. 1953. Zur Unterscheidung fossiler Spaltöffnungen der mitteleuropäischen Coniferen. *Flora* 140:523–33.

United Nations Environment Programme (UNEP). 1994. *Environmental Data Report 1993–94*. Oxford: Blackwell.

Urey, H. C. 1947. The thermodynamic properties of isotopic substances. *Journal of the Chemical Society* 1947:562–81.

Vesajoki, H., and M. Tornberg. 1994. Outlining the climate in Finland during the pre-instrumental period on the basis of documentary sources. In *Climate in Europe 1675–1715*, eds. B. Frenzel, C. Pfister, and B. Gläser, 51–60. Stuttgart, Germany: Gustav Fischer Verlag.

Vonder Mühll, D., M. Hoelzle, and S. Wagner. 1994. Permafrost in den Alpen. *Geowissenschaften* 12(5–6):149–53.

von Gravenstein, U., H. Erlenkenhauser, A. Kleinmann, J. Müller, and P. Trimborn. 1994. High frequency climate oscillations during the last deglaciation as revealed by oxygen-isotope records of benthic organisms (Ammersee, southern Germany). *Journal of Palaeolimnology* 11:349–57.

Vyverman, W., and K. Sabbe. 1995. Diatom-temperature transfer functions based on the altitudinal zonation of diatom assemblages in Papua New Guinea: A possible tool in the reconstruction of regional palaeoclimatic changes. *Journal of Paleolimnology* 13:65–77.

Wagenbach, D. 1994a. Special problems of mid-latitude glacier ice-core research. In *Greenhouse Gases, Isotopes and Trace Elements in Glaciers as Climatic Evidence of the Holocene*, eds. W. Haeberli and B. Stauffer, 10–4. Zurich: Versuchsanstalt für Wasserbau, Hydrologie und Glaziologie der Eidgenössischen Technischen Hochschule Zürich, Arbeitsheft 14.

Wagenbach, D. 1994b. Results from the Colle Gnifetti ice-core programme. In *Greenhouse Gases, Isotopes and Trace Elements in Glaciers as Climatic Evidence of the Holocene*, eds. W. Haeberli and

B. Stauffer, 19–22. Zurich: Versuchsanstalt für Wasserbau, Hydrologie und Glaziologie der Eidgenössischen Technischen Hochschule Zürich, Arbeitsheft 14.

Wagenbach, D. 1989. Environmental record in Alpine glaciers. In *The Environmental Record in Glaciers and Ice Sheets*, eds. H. Oeschger and C. C. Langway, 69–83. Chichester, U.K.: Wiley.

Wagner, S. 1994. Three-dimensional flow and age distribution at a high altitude ice-core drilling site. In *Greenhouse Gases, Isotopes and Trace Elements in Glaciers as Climatic Evidence of the Holocene*, eds. W. Haeberli and B. Stauffer, 33–9. Zurich: Versuchsanstalt für Wasserbau, Hydrologie und Glaziologie der Eidgenössischen Technischen Hochschule Zürich, Arbeitsheft 14.

Walker, I. R., R. J. Mott, and J. P. Smol. 1991. Alleröd-Younger Dryas lake temperatures from midge fossils in Atlantic Canada. *Science* 253:1010–12.

Walker, I. R., J. P. Smol, D. R. Engstrom, and H. J. B. Birks. 1991. An assessment of Chironomidae as quantitative indicators of past climate change. *Canadian Journal of Fisheries and Aquatic Sciences* 48:975–87.

Wang, W.-C., Z. Zeng, and T. R. Karl. 1990. Urban heat islands in China. *Geophysical Research Letters* 17(12):2377–80.

Weber, R. O., P. Talkner, and G. Stefanicki. 1994. Asymmetric diurnal temperature change in the Alpine region. *Geophysical Research Letters* 21:673–7.

Welten, M. 1982. Stand der Palynologische Quartär forschung am Schweizerischen Nordalpenrand. Geographica Helvetica 2: 75–84.

West, R. G. 1977. *Pleistocene Geology and Biology*. London: Longman.

Wick, L., and W. Tinner. 1997. Vegetation changes at timberline as indicators of Holocene climate oscillations in the Alps. *Arctic and Alpine Research*.

Wright, H. E., J. E. Kutzbach, T. Webb, W. F. Ruddiman, F. A. Street-Perrott, and P. J. Bartlein, eds. 1993. *Global Climates since the Last Glacial Maximum*. Minneapolis: University of Minnesota Press.

Yurtsever, Y. 1975. *Worldwide Survey of Stable Isotopes in Precipitation*. Internal Report, Isotope Hydrology Section. Vienna: International Atomic Energy Agency.

Yurtsever, Y., and J. R. Gat. 1981. Atmospheric waters. In *Stable Isotope Hydrology: Deuterium and Oxygen-18 in the Water Cycle*, eds. J. R. Gat and R. Gonfiantini, 103–39. Vienna: International Atomic Energy Agency.

Zoller, H. 1977a. Alter und Ausmass postglazialer Klimaschwankungen in den Schweizer Alpen. In *Dendrochronologie und postglaziale Klimaschwankungen in Europa*, ed. B. Frenzel, 271–81. Wiesbaden, Germany: Steiner.

Zoller, H. 1977b. Alter und Ausmass postglazialer Klimaschwankungen und ihr Einfluss auf die Waldentwicklung Mitteleuropas einschliesslich der Alpen. *Berichte der deutschen Botanischen Gesellschaft* 80(10):690–96.

Zoller, H. 1960. Pollenanalytische Untersuchungen zur Vegetationsgeschichte der insubrischen Schweiz. *Denkschrift der schweizerischen naturforschenden Gesellschaft* 83:45–156.

Züllig, H. 1988. Waren unsere Seen früher wirklich "rein"? Anzeichen von Früheutrophierung gewisser Seen im Spiegel jahrtausendealter Seeablagerungen. *Gas, Wasser, Abwasser* 68(1):17–32.

4 Future Alpine Climate

Dimitrios Gyalistras, Christoph Schär, Huw C. Davies, and Heinz Wanner

4.1 INTRODUCTION

Our current understanding of human interference with the climate system indicates the possibility of rapid global-scale warming accompanied by changes in planetary-scale flow patterns and significant continental and regional-scale effects. However, the climate's evolution is substantially uncertain. Because of the complexity and nonlinear nature of the climate system itself, the error bars associated with the uncertainties of the future level of climate-modifying GHGs and aerosols are considerable, and the limitations in our scientific understanding severe. The inadequacy of the observational database for the climate system and the computer resources required to effect a single prediction or an ensemble of predictions compound these difficulties. Nevertheless, a significant anthropogenically induced climate change is sufficiently likely to warrant an assessment of its possible nature, amplitude, and global and regional character.

In this context, the term "scenario" has entered the discussion on climate change. Climate change scenarios are viewed as formal prescriptions of a plausible space-time evolution of the system. They are usually either physically or statistically based descriptions of future climatic conditions designed for a variety of purposes. For example, they can be geared to providing insight on the climate system's sensitivity to prespecified changes in the future climate forcing, to assessing climate change's relative importance compared to other aspects of global and regional change, and to exploring possible interactions of climate change with other anticipated environmental and socioeconomic trends.

This chapter focuses on regional scenarios for the future Alpine climate. Chapter 2 presented an overview of Alpine climate itself, and chapter 3 discussed aspects of its history. The next two sections outline the need for, and the framework of, scenario construction. Thereafter, section 4.4 provides a critical overview of the available techniques, followed in section 4.5 by a presentation and intercomparison of a series of Alpine scenarios associated with increased atmospheric GHG concentrations.

4.2 THE NEEDS

Alpine climate change scenarios have a wide range of purposes (see also the discussion in the concluding chapter). One common use, and the one pursued here, is as input information for models developed to examine the impact(s) of a hypothesized climate change. Consideration of different impacts entails using models with different methodological approaches, data requirements, and objectives. The following list catalogues some applications concerned with impacts and provides a resume of scenario needs in the respective fields.

1. Social science: Some social science issues relating to the general perception of climate change require a broad-brush qualitative assessment of possible change rather than a quantitative scenario (see, e.g., Stern, Young, and Druckman 1992). In particular, they may require information on climatic variability and the degree of uncertainty rather than quantitative estimates of mean changes (e.g., Bailly 1994).

2. Economics: The wide range of economic issues includes the redistribution of resources, the cost of abatement and adaptation, and fiscal adjustments. These issues in turn impinge on the insurance, energy, agricultural, transport, and tourism sectors. (Chapter 8 pursues some of these themes further.) Examining economic issues usually requires quantitative scenarios of key parameters' time evolution, such as the expected value of the annually averaged temperature (e.g., Tahvonen, von Storch, and von Storch 1994), the frequency distributions of wind gusts over areas of several to tens of thousands of square kilometers, parameters related to heating requirements and energy consumption under a changed climate, or regional or local information on the variability and duration of snow cover (e.g., Abegg and Froesch 1994).

3. Glaciology and geomorphology: These disciplines concern climate change's effects on glaciers, mountain permafrost, and the stability of slopes, requiring information on seasonal changes in temperature and the type and total amount of precipitation (rain or snow) for the particular local (spatial scale of one to ten kilometers) cryospheric features under study. They may also require supplementary information on changes in wind, snow cover, global radiation, and humidity (W. Häberli, personal communication).

4. Hydrology: Examining climate change's impact on water quantity and quality usually demands high-resolution data. For example, studying a particular catchment requires temperature and precipitation data with a monthly or even hourly temporal resolution and a one- to ten-kilometer spatial scale. Supplementary information can include changes in wind, global radiation, and humidity, and the probabilities for dry periods and extreme precipitation events (H. Lang, personal communication; see also Lettenmaier 1995).

5. Ecology: Models used to study impacts of climate change on organisms, populations, or whole ecosystems require particularly demanding and complex input data. For example, table 4.1 lists the specific requirements for four

Table 4.1 Climatic input requirements of a selection of Alpine case studies for climate change impact

Case study	Model type of impact system	Spatial resolution	Time window of projections	Temporal resolution	Time of the year	Input data
Distribution of plant species in the alpine belt	Statistical	Test regions, 100 m	Future time period, e.g., 2030–50	1 season (1 month)	Winter... autumn	$E[T](s), E[R](s),$ Min$[T]$(Win), Max$[T]$(Sum), $N \geq 0\,\mathrm{cm}[H](m)$
Distribution of potential natural forest vegetation	Statistical	Switzerland, 1 km	Future time period, e.g., 2030–50	1 month	Jan...Dec	$E[T](m), E[R](m)$
Forest succession and soil dynamics	Dynamic	Representative locations, Europe	Present... 2100+ (3000)	1 month	Jan...Dec	$T, R_{(y,m)}$ $(E[T](m), E[R](m),$ Cov$[T,R](m), \ldots)$
Low-elevation grassland ecosystems	Dynamic	Representative locations, Switzerland	Present... 2030+	1 hour	Mar...Nov	$T, R, S, W, U_{(y,m,d,h)}$ $(E[T](m), E[R](m), \ldots$ Cov$[T,S,W,U](d),$ $E[T](h), E[S](h), \ldots)$

Note: T = Temperature, R = Precipitation, S = Radiation, W = Windspeed, U = Humidity, H = Snow Height; y = year, s = season, m = month, d = day, h = hour; $X(y, m, \ldots)$ = realization of random variable X for year y, month m, etc. $E[X](i)$ = expected value of X for period i; $N \geq z[X](i)$ = number of days within period i at which X exceeds the threshold value z; Min/Max$[X](i)$ = absolute minimum/maximum of X within month or season i. Cov$[X, Y, \ldots]$ = covariance matrix of $X, Y, \ldots$. Main sources (by table row, from top to bottom): A. Guisan (personal communication; see also chapter 6); Brzeziecki, Kienast, and Wildi 1993; Fischlin, Bugmann, and Gyalistras 1995 (see also chapter 6); Fuhrer 1996.

particular case studies of Alpine ecosystems. Each application obviously requires specific combinations of climatic input data, and taken together they comprise all the major climatic elements.

All studies require a high spatial resolution and several applications require some measures of variability (e.g., standard deviations or long-term extremes), or even a proxy of a time series for particular variables. Moreover, some of the systems studied respond to climate on very long time scales and hence ideally require input on the natural variability and changes of climate on extended time scales (tens to hundreds of years).

From the foregoing remarks, it is apparent, first, that no single scenario can satisfy all potential users (see also Robinson and Finkelstein 1991; Robock et al. 1993). Second, developing a rationale for using simpler impact models with lowered input data demands would be helpful. Indeed, the complex, multiparameter dependency of the climate-impact relationship might make it appropriate to adopt a heuristic stepwise procedure. For example, a tentative scenario could assume that some parameters are of secondary importance and can be neglected or assigned prespecifiable approximate values. Thereafter, a study could explore their contribution and sensitivity. Such a procedure could concomitantly help identify aspects of climate change that merit particular consideration.

The Alpine region certainly offers the scenario developer a significant challenge, because the terrain induces a rich range of regional- and subregional-scale phenomena and related climatic features (see chapter 2). However, the developer also has at his disposal an almost unequaled historical and proxy climatic database (chapters 2 and 3) and substantial related information (chapters 3, 5, and 6).

4.3 FRAMEWORK FOR SCENARIO CONSTRUCTION

A generally valid climate change scenario taking into account the myriad interrelated physical, chemical, biological, and socioeconomic factors is clearly unattainable. Chapter 8 delineates aspects of this challenge and discusses the "integrated assessment" approach to the overall problem. This chapter focuses on developing one limited type of scenario that deals expressly with the physical component of the climate system in isolation. In effect, this approach yields first-order estimates of possible regional climate changes and, if suitably modified, could form the basis for one of the interactive components of a more general assessment procedure.

In principle, a scenario of this limited form should have a firm, physically based foundation and be consonant with current knowledge and understanding. Its relation to the entire global climate system should both reflect that system's physics and provide estimates of regional change consistent with projections of global climate change. On the regional scale itself, the scenario should exhibit spatial and temporal consistency and compatible interrelationships among the different climate parameters and weather variables.

Figure 4.1 outlines a simplified framework for deriving a regional climate scenario of this type. This sequential framework reduces the physical system's multiscale interactions to a linear chain of transfer functions, each of which causally maps a set of inputs from one step to a set of outputs that enter the next. Certainly, this approach presents a simplification, since it neglects the observed system's two-way couplings and feedbacks many of and in particular assumes a one-way transfer of information from the global to the regional scale.

Consider now each step in turn. The function f_1 (figure 4.1) specifies human behavior's effect on large-scale forcings of the global climate system. This could be derived from a socioeconomic model whose inputs would be the initial scenarios for human activities such as population growth, energy consumption, and agriculture (e.g., Legget, Pepper, and Swart 1992). Estimates are required on emissions of GHGs and aerosols and also conceivably on changes in land surface boundary conditions. A simple emissions scenario for GHGs, often used to test climate models' sensitivity, is to assume that a new net radiative forcing of all GHGs corresponds to a doubling of the CO_2 concentration relative to the preindustrial value (the basis for the so-called $2 \times CO_2$ scenarios). An alternative to a simple step doubling of CO_2 is to prescribe a time-dependent change of say a 1 percent increase in GHGs per

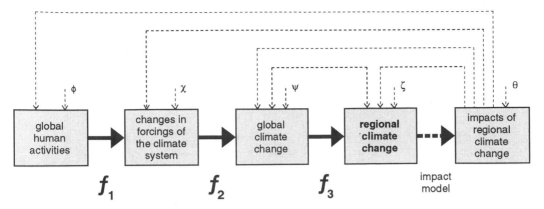

Figure 4.1 A simplified framework for deriving regional climate change scenarios consistent with stipulated anthropogenic climate change. The f_i refer to the required steps; thick lines indicate the flow of information; thick broken line refers to impact models (not discussed in this chapter); thin lines are some possible but neglected feedbacks; $\varphi, \chi, \psi, \zeta$, and θ are additional, not explicitly considered, or unknown factors.

year. (Other alternatives are the so-called business-as-usual and accelerated policies scenarios [Houghton, Jenkins, and Ephraums 1990], and the "IS92a-f" scenarios of Houghton et al. [1994]).

The function f_2 represents a climate model that quantifies the altered large-scale forcings' effect on the global climate. There are a number of these global climate models, ranging in complexity from zero- and one-dimensional energy balance models (e.g., North, Cahalan, and Coakley 1981) to three-dimensional coupled atmosphere-ocean models, the so-called general circulation models or GCMs. Overviews of these various types are available (e.g., Bach, Jung, and Knottenberg 1985; Schlesinger and Mitchell 1985, Henderson-Sellers and McGuffie 1987).

The function f_3 seeks to infer the regional climate's nature from larger-scale climate information (usually derived from a GCM). The process has been termed "downscaling" or "regionalization," and its objective is to deliver the kind of detailed regional climatic information discussed earlier. The need for such a step arises in part from GCMs' limitations. Section 4.4 provides an overview of the various approaches to downscaling alongside an account of GCMs' capability and limitations.

The choice and indeed the sophistication of the method adopted for scenario construction clearly depends on the nature of the application. For example, a simple empirical approach may suffice for initial sensitivity studies. More sophisticated applications may require strict global-regional consistency of the putative climate change, spatially consistent regional-scale distributions of climatic parameters (e.g., Brzeziecki et al. 1993), or temporal consistency between local weather variables (e.g., Fischlin, Bugmann, and Gyalistras 1995; Fuhrer 1997). Figure 4.2 shows an example of intervariable temporal relationships that may have to be considered in scenario construction.

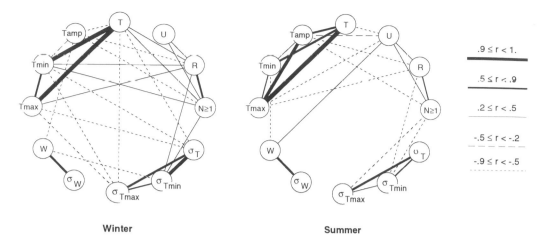

Figure 4.2 An indication of weather variables' interdependency. The values refer to the observed interannual correlations of seasonal statistics for weather variables for a site at Bever in Switzerland's Oberengadine valley during the period 1901–80. $T/T_{min}/T_{max}$ signify the seasonal mean daily mean/minimum/maximum temperature; T_{amp} = seasonal mean daily temperature amplitude; R = seasonal precipitation total; $N \geq 1$ represents the number of days within season with a total of more than one millimeter of precipitation; U = seasonal mean relative humidity; W = seasonal mean daily wind speed; σ_i = within-season standard deviation of variable i.

Again it may be necessary to adopt auxiliary techniques to supplement the downscaling to help meet a particular application's data requirements. For example, time series analysis can help identify weather variables' key statistical parameters and simulate mutually consistent time series of these variables for presumed changes in climate, as in "stochastic weather generators" (e.g., Richardson 1981; Mearns, Katz, and Schneider 1984; Woo 1992; Wilks 1992; Gyalistras, Fischlin, and Riedo 1997). EOF analysis (e.g., Preisendorfer 1988) can be used to estimate the main patterns of joint variability among several variables measured at one or several locations, and these patterns can be recombined to produce, under scenario assumptions, new internally self-consistent states (e.g., Wilks 1989; Wigley et al. 1990). Finally, spatial interpolation techniques based on empirical relationships that govern climate parameters' regional distributions and statistical models that describe the residual variability can estimate present climate and climatic changes at locations with few or no measurements (e.g., Running, Nemani, and Hungerford 1987; Phillips, Dolph, and Marks 1992; Barancourt, Creutin, and Rivoirard 1992; Ishida and Kawashima 1993, Daly, Neilson, and Phillips 1994; Gyalistras and Fischlin 1996).

4.4 THE TECHNIQUES: DESCRIPTION AND CRITIQUE

Several approaches for constructing regional climate scenarios (see, e.g., Lamb 1987; Robinson and Finkelstein 1991; Giorgi and Mearns 1991;

Table 4.2 Three approaches for constructing regional climate change scenarios

Approach	Emphasis	Mathematical foundation	Main tools	Resolution of annual cycle	Resolved processes	Examples
Empirical	Empirical (observation-based)	Statistics (differences of distributions, trend analyses, classifications, etc.)	Statistical and time series models, eigen techniques	Year to seasons	Century time scale to inter-annual variability	Paleoclimatic analogues, extrapolation of trends
Semi-empirical				Seasons to 1–5 days	Decadal-scale variability to planetary scale waves	Application of weather types to global climate model output
Modeling	Theoretical (process-based)	Coupled nonlinear differential equations	Deterministic dynamic simulation models	Hours to minutes	Cyclones, fronts, regional-scale circulations and interactions with topography	Nesting of high-resolution in low-resolution climate models

Robock et al. 1993; Pittock 1993; Viner and Hulme 1994; Carter et al. 1994; von Storch 1995) differ in their methodology, degree of complexity, and the form and detail of the resulting scenarios but can be categorized into three general classes: empirical approaches, semiempirical approaches, and regional process–based models (cf. Giorgi and Mearns 1991; Viner and Hulme 1994).

Table 4.2 juxtaposes the characteristics of these classes. The empirical approach is based entirely upon interpreting data from past and present climates, whereas the model approach is based upon a physicomathematical representation of the system. The semiempirical approach is intermediate in that it combines observations and model-based results. Note further that GCMs are themselves physicomathematical representations of the climate system. On the one hand, they form a primary information source for global climate change, and on the other, their limitations provide the rationale for developing regionalization methods. Therefore we first overview GCMs' capabilities and limitations and thereafter discuss with particular reference to the Alpine region the basis for, and the specific advantages and limitations of, each of the three approaches listed above.

4.4.1 General Circulation Models

4.4.1.1 Nature of Models Solar radiation drives the earth's climate system, which involves processes on a wide range of temporal and spatial scales in the atmosphere, oceans, soil, biosphere and cryosphere. GCMs effect the most detailed and comprehensive representation of these processes and their mutual interactions. These models constitute an internally consistent physicomathematical description of the climate system. They include a sophisticated representation of the atmosphere and are coupled with representations of the land surface and the oceans whose complexity can range from a prescription of sea surface temperature to fully coupled ocean circulation models including a formulation of sea ice.

Three-dimensional, time-dependent partial differential equations govern the distribution of heat, momentum, and moisture in the model's atmospheric component. They describe the continuum's evolution, and to be solved numerically using a computer they must be discretized onto a computational grid (with a finite number of degrees of freedom). The vertical discretization involves typically ten to thirty computational layers. For fully coupled atmosphere-ocean models the highest realizable horizontal resolution on current high-speed computers is of the order of 300 kilometers, although atmospheric models with prescribed sea surface temperature can have higher resolutions. Thus the flow phenomena explicitly represented include cyclones and anticyclones, the meandering midlatitude jet stream, and continent-scale monsoon circulations.

Smaller-scale phenomena unresolved by the computational grid can nevertheless in reality feed back to the larger scales via their induced fluxes of heat, moisture, and momentum and thereby critically influence the climate. A key example is the vertical exchange of energy and moisture through small-scale (unresolved) deep tropical convective cloud activity. The influence of such small-scale processes must hence be approximated physically in terms of the resolved circulations. This so-called sub–grid scale parameterization procedure captures the effect of boundary layer, cloud and precipitation, radiative, and topographical processes.

These parameterization schemes, although often highly sophisticated, are nevertheless approximations and are, to a measure, tuned to match the current climate. This restricts their domain of validity for climate change studies. A similar restriction applies to the "flux-adjustment" procedure introduced at the atmosphere-ocean interface (see, e.g., Sausen, Barthel, and Hasselmann 1988): This procedure employes empirical adjustments of heat, momentum and fresh water fluxes to compensate for inconsistencies between the fluxes simulated by the fast (atmospheric) and slow (oceanic) components of coupled models, and is also based on current climatic conditions.

4.4.1.2 Model Performance Simulations of GCMs can be assessed by examining the degree to which they reproduce either the current-day climate (including its variability) or documented past climates and climatic changes. In effect, the latter assessment tests the GCMs' semiempirically tuned components. Examples include examining orbital parameters' (e.g., Phillipps and Held 1994), land-sea distribution's (e.g., Oglesby 1991), and vegetation cover's (e.g., Foley et al. 1994) impact on climate. However, the lack of data during paleoclimatic episodes hampers the continental- and regional-scale validation of such simulatoins.

Comparing GCM simulations with the current climate indicates that the models reproduce relatively well the zonal mean atmospheric circulation and

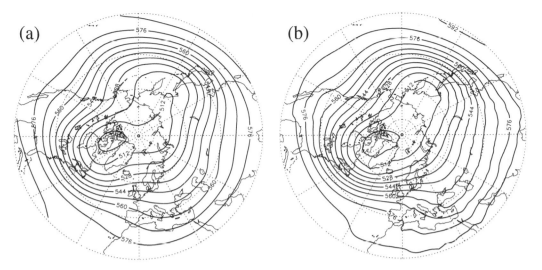

Figure 4.3 Comparison of (a) observed and (b) simulated distributions of the current mean winter 500 hPa geopotential height. Simulated results are for a GCM control experiment with prescribed climatological sea surface temperature (from Beniston et al. 1995).

the planetary-scale geographical distribution of climatic zones (e.g., Gates et al. 1996). This is illustrated in figure 4.3 which juxtaposes the observed mean winter height of the 500 hPa surface with that derived from a high-resolution model simulation. The spacing of the height contours inversely measures the strength of the midlatitude westerly flow, and the wave pattern itself provides information related to the continental-scale extratropical climate and the location and amplitude of the major storm tracks (see chapter 2). Note that the model simulates the midlatitude jet's mean meridional location fairly well, but shows some deviations in the amplitude and location of the stationary planetary-scale wave patterns.

There are several indications that GCMs can also simulate planetary-scale natural and anthropogenic climatic changes. First, simulations of the El Niño phenomenon in the Pacific usefully predict gross indices of tropical climate on time scales up to one year or more (see, e.g., Latif et al. 1994). This success is particularly remarkable because El Niño relies on the dynamical interaction of the atmospheric and oceanic circulations. Second, simulations of the aerosol effects of the 1991 Pinatubo volcanic eruption on the global mean surface air temperature (e.g., Hansen et al. 1992) successfully mirrored the aerosol-induced transient global cooling of 1992–94 and the subsequent return to the prevailing climate evolution of this century. Third, recent simulations of this century's climate have estimated the global mean trend in the temperature in reasonable agreement with the observed warming (Mitchell, Davis, et al. 1995; Mitchell, Johns, et al. 1995). The latter point is illustrated in figure 4.4. It can be seen that the simulated long-term changes in annual mean surface air temperature correspond reasonably to the actual observations. However, as consequence of the uncertainties natural variability, the

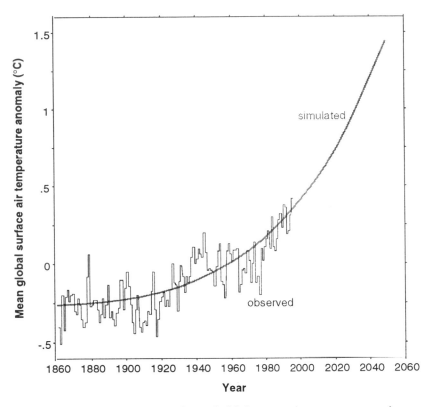

Figure 4.4 A comparison of the observed global mean surface air temperature from 1861 to 1995 (thin line, based on data from Jones 1988, updated) with the value derived from a transient coupled atmosphere-ocean simulation. The observations are given as anomalies relative to the 1950–79 mean. The smooth curve was derived from differences simulated in a climate change experiment with the Hadley Centre GCM relative to the long-term mean state from a control simulation under constant (circa 1860) GHG concentrations. The climate change experiment was started in the year 1860 and was driven by observed and projected increases of GHGs and aerosol concentrations (after Mitchell, Johns, et al. 1995).

history of the forcing, and the sensitivity of the model introduce, it is difficult to attribute the recent warming unambiguously to increases in GHGs. (Recent notable attempts are those of Santer et al. 1996 and Hegerl et al. 1996.) Figure 4.4 also includes the same numerical model's projections for the global mean surface air temperature until 2050. The projected warming exceeds the natural climate variations over many thousands of years.

4.4.1.3 Limitations of the Model GCMs' ability to model continental- and regional-scale effects has significant limitations, as is particularly evident in the climate scenarios derived from GCMs for specific regions (e.g., Robock et al. 1993; Risbey and Stone 1996; Cubasch et al. 1996) and the models' representation of synoptic-scale variability (cyclone frequency and weather types). For example, Risbey and Stone (1996) found the simulation of the regional climate of the Sacramento basin in the western United States serious deficient because of shortcomings in the GCM synoptic climatologies. Likewise Hulme et al. (1993) concluded that the models did not adequately reproduce the observed frequency of daily airflow types over the British Isles. Difficulties in simulating cyclones' frequency, amplitude, and character also have some global-scale repercussions because cyclones are central to the poleward atmospheric transport of heat (Stone and Risbey 1990).

A further problem relates to the models' ability to simulate the hydrological cycle. Even on the global scale, current climate models show marked differences (Hulme 1994). Global mean annual precipitation varies among GCMs by a factor of up to 1.5. (Note, however, that there are also substantial uncertainties in the observed precipitation signal.) The inadequate representation of the hydrological cycle, partially related to coarse spatial resolution, has important repercussions for coupled atmosphere-ocean studies because the rainfall freshwater fluxes significantly force the oceanic circulation. Precipitation and evapotranspiration depend heavily on regional details of the topography, soil type, and even vegetation, and adequate representation of these processes requires much higher resolution than currently used in coupled GCMs (approximately, 300 kilometers).

GCMs' success on the global scale does not contradict their shortcomings on the synoptic and regional scales (von Storch 1995). The models have been designed and tuned to reproduce global climate accurately in response to the large-scale forcing (e.g., insolation and the land-sea distribution). Regional climates, on the other hand, depend strongly on the interaction of the large-scale flow with regional scale features and this interaction is generally less well represented in GCMs.

Significant improvement in GCMs depends critically on whether the current deficiencies are related to model resolution (cf. Boville 1991; Boyle 1993) or to the lack of appropriate parameterizations of physical processes (cf. Senior and Mitchell 1993; Stone and Risbey 1990). In the former case an increase of computing power alone would suffice to improve the model results, whereas in the latter case further improvements of the model formulation may require detailed investigations of the parameterized processes and their feedbacks with the climate system.

In spite of their shortcomings, GCMs represent the only currently foreseeable approach to projecting future climate change in a manner that takes into account the fundamental nonlinear dynamical and thermodynamical interactions of the coupled atmosphere-ocean system. Nevertheless, conceptional and technical limitations indicate that, at least in the medium term, purely the improvement of the global models will not solve the problem of regionalization.

4.4.2 Empirical Approach

In the empirical approaches, regional climatic scenarios are derived on the basis of past climate data—instrumental data, historical information from the last few centuries, or reconstructions of past climates at geological timescales. (Chapter 3 discussed the latter two data sources.) Empirical approaches include:

• **Temporal analogs:** A previous period of global or hemispherical warm temperatures (or warming trend) is identified, and the accompanying shifts in regional climate are viewed as analogs for future changes associated with a GHG-induced global warming (see, e.g., Flohn 1979; Palutikof, Wigley, and Lough 1984; Crowley 1990; Giorgi and Mearns 1991). For analogs derived from the instrumental period, a globally warm state can be defined in several ways, such as departure of averages from consecutive years relative to the long-term average or ensembles of the warmest and coldest years in the available record.

• **Spatial analogs:** A hypothesis on future climate change (e.g., "warmer and drier") is formulated for the region of interest, and observations for another region or location currently experiencing a climate similar to that hypothesized are adopted as the basis for a scenario (e.g., Brown & Katz 1995). For example, for the Alpine region, a "warmer-world" analog can be estimated for a specific location based on measurements from nearby sites at a lower elevation.

• **Ad hoc adjustments:** Meteorological observations from the region of interest are plausibly modified to produce changes for individual climate parameters or for a combination of them. A simple variant is to shift all elements of a measured time series by a fixed amount (e.g., $\pm 2°C$). An alternative is to modify the daily weather record algorithmically to obtain a prescribed change in a key parameter, such as the number of rainfall events (Robock et al. 1993). Coherent multiparameter adjustments should conform to the known properties of the spatiotemporal behavior of the variables under consideration, their relationship to regional weather patterns, and indications from GCMs.

One typical approach uses temporal analogs based on a low-order description of the global system (e.g., by means of a single variable such as the mean Northern Hemispheric surface air temperature). The results shown in figure 4.5, depicting European-scale changes in winter mean sea level pressure observed during two different episodes of Northern Hemispheric warming, illustrate the approach's simplicity and limitations. The two panels display the difference between the warmest (1934–53) and the coldest (1901–20) sequence of twenty consecutive years prior to 1970 (Palutikof, Wigley, and Lough 1984; see also Lough, Wigley, and Palutikof 1983), and the trend in surface pressure during a more recent episode of Northern Hemispheric warming, namely 1961–90 (Schönwiese et al. 1994; see also Flohn et al. 1992). There are notable differences between these two cases. Panel (a) indicates that a shift toward anomalous high pressure over Scandinavia and low pressure over the Mediterranean accompanied the Northern Hemispheric warming during the first part of this century, whereas a trend

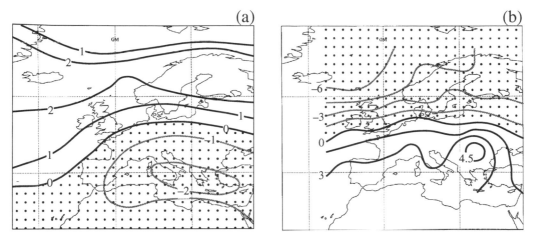

Figure 4.5 Effects of Northern Hemispheric warming as reflected in the winter European mean sea level pressure (in hPa). Panel (a) is the difference between the warmest (1934–53) and the coldest (1901–20) sequence of twenty years prior to 1970. Panel (b) shows the trend between 1961 and 1990 in hPa/(30 years). Grey contours and stippled regions indicate negative pressure anomalies for warm episodes. Redrawn after Palutikof, Wigley, and Lough (1984) and Schönwiese et al. (1994). Data are averaged over the period December to February.

with almost the reverse pattern accompanied the warming after 1960. This dichotomy reflects two distinctly different shifts in the European winter circulation. In the first case blocking events increased in frequency and persistence, whereas the second period coincided with increased westerlies over the North Atlantic and European sectors.

In effect, representing the global climate system by one single variable is an oversimplification. One inference is that different periods of global warming might be attributable to different causes and result in a variety of continental and regional scale changes. This clearly limits the usefulness of this form of temporal analogs (see also Jones and Kelly 1983).

Generally, empirical approaches' root problem is that they do not account physically for the forcing's effects on the global climate, making it difficult to assess the resulting scenarios' domain of validity or representativity. Associated problems include the limited time span and small range of changes during the instrumental period in comparison with the larger or more rapid changes that might occur in the future, the limited spatiotemporal resolution and nature of the available variables for the paleoclimatic analogs, and the multiple nature of and changes in the forcings of global and regional climate. (In the latter context, the temporal analog approach can be refined to take into account, at least partly, such changes. For example, Bayer et al. (1994) attempted to isolate the effects of volcanic indices, solar flux, the So index, and atmospheric CO_2 concentration. Such studies can yield comparative estimates of the effects of natural and anthropogenic forcings on global climate.)

In spite of these drawbacks, the empirical approaches can provide important information on climatic changes and represent spatiotemporally consistent pictures of altered regional climates. Furthermore they are technically simple and transparent and require modest amounts of data. They provide, at comparatively small expense, a first iteration of regional climate scenarios and as such can be useful in the study of sensitivity, threshold effects, and tolerance limits of systems influenced by climate. Indeed, given the uncertainties associated with future climate behavior, they provide an important complement to the more mathematical and physical techniques we now consider.

4.4.3 Semiempirical Approach

Semiempirical downscaling techniques invoke a scale separation that assumes that larger-scale atmospheric circulations provide the environment and primary forcing for smaller-scale climate components. Accordingly, semiempirical downscaling assumes that the variability regional- and local-scale processes actually introduce either remains statistically unchanged (when estimating long-term climatic shifts) or can be empirically "added" to the downscaled signal to generate possible regional weather sequences. The basic procedure is as follows:

1. Select a set of predictor variables (predictors) as simulated by a GCM.

2. Select a set of regional or local variables of interest.

3. Use simultaneous observations of the two sets of variables to establish an empirical transfer function f (the function $f3$ in figure 4.1) to predict the regional variables from the predictors.

4. Apply this function to output from GCM simulations to infer the associated climatic shifts or to generate possible weather sequences for prescribed large-scale regimes.

The methods can differ in the predictors, the regional variables, the transfer functions f, and the temporal resolutions. To date most studies have focused on temperature and precipitation at a seasonal, monthly, or daily resolution. The predictors can range from regional averages of a few surface variables to several atmospheric fields. The transfer function can be derived with any method that maps a set of inputs (predictors) to a set of outputs (regional variables) and can be linear (von Storch, Zorita, and Cubasch 1993; Gyalistras et al. 1994; see also Bretherton, Smith, and Wallace 1992), piecewise linear (e.g., Karl et al. 1990), fully nonlinear (e.g., Hewitson 1994; McGinnis 1995), or classification based (e.g., Hay et al. 1992; Bardossy and Plate 1992). Regional weather variability has been simulated with stochastic modeling techniques. A common approach is to sample at random from the empirical distributions of the local weather variables conditional on the large-scale state (see, e.g., Bardossy and Plate 1992; Hughes and Guttorp 1994; Matyasovszky and Bogardi 1994; Wilby 1994; Zorita et al. 1995).

Three major types of semiempirical downscaling methods can be distinguished based on the kind of predictors used:

- **Grid point-based methods:** These methods predict changes in regional climates based on the changes simulated by a GCM at one or a few grid points in the vicinity of the region of interest (e.g., Kim et al. 1984; Bach, Jung, and Knottenberg 1985; Wilks 1989; Karl et al. 1990; Santer et al. 1990; Wigley et al. 1990; Viner and Hulme 1993). The methods are usually relatively simple to implement, but the low reliability of climate and climate changes simulated at individual GCM grid points constrains them. The latter restriction is particularly stringent for regions of complex topography such as the Alpine region.

- **Synoptical (or weather type–based) downscaling:** These methods are based on weather systems' tendency to exhibit distinct, relatively stable patterns that typically persist from one to five days. These patterns can be classified into a limited number of synoptic weather types or Grosswetterlagen (see chapter 2). They are classified based on daily weather maps for one or more reference levels, for example, the 1,000, 850 and 500 hPa surfaces (Yarnal 1993; see also Wanner 1980; Perry 1983). Local variables are estimated in terms of the frequency and occurrence of weather types within a given reference period, such as a season (see, e.g., table 2.1) or a day (e.g., Bardossy and Plate 1992). Regional climate and weather scenarios can be constructed based on calendars of weather types obtained from GCM runs.

- **Regionally optimized downscaling:** Large-scale patterns designed to be optimally (statistically) predictive in terms of the regional variables can be employed as an alternative to a priori defined weather types. The patterns are normally defined based on a simultaneous statistical analysis of large-scale and regional weather states at a seasonal (e.g., von Storch, Zorita, and Cubasch 1993) to daily (e.g., Hewitson 1994; Zorita et al. 1995) resolution. As in synoptical downscaling, one or several atmospheric fields over a sector containing the region of interest typically provides the large-scale information. GCM-simulated fields are then used to project regional climate and weather scenarios. Gyalistras et al. (1994) and Fischlin and Gyalistras (1997) have applied the regionally optimized technique to the Alpine region.

These semiempirical methods' reliability and applicability depends critically on the appropriate choice of the large-scale predictors. Not only should suitable predictors describe the relevant processes that affect the regional variables (see, e.g., Giorgi and Mearns 1991), but ideally GCMs should also simulate them reliably. These criteria are usually more easily satisfied for the anomalies of atmospheric variables relative to their long-term mean rather than for absolute values, free-atmosphere variables rather than for surface variables, and smoothly varying fields (such as for sea level pressure) rather than for discontinuous fields and fields of derived variables (such as precipitation). The further trenchant assumption is that the predictors will also be realistically simulated in a climate change simulation conducted with a GCM.

A further step in the procedure derives from the fact that though present GCMs simulate some of the major large-scale patterns of atmospheric variability fairly well (e.g., Hewitson and Crane 1992; von Storch, Zorita, and Cubasch 1993; Zorita et al. 1995), synoptic weather types that depend on grid point–scale information are not yet well reproduced (e.g., Hulme et al. 1993). To compensate for this deficiency, the GCMs' gridpoint–scale detail is often filtered out from the description of the large-scale state by projecting the atmospheric-state vector (consisting typically of tens to hundreds of gridded variables) onto a lower-dimensional subspace spanned by the system's first few (say, fiver to ten) EOFs, determined from the analysis of the observed fields. This procedure may, however, miss potentially important detail related, for example, to rare or extreme events.

To illustrate the semiempirical downscaling approach, we outline one application for the inner-Alpine site of Sion. Following Gyalistras et al. (1994), the predictors were taken to be the gridded anomalies of seasonal mean sea level pressure and near-surface temperature for the European and North Atlantic sector. The regional anomalies examined were the seasonal mean temperature and precipitation at Sion. The transfer function f consisted of sets of multivariate linear regression equations determined separately for each season with the Canonical Correlation Analysis in EOF space (e.g., Barnett and Preisendorfer 1987).

Figure 4.6 shows the large-scale pressure patterns derived to account for interannual variations of seasonal mean temperatures at Sion. The numbers next to the dots denote the local temperature anomalies associated with a given pattern. The displayed pressure-temperature relationships have a physically plausible interpretation: Negative (positive) pressure anomalies indicate predominance of counterclockwise (clockwise) flow around the center of a trough (ridge), and hence the local temperature anomalies at Sion relate to the mean strength of southwesterly flow in winter, more meridional (i.e., north-south) flow in the transition seasons, and large-scale subsidence associated with counterclockwise circulation over Europe in summer. For prediction of local temperatures the temperature anomalies shown in figure 4.6 are scaled depending on the similarity of a given (observed or GCM-simulated) pressure anomaly with the respective pattern.

Figure 4.7 measures the technique's success in predicting the interannual temperature variability from large-scale observations for the reference interval 1901–47 and the subsequent verification period 1948–77. The correlation coefficients (r) between the observed and the predicted time series during the verification period were significant for all seasons ($r \geq 0.63$). However, the technique did not reproduce well the observed upward temperature trend in winter and the temperature levels for summer and autumn after 1965. This could be due to systematic changes in large-scale flow features that the technique did not take into account (e.g., the upper-level flow patterns) or local effects (such as urban warming trends or increases in air pollution). For precipitation, the technique yielded good results (not shown)

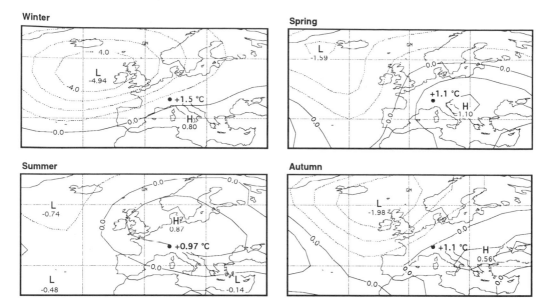

Figure 4.6 The regionally optimized downscaling approach applied to the alpine region. Shown, for each season, are statistically detected patterns of seasonal mean sea level pressure anomalies that account for the year-to-year variability of seasonal mean temperature at the inner-alpine site of Sion (analysis interval 1901–47, anomalies in hPa). The patterns were obtained from a Canonical Correlation Analysis between a large-scale signal, comprising a vector consisting of the seasonal mean sea level pressure and the near-surface temperature field, and a regional-scale vector consisting of seasonal mean temperature and square-root transformed total seasonal precipitation. The dots denote Sion's approximate position, and the temperatures refer to the local anomalies associated with the sea level pressure anomalies.

for winter ($r = 0.74$) but weaker results for the transition seasons ($r \geq 0.35$) and summer ($r = 0.29$), when smaller-scale processes control the precipitation more closely.

A major restriction of semiempirical approaches is that they do not base the link between large-scale and regional climate on physical principles. Statistical relationships established for the present climate need not necessarily remain valid under a changed climate. Another disadvantage is that they do not consider systematic changes in regional or local climate forcings (such as changes in vegetation cover or soil properties) not accounted for in the parent GCM. This is crucial if the empirical link between a regional variable and the large-scale climate is inadequate (as, for example, for summertime precipitation).

Nevertheless, semiempirical downscaling methods enable the effective combination of physically consistent results from global or regional climate models with observations, and they can attain a very high spatial resolution—in principle limited only by the density of regional measurements. Moreover, the approach requires only modest computing resources and is therefore suitable to derive a wide range of scenarios, or estimate time-dependent regional changes from corresponding GCM simulations.

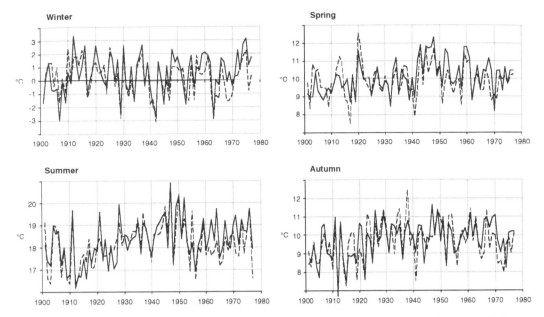

Figure 4.7 Comparison between observed (solid lines) and reconstructed, statistically downscaled (dashed lines) seasonal mean temperatures at the inner-alpine site of Sion. The local temperatures were predicted from observed seasonal mean sea level pressure and near-surface temperature anomalies over the European and North Atlantic sector based on a linear regression model. (A Canonical Correlation Analysis was performed using global and local measurements for 1901–47; c.f. figure 4.6.) The explained variances ($100 \cdot r^2$) in the verification interval 1948–77 are 60 percent for winter, 55 percent for spring, 54 percent for summer, and 40 percent for autumn.

Forerunner studies related to semiempirical downscaling for the Alpine region are those of Schüepp and Fliri (1967) and Kirchhofer (1974). However, the application of this approach to derive Alpine climate change scenarios is still limited, and as yet no systematic comparisons of the various methods have been undertaken for the Alpine region.

4.4.4 Modeling Approach

It was noted earlier that GCMs' heavy computational needs and coarse horizontal resolution restrict their use for regional studies. The speed of supercomputers increases with an impressive doubling time of around two years (Bengtsson 1990), but numerical resolution improves much more slowly because the computational requirements grow with the fourth power of the spatial resolution (assuming congruent horizontal and vertical resolutions).

Yet a high horizontal resolution is crucial for modeling the regional-scale climate because the latter is highly influenced by mesoscale processes and phenomena operating at spatial scales of a few to a few hundred kilometers (see chapter 2). Key processes in this category are those that generate Alpine-scale circulations (such as lee cyclogenesis events) or extreme events (such as the triggering of convection by approaching fronts, which may lead

(a)   (b)

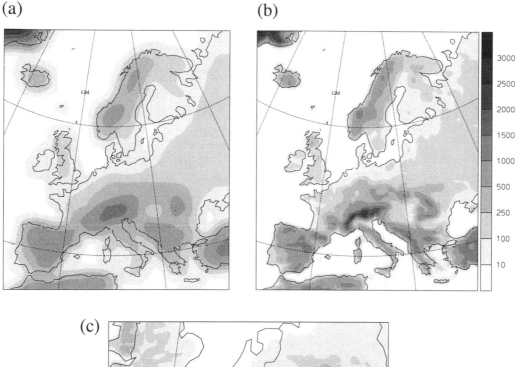

(c)

Figure 4.8 The European and Alpine orography for three spatial resolutions: (a) a typical coupled atmosphere-ocean climate GCM resolution (350 kilometers), (b) a typical mesoscale resolution (56 kilometers), and (c) a fine-mesh mesoscale resolution (14 kilometers).

to severe thunderstorms and flooding; see, e.g., Cacciamani et al. 1995; Binder and Schär 1995). Such phenomena are controlled by the Alpine topography.

Figure 4.8 illustrates the link between resolution and orography using representations of the European and Alpine relief with three resolutions, namely that of a typical coupled atmosphere-ocean GCM (horizontal resolution 350 kilometers), a mesoscale model (56 kilometers), and finally a fine-mesh model (14 kilometers). The intermediate resolution of 56 kilometers is sufficient to resolve the interaction of approaching synoptic systems with the Alpine massif and some key dynamical effects (such as lee cyclogenesis). Several weather services now use models with 14-kilometers resolution operationally for forecasting. Explicitly resolving the important cloud- and rain-producing convective processes mentioned above would require even higher resolution, and these are usually parameterized in regional climate models.

Three distinct physically based modeling techniques, all initially developed in the context of weather forecasting, and in essence relying on "atmosphere only" models using externally prescribed sea surface temperatures, explicitly simulate regional climates:

• **High-resolution GCMs:** Scenario calculations and sensitivity studies with high-resolution GCMs having a globally uniform horizontal resolution equivalent to approximately 120 kilometers are currently underway at the MPI/ETH (Hamburg/Zurich; see Marinucci et al. 1995; Wild et al. 1995; Beniston et al. 1995), the NCAR (Boulder, Colorado; see Boville 1991; Boyle 1993), the Hadley Center (Bracknell), and at Météo-France (Toulouse; see Déqué and Piedelievre 1995). This method's advantage is the globally homogenous treatment of all processes. However, at present, the method's high computational requirements allow for only comparatively short simulations, and the resolution is insufficient to reproduce Alpine precipitation patterns.

• **Variable-resolution GCMs:** A procedure developed at the French Meteorological Service uses spectral techniques with nonisotropic base functions to enhance GCMs' resolution in the region of interest (see Déqué and Piedelievre 1995). This method yields resolutions around 100 kilometers in the desired region and in principle is cost effective. However, parameterization schemes to cope with variable resolution have not yet been fully formulated.

• **Nesting of a regional climate model into a GCM:** Here a high-resolution regional climate model (RegCM, also referred to as a "limited area model," or LAM) is embedded within a GCM. The nesting is usually one way, that is, the information flows exclusively from the larger- to the smaller-scale model. This approach is commonly and successfully used in numerical weather prediction, and Dickinson et al. (1989) and Giorgi (1990) have pioneered its use in regional climate simulations. Regional climate models of various resolutions have been tested and used for process studies over Europe by Giorgi and Marinucci (1991), Marinucci and Giorgi (1992), Cress et al. (1994), Cress et al. (1995), Podzun et al. (1995), Beniston et al. (1995), Marinucci et al. (1995), Jones, Murphy, and Noguer (1995), Lüthi et al. (1996), and Schär et al. (1996). Scenario calculations over the Alps under enhanced concentrations of GHGs have been reported by Giorgi, Marinucci, and Visconti (1992), Beniston et al. (1995), Rotach et al. (1997), Giorgi et al. (1997), and Jones et al. (1997).

A climate scenario employing one of the above approaches requires two simulations with some low-resolution GCM to furnish the distribution of sea surface temperature and sea ice as well as the initial and lateral boundary conditions. The first of these simulations is the control climate ($1 \times CO_2$), and the second relates to a putative future climate valid for either an equilibrium climate ($2 \times CO_2$ equilibrium) or a time slice of five to ten years from a transient climate experiment (corresponding to the time of CO_2 doubling). These simulations then form the database for two simulations undertaken with a RegCM or a high-resolution GCM. The difference between the latter two high-resolution simulations measures the climate change signal. This approach delivers a signal that excludes the systematic mean model error, but there may be limited justification for adopting such a procedure because the model bias often has a magnitude similar to that of the climate change signal. (Most downscaling techniques apply variants of this procedure implicitly or explicitly.)

Testing high- and variable-resolution GCMs is only possible in a climatological sense, that is, by driving the global models with observed surface conditions (mainly sea surface temperatures and sea ice) over many years, then comparing the simulated and observed mean climatology over the region of interest. The inability to validate individual years or seasons of a GCM experiment and the inappropriateness of doing so arise from the climate system's chaotic and noisy nature, which in turn implies that the regional weather is not deterministically predictable beyond a week or so, even for a known sea surface state.

In contrast, in the one-way nesting approach to regional climate modeling, the lateral boundary conditions and the initial data deterministically control the RegCM's regional climate (provided the computational domain is not too large). Thus a RegCM can be driven continously at its lateral boundaries with observed atmospheric data, and comparatively short (e.g., month-long) numerical simulations can be compared against the observed regional climate of a particular month (e.g., Anthes et al. 1989; Giorgi and Marinucci 1991; Lüthi et al. 1996).

Results from Cress et al. 1994 and Lüthi et al. 1996 provide some indication of a RegCM's performance. The model has a resolution of fifty-six kilometers in the horizontal and thirty computational levels in the vertical. It includes parameterizations of cloud microphysics, moist convection, surface hydrology, and other processes (for a detailed description, see Majewski 1991). At the lateral boundaries it is driven with observed analysis fields (updated every six hours); the interior flow is simulated and includes surface hydrological and precipitation processes. Results are presented for month-long simulations of January and July 1993.

Figure 4.9 shows the simulated mean sea level pressure, along with the atmosphere's observed state; the model and observations differ by less than a few hPa. The simulated dynamical response is a superposition of the driving analysis fields with smaller-scale circulations resolved by the RegCM.

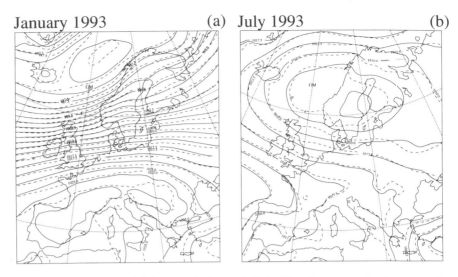

January 1993 (a) July 1993 (b)

Figure 4.9 Simulated (full lines) and observed (dashed lines) mean sea level pressure for regional climate simulations of (a) January 1993 and (b) July 1993. After Cress et al. 1994.

The latter result from better resolution of topography, land-sea contrast, surface inhomogeneities and dynamically generated features (such as cold and warm fronts). Because much precipitation falls in conjunction with such mesoscale circulations, the simulated precipitation response from a RegCM may be significantly different than that of the driving GCM, even on the scales resolved by the GCM (see, e.g., Marinucci and Giorgi 1992).

Figure 4.10 shows simulated precipitation totals for the two individual months along with a verifying analysis based on observations at approximately 1,400 rain gauge stations. (Note that the observations cover only a part of the domain.) January 1993 was a rather dry month in central Europe, and the simulation captures the precipitation distribution remarkably well (figure 4.10 a,b). The borderline between the dry Mediterranean climate and the wet climate in central and northern Europe is captured very well, but precipitation appears to be slightly overestimated. During July 1993 (figure 4.10c) the midlatitude temperate climate zone of Europe received precipitation amounts of between 20 and 150 mm, with parts of the Mediterranean receiving lesser amounts. In the near-Alpine domain, widespread areas exceeded 300 millimeters, primarily as a result of convective precipitation. Comparison with the simulated values (figure 4.10d) shows some obvious deficiencies with overestimation in Scandinavia and underestimation over much of central Europe. The Alpine anomaly is captured but underestimated.

Careful further diagnosis of the results (Lüthi et al. 1996) demonstrated that the modeling approach can simulate the primary dynamical fields over the European sector with a quality comparable, throughout the entire month-long integration period, to that of a two- to three-day forecast. The rainfall distribution is reproduced quite successfully in winter (and the transition seasons), and the regional-scale effects of the interannual variability appear

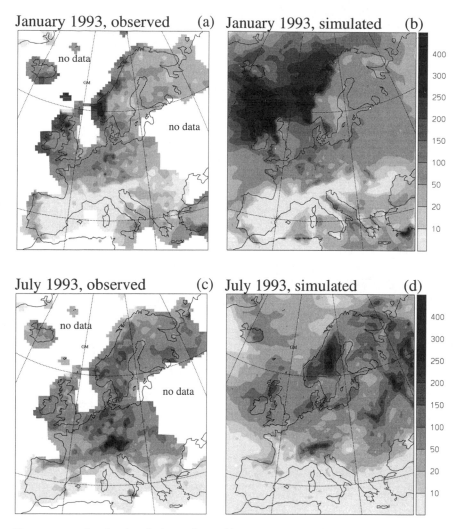

Figure 4.10 Simulated and observed monthly precipitation totals (in mm per month) for regional climate simulations for January 1993 (panels a and b) and July 1993 (panels c and d). The observations in (b) and (d) are from a surface-based network of approximately 1,400 rain gauges covering a part of the computational domain. (Note there are no observations in the white areas.) After Lüthi et al. 1996.

predictable from their large-scale counterparts, indicating that the regional model could also realistically simulate other (say, anthropogenic) climate changes given realistic driving data. The simulated rainfall in summer, when convective precipitation and soil-atmosphere interactions become more important, has the qualitatively correct spatial pattern but is quantitatively unsatisfactory, and the approach requires further improvements.

The modeling errors and uncertainties increase substantially when output from a GCM drives a fine-mesh model. The GCM simulation's deficiencies normally combine with and add to the regional model's errors (see, e.g., Dickinson et al. 1989). Indeed for domain sizes comparable to those shown in figure 4.9, a RegCM predicts a large-scale dynamical mean climatology generally very close to that of the driving GCM simulations (see Podzun et al. 1995; Jones, Murphy, and Noguer 1995). Hence, improvements in regional climate projections with the combined GCM/RegCM approach require the improvement of both components.

The modeling approach can also be combined with other methods to give mixed downscaling procedures. For example, the approach proposed by Frey-Buness (1993) and Frey-Buness, Heimann, and Sausen (1995) involves (cf. the synoptic downscaling approach) defining a set of large-scale weather regimes and eliciting the regional climate associated for each class using a RegCM model. In this approach, the RegCM is not driven conventionally, with continously varying data at the lateral boundaries, but is instead run separately for each weather regime, with vertical profiles of atmospheric variables corresponding to the temporal and horizontal mean values derived, for example, from GCM runs. This procedure excludes the simulation of transient disturbances' passage into the regional-model domain but reduces the modeling approach's very high computational costs. It therefore constitutes a viable approach for constructing more extensive (e.g., time-dependent) climate scenarios. Further it has proven feasible to use a RegCM to generate regional surrogate scenarios of a region's sensitivity to climate change, for example, to examine the potential precipitation changes resulting from a warmer global atmosphere and ocean, assuming that the frequencies of weather types remained unchanged (Schär et al. 1996).

In addition to their application for downscaling, high-resolution climate simulations based on any of the above approaches can also be used to study in detail various processes related to regional climate change. Important issues such as the sensitivity of summer precipitation to soil-moisture content and ambient moisture transport, or the feedbacks of an elevated snowline (e.g., Giorgi et al. 1997), can be investigated.

4.5 SOME ALPINE SCENARIOS

A series of climate change scenarios related to the Alps or individual Alpine locations have been developed in the last few years. This section presents and discusses a sample of these scenarios.

4.5.1 Description and Discussion

4.5.1.1 A Paleoclimatic Analogue Scenario Guiot, Harrison, and Prentice (1993) reconstructed annual mean temperature and precipitation for the Holocene warm phase (6,000 yr B.P.) over Europe and the Alps at a $1° \times 1°$ latitude-longitude resolution. They performed a principal component analysis on fossil pollen spectra to identify a set of closest modern analog locations for each fossil sample subject to a series of constraints provided by historical lake level data. The climates of the analog locations were then used to derive the corresponding paleoclimatic maps.

In the Alpine region, the annual mean temperature at the northern and western Alps was found to be at least 3°C higher than the present climate. In contrast, no marked changes, or even a temperature decrease of 1–2°C, were found for the southeastern Alps. Reconstruction of the precipitation indicated significant decreases (more than 30 percent) in the eastern and Mediterranean Alps and smaller decreases at other Alpine locations.

Changes in insolation caused by changes in the Earth's orbital characteristics most likely drove the warming in the paleoclimatic setting considered. Indeed the changes are consistent with higher annual insolation at high latitudes and higher summer and lower winter insolation in midlatitudes compared to the present. The authors also argue that the results conform with warmer high-latitude North Atlantic sea surface temperatures and a strong poleward shift of the subtropical anticyclone compared to the present. This warm analog's origin does not necessarily suggest a close correspondence with a GHG-induced climatic change.

4.5.1.2 An Instrumental Analogue Scenario Table 4.3 provides an instrumental analog scenario for eight representative locations in the Swiss Alps. The scenario was constructed by determining the local temperature and precipitation changes between the warmest (1981–90) and coolest (1901–10) decades of globally and annually averaged near-surface temperature recorded since 1861 (Jones and Briffa 1992). The Swiss temperatures show positive anomalies for all seasons and locations with the exception of Davos in summer. The strongest warming occurred in winter and autumn, in particular on the Alpine north side (Basel, Bern, Neuchatel, Saentis, Zuerich). Annual precipitation totals remained unchanged but were redistributed within the year: Winters were generally wetter and summers drier during 1981–90 than during the reference decade. The strongest winter precipita-

Table 4.3 Differences in the mean temperature and precipitation totals for 1981–90 relative to 1901–10 for eight locations in Switzerland, derived using the analog approach.

| Location | m.a.s.l. | ΔTemperature (in °C, relative to the 1901–10 mean) | | | | | ΔPrecipitation (in % of 1901–10 mean) | | | | |
		Year	Winter	Spring	Summer	Autumn	Year	Winter	Spring	Summer	Autumn
Lugano	276	0.92	1.61	0.48	0.52	1.28	−5.8	11.2	23.8	−21.6	−15.3
Basel Binningen	317	1.64	1.82	1.37	1.63	1.92	4.2	18.5	25.9	−11.7	−1.1
Altdorf	451	0.56	0.73	0.23	0.41	1.03	1.0	35.9	−2.3	−11.5	5.1
Neuchatel	487	1.04	1.53	0.51	0.79	1.47	4.7	9.9	20.0	−13.8	15.1
Zuerich SMA	556	1.37	1.80	1.02	1.13	1.71	5.4	18.3	1.6	−6.1	22.6
Bern-Liebefeld	570	1.05	1.70	0.56	0.74	1.35	7.0	8.9	18.4	−6.6	18.1
Davos	1,590	0.78	2.27	0.20	−0.35	1.26	−5.1	28.4	−8.4	−12.8	−7.7
Saentis	2,500	1.41	1.72	1.18	1.07	1.84	−11.2	18.2	−15.5	−19.1	−16.7
Average		1.09	1.65	0.69	0.74	1.48	0.0	18.7	7.9	−12.9	2.5

tion increases were in the central Alps, whereas the largest summer decreases were recorded on the Alpine south side.

Between the two climatic periods measured, atmospheric CO_2-concentrations steadily increased, from about 295 ppmv at the beginning of the century to approximately 354 ppmv in 1990. In the same period, the global and Northern Hemispheric temperatures are estimated to have warmed by respectively 0.58°C and 0.53°C. This scenario does not necessarily indicate the trend of future Alpine climate change. Caveats are that the signals reflect the climate's high natural variability and thus depend on the choice of the two intervals, and that they are derived from data possibly contaminated by measurement errors and systematic non–CO_2-induced changes in regional climate forcings (e.g., the urban heat island effect).

4.5.1.3 A Scenario from Extrapolation of Synoptical Trends Table 4.4 depicts a scenario based on the continuation of observed trends in synoptic circulation patterns according to Schüepp (1968). The Schüepp classification scheme comprises forty Alpine weather types (fifteen convective, twenty advective, and five mixed) and is semiobjectively based on five weather parameters (defined using observational data within a circular region of 222-kilometers radius centered over the eastern Swiss Alps): the direction and speed of the upper level wind (500 hPa), the direction of surface wind (1000 hPa), the height of the 500 hPa surface, and the baroclinicity.

In the first step of constructing the scenario, the frequencies of the forty weather types were calculated for two periods with more or less constant frequency spectra (1945–74 and 1975–91). Then the effects of the most important frequency changes were taken into account based on calculated temperature and precipitation maps of all the important weather types given by Fliri (1984).

From the most frequent weather types the four most pronounced frequency increases in the 1975–91 interval were found for (Rickli and Salvisberg, personal communication) (a) the anticyclonic type with 500 hPa winds from the west, (b) the anticyclonic type with 500 hPa winds from northwest to north, (c) the type with weak pressure gradients and upper-level winds from northwest to north, and (d) the type with weak pressure gradients and upper-level winds from southwest to south. The frequencies of (a) and (c)

Table 4.4 Estimate of alpine climate for the next ten to twenty years based on the linear extrapolation of synoptical trends during 1945–91.

	Temperature	Precipitation
Winter	Warmer in the mountains	Relatively dry
Spring	Cooler in the southern and warmer in the northern alpine valleys	Wet in the southern Alps
Summer	Average	Average
Autumn	A little warmer, especially in the southern Alps	Rather wet in the northeastern Alps

increased substantially in summer and autumn, whereas the frequency of type (b) increased in autumn and winter and of type (d) in spring. Furthermore, two weather types showed a clear decline. The type with weak pressure gradients and weak upper winds (500 hPa level) declined in spring and autumn, and the northerly type with warm air advection exhibited a decreasing frequency in winter and spring.

The extrapolation of these changes into the future (table 4.4) yields a tendency toward warmer, relatively dry winters, especially at higher Alpine altitudes. For summer a less-marked increase is predicted in the tendency for prevailing southwesterly flows and a concomitant increase in precipitation on the Alpine south side.

These Alpine circulation changes are consistent with the tendencies in the Atlantic and European area, which show a marked deepening of pressure over the Northern Atlantic Ocean, a strengthening of the north-south pressure gradient, and an increase of anticyclones in winter in the southern downstream branch of the reinforced westerlies (Flohn et al. 1992). During summer, a less accentuated tendency was observed for trough formation southwest of the Alps. The extent to which these changes reflect increasing concentrations of radiatively active gases in the atmosphere is, however, uncertain.

4.5.1.4 Scenarios from Semiempirical Downscaling The semiempirical downscaling method outlined in section 4.3 has been used to derive several Alpine-related climate change scenarios. Three are discussed here.

The first scenario, derived by Gyalistras et al. (1994) for five representative Swiss locations, provides information on possible changes in seventeen seasonal statistics related to daily temperatures, precipitation, sunshine duration, relative humidity, and wind speed. The scenario was downscaled from an IPCC Business as Usual simulation performed with the MPI/Hamburg ECHAM1-T21/LSG fully coupled Oceanic-Atmospheric GCM (Cubasch et al. 1992). Figure 4.11 summarizes the changes obtained for six seasonal statistics in the last decade (scenario years 2075–84) of the GCM-experiment relative to the 1901–80 baseline. Figure 4.12 shows the projected changes for winter mean temperature and total precipitation at Bern and Lugano, on the Alpine north and south sides, respectively. In winter, the scenario depicts a shift toward milder, wetter conditions with reduced daily temperature amplitudes. For summer, it specifies hotter, wetter conditions with substantially increased relative sunshine durations and daily temperature amplitudes. (For further details, see Gyalistras et al. 1994.)

The second and third examples refer to work by Gyalistras (1994) and Gyalistras and Fischlin (1995) who used the same method as in the first example to estimate changes in monthly mean temperature and precipitation at forty Swiss locations. In addition, a random resampling procedure (Efron 1979; DiCiccio and Tibshirani 1987) was used to account for uncertainties in the formulation of the statistical downscaling models and the definition of the baseline climate. The background data for the two scenarios were

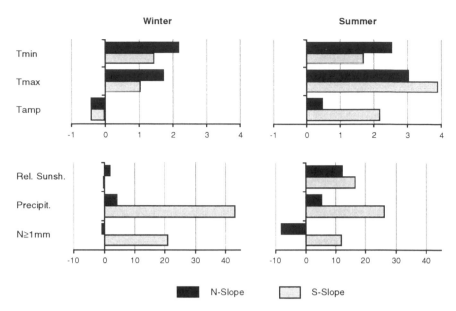

Figure 4.11 A Business as Usual climate change scenario for the Swiss Alps for 2075–84 derived with the semiempirical downscaling technique. The scenario is based on a global simulation with the ECHAM1-T21/LSG GCM (Cubasch et al. 1992) under the time-dependent IPCC Business As Usual scenario for future atmospheric GHG concentrations. N-Slope and S-slope refer respectively to mean changes from three locations on alpine north side (Bern, Davos, Saentis) and two locations on the alpine south side (Bever, Lugano). Rel. Sunsh. = seasonal mean daily relative sunshine duration; Precipit. = total precipitation; N ≥ 1 = number of days within total precipitation greater than one millimeter; T_{min}/T_{max} = seasonal mean daily minimum/maximum temperature; and T_{amp} = seasonal mean daily temperature amplitude. Temperatures are in °C; all other weather statistics are in percentage of the 1901–80 long-term means. (The N-slope values for sunshine and precipitation exclude Saentis.) After Gyalistras et al. 1994.

respectively from $2 \times CO_2$ simulations performed with the ECHAM1-T21/LSG GCM (Cubasch et al. 1992) and the Canadian Climate Centre (CCC) GCMII (Boer, McFarlane, and Lazare 1992). Figure 4.13 shows the obtained changes in the annual cycles of temperature and precipitation averaged for the forty locations considered.

In terms of the annual and areal (i.e., all forty locations) averages, the projected temperature (precipitation) changes amounted to +1.3°C(−3 percent) for the ECHAM1 and to +2.5°C (+16 percent) for the CCC simulation. For temperature, the empirical 90 percent confidence intervals (the grey areas in figure 4.13) straddling these best-estimate values amount to ±0.3°C (ECHAM1) and ±0.5°C (CCC). For precipitation, the corresponding values are ±10 percent and ±20 percent, respectively (Gyalistras and Fischlin 1995). The uncertainties tend to be largest in the summer and autumn months (figure 4.13) and are larger for the individual locations than for the areal averages (not shown).

The signals downscaled from the two GCM simulations show marked differences in strength, and the changes in the regional temperatures and pre-

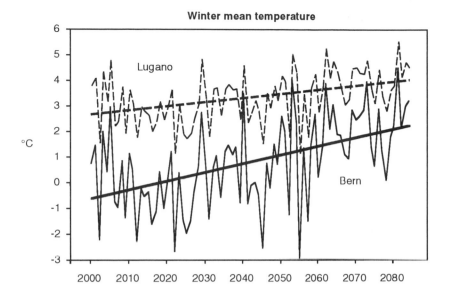

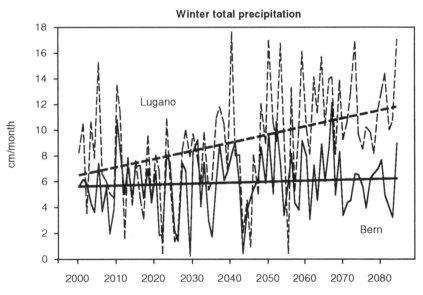

Figure 4.12 A time-dependent Business as Usual climate change scenario for winter mean temperature and winter total precipitation on the Alps' north (Bern) and south (Lugano). The scenario was derived by semiempirical downscaling (Gyalistras et al. 1994) from a transient climate change simulation with the ECHAM1-T21/LSG GCM (Cubasch et al. 1992) and refers to the 1901–80 baseline. Year-to-year variations relative to the linear trend lines were rescaled to have the same standard deviation as the 1901–80 observations.

Mean temperature change (°C)　　　　　　　Total precipitation change (%)

ECHAM1/LSG

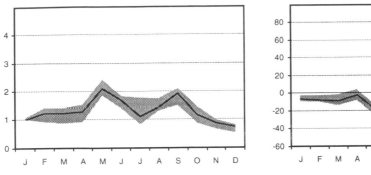

CCC-GCMII

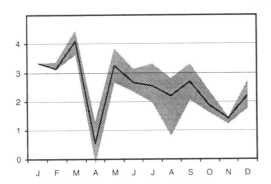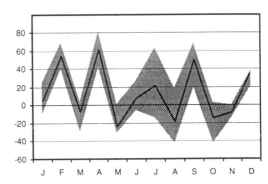

Figure 4.13 Two $2 \times CO_2$ equilibrium climate change scenarios for the Swiss region derived with the semiempirical downscaling technique. Shown are regionally averaged changes in monthly mean temperature (left) and total precipitation (right) downscaled from $2 \times CO_2$ experiments with the Hamburg ECHAM1-T21/LSG GCM (Cubasch et al. 1992, top) and the Canadian CCC-GCMII (Boer, McFarlane, and Lazare 1992, bottom). Regional averages were computed from separately downscaled changes at forty irregularly distributed locations (see figure 4.14). Grey areas correspond to the empirical 90 percent confidence interval of the regionally averaged change. Changes are given for temperature in °C and for precipitation in percentage of the 1931–80 long-term mean. After Gyalistras 1994.

cipitation totals exhibit strong month-to-month variability (figure 4.13). The latter is particularly large for the CCC scenario which was based on only five years of GCM-simulated data, as opposed to twenty years for the ECHAM1-scenario.

　　Figure 4.14 shows the spatial distribution of the annual mean temperature change for Switzerland under the ECHAM1 $2 \times CO_2$ scenario. The map was produced by first using linear regression to interpolate the downscaled changes at the aforementioned forty Swiss climate stations to an additional sixty-eight long-term (at least twenty years of data) "secondary" stations; then, inverse-distance weighted means were applied to estimate temperature anomalies on a 2.5 km $\times$ 2.5 km latitude-longitude grid (Gyalistras and Fischlin 1995, 1996). The resulting signal has a strong north-south gradient

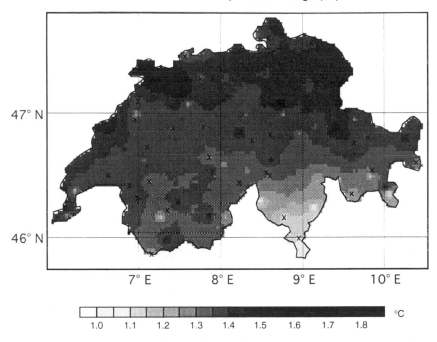

Annual mean temperature change (°C)

Figure 4.14 A $2 \times CO_2$ equilibrium climate change scenario for the Swiss annual mean temperature field constructed by combining semiemprical downscaling with a spatial interpolation procedure. Changes are given relative to the 1931–80 baseline. The scenario was based on monthly temperature anomalies (see also figure 4.13) downscaled at forty long-term climatological stations (shown as x_s) from a $2 \times CO_2$ simulation with the ECHAM1-T21/LSG GCM (Cubasch et al. 1992). For spatial interpolation sixty-eight additional stations with at least twenty years of data available were used (not shown). The main ridge of the Alps crosses Switzerland from west-southwest to east-northeast between 46 and 47°N (cf. figure 4.8c). From Gyalistras and Fischlin 1995.

across the main Alpine ridge and provides a first-order estimate of possible change under the given global scenario. Clearly, the validity of the applied regressions between climate stations under a changed climate and the fact that the procedure did not explicitly account for orographic effects both limit the spatial precision of this scenario.

Qualitatively similar maps were obtained for the two other examples (not shown). All three scenarios presented also have in common the prediction of a tendency for increased precipitation on the Alpine south side (cf. figures 4.11 to 4.13). To a measure, the similarities reflect similar large-scale patterns of change in the GCMs, but might also be linked to the downscaling procedure's inability to pick up subtle differences among the different GCM simulations. Note, however, the variations in the details of the scenarios (figure 4.13).

These scenarios are of course subject to the limitations discussed in section 4.4.3, but they do provide data with high spatial and temporal resolution and

Winter mean temperature change (˚C)

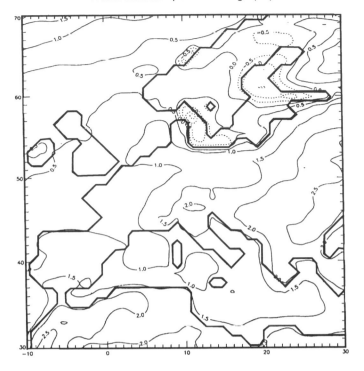

Winter total precipitation change (mm/day)

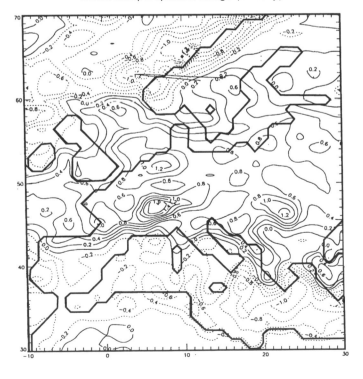

are convenient starting points for regional or even local sensitivity studies with impact models (e.g., Riedo et al. 1997, Fischlin and Gyalistras 1997).

4.5.1.5 A Scenario from a High-Resolution GCM

Beniston et al. (1995; see also Cubasch et al. 1996) have carried out a global simulation at a $1.1° \times 1.1°$ latitude-longitude (approximately 120×80 kilometers over Europe) resolution with the MPI/Hamburg ECHAM3-T106 spectral GCM (DKRZ 1993). The simulation was performed as a time slice experiment, with sea surface temperature and sea-ice boundary conditions taken from the same transient IPCC Business as Usual run also used to construct the first scenario reported in section 4.5.1.4 (ECHAM1-T21/LSG GCM $5.6° \times 5.6°$ latitude-longitude resolution; Cubasch et al. 1992). The scenarios are given relative to the climate simulated under 1985 conditions (i.e., CO_2 values of 344 ppmv) and pertain to a time window of five years around 2045 (corresponding to an equivalent of approximately 750 ppmv of CO_2).

Figure 4.15 shows the projected changes in winter mean temperature and precipitation over Europe relative to the values simulated for present climate. In the Alpine region, temperature increased by 1.5–2°C with an indication of stronger warming in the eastern Alps. Winter Alpine precipitation increased by 30–40 percent, with the largest increases in the southwestern Alps. Uncertainties in the precipitation signal are linked to precipation's high spatiotemporal variability, the relatively short sampling period, and the model's still relatively coarse resolution. For summer (not shown) the modeled Alpine temperature increased by as much as 4–6°C and was accompanied by a strong decrease in precipitation (on the order of 40–50 percent). These changes may overestimate the regional climate's true sensitivity because the model systematically underpredicts the strength of the eastern Mediteranean branch of the Atlantic jet as well as the low tropospheric humidity content. This in turn yields a systematic underprediction of the present summer precipitation throughout Europe (Marinucci et al. 1995). An inadequate representation of regional soil moisture, which can lead to an excessively rapid evaporation during spring and thus enhanced soil drought and enhanced warming in summer probably further amplified this inconsistency (Beniston et al. 1995).

4.5.1.6 Scenarios from Regional Climate Models Nested in GCMs

This section presents results from three studies using high-resolution regional climate models that pertain to the Alps. In the first, a

Figure 4.15 A Business as Usual climate change scenario for Europe by the year 2045 obtained from simulations with the ECHAM3-T106 high-resolution GCM. Shown are average changes from a five-year time window from the year 2045 relative to the means from a five-year control run. Boundary conditions for the two high-resolution GCM runs were taken from a control and a transient climate change simulation with the ECHAM1-T21/LSG GCM (Cubasch et al. 1992), respectively. Temperatures are in °C, precipitation in mm/day. From Beniston et al. 1995.

Table 4.5 Comparison of mean observed near-surface temperature and precipitation over the Alpine region with corresponding values from model simulations of current $1\times CO_2$ and putative $2\times CO_2$ climate.

		Temperature (°C)				Precipitation			
		Jan	Apr	Jul	Oct	Jan	Apr	Jul	Oct
Observed	OBS	−1.0	8.0	18.0	9.1	3.10	3.57	4.00	3.87
$1\times CO_2$ bias	CCM1−OBS	+3.0	−1.0	+9.1	+2.1	+15%	+12%	−83%	−12%
	MM4−OBS	−3.0	−4.9	+5.0	−5.2	−38%	−29%	−65%	−43%
$2\times CO_2$ effect	CCM1	+3.5	+3.0	+2.6	+4.2	+20%	+20%	+21%	+13%
	MM4	+3.7	+3.2	+2.4	+4.7	+1%	+6%	+16%	−12%

Note: The labels OBS, CCM1, and MM4 refer respectively to observations, the CCM1 general circulation model, and the nested MM4 regional climate model. The $2\times CO_2$ effect is defined as $2\times CO_2 - 1\times CO_2$. Precipitation is given in millimeter per day for the observations, and in percentage of the observed values for the model-derived results. Compiled from data in Marinucci and Giorgi 1992 and Giorgi, Marinucci, and Visconti 1992.

$2 \times CO_2$ simulation for the European sector using the GCM/RegCM nesting strategy (Marinucci and Giorgi 1992; Giorgi, Marinucci, and Visconti 1992), lateral boundary conditions for the regional climate model were taken from the global climate model CCM1 of NCAR run with a $4.5° \times 7.5°$ latitude-longitude resolution (Washington and Meehl 1984). The RegCM was itself the mesoscale model MM4 of the NCAR/Pennsylvania State University (Anthes, Hsie, and Kuo 1987), and was run at a 70 km $\times 70$ km latitude-longitude resolution. Table 4.5 summarizes the results for the Alpine region. Both the GCM and the RegCM simulated a warming, with the strongest changes (on the order of 3.5–4.5°C) in January and October. Model comparison of the precipitation values shows less agreement, but both models projected wetter conditions in all seasons (an exception is the October precipitation in the RegCM). The control experiments' deviations from the observed climate generally exceed the $2 \times CO_2$ effects, even for the mean surface temperature (table 4.5), so that only limited confidence can be attached to the projected regional change.

The second example refers to a simulation with a twenty-kilometers resolution version of a newer version of the MM4 model, RegCM2 (Giorgi Marinucci, and Bates 1993; Giorgi et al. 1993; Marinucci et al. 1995) reported by Rotach et al. (1997). The lateral boundary data for the RegCM2 was derived from a $2 \times CO_2$ time slice experiment with the high-resolution ECHAM3-T106 GCM (Beniston et al. 1995, see section 4.5.1.5). Table 4.6 gives some results. In general, the RegCM2 performs less well than the driving model and is generally too warm and too wet. Compared to the ECHAM3 GCM, the RegCM2 projected, under the hypothetically warmer atmosphere, a smaller temperature increase in both January and July and a smaller precipitation increase in January. The areal-averaged precipitation for July showed a strong decrease similar to that of the ECHAM3 GCM, again suggesting that the problematic soil moisture deficit in the high-resolution GCM dominates the RegCM's climate change signal. Projected changes

Table 4.6 Comparison of mean observed near-surface temperature and precipitation over the western alpine region with corresponding values derived from model simulations of current $1\times CO_2$ and putative $2\times CO_2$ climate.

		Temperature (°C)		Precipitation	
		Jan	Jul	Jan	Jul
Observed	OBS	−0.8	17.8	2.91	3.87
$1\times CO_2$ bias .	ECHAM–OBS	+0.7	+1.2	+12%	−61%
	RegCM2–OBS	+2.7	+1.8	+41%	−34%
$2\times CO_2$ effect	ECHAM	+1.8	+5.5	+30%	−40%
	RegCM2	+0.9	+3.9	+6%	−38%

Note: The labels OBS, ECHAM, and RegCM2 refer to observations, the ECHAM3-T106 high-resolution general circulation model, and the nested RegCM2 regional climate model, respectively. The $2\times CO_2$ effect is defined as $2\times CO_2 - 1\times CO_2$. Precipitation is given in millimeters per day for the observations, otherwise in percentage of the observed values. Compiled from data in Marinucci et al. 1995 and Beniston et al. 1995.

within the regional model domain were, however, found to be spatially variable, in particular over the Alpine chain. For instance, simulated January precipitation increased most markedly over the model southwestern Alps, but—in contrast to the driving GCM (figure 4.15)—decreased over France. In July a large precipitation increase was projected in a few small regions, in particular in the eastern Alps.

As in the previous example, the regional climate changes were again projected to be small compared to the model bias (table 4.6). Furthermore, the most pronounced changes tended to be projected for regions of largest errors (Beniston et al. 1995, Rotach et al. 1997). These results suggest that further improvement in both the global and the regional model components are needed to gain more confidence in the projected regional patterns of change.

The third example is a nested simulation carried out at a 21 × 29 kilometers latitude-longitude resolution with the REWI3HD model (Heimann 1993) using Frey-Buness's (1993) nonnested technique (see section 4.4). Weather regimes pertinent to the Alpine region for present and possible future conditions were derived from two different simulations with the ECHAM3-T42 GCM (Roeckner et al. 1992). The first GCM simulation was driven with the constant 1979–88 annual mean cycle for the sea surface temperature, and for the second simulation the sea surface temperatures were adjusted according to the mean sea surface temperature anomalies simulated during the years 2075-84 of a business-as-usual experiment with the ECHAM1-T21/LSG GCM (Cubasch et al. 1992).

Figure 4.16 shows the estimated changes in January and July mean temperature. In both cases, a warming was distributed irregularly over the Alps. In January, the modeled signal was less pronounced at the Alpine foothills (about +2°C) than at the mountain ridge and the rest of central Europe

January mean temperature change (°C)

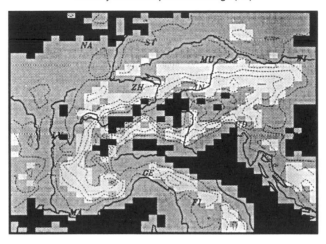

July mean temperature change (°C)

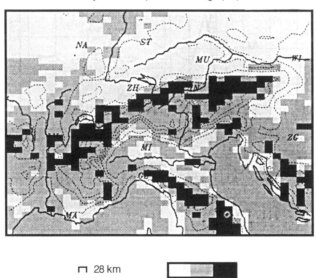

⊓ 28 km

0   2.5   2.8   3.1   °C

Figure 4.16 A Business as Usual climate change scenario for the Alpine region for 2075–84 derived from simulations with the REWIH3D-RegCM. Shown are changes derived from day-long simulations for a series of large-scale weather classes under present and hypothetical future conditions. Boundary conditions for each weather class were obtained from simulations with the ECHAM3-T42 GCM, which for the definition of the present climate was driven with 1979–88 observed sea surface temperatures. For construction of the scenario, the sea surface temperatures were adjusted according to the changes simulated in a transient climate change simulation with the ECHAM1-T21/LSG GCM (Cubasch et al. 1992). From Frey-Buness 1993.

(values up to 3.1°C). In July, the simulated warming increased toward the main Alpine ridge, but the signal on the southern slope (+2.1°C) was less than that on the northern slope (+2.4°C). Projected precipitation increased in January over the entire Alpine region by circa 40–50 percent. For July, a small increase or none at all was simulated at the model northern slope and an increase of on average circa 30 percent at the southern slope (not shown). These projected major changes are beset with uncertainty because the model contained no formulation of convective precipitation and because the results to some extent also depend on the exact definition of the weather classes (Frey-Buness 1993).

4.5.2   Intercomparison

In comparing the main features of the scenarios presented in the section 4.5.1, we aim not to extract a most probable scenario but rather to explore the Alpine climate's possible sensitivity to a CO_2 doubling. Table 4.7 lists aspects of the various scenarios. The compilation is restricted to changes in regionally averaged seasonal mean temperatures and seasonal precipitation totals. Further the spatial domain is subdivided into northern and southern Alpine subregions corresponding to the regions northwest and southeast of the main Alpine ridge (cf. figure 4.8c).

For the discussion, it is useful to view climate variables (e.g., seasonal mean temperature) as random variables. The description of the probability distributions of these variables requires knowledge of their expected value, variance, and other statistical characteristics. Here only changes in the expected values are estimated for the various scenarios from the arithmetic mean of a limited sample of measurements and projected values. Thus, because of the natural climatic variability, considerable uncertainties are present even within the individual scenarios (cf. figure 4.13), and table 4.7 does not address this aspect. These uncertainties typically increase with smaller averaging regions and with an increasing resolution of the annual cycle.

The scenarios do not provide direct estimates of changes in climatic variability. Nevertheless some information can be inferred on the possible changes in the frequencies of extreme events, since in effect a shift in the expected value of a climate variable will also affect the tails of its distribution function. For example, consider winter mean temperature in Bern, Switzerland. To a good approximation, it is normally distributed with an expected value of −0.2°C and a standard deviation of 1.56°C. The present theoretical probability for a winter warmer than 2.5°C is 4.2/100 yr. (The empirical value for the 1901–80 analysis interval is 1.25/100 yr). Thus, assuming that the variance of the temperature remains unchanged, a relatively modest increase of the expected value by 0.8°C (scenario 9 in table 4.7) connotes an increase of the probability for a "warm" winter to almost one per decade; and under the 3.7°C increase of scenario 7, to a value larger than 70/100 yr (see also figure 4.12). These estimates depend sensitively on the constancy of

Table 4.7 Selected climate change scenarios for the alpine region

Type of study	No.	Applied technique	Major historical, or assumed future forcing	Climate model(s)	ΔTemperature REG	Win	Spr	Sum	Aut	ΔPrecipitation (in %) REG	Win	Spr	Sum	Aut
Empirical	(1)	Paleoclimatic analog based on pollen and lake level data (Guiot, Harrison, and Prentice 1993)	Orbital parameters of Holocene warm phase (6,000 yr B.P.)	—	**N, W**	≥ +3°C				**E, SW**	≤ −30%			
					SE	< −1–2°C				**other**		(−)		
	(2)	Instrumental analog: change of means 1981–90 relative to 1901–10 (see section 4.5.1.2)	Historical GHG increase (CO_2 from 295 to 354 ppmv)	—	N	+1.7	+0.9	+1.1	+1.7	N	+15	+10	−11	+8
					S	+1.6	+0.5	+0.5	+1.3	S	+11	+24	−22	−15
	(3)	Extrapolation of frequencies of alpine weather types: 1975–91 vs. 1945–74 (section 4.5.1.3)	Historical GHG increase (CO_2 from 310 to 356 ppmv)	—	N	(+)	(+)	(o)	(o)	N	(−)	(o)	(o)	(+)
					S	(+)	(−)	(o)	(+)	S	(−)	(+)	(o)	(o)
Semi-empirical	(4)	Linear statistical downscaling using anomalies of large-scale atmospheric fields (Gyalistras et al. 1994; Gyalistras 1994)	2×CO_2 (from 330 to 660 ppmv)	CCC/GCMII	N	+3.1	+2.6	+2.6	+2.2	N	+30	+4	−4	+4
					S	+2.2	+2.0	+2.2	+1.9	S	+36	+14	+1	+13
	(5)		2×CO_2 (from 344 to 720 ppmv)	ECHAM1-T21/LSG	N	+1.1	+1.6	+1.5	+1.4	N	−3	−14	−2	−2
					S	+0.9	+1.3	+1.3	+1.2	S	+2	+2	+11	+35
	(6)		IPCC BaU, years 2075–84 (from 344 to 1100 ppmv CO_2 equivalent)	ECHAM1-T21/LSG	N	+2.0	•	+2.7	•	N	+4	•	+6	•
					S	+1.2	•	+2.4	•	S	+43	•	+27	•

		Model/GCM	Scenario					N/S				
Model based	(7)	CCM1, MM4-RegCM	2×CO_2 (from 330 to 660 ppmv)	Nesting of RegCM in GCM (Giorgi, Marinucci, and Visconti 1992)	+3.7	+3.2	+2.4	+4.7		+1	+6	+16	−12
	(8)	ECHAM1 T21/LSG, ECHAM3-T106 high res. GCM	IPCC BaU, years 2045–49 (from 344 to 750 ppmv CO_2 equivalent)	High-resolution GCM driven by coarse-resolution GCM (Beniston et al. 1995)	+1.8	•	+5.5	•		+30	•	−40	•
	(9)	As above, but in addition RegCM2-RegCM	"	As (8), but in addition RegCM nested in the high-resolution GCM (after Beniston et al. 1995, Marinucci et al. 1995)	+0.8	•	+3.5	•	N	+5	•	−30	•
					+1.2	•	+2.5	•	S	+15	•	−5	•
	(10)	ECHAM1-T21/LSG & ECHAM3-T42, REWIH3D-RegCM	IPCC BaU, years 2075–84 (from 344 to 1100 ppmv CO_2 equivalent)	As (9), but RegCM simulations stratified by large-scale weather classes (after Frey-Buness 1993; Frey-Buness, Heimann, and Sausen 1995)	+2.0	•	+2.1	•	N	≥ +50	•	≥ 0	•
					+2.3	•	+2.4	•	S	≥ +45	•	≥ +30	•

Note: REG refers to alpine subregions such that (W/SW/SE) denotes the western/southwestern/southeastern Alps and N and S the northern and southern alpine slopes. The symbols (+/o/−) refer respectively to a positive trend/no trend/negative trend; BaU = business as usual. Bullets denote missing values. For scenario 1, the data are annual averages; for Scenario 2, the N-slope averages are for Basel, Bern, Neuchatel, Saentis, and Zuerich, and the S-slope data are for Lugano only; for Scenario 3, the temperature trends apply in winter for mountain regions, in spring to valleys; for scenarios 4 and 5 the data are averages for thirty-two locations on the north and eight locations on the south slopes of the Swiss Alps; for Scenario 6, the temperatures signify averages for the locations Bern, Davos, and Saentis (N-slope) and Bever and Lugano (S-slope), and likewise for the precipitation (with Saentis excluded); for Scenarios 7–10, the data are temporal and spatial averages and apply to the entire alpine region. The data are derived for scenario 7 from five simulations of the months of January, April, August, and November; for scenario 8 from a five-year simulation; for scenario 9 from five simulated Januarys and Julys for the western/central Alpine region northward of 44°N and westward of 13°E; and for scenario 10 from ten simulated Januarys and Julys. Changes for the N- and S-slope of the Alps in the scenarios 9 and 10 were obtained from our own analyses of the results provided by Beniston et al. (1995) and Frey-Buness (1993).

the variance (cf. Katz and Brown 1992). More sophisticated assessments of possible changes in the probabilities of extreme events can be derived from time series modeling techniques (e.g., Mearns, Katz, and Schneider 1984).

As can be seen from table 4.7, the various scenarios refer to differing baseline climates, CO_2 concentrations, and forcing of global climate. In particular, scenario 1 reflects changes in the earth's orbital parameters and was therefore not considered further. The empirical scenarios 2 and 3 refer to the historical observed CO_2 increase; the GCM-based scenarios 4, 5, and 7 to a $2 \times CO_2$ equilibrium climate change; and the remaining scenarios to a future time window under the IPCC Business as Usual emissions scenario. To compare these scenarios, consider changes to the annually and spatially (Alpinewide) averaged expected values of temperature and precipitation. These changes are assessed in terms of the following standardized sensitivity measure:

$$(X_{new} - X_o) \cdot \ln(2)/\ln(GHG_{new}/GHG_o)$$

for an increase of the global atmospheric GHG content. Here X is the chosen parameter, GHG refers to the atmospheric GHG content in ppmv-equivalent CO_2 concentration, and the subscripts o and new to respectively the baseline and the perturbed values. (The logarithm is introduced to account for the dimunition in the rate of increase of the global radiative forcing at higer levels of atmospheric CO_2; cf. Shine et al. 1990.) The change $[X_{new} - X_o]$ is estimated for each scenario by averaging with equal weights all available seasonal values from the northern and southern subregions of the Alps (table 4.7).

Figure 4.17 plots the values derived for the index. Semiempirically derived scenarios 4–6 gave comparatively small temperature sensitivities and positive precipitation sensitivities. Scenarios 7–10, derived from regional climate models, yielded generally larger temperature sensitivities (scenarios 7 and 8), and in two cases negative precipitation sensitivities (scenarios 8 and 9). Note however that estimates 6, and 8–10 are not independent, because they were all based on the same global simulation with the ECHAM1-GCM; in particular estimates 8 and 9 (and respectively 6 and 10) were derived from the same information taken from the ECHAM1-GCM at around the simulation years 2045 (and respectively 2080; see also table 4.7).

The largest temperature sensitivity (4.4°C) is that from purely empirical scenario 2. All other GCM-based scenarios 4–10 yielded between 1.2°C (scenario 5) and 3.5°C (scenario 7; figure 4.17). In scenarios 2, 4–6 and, for summer only, 9 and 10, warming was found to be stronger at the north slope than at the south slope of the Alps (table 4.7).

The relatively wide range of values for temperature is at least in part related to the different character of the climate simulated by the driving GCMs. Note, for example, that the increase in the globally and annually averaged near-surface air temperature under $2 \times CO_2$ conditions was 1.7°C in the ECHAM1-GCM, 3.5°C in the CCC-GCMII, and 4°C in the CCM-

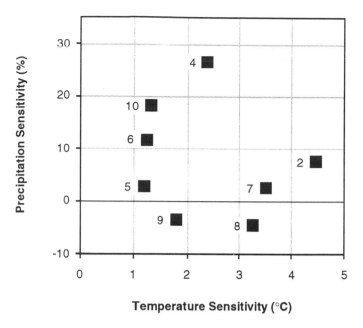

Figure 4.17 Comparison of estimates for the $2 \times CO_2$ sensitivity of the annually and regionally averaged temperature and precipitation over the alpine region, as derived from different climate change scenarios. The numbers refer to the scenarios from table 4.7—(2): instrumental analogue scenario; (4)–(6): scenarios derived by semiempirical downscaling; (8): scenario from a high-resolution GCM; (7), (9), (10): scenarios from RegCM simulations. The estimates (4)–(10) were obtained from global climate change simulations with three different driving GCMs. Several estimates were obtained from a limited number of months and subregions of the Alps (see table 4.7 and explanations in the text).

GCM. The average temperature sensitivity obtained from all seven GCM-based scenarios for the Alpine region was 2.1°C, relatively small compared to the mean global sensitivity of the three GCMs (3.1°C). One possible explanation lies in the moderating effect of the nearby Atlantic. The larger warming found at the north slope of the Alps in several of the scenarios probably reflects the influence of particularly strong warming at higher latitudes and over the interiors of the continent in the driving GCMs.

For annual precipitation the results obtained from the $2 \times CO_2$ sensitivity study range from −4 percent to +27 percent (scenarios 8 and 4, respectively; figure 4.17). The average is approximately 8 percent. In scenarios 4–6, 9, and 10, the precipitation increased on the southern slope of the Alps—the one exception is for summer in scenario 9.

The tendency toward wetter conditions agrees with an increase in global precipitation of 3–15 percent climate models project under a CO_2 doubling. In the Alpine region, a precipitation increase would be linked to regions favorable for orographic uplifting and convective activity (see also section 2.3).

It follows from the above that the scenarios indicate an increase in both temperature and precipitation under putative $2 \times CO_2$-conditions. However, no other coherent pattern can be discerned in figure 4.17. Several factors may have contributed to this result:

First, as section 4.4 has already discussed, each scenario suffers from specific problems and limitations, and these may well lead to a misrepresentation of the sensitivity of Alpine climate in each individual case. For instance, purely empirical scenario 2 may well overestimate the temperature sensitivity because of sampling uncertainty or purely regional climate trends; semiempirical scenario 4 may overestimate the precipitation sensitivity because it uses empirical extrapolations; and scenarios 8 and 9 may depict a too hot, too dry climate because they inadequately represent soil moisture in summer (see section 4.5.1.5). Second, estimates of several sensitivities were based on a limited number of months or seasons (scenarios 6–10) or on a relatively small number of irregularly distributed locations that covered only a small part of the Alpine region (scenarios 4–6). This renders the evaluation of the annual and regional averages particularly uncertain. Finally, scenarios 6, 8, and 10 refer to future time windows under a time-dependent climate change simulation and our simple sensitivity measure did not take into account the coupled ocean-atmosphere system's time-delayed response.

Recall also that the present study has considered only a limited number of global climate simulations, and moreover the GCMs used in this study did not fully cover the range of 1.7°C–5.3°C obtained from $2 \times CO_2$ simulations with state-of-the-art GCMs (Mitchell et al. 1990; Gates et al. 1992; Kattenberg et al. 1996). Finally, note that future Alpine climate will also depend on factors other than those associated with the anticipated increases in GHGs including the effects of atmospheric aerosols, which may not combine linearly with the above scenarios, and climate's natural variability on decadal and larger timescales.

In view of these uncertainties, there is at present no compelling evidence to select one particular scenario among those presented. Indeed, the wide spectrum of the results reflects the approximations that underlie, and the uncertainties associated with, the generation of Alpine climate scenarios. In effect, the spectrum of values is a salutary reminder of the care and caveats that must be associated with any application. However, the results do indicate to some degree the more trustworthy components of the scenarios. For example, under particular assumptions on the future forcings of global climate, projections for winter and for temperature are more likely to be reliable than for summer and for precipitation amounts. In turn, these tentative indicators influence the measure of uncertainty to be attached to estimates of specific impacts.

4.6 SUMMARY AND CONCLUSION

The response to global climate change will probably exhibit significant regional variations. This chapter has focused on the Alpine region and provided an overview of the possible future evolution of the region's climate.

The distinctiveness of Alpine weather systems and their relationship to the global-scale circulation (chapter 2) implies that a change in the global climate could also generate special features in the Alpine region.

Our consideration of Alpine climate focused on physically induced atmospheric changes operating in isolation from other environmental and human components of the system. The main thrust was to derive regional climate scenarios. The competing constraints placed on the one hand by the inherent uncertainties associated with climate prediction and on the other by the desirability of having available assessments of the nature of possible climate change influenced the methods adopted. The discussion has emphasized that the derived scenarios are not forecasts but applications-oriented descriptions of conceivable future climate changes geared for the examination of one or a range of possible impacts.

Our discussion involved a study of the approaches currently available for generating climate scenarios, providing examples for the Alpine region based on these approaches, and commenting on the range of their application. The scenarios provided for the Alpine region were derived using the extant range of approaches: the analyses of past climates, the continuation of recent trends, the semiempirical downscaling of future global climates as simulated by coarse-resolution climate models, and the simulation of future climate with high-resolution global and nested regional climate models. Attention was drawn to the specific advantages and limitations of each approach.

The techniques discussed are subject to a series of basic limitations that introduce large uncertainties into any assessments of future Alpine climate. Nevertheless, taken together, they provide a set of complementary tools for studying Alpine climate change and its possible implications. In the following, we discuss these two aspects of basic limitations and practical utility.

In relation to the limitations, several points need to be stressed: First, the limitations arise because projections of future climate change inevitably assume future anthropogenic forcings and are unavoidably obscured by the stochastic component of the variability associated with natural forcing (e.g. solar irradiance and volcanic activity) and the climate system's chaotic nature (see chapter 7). These limitations are mitigated to the extent that current GCM results suggest anthropogenic effects with an amplitude exceeding that of the natural variability experienced in the past several thousand years. More trenchantly, studies of the combined effects of several, possibly counteracting, anthropogenic forcings (such as the increasing concentration of both aerosols and GHGs) are in their infancy. Aerosol effects, not considered in the present study, could not only reduce future warming over central Europe (at least temporarily) by several tenths of a degree Celsius but also modify the presently projected changes in the general circulation (e.g., Taylor and Penner 1994; Mitchell, Davis, et al. 1995; Mitchell, Johns, et al. 1995)

Second, present climate models, in spite of their sophistication, inadequately represent several important processes and interactions. Feedbacks related to the biosphere and the global carbon cycle can be important particularly on longer timescales (decades and beyond). The oceans are huge reservoirs for heat and CO_2, and simulation studies addressing the role of oceanic variability have only just started (e.g., Delworth, Manabe, and Stouffer 1993; Manabe and Stouffer 1994). In this context the oceanic subsystem might exhibit multiple steady states (e.g., Marotzke and Willebrand 1991), and possible future subtle shifts in the boundary conditions of the oceanic circulation related to meltwater inflow and precipitation could influence the transitions between the different modes (see also chapter 7). These unrepresented or partially represented processes influence the comparative evolution of the real and modeled atmosphere-ocean system.

Third, the amplitude of future climate change could exceed the range of the data available to construct statistical and physical models of climate. This in turn limits the range of applicability of empirical and semiempirical regionalization approaches and the validity of the parameterizations in the physical models, because these have been tuned for current climatic conditions.

Fourth, the departure of GCM control simulations of current-day climate from the observed climate limits the credibility of scenarios based on such models. Removal of the bias by comparison of $1 \times CO_2$ and perturbed states (rather than a comparison of the perturbed states with observed climate) becomes questionable if the bias's relative amplitude is large compared to the climate change signal. Indeed, at present, GCMs typically show large biases over the European sector and the Alpine region, and thus care must be taken in interpreting the resulting scenarios.

Finally, we comment on the scenarios' utility. The results demonstrate the potential for complex and significant climate change in the Alpine region. Comparison of the scenarios provides initial estimates of the sensitivity range of Alpine temperature and precipitation to a doubling of atmospheric GHG forcing. The estimates suggest a general warming of the Alps under increasing GHG concentrations and indicate that in most subregions of the Alps an increase rather than a decrease in precipitation could accompany such a warming. Additional studies are needed to understand better the differences between the present estimates, to investigate a broader range of conditions, such as combined sensitivities due to a simultaneous increase of GHGs and aerosols, and to assess more reliably the relative importance of regional-scale feedbacks for the region's climate.

Several of the scenarios this chapter presented are comprehensive enough to allow for detailed impact and sensitivity studies in a range of sectors affected by climate. They provide first-order estimates of possible climatic changes and can be used to test and simplify impact models, to detect sensitive areas, and to explore the possible interactions of climate change with other aspects of global and regional change. The scenarios have already proven useful for initial sensitivity studies in the ecological, agricultural, and touristic sectors. (For ecological applications, see chapter 6 and Bugmann and Fischlin 1994, Brzeziecki, Kienast, and Wildi 1995, and Fischlin and Gyalistras 1997; for agriculture, see Flückiger 1995, Fuhrer 1997 and Riedo et al. 1997; for tourism, see Abegg 1996.)

ACKNOWLEDGMENTS

We thank the Swiss Meteorological Institute (SMA) for the generous access to the their meteorological data. We also thank Drs. M. Rottach (ETH Zurich) and D. Heimann (DLR Oberpfaffenhofen) for providing and discussing the results of the RegCM2 and REWI3HD models. We acknowledge useful comments from Prof. M. Beniston (University of Fribourg, Switzerland) and three anonymous reviewers, who contributed substantially to an improvement of the manuscript.

REFERENCES

Abegg, B. 1996. *Klimaänderung und Tourismus: Klimafolgenforschung am Beispiel des Wintertourismus in den Schweizer Alpen.* Zurich: Verlag der Fachvereine.

Abegg, B., and R. Froesch. 1994. Climate change and winter tourism: Impact on transport companies in the Swiss canton of Graubünden. In *Mountain Environments in Changing Climates*, ed. M. Beniston, 328–40. London: Routledge.

Anthes, R. A., E. Y. Hsie, and Y. H. Kuo. 1987. *Description of the Penn State/NCAR Mesoscale Model Version 4 (MM4).* NCAR Tech. Note TN-282+STR, National Center for Atmospheric Research, Boulder, Colorado.

Anthes, R. A., W. Y. Kuo, E. Y. Hsie, S. Low-Nam, and T. W. Bettge. 1989. Estimation of skill and uncertainty in regional numerical models. *Quart. J. Roy. Met. Soc.* 115:763–806.

Bach, W., H. J. Jung, and H. Knottenberg. 1985. *Modeling the Influence of Carbon Dioxide on the Global and Regional Climate: Methodology and Results.* Münstersche geographische Arbeiten 21, Schoeningh, Paderborn.

Bailly, A. S. 1994. Environmental perception, climate change and tourism. In *Mountain Environments in Changing Climates*, ed. M. Beniston, 318–27. London: Routledge.

Barancourt, C., J. D. Creutin, and J. Rivoirard. 1992. A method for delineating and estimating rainfall fields. *Water Resour. Res.* 28:1133–44.

Bardossy, A., and E. J. Plate. 1992. Space-time model for daily rainfall using atmospheric circulation patterns. *Water Resour. Res.* 28:1247–59.

Barnett, T., and R. Preisendorfer. 1987. Origins and levels of monthly and seasonal forecast skills for the United States surface air temperatures determined by canonical correlation analysis. *Mon. Wea. Rev.* 115:1825–50.

Bayer, D., M. Denhard, S. Meyhöfer, J. Rapp, and C.-D. Schönwiese. 1994. *Trend- und multiple Signalanalyse globaler bzw. europäischer Klimavariationen.* Berichte des Instituts für Meteorologie und Geophysik 98, Universität Frankfurt am Main, Frankfurt.

Bengtsson, L. 1990. Advances in numerical prediction of the atmospheric circulation in the extra-tropics. In *Extratropical Cyclones: The Erik Palmén Memorial Volume*, eds. C. Newton and E. Holopainen, 194–220. Boston: American Meteorological Society.

Beniston, M., A. Ohmura, M. Rotach, P. Tschuck, M. Wild, and R. M. Marinucci. 1995. *Simulation of Climate Trends over the Alpine Region: Development of a Physically-Based Modeling System for Application to Regional Studies of Current and Future Climate.* Internal Report, Department of Geography, Swiss Federal Institute of Technology (ETH), Zurich.

Binder, P., and C. Schär. 1995. *The Mesoscale Alpine Programme: Design Proposal.* Swiss Meteorological Institute, Zürich.

Boer, G. J., N. A. McFarlane, and M. Lazare. 1992. Greenhouse gas–induced climate change simulated with the CCC second-generation general circulation model. *J. Clim.* 5:1045–77.

Boville, B. A. 1991. Sensitivity of simulated climate to model resolution. *J. Clim.* 4:469–85.

Boyle, J. S. 1993. Sensitivity of dynamical quantities to horizontal resolution for a climate simulation using the ECMWF (Cycle 33) model. *J. Clim.* 6:796–815.

Bretherton, C. S., C. S. Smith, and J. M. Wallace. 1992. An intercomparison of methods for finding coupled patterns in climate data. *J. Clim.* 5:541–60.

Brown, B. G, and R. W. Katz 1995. Regional analysis of temperature extremes: Spatial analogue for climate change? *J. Clim.* 8:108–119.

Brzeziecki, B., F. Kienast, and O. Wildi. 1995. Modelling potential impacts of climate change on the spatial distribution of zonal forest communities in Switzerland. *J. Veg. Sci.* 6:257–268.

Brzeziecki, B., F. Kienast, and O. Wildi. 1993. A simulated map of the potential natural forest vegetation of Switzerland. *J. Veg. Sci.* 4:499–508.

Bugmann, H. and A. Fischlin, 1994. Comparing the behaviour of mountainous forest succession models in a changing climate. In *Mountain Environments in Changing Climates*, ed. M. Beniston, 237 55. London: Routledge.

Cacciamani, C., F. Battaglia, P. Patruno, L. Pomi, A. Selvini, and St. Tibaldi. 1995. A climatological study of thunderstorm activity in the Po Valley. *Theor. Appl. Climatol.* 50:185–203.

Carter, T. R., M. L. Parry, H. Harasawa, and S. Nishioka 1994. *IPCC Technical Guidelines for Assessing Climate Change Impacts and Adaptations.* London and Tsukuba: Dept. of Geography, University College of London, and National Institute for Environmental Studies, Japan.

Cress, A., H. C. Davies, C. Frei, D. Lüthi, and C. Schär. 1994. *Regional Climate Simulations in the Alpine Region.* LAPETH Report 32, Atmospheric Physics ETH, Zürich.

Cress, A., D. Majewski, R. Podzun, and V. Renner. 1995. Simulation of European climate with a limited area model. Part I: Observed boundary conditions. *Contrib. Atmosph. Phys.* 68:161–78.

Crowley, T. J. 1990. Are there any satisfactory geologic analogs for a future greenhouse warming? *J. Clim.* 3:1282–92.

Cubasch, U., K. Hasselmann, H. Höck, E. Maier-Reimer, U. Mikolajewicz, B. D. Santer, and R. Sausen. 1992. Time-dependent greenhouse warming computations with a coupled ocean-atmosphere model. *Clim. Dyn.* 8:55–69.

Cubasch, U., H. von Storch, J. Waszkewitz, E. Zorita. 1996. Estimates of climate change in Southern Europe derived from dynamical climate model output. *Clim. Res.* 7:129–149.

Daly, C., R. P. Neilson, and D. L. Phillips. 1994. A statistical-topographic model for mapping climatological precipitation over mountainous terrain. *J. Appl. Meteorol.* 33:140–58.

Delworth, T., S. Manabe, R. J. Stouffer. 1993. Interdecadal variations of the thermohaline circulation in a coupled ocean-atmosphere model. *J. Clim.* 6:1993–2011.

Déqué, M., and J. P. Piedelievre. 1995. High resolution climate simulation over Europe. *Clim. Dyn.* 11:321–39.

DiCiccio, T., and R. Tibshirani. 1987. Bootstrap confidence intervals and bootstrap approximations. *J. Am. Stat. Assoc.* 82:163–70.

Dickinson, R. E., R. M. Errico, F. Giorgi, and G. T. Bates. 1989. A regional climate model for the western United States. *Clim. Change* 15:383–422.

DKRZ (Deutsches Klimarechenzentrum) 1993. *The ECHAM3 Atmospheric General Circulation Model.* DKRZ Tech. Report 6, Deutsches Klimarechenzentrum, Hamburg.

Efron, B. 1979. Bootstrap methods: Another look at the jacknife. *The Annals of Statistics* 7:1–26.

Fischlin, A., H. Bugmann, and D. Gyalistras. 1995. Sensitivity of a forest ecosystem model to climate parametrization schemes. *Environ. Pollut.* 87:267–82.

Fischlin, A., and D. Gyalistras. 1997. Assessing impacts of climatic change on forests in the Alps. *Global Ecol. Biogeogr. Lett.* 6:19–37.

Fliri, F. 1984. *Synoptische Klimatographie der Alpen zwischen Mont Blanc und Hohen Tauern (Schweiz-Tirol-Oberitalien).* Wissensch. Alpenvereinshefte 29, Innsbruck.

Flohn, H. 1979. Can climate history repeat itself? Possible climatic warming and the case of paleoclimatic warm phases. In *Man's Impact on Climate*, ed. W. Bach, Y. Pankrath, and W. W. Kellogg, 15–28. New York: Elsevier Science.

Flohn, H., A. Kapala, H. R. Knoche, and H. Mächel. 1992. Water vapour as an amplifier of the greenhouse effect: New aspects. *Meteorol. Z.*, 1:122–38.

Flückiger, S. D. 1995. *Klimaänderungen: Ökonomische Implikationen innerhalb der Landwirtschaft und ihrem Umfeld aus globaler, nationaler und regionaler Sicht.* Ph.D. diss. no. 11276, Swiss Federal Institute of Technology (ETHZ): Zürich.

Foley, J. A., J. E. Kutzbach, M. T. Coe, and C. Levis. 1994. Feedbacks between climate and boreal forests during the Holocene epoch. *Nature* 371:52–4.

Frey-Buness, A. 1993. *Ein statistisch-dynamisches Verfahren zur Regionalisierung globaler Klimasimulationen.* Forschungsbericht DLR DLR-FB 93-47, Deutsche Forschungsanstalt für Luft- und Raumfahrt, Oberpfaffenhofen, Germany.

Frey-Buness, A., D. Heimann, and R. Sausen. 1995. A statistical-dynamical downscaling procedure for global climate change simulations. *Theor. Appl. Climatol.* 50:117–31.

Fuhrer, J., ed. 1997. *Klimaveränderungen und Grünland—Eine Modellstudie über die Auswirkungen zukünftiger Klimaveränderungen auf das Dauergrünland in der Schweiz.*, Zürich: Verlag der Fachvereine.

Gates, W. L., A. Henderson-Sellers, G. J. Boer, C. K. Folland, A. Kitoh, B. J. McAvaney, F. Semazzi, N. Smith, A. J. Weaver, and Q.-C. Zeng. 1996. Climate models: Evaluation. In *Climate Change 1995: The Science of Climate Change. Contribution of Working Group I to the Second Assessment Report of the Intergovernmental Panel on Climate Change*, ed. J. T. Houghton, L. G. Meira Filho, B. A. Callander, N. Harris, A. Kattenberg, and K. Maskell, 228–84. Cambridge: Cambridge University Press.

Gates, W. L., J. F. B. Mitchell, G. J. Boer, U. Cubasch, and V. P. Meleshko. 1992. Climate modelling, climate prediction and model validation. In *Climate Change 1992: The Supplementary Report to the IPCC scientific assessment*, ed. J. T. Houghton, B. A. Callander, and S. K. Varney, 96–134. Cambridge: Cambridge University Press.

Giorgi, F. 1990. On the simulation of regional climate using a limited area model nested in a general circulation model. *J. Clim.* 3:941–63.

Giorgi, F., J. W. Hurrell, M. R. Marinucci, and M. Beniston. 1997. Elevation dependency of the surface climate signal: a model study. *J. Clim.* 10:288–296.

Giorgi, F., and M. R. Marinucci. 1991. Validation of a regional atmospheric model over Europe: Sensitivity of wintertime and summertime simulations to selected physics parameterizations and lower boundary conditions. *Quart. J. Roy. Met. Soc.* 117:1171–206.

Giorgi, F., M. R. Marinucci, and G. T. Bates. 1993. Development of a second generation regional climate model. I: Boundary layer and radiative transfer processes. *Monthly Weather Review* 121:2794–813.

Giorgi, F., M. R. Marinucci, G. T. Bates, and G. Decanio. 1993. Development of a second generation regional climate model. II: Convective processes and assimilation of lateral boundary conditions. *Monthly Weather Review* 121:2814–32.

Giorgi, F., M. R. Marinucci, and G. Visconti. 1992. A $2 \times CO_2$ climate change scenario over Europe generated using a limited area model nested in a General Circulation Model. Part II: Climate change scenario. *J. Geophys. Res. D*, 97:10011–28.

Giorgi, F., and L. O. Mearns. 1991. Approaches to the simulation of regional climate change: A review. *Rev. Geophys.* 29:191–216.

Guiot, J., S. P. Harrison, and C. I. Prentice. 1993. Reconstruction of Holocene precipitation patterns in Europe using pollen and lake-level data. *Quaternary Res.* 40:139–49.

Gyalistras, D. 1994. *Case Studies in Bioclimatic Scenario Derivation.* Internal Report, Systems Ecology, Swiss Federal Institute of Technology ETHZ, Zürich.

Gyalistras, D., and A. Fischlin. 1996. *Derivation of Climate Change Scenarios for Mountainous Ecosystems: A GCM-based Method and the Case Study of Valais, Switzerland.* Systems Ecology Report 22, Swiss Federal Institute of Technology ETHZ, Zürich.

Gyalistras, D., and A. Fischlin. 1995. Downscaling: Applications to ecosystems modeling. In *Proceedings of the Sixth International Meeting on Statistical Climatology*, Galway, Ireland, ed. I. Muirterchaigh, 189–92. Galway: Univ. College.

Gyalistras, D., A. Fischlin, and M. Riedo. 1997. Herleitung stündlicher Wetterszenarien unter zukünftigen Klimabedingungen. In *Klimaveränderungen und Grünland—Eine Modellstudie über die Auswirkungen zukünftiger Klimaveränderungen auf das Dauergrünland in der Schweiz*, ed. J. Fuhrer, 207–63. Zürich: Verlag der Fachvereine.

Gyalistras, D., H. von Storch, A. Fischlin, and M. Beniston. 1994. Linking GCM-simulated climatic changes to ecosystem models: Case studies of statistical downscaling in the Alps. *Clim. Res.* 4:167–89.

Hansen, J., A. Lacis, R. Ruedy, and M. Sato. 1992. Potential climate impact of Mount Pinatubo eruption. *Geophys. Res. Lett.* 19:215–8.

Hay, L. E., G. J. McCabe Jr., D. M. Wolock, and M. A. Ayers. 1992. Use of weather types to disaggregate General Circulation Model predictions. *J. Geophys. Res. D* 97:2781–90.

Hegerl, G. C., H. von Storch, K. Hasselmann, B. D. Santer, U. Cubasch, and P. D. Jones. 1996. Detecting greenhouse gas induced climate change with an optimal fingerprint method. *J. Clim.* 9:2281–2306.

Heimann, D. 1993. *REWIH3D-1.0 Modell und Programmbeschreibung.* Forschungsbericht DLR 3, Deutsche Forschungsanstalt für Luft- und Raumfahrt, Oberpfaffenhofen, Germany.

Henderson-Sellers, A., and K. McGuffie. 1987. *A Climate Modelling Primer.* Chichesterm, U.K.: John Wiley & Sons.

Hewitson, B. 1994. Regional climates in the GISS General Circulation Model: Air surface temperature. *J. Clim.* 7:283–303.

Hewitson, B., and R. G. Crane. 1992. Regional climates in the GISS GCM: Synoptic-scale circulation. *J. Clim.* 5:1002–11.

Houghton, J. T., G. J. Jenkins, and J. J. Ephraums, eds. 1990. *Climate Change: The IPCC Scientific Assessment.* Cambridge: Cambridge University Press.

Houghton, J. T., L. G. Meira Filho, J. Bruce, H. Lee, B. A. Callander, E. Haites, N. Harris, and K. Maskell. 1994. *Climate Change 1994: Radiative Forcing of Climate Change and an Evaluation of the IPCC 1992 IS92 Emission Scenarios*. Cambridge: Cambridge University Press.

Hughes, J. P., and P. Guttorp. 1994. A class of stochastic models for relating synoptic atmospheric patterns to regional phenomena. *Water Resour. Res.* 30:1535–46.

Hulme, M. 1994. Validation of large-scale precipitation fields in General Circulation Models. In *Global Precipitations and Climate Change*, ed. M. Desbois and F. Desalmand, 387–405. Berlin (etc.): Springer.

Hulme, M., K. R. Briffa, P. D. Jones, and C. A. Senior. 1993. Validation of GCM control simulations using indices of daily airflow types over the British Isles. *Clim. Dyn.* 9:95–105.

Ishida, T., and S. Kawashima. 1993. Use of cokriging to estimate surface air temperature from elevation. *Theor. Appl. Climatol.* 47:147–57.

Jones, P. D. 1988. Hemispheric surface air temperature variation. *J. Clim.* 1:654–60.

Jones, P. D., and K. R. Briffa. 1992. Global surface air temperature variations during the twentieth century. Part 1: Spatial, temporal and seasonal details. *The Holocene* 2:165–79.

Jones, P. D., and P. M. Kelly. 1983. The spatial and temporal characteristics of Northern Hemisphere surface air temperature variations. *J. Climatol.* 3:243–52.

Jones, R. G., J. M. Murphy, and M. Noguer. 1995. Simulation of climate change over Europe using a nested regional climate model. I: Assessment of control climate, including sensitivity to location of lateral boundaries. *Quart. J. Roy. Met. Soc.* 121:1413–49.

Jones, R. G., J. M. Murphy, M. Noguer, and A. B. Keen 1997. Simulation of climate change over Europe using a nested regional climate model. II: Comparison of driving and regional model responses to a doubling of carbon dioxide concentration. *Quart. J. Roy. Met. Soc.* 123:265–292.

Karl, T. R., W.-C. Wang, M. E. Schlesinger, R. W. Knight, and D. Portman. 1990. A method relating General Circulation Model simulated climate to the observed local climate. Part I: Seasonal statistics. *J. Clim.* 3:1053–79.

Kattenberg, A., F. Giorgi, H. Grassl, G. A. Meehl, J. F. B. Mitchell, R. J. Stouffer, T. Tokioka, A. J. Weaver, and T. M. L. Wigley. 1996. Climate models: Projections of future climate. In *Climate Change 1995: The Science of Climate Change. Contribution of Working Group I to the Second Assessment Report of the Intergovernmental Panel on Climate Change*, ed. J. T. Houghton, L. G. Meira Filho, B. A. Callander, N. Harris, A. Kattenberg, and K. Maskell, 289–357. Cambridge: Cambridge University Press.

Katz, R. W., and B. G. Brown. 1992. Extreme events in a changing climate: Variability is more important than averages. *Clim. Change* 21:289–302.

Kim, J. W., J. T. Chang, N. L. Baker, D. S. Wilks, and W. L. Gates. 1984. The statistical problem of climate inversion: Determination of the relationship between local and large-scale climate. *Mon. Wea. Rev.* 112:2069–77.

Kirchhofer, W. 1974. *Classification of European 500 mb Patterns*. Arbeitsbericht SMA 43, Schweizerische Meteorol. Zentralanstalt, Zürich.

Lamb, P. J. 1987. On the development of regional climatic scenarios for policy-orientated climatic-impact assessment. *Bull. Amer. Met. Soc.* 68:1116–23.

Latif, M., T. P. Barnett, M. A. Cane, M. Flügel, N. E. Graham, H. von Storch, J.-S. Xu, and S. E. Zebiak. 1994. A review of ENSO prediction studies. *Clim. Dyn.* 9:167–79.

Legget, J. A., W. J. Pepper, and R. J. Swart. 1992. Emissions scenarios for the IPCC: An update. In *Climate Change 1992: The Supplementary Report to the IPCC Scientific Assessment*, ed. J. T. Houghton, B. A. Callendar, and S. K. Varney, 69–95. Cambridge: Cambridge University Press.

Lettenmaier, D. 1995. Stochastic modeling of precipitation with applications to climate model downscaling. In *Analysis of Climate Variability: Applications of Statistical Techniques*, ed. H. von Storch and A. Navarra, 197–212. Berlin: Springer.

Lough, J. M., T. M. L. Wigley, and J. P. Palutikof. 1983. Climate and climate impact scenarios for Europe in a warmer world. *J. Clim. and Appl. Met.* 22.1673–84.

Lüthi, D., A. Cress, H. C. Davies, C. Frei, and C. Schär. 1996. Interannual variability and regional climate simulations. *Theor. Appl. Climatol.* 53:185–209.

Majewski, D. 1991. The Europa-Modell of the Deutscher Wetterdienst. In *ECMWF Proceedings, "Numerical Methods in Atmospheric Models,"* 147–91. Reading, UK: ECMWF (European Centre for Medium-Range Weather Forecasts).

Manabe, S., and R. J. Stouffer. 1994. Multiple-century response of a coupled ocean-atmosphere model to an increase of atmospheric carbon dioxide. *J. Clim.* 7:5–23.

Marinucci, M. R. and F. Giorgi. 1992. A $2 \times CO_2$ climate change scenario over Europe generated using a limited area model nested in a General Circulation Model. Part I: Present-day seasonal climate simulation. *J. Geophys. Res. D* 97:9989–10009.

Marinucci, M. R., F. Giorgi, M. Beniston, M. Wild, P. Tschuck and A. Bernasconi. 1995. High resolution simulations of January and July climate over the western Alpine region with a nested regional modeling system. *Theor. Appl. Climatol.* 51:119–38.

Marotzke, J., and J. Willebrand. 1991. Multiple equilibria of the global thermohaline cicrculation. *J. Phys. Oceanogr.* 21:1372–85.

Matyasovszky, I., and I. Bogardi. 1994. Comparison of two general circulation models to downscale temperature and precipitation under climate change. *Water Resour. Res.* 30:3437–48.

McGinnis, D. L., 1995. Downscaling techniques for snowfall prediction in global change studies. In *Proceedings of the Sixth International Meeting on Statistical Climatology*, Galway, Ireland, ed. I. Muirterchaigh, 335–8. Galway: Univ. College.

Mearns, L. O., R. W. Katz, and S. S. Schneider. 1984. Extreme high-temperature events: Changes in their probabilities with changes in mean temperature. *J. Clim. Appl. Meteorol.* 23:1601–13.

Mitchell, J. F. B., R. A. Davis, W. J. Ingram, and C. A. Senior. 1995. On surface temperature, greenhouse gases, and aerosols: Models and observations. *J. Clim.* 8:2364–86.

Mitchell, J. F. B., T. C. Johns, J. M. Gregory, and S. F. B. Tett. 1995. Climate responses to increasing levels of greenhouse gases and sulphate aerosols. *Nature* 376:501–4.

Mitchell, J. F. B., S. Manabe, V. Meleshko, and T. Tokioka. 1990. Equilibrium climate change—and its implications for the future. In *Climate Change—The IPCC Scientific Assessment*, Report prepared for IPCC by Working Group 1, ed. J. T. Houghton, G. J. Jenkins, and J. J. Ephraums, 139–73. Cambridge: Cambridge University Press.

North, G. R., R. F. Cahalan, and J. A. Coakley. 1981. Energy balance climate models. *Rev. Geophys. Space Phys.* 19:91–121.

Oglesby, R. J. 1991. Joining Australia to Antarctica: GCM implications for the Cenozoic record of Antarctic glaciation. *Clim. Dyn.* 6:13–22.

Palutikof, J. P., T. M. L. Wigley, and J. M. Lough. 1984. *Scenarios for Europe and North America in a High CO_2, Warmer world*. Technical Report, Carbon Dioxide Res. Div. TR012, United States Department of Energy, Washington, DC:

Perry, A. 1983. Growth points in synoptic climatology. *Prog. Phys. Geogr.* 7:90–6.

Phillips, D. L., J. Dolph, and D. Marks. 1992. A comparison of geostatistical procedures for spatial analysis of precipitation in mountainous terrain. *Agric. For. Meteorol.* 58:119–41.

Phillipps, P. J., and I. M. Held. 1994. The response to orbital perturbations in an atmospheric model coupled to a slab ocean. *J. Clim.* 7:767–82.

Pittock, A. B. 1993. Climate scenario development. In *Modelling change in environmental systems*, ed. A. J. Jakeman, M. B. Beck, and M. J. McAleer, 481–503. Chichester, U.K.: Wiley.

Podzun, R., A. Cress, D. Majewski, and V. Renner. 1995. Simulation of European climate with a limited area model. Part II: AGCM boundary conditions. *Contrib. Atmosph. Phys.* 68:179–92.

Preisendorfer, R. W. 1988. *Principal Component Analysis in Meteorology and Oceanography.* Amsterdam: Elsevier.

Richardson, C. W. 1981. Stochastic simulation of daily precipitation, temperature, and solar radiation. *Water Resour. Res.* 17:182–90.

Riedo, M., D. Gyalistras, A. Grub, M. Rosset, and J. Fuhrer. 1997. Modelling grassland responses to climate change and elerated CO_2 *Acta Oecol. Appl.* 18: 305–11. (in press).

Risbey, J. S. and P. H. Stone. 1996. A case study of the adequacy of GCM simulations for input to regional climate change assessments. *J. Climate* 9:1441–67.

Robinson, P. J. and Finkelstein P. L., 1991. The development of impact-oriented climate scenarios. *Bull. Amer. Meteor. Soc.* 72:481–90.

Robock, A., R. P. Turco, M. A. Harwell, T. P. Ackerman, R. Andressen, H. S. Chang, and M. V. K. Sivakumar. 1993. Use of General Circulation Model output in the creation of climate change scenarios for impact analysis. *Clim. Change* 23:293–335.

Roeckner, E., K. Arpe, L. Bengtsson, S. Brinkop, L. Dümenil, M. Esch, E. Kirk, F. Lunkeit, M. Ponater, B. Rockel, R. Sausen, U. Schlese, S. Schubert, and M. Windelband. 1992. *Simulation of the Present-Day Climate with the ECHAM Model: Impact of Model Physics and Resolution.* MPI Report 93, Max-Planck-Institut für Meteorologie, Hamburg.

Rotach, M. W., M. R. Marinncci, M. Wild, P. Tschnck, A. Ohmura, and M. Beniston. 1997. Nested regional simulation of climate change over the Alps for the scenario of a doubled greenhouse forcing. *Theor. Appl. Climatol.* 57:209–227.

Running, S. W., R. R. Nemani, and R. D. Hungerford. 1987. Extrapolation of synoptic meteorological data in mountainous terrain and its use for simulating forest evapotranspiration and photosynthesis. *Can. J. For. Res.* 17:472–83.

Santer, B. D., K. E. Taylor, T. M. L. Wigley, T. C. Johns, P. D. Jones, D. J. Karoly, J. F. B. Mitchell, A. H. Oort, J. E. Penner, V. Ramaswamy, M. D. Schwarzkopf, R. J. Stouffer, and S. Tett. 1996. A search for human influences on the thermal structure of the atmosphere. *Nature* 382:39–46.

Santer, B. D., T. M. L. Wigley, M. E. Schlesinger, and J. F. B. Mitchell. 1990. *Developing Climate Scenarios from Equilibrium GCM-Results.* MPI Report 47, Max-Planck-Institut für Meteorologie, Hamburg.

Sausen, R., K. Barthel, and K. Hasselmann, 1988. Coupled ocean-atmosphere models with flux corrections. *Clim. Dyn.* 2:154–63.

Schär, C., C. Frei, D. Lüthi, and H. C. Davies. 1996. Surrogate climate-change scenarios for regional climate models. *Geophys. Res. Lett.* 23:669–72.

Schlesinger, M. E., and J. F. B. Mitchell. 1985. *Model Projections of Equilibrium Climatic Response to Increased CO_2-Concentration.* Department of Atmospheric Sciences and Climatic Research Institute, Oregon State University, Corvallis.

Schönwiese, C.-D., J. Rapp, T. Fuchs, and M. Denhard. 1994. Observed climate trends in Europe 1891–1990. *Meteorol. Z.* 3:22–8.

Schüepp, M. 1968. *Kalender der Wetter- und Wittterungslagen im zentralen Alpengebiet.* Veröffentl. Schweiz. Meteorol. Zentralanstalt 11, Schweiz. Meteorol. Zentralanstalt, Zürich, Switzerland.

Schüepp, M., & F. Fliri. 1967. *Witterungsklimatologie.* In K. Schram and J. C. Thams, eds., Abhandlungen der 9 Internationalen Tagung für Alpine Meteorologie, Zürich, Switzerland, Schweizerische MeteorologischeAnstalt, 369.

Senior, C. A., and J. F. B. Mitchell. 1993. Carbon dioxide and climate: The impact of cloud parameterization. *J. Clim.* 6:393–418.

Shine, K., R. G. Derwent, D. J. Wuebbles, and J.-J. Morcrette. 1990. *Radiative Forcing of Climate.* In *Climate Change—The IPCC scientific assessment,* ed. J. T. Houghton, G. J. Jenkins, and J. J. Ephraums, 45–73. Report prepared for IPCC by Working Group 1. Cambridge: Cambridge University Press.

Stern, P. C., O. R. Young, and D. Druckman, eds. 1992. *Global Environmental Change: Understanding the Human Dimensions.* Washington DC: National Academy Press.

Stone, P. H., and J. S. Risbey. 1990. On the limitations of general circulation climate models. *Geophys. Res. Lett.* 17:2173–6.

Tahvonen, O., H. von Storch, and J. von Storch. 1994. Economic efficiency of CO_2 reduction programs. *Clim. Res.* 4:127–41.

Taylor, K. E., and J. E. Penner. 1994. Response of the climate system to atmospheric aerosols and greenhouse gases. *Nature* 369:734–7.

Viner, E., and P. Hulme. 1994. *The Climate Impacts LINK Project: Providing Climate Change Scenarios for Impacts Assessment in the UK.* Report prepared for the UK Department of the Environment Climatic Research Unit, Norwich.

Viner, E., and P. Hulme. 1993. *Construction of Climate Change Scenarios by Linking GCM and STUGE2 Output.* Climate Impacts LINK Project, Technical Note 2. Norwich, U.K.: University of East Anglia.

von Storch, H. 1995. Inconsistencies at the interface of climate impact studies and global climate research. *Meteorol. Z.* 4:72–80.

von Storch, H., E. Zorita, and U. Cubasch. 1993. Downscaling of global climate change estimates to regional scales: An applicaton to Iberian rainfall in wintertime. *J. Clim.* 6:1161–71.

Wanner, H. 1980. Grundzüge der Zirkulation der mittleren Breiten und ihre Bedeutung für die Wetterlagenanalyse im Alpenraum. In *Das Klima—Analysen und Modelle, Geschichte und Zukunft,* 117–24. Berlin: Springer.

Washington, W. M., and G. A. Meehl. 1984. Seasonal cycle experiment on the climate sensitivity due to a doubling of CO_2 with an atmospheric general circulaton model coupled to a simple mixed layer ocean model. *J. Geophys. Res. D* 89:9475–503.

Wigley, T. M. L., P. D. Jones, K. R. Briffa, and G. Smith. 1990. Obtaining sub-grid scale information from coarse-resolution general circulation model output. *J. Geophys. Res. D* 95:1943–53.

Wilby, R. L. 1994. Stochastic weather type simulation for regional climate change impact assessment. *Water Resour. Res.* 30:3395–403.

Wild, M., A. Ohmura, H. Gilgen, and E. Roeckner. 1995. Validation of GCM simulated radiative fluxes using surface observations. *J. Clim.* 8:1309–24.

Wilks, D. S. 1992. Adapting stochastic weather generation algorithms for climate change studies. *Clim. Change* 22:67–84.

Wilks, D. S. 1989. Statistical specification of local surface weather elements from large-scale information. *Theor. Appl. Climatol.* 40:119–34.

Woo, M.-K. 1992. Application of stochastic simulation to climatic-change studies. *Clim. Change* 20:313–30.

Yarnal, B. 1993. *Synoptic Climatology in Environmental Analysis: A Primer*. London: Belhaven Press.

Zorita, E., J. P. Hughes, D. P. Lettemaier, and H. von Storch. 1995. Stochastic characterization of regional circulation patterns for climate model diagnosis and estimation of local precipitation. *J. Clim.* 8:1023–42.

5

Sensitivity of Plant and Soil Ecosystems of the Alps to Climate Change

Jean-Paul Theurillat, François Felber, Patricia Geissler,
Jean-Michel Gobat, Marlyse Fierz, Andreas Fischlin,
Philippe Küpfer, André Schlüssel, Caterina Velluti,
Gui-Fang Zhao, and Jann Williams

5.1 CLIMATE AND VEGETATION

Climate is a major determinant of the distribution of ecosystems on earth (e.g., Woodward 1987), as the latitudinal zonation of the vegetation biomes, such as evergreen rain forest, boreal forest, and tundra, as belts or zonobiomes from the equator to the poles illustrates. Climate fundamentally influences the distribution of both plant species and vegetation (e.g., Walter 1984, 1985; Walter and Breckle 1991; Ellenberg 1986, 1988, 1996). In high mountains, climate influences not only ecosystems' latitudinal distribution but also their altitudinal distribution. Indeed, altitudinal zonation characterizes the vegetation of all high mountains that form separate biomes, called orobiomes, that occur within the zonobiomes. Thus, an orobiome consists of all the vegetation zonation from the bottom to the top of the mountain range. The gradual decrease of mean air temperature with increasing altitude (0.55 K per 100 meters' elevation at middle latitudes) is about 1,000 times steeper in Europe and northern Asia (about 1 K per 200 kilometers) than the latitudinal decrease toward the poles (Walter 1984, 1985). This is half the gradient predicted for North America by Hopkin's law (MacArthur 1972). Hence, vegetation belts of orobiomes are 500 to 1,000 times narrower than the vegetation zones of zonobiomes, and a modest change in mean annual air temperature may significantly affect the altitudinal distribution of plants and vegetation (e.g., Halpin 1994a, 1994b; Ozenda and Borel 1991, 1994). Projections for global warming by the latter half of the next century should modify biomes' present ranges. According to Halpin (1994a, 1994b), high mountain systems such as the European Alps are likely to be particularly vulnerable to climate change, and this may have severe biological and economic consequences.

Considering the biological and socioeconomic importance of both the forests and the alpine zone of the Alps, several questions should be addressed concerning global warming's potential impacts:

- To what extent is climate influencing these ecosystems?

- In what ways are these ecosystems likely to respond to climate change?

• How may different components of these systems respond?

• What will be the rate of change and the limits of response of these ecosystems and their different components?

• What will be the mechanisms of response?

• How can we best assess the impacts of climate change on forests and alpine ecosystems?

It is an exceedingly complex task to assess climate change's potential impacts on ecosystems. More fundamental research is required before a reliable evaluation of the major trends can be made. Still, the defining features of the subalpine-alpine ecocline (i.e., relatively unmodified, marked visual difference between the upper subalpine forests and alpine meadows) make it a useful system for studying climatic change's potential impact on mountain ecosystems. This raises a further question: Is it possible to find markers in the biosphere and the pedosphere to monitor the evolution of climate change's impact on ecosystems?

A comprehensive review of the sensitivity to climate change of the diverse range of ecosystems in the Alpine region is beyond the scope of this chapter. For this reason, we focus mainly on the potential response of soil ecosystems and the alpine zone to changed temperature and precipitation regimes predicted for the coming century. Other important aspects of global change (enhancement of atmospheric CO_2 concentration, eutrophication, acidification, increase in UV-B, changes in land use) will be mentioned only where they interact synergistically with climatic factors.

Here we analyze the potential response of plants of the alpine zone to climate change, using published literature and studies in progress at two sites in the Alps of the Valais, where soils and vegetation (bryophytes, vascular plants) are being investigated at several levels of complexity. These studies are referred to as the "Ecocline project." Figure 5.1 renders the organizational levels and interactions considered in the Ecocline project.

In this chapter, we

1. briefly review the biodiversity of bryophytes and vascular plants in the European Alps

2. outline the main ecological and biological factors prevailing for both soils and vegetation in the subalpine and alpine zones of the Alps, emphasizing the importance of climatic factors

3. present an integrated assessment, based on examples from the past, of how alpine ecosystems may respond differentially to climate change and what the consequences of this may be for biodiversity.

Exploration of these points is coupled with a summary of key assessments of vegetation and soil ecosystems' response to climate change. The chapter concludes by proposing an integrated vision about the paradigm of species' individual response.

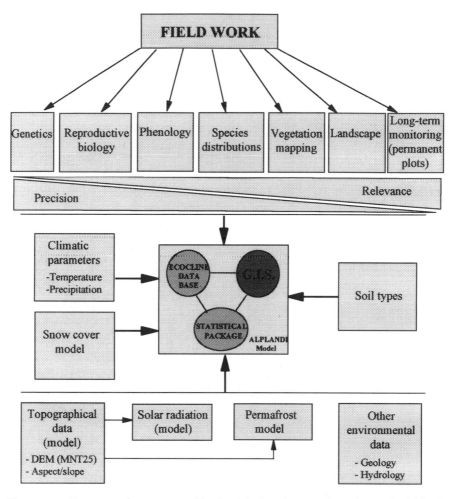

Figure 5.1 Domains and organizational levels of the biosphere in relationship to the fields of study, methods of investigation, and timescale. Relevance increases with both spatial scaling-up and time. Precision increases in the opposite direction.

5.2 THE DISTRIBUTION OF BRYOPHYTES AND VASCULAR PLANTS IN THE ALPS

5.2.1 Vegetation Belts, Forests, and Alpine Ecosystems of the Alps

Changes in elevation result in the formation of more or less homogeneous sections of vegetation extending over 700 meters in elevation called *vegetation belts* (figure 5.2). Given the regular temperature decrease with elevation, a vegetation belt corresponds to a range in mean annual temperature of around 3.8 K. According to Schröter (1923), 100 meters of elevation shortens the growing season by nine days, on average, annually (six in spring and three in autumn).

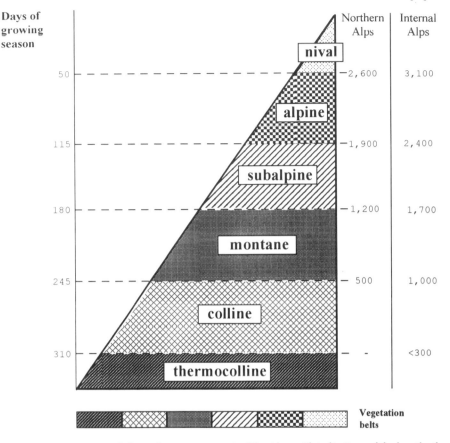

Figure 5.2 Vegetation belts in the temperate part of the Alps, with indications of the length of the growing season (days with mean daily air temperature of at least 5°C).

The highest belt where forest can potentially still grow is termed the *sub-alpine vegetation belt*. Generally, this belt's lower limit in the European Alps is between 1,200 and 1,700 m a.s.l., and its upper limit is between 1,900 and 2,400 m (Landolt and Aeschimann 1986; Landolt 1992; Favarger 1995b). Formations of phanerophytes (woody plants taller than 50 centimeters—trees and shrubs) characterize this vegetation belt, which supports boreal coniferous forests of species such as Norway spruce (*Picea abies* (L.) Karst.), larch (*Larix decidua* Miller), Arolla pine (*Pinus cembra* L.) and mountain pine (*Pinus mugo* subsp. *uncinata* (DC.) Domin). The subalpine deciduous forests, particularly beechwoods, grow in the lower part of this vegetation belt on the external border of the Alps and are subject to a more Atlantic climate (Pache, Michalet, and Aimé 1996). The subalpine vegetation belt's mean annual temperature is between 0.5°C and 4°C, and the vegetation growth period extends over 110 to 180 days (Ozenda 1985; Theurillat 1991).

The *alpine vegetation belt* begins at around 1,900 to 2,400 m a.s.l. and extends up to 2,600 to 3,100 m. Formations of swards and low heaths, the absence of phanerophytes, a decrease in abundance of annual plants (therophytes) and of bulbar plants (geophytes), and the presence of cushion plants characterize this belt (e.g., Landolt and Aeschimann 1986; Landolt 1992; Ozenda 1985; Favarger 1989, 1995a). All alpine plant communities can be considered "specialized" (i.e., limited to particular habitats; see section 5.4.3) with geomorphology, microrelief, and site exposure strongly determining their distribution. These factors determine the microclimate and the related length of snow cover (Thornwaite 1954).

Above 2,600–3,100 meters is found the *nival belt*, where the growing season is less than 50 days. At these elevations, snow remains on flat ground throughout the summer. These environmental conditions preclude the development of closed plant communities such as swards. Vascular plants of the alpine belt form loose plant communities in screes and rock fissures in the lower part of the nival belt. Higher up, these plants grow only sporadically in favorable, rocky places.

5.2.1.1 Forests The forests of the European Alps constitute a major ecosystem and fulfill several important functions. They regulate runoff, thereby moderating the hydrological regime and preventing erosion of steep slopes (Brooks et al. 1991). They protect human settlements from avalanches and landslides. They provide fuel, pulpwood, and timber for human use and, in conjunction with alpine meadows, are the dominant component of an attractive landscape for recreational activities. In addition to their economic roles, Alpine forests are among the most natural remaining ecosystems in the Alps, other than alpine ecosystems, and thus they play an important role in biodiversity conservation. On a larger scale, understanding Alpine forests' climate sensitivity may provide important insights into the potential responses of mountainous forest ecosystems throughout the world.

Forests and the atmosphere form an intertwined system in which temperature, precipitation, and winds influence the structure of forests, and forests in turn feed back to the climatic system via changes in albedo, surface roughness, and evapotranspiration (e.g., Woodward 1987; Running and Nemani 1991). On a regional scale, studies have shown the potential of these feedbacks to affect regional climate (e.g., Lean and Warrilow 1989; Pielke et al. 1994).

In the context of climate change, forests are important terrestrial ecosystems because they can fix, through photosynthesis, and store a significant amount of carbon. For instance, on a global scale, forest ecosystems store 80 percent of the above-ground and 40 percent of the below-ground nonfossil carbon, which amounts to 62–78 percent of all carbon stored in the terrestrial biosphere and 61–77 percent percent of all carbon in the world's biomass (e.g., Dixon and Turner 1991; Wisniewski et al. 1993; Dixon et al. 1994; Perruchoud and Fischlin 1995). Therefore significant changes in the extent of the world's forests could exacerbate changes in climate by releasing large amounts of carbon into the atmosphere (King and Neilson 1992; Neilson 1993; Smith and Shugart 1993; Watson et al. 1996).

5.2.1.2 Alpine and Nival Zones In a phytogeographical sense, the alpine zone, or alpine vegetation belt, is the area between the tree line and the snow line. About one-fourth (1,000–1,100 species) of the Alps' flora is found in the alpine zone, in conjunction with the upper forest zone and the nival zone. These species, called orophytes, can be considered the Alps' "primary diversity" that is, those species native to the Alps as opposed to those (mainly weeds, ruderal, and grassland species) whose natural distribution has been widely extended through human activity. Most orophytes evolved from the surrounding flora of the warm lowlands in the late Tertiary (Pliocene). Their distribution in the alpine zone is more a factor of microclimates generated by microrelief (topoclimates) than of the general climate. The distribution of these orophytes is also considered to be in equilibrium with natural conditions because the alpine belt's upper zone, at least, is considered to have experienced minimal impact from human use. In this respect, the alpine and nival belts are the largest relatively unmodified ecosystem in middle Europe.

The tree line, the contact between the forest and Alpine belt, is a striking bioclimatic boundary (e.g., Tranquillini 1979; Wardle 1983; Holtmeier 1989; Tallis 1991; Slatyer and Noble 1992). Where edaphic and orographic factors abruptly prevent the growth of trees, the treeline can be a sharp boundary (an ecotone); where climate determines the boundary, a transitional zone (an ecocline) of heaths, shrubs, stunted trees, and meadows can occur between the forest limit, or timber line, and the tree limit, or tree line. According to Tallis (1991) and Slatyer and Noble (1992), the tree line is determined by "trees" taller than two meters (i.e., microphanerophytes; see also sections 5.2.1, 5.4.6)

This subalpine-alpine ecocline is called the "kampfzone" or the "krummholz zone." Its distribution fluctuates in response to local factors such as topography and site exposure. The kampfzone now rarely exists under natural conditions in the Alps because human activities have lowered the upper limit of forest almost everywhere, thus broadening the subalpine-alpine ecocline (see also section 5.4.6).

5.2.2 Bryophytes

5.2.2.1 Characteristics Mosses, liverworts, and hornworts are collectively known as bryophytes, the second largest group of land plants, comprising some 23,000 species worldwide. They are poikilohydric plants, that is, plants whose water status completely depends on that of the aerial environment (Walter 1931, 1985; Walter and Breckle 1985, 1991). Such plants tolerate drought by reducing photosynthesis and respiration and withstanding complete tissue dehydration. In contrast to mosses, the homeohydric vascular plants have developed morphological adaptations and internal conducting systems that enable them to maintain relatively constant water contents independent of their environment. Poikilohydry provides the physiological ability to survive, to the point of desiccation, a high variation in water content. In dry conditions, bryophytes strongly resist frost and heat (Longton 1980). Growth is restricted to periods only when the moss is wet. Some bryophytes can photosynthesize at very low internal water contents (Proctor 1982), an important ability for competition, particularly with changing growth conditions.

Bryophytes live in microenvironments whose conditions may differ from those of the surrounding vegetation, determined by (macro)climate, and thus they show characteristic substrate specificity. Bryophytes can colonize bare soil after landslides or other disturbances (Slack 1988). In other situations, they constitute the edaphic climax (see section 5.4.3) on rocks, in crevices, and in screes, fens, and streams. Various bryophyte communities are present as synusiae in plant communities dominated by vascular plants.

In bryophytes, asexual reproduction is often a more important means of propagation than sexual reproduction (During and van Tooren 1987). They have a variety of modes of vegetative propagation. In harsh environments, bryophytes may persist at a vegetative stage for long periods. In dioecious taxa (those with sexes on separate plants), successful sexual reproduction depends on the simultaneous presence of both sexes, which geography may restrict (Longton and Greene 1969a). Genetic variability is especially high in mosses (Wyatt, Stoneburner, and Odrzykosski 1989; Wyatt 1994), as is their phenotypic plasticity (Longton 1974; Busby, Bliss, and Hamilton 1978).

Unfortunately, ecophysiological data are unavailable for alpine bryophytes, although extensive ecophysiological work has been carried out in polar regions. Because arctic (and antarctic) environments may to some extent be comparable to alpine conditions, these results may provide some insights relevant to the European Alps. In addition, the species investigated also occur in central Europe.

The productivity of bryophytes is low, but their aboveground production may be similar to that of vascular plants in arctic areas (Oechel and Sveinbjörnsson 1978). In Canada, Busby, Bliss, and Hamilton (1978) and Robinson, Vitt, and Timoney (1989) have studied bryophytes' response to microclimatic and edaphic gradients, especially water. Bryophytes' role in ecosystems' processes such as nutrient and water interception and retention is still poorly understood, at least for temperate and tropical regions (Longton 1992). Again, more data are available for polar regions (Longton 1984, 1988). Studies on nutrient cycling in the Alaskan taiga (Oechel and Van Cleve 1986; Chapin et al. 1987) showed that the bryophyte layer functions as a selective filter, acquiring some nutrients rapidly and losing them slowly.

Bryophyte tissue decomposes much more slowly than that of vascular plants because it has a high concentration of lignin-like compounds, thus accumulating acidifying organic matter in the soil (e.g., Hobbie 1996).

Herbivorous animals normally do not graze mosses because of their low digestibility.

5.2.2.2 Patterns of Distribution Bryophytes have an ancient origin, with their oldest fossils dating to the upper Devonian. Extant species have been found in Eocene deposits. In Great Britain and North America, fossil alpine species were discovered at sites far from the present distribution range (Miller 1983). For the European Alps, paleoecological data support the presence of bryophytes in postglacial deposits, as in peat profiles, but unfortunately without no further identification than just "mosses." However, we may assume that the present alpine bryophyte vegetation established itself in the early Holocene.

In terms of species, 1,154 mosses and 423 liverworts are known for Europe, including the Azores (Corley and Crundwell 1991; Grolle 1983). More than two-thirds of the European species are found in the Alps, of which about 40 percent can be considered orophytes, based on the Swiss flora, for which 776 mosses and 252 liverwort species are reported (Urmi et al. 1992). The highest-growing moss in the Alps is an indeterminate *Grimmia* Hedw. species at 4,638 meters a.s.l. at the top of the Dufourspitze (Monte Rosa) on the Swiss-Italian border (Vaccari 1911). *Grimmia recurva* Schwägr. (= *G.*, cf. *sessitana* De Not.; E. Maier, personal communication) is the highest-growing known species, at 4,554 meters a.s.l. at the top of Punta Gnifetti (Monte Rosa) in Italys Aosta Valley (Vaccari 1913).

In contrast to the abundance of vascular plants, there are very few endemic bryophytes in the Alps, with only the liverworts *Riccia breidleri* Steph. and *Herbertus sendtneri* (Nees) Lindb. recognized as such. Moss species once considered endemic have since either been discovered on another continent or reduced to synonymy after taxonomic revision. Many rare alpine species are more frequent in arctic regions. In contrast to the lowlands, where the significant human impact has consequently impoverished the diversity, the alpine bryophyte vegetation is still quite rich and well developed, mainly in subalpine forest and heath communities and in humid vegetation types such as fens, springs, and snowbeds, where the bryophyte layer may show considerable productivity and diversity.

The great majority of bryophytes have a large distribution range, circumboreal for the majority of alpine species. Throughout the Alps, each species' ecological requirements varies little compared to those of vascular plants. The results of bryoecological research are therefore applicable to large geographical areas.

In the Alps, some genera of bryophytes are highly diversified and also possess great phenotypic plasticity. Ecological modifications that reflect unfavorable growth conditions may hamper taxonomic evaluation, yet they may be useful for environmental monitoring. For example, Frisvoll (1988) has recently revised the complex of *Racomitrium heterostichum* (Hedw.) Brid., considered to date one species in central Europe, for the Northern Hemisphere. At least five well-delimited taxa, each with a particular ecology, are reported for the Alps. A similar program has been set up for the genus *Grimmia* Hedw, whose taxonomy in the Alps was poorly understood until recently. As knowledge on habitat requirements has improved, it has been suggested that some species might be used as indicator species (Maier and Geissler 1995).

5.2.3 Vascular Plants

5.2.3.1 Characteristics Orophytes (see section 5.2.1.2) are specifically adapted to cold conditions through physiological and morphological characteristics (e.g., Atkin, Botman, and Lambers 1996; Friend and Woodward 1990; Körner 1989, 1991, 1992, 1994, 1995; Körner and Diemer 1987; Körner et al. 1989; Körner and Larcher 1988; Larcher 1980, 1994). These species have developed several mechanisms of resistance to frost. They assimilate efficiently at temperatures near 0°C, despite the lower partial pressure of CO_2, exhibit a high mitochondrial respiration (dark respiration) rate in comparison to plants from the lowlands (see also section 5.4.2.1), and accumulate large amounts of soluble sugars. They also have a high nitrogen content and abundant secondary compounds. For example, they possess a high level of antioxidants, such as ascorbic acid, as an adaptation to the intense irradiation that may generate oxidative substances (Wildi and Lütz 1996). Plants in the alpine belt also exhibit morphological and anatomical adaptations to low temperatures and hereditary features such as thick leaves, small size, and slow growth. The tillers of the important alpine sedge *Carex curvula* All., for example, may grow on the order of 0.5–1 millimeter per year (Grabherr, Mähr, and Reisigl 1978). However, this occurs in conjunction with daytime canopy temperatures that may reach 20–30°C because of strong radiation. Alpine plants' inherently slow growth is related to a small leaf area per leaf biomass, a limited number of dividing cells at the growing points, and low night temperatures' inhibiting mitotic division and therefore cell growth. Based on the environmental conditions at the alpine belt, plant morphology, and plant longevity, orophytes are considered "stress tolerators" according to the "competitive—stress tolerant—ruderal" scheme Grime (1979) developed. However, closer observations in the high-alpine and nival belts indicate that orophytes do not completely fit the concept of stress tolerators, because they also show competitive and ruderal traits, such as supporting considerable herbivory by small mammals (Diemer 1996).

5.2.3.2 Patterns of Distribution The European Alps support a high level of plant biodiversity in a relatively small area, with about 4,500 vascular plant species, more than a third of the entire European flora (west of the Urals). Favarger (1972) identified 1,049 orophytic taxa in the Alps (about 23 percent of the Alps' entire flora). The highest growing vascular plant observed in the Alps is a saxifrage (*Saxifraga biflora* All.) at 4450 meters a.s.l. at the Dom in Valais (Anchisi 1985).

According to Pawlowski (1970), 397 taxa, in the narrow sense, are endemic to the Alps. In addition, there are 55 subendemic taxa (plants having their main distribution in the Alps but extending also to other mountain ranges). Among the endemics and subendemics, more than half are orophytes. Based on Pawlowski's figures, Favarger (1972) mentioned 331 (32 percent) endemics among his 1,049 species of orophytes. However, the status of endemic taxa remains unclear because of the lack of a precise assessment of the Alps' flora as well as taxonomic divergences. In a broader, conservative sense, only 15 percent of the plant species of the Alps may be endemics (Favarger personal communication).

Orophytes' distribution patterns result mainly from the integration of two dominant factors: climate and macrorelief. Glaciation during the Pleistocene played a key role in shaping the present distribution patterns of species according to intrinsic biological characteristics (dispersion capacity, competitive ability, adaptation, and persistance). Species' individualistic response to climatic vicissitudes makes it difficult to use their distribution to define phytogeographical regions in the European Alps, and no agreement exists between authors even for the major divisions (Theurillat et al. 1994). However, some general patterns exist (Ozenda 1985, 1988, 1994; Favarger 1989, 1995a, 1995b; Theurillat 1995).

Four main distribution patterns of orophytes can be distinguished:

1. Many species are distributed through the whole range

2. Many species occur only in the western or eastern part of the Alps, on either side of a line from Bodensee to Lago di Como, or a line from Lake Geneva to Lago Maggiore

3. Many species exhibit a north-south disjunction in their distribution as a result of the absence of limestone in the central part of the Alps

4. Several species, including both those distributed through the entire range of the Alps as well as eastern or western distributions, show wide disjunction or fragmented distributions, mainly due to glacial events.

In addition to these four patterns, several species possess chromosomic races (diploid, polyploid) that show a west-east distribution (phytogeographical pseudovicariance; Favarger 1962).

Glacial expansion had four other effects on the distribution of orophytes, in addition to those on general distribution patterns:

1. the ancestors of the Arcto-Tertiary flora were almost completely extirpated

2. many orophytes were limited to refugia in the northeastern, southeastern, and southwestern margins of the Alps, where they are now distributed; therefore, the main centers of endemism are the Maritime Alps, the southern Alps between Lago di Como and Lago di Garda, and the low eastern Alps

3. some orophytes moved to lowland areas surrounding the Alps, where they have persisted until the present

4. some orophytes moved to neighboring high mountains and the Arctic; according to figures derived from Jerosch (1903) for Switzerland, about 30 percent of orophytes also occur in the Arctic, and more than 50 percent occur in the mountains of central and southern Europe.

5.2.4 Summary

• Altitudinal vegetation belts, each spanning about 700 meters in elevation, can be distinguished in the European Alps. The subalpine belt is the highest potential limit for forest. The alpine belt above it is composed of low heaths and swards, and the nival belt above that of sparse vegetation.

• The subalpine and alpine vegetation belts are mainly composed of plants, called orophytes, that are specially adapted to cold conditions through several mechanisms.

• In the Alps, about 40 percent of bryophytes and about 25 percent of vascular plants are orophytes.

• Bryophytes tolerate almost complete dehydration, in contrast to most vascular plants.

• The microclimate of bryophytes' habitats usually differs significantly from the mesoclimate.

• Most bryophytes are distributed over the entire Alps, whereas vascular plants have a wider array of distribution patterns.

• Four main distribution patterns for vascular plants result from past climate change, orography, dispersal and/or competitive abilities: entire range, eastern-western distribution, north-south, and east-west disjunction.

• The Maritime Alps, the southern Alps, and the low eastern Alps are the main centers of endemism for vascular plants because of their role as refugia during Pleistocene glaciations.

• Only a few bryophytes are endemic to the entire Alps, whereas there is about 15 percent endemism among orophytic vascular plants in the Alps.

5.3 SUBALPINE AND ALPINE SOILS: GENERAL CHARACTERISTICS

Scientific understanding of subalpine and alpine soils in the European Alps is limited. In the Swiss canton of the Valais, for example, where the Ecocline studies are in progress, few studies have been published on the formation and evolution of soils (e.g., Paternoster 1981; Spaltenstein 1984; Keller 1991). Soil formation and evolution in the French Alps are slightly better known (e.g., Bottner 1972; Legros and Cabidoche 1977; Bartoli and Burtin 1979; Robert, Cabidoche, and Berrier 1980; Dambrine 1985), as are those of the eastern Alps.

The classical definitions of soil types distinguish two main categories according to the minerals in the parent rock: those on carbonate rock and those on acidic siliceous rock. This distinction is arbitrary because there are many intermediate types of rock (e.g., silica-rich calcschists). Furthermore, decarbonated loess can cover limestone and give rise to acidic soils. The soil's present physicochemical characteristics are likely to determine its reaction to climate change.

Eight main types of subalpine and alpine soils can be identified, outside of wet zones, based on their degree of evolution (table 5.1).

5.3.1 Time Scales in Soil Development and Evolution

Soils bring together many ecosystem processes, integrating mineral and organic processes; solid, liquid, and gas processes; and biological, physical, and chemical processes (Arnold et al. 1990; Yaalon 1990). Soil may respond more slowly to environmental changes than other elements of the ecosystem, such as, for example, the plant and animal components. In subalpine and alpine climates, the time needed to form and stabilize soil is estimated to be several thousand years. So how could soil development follow rapid climate changes, and how could we use soils to detect these changes?

This question belittles the complexity of soil, which develops through processes operating at differential rates. Only the soil's complete evolution, to the formation a stable soil "type," requires the entire period. Within the soil ecosystem, many processes operate much faster than the evolution of mature soil. Moulds begin to decompose Arolla pine needles internally, for example, only a few weeks after they fall, and podzolized soil layers can form in only a few centuries. These changes take place on a wholly different scale from the millennia required for soil ecosystems to form in situ.

The different timescales at which soil components evolve challenges another common misconception, namely that soils evolve more and more slowly as the climate becomes harsher. Soils in cold regions are assumed to take longer to change over time than those in hot regions, where the vegetation has a high rate of productivity and a rapid biogeochemical turnover. But this assumption neglects soil development's dependence not only on climate, but also on local conditions, including topography, which allows constant rejuvenation through colluvial movement; calcium content, which prevents acidification; and coarse soil texture, which promotes rapid water transit. Under these conditions, alpine soils can evolve rapidly and may help reveal climate changes at an early stage (Labroue and Tosca 1977). Keller (1992, 187) observed: "These ecosystems [at high altitudes], long thought of as little evolved, in fact constitute very active systems, which can be used as models to study the different biological and biochemical factors in the first stages of soil development."

Sensitivity of Plant and Soil Ecosystems of the Alps to Climate Change

Table 5.1 Comparison of eight subalpine and alpine soil types

Reference Baize and Girard 1992	Soil type Duchaufour 1977	Soil type Soil taxonomy 77	Horizonts Baize and Girard 1992	pH range (average)	Dominant processes	Humus forms	Vegetation (examples)
Cryosol	Cryosol	Pergelic Cryorthent	O-J-C	4.0–8.0	Frost	Moder, mor	Drabion hoppeanae Androsacion alpinae
Regosol	Régosol	Lithic Entisol	O-A-C	4.0–8.0	Rock alteration	Moder	Thlaspion rotundifolii Petasition paradoxi
Organosol	Sol lithocalcique	Lithic Cryumbrept	OL-OF-OH-(A)-R	3.5–6.0	Organic matter accumulation	Mor	Piceion abietis
Rankosol	Ranker	Lithic Humitropept	OL-OF-A-R	3.5–5.0	Weak humus-clay aggregation	Moder	Caricion curvulae
Rendisol	Sol humocalcique	Entic Cryumbrept	OL-OF-Aci-R	5.5–7.5	Humus-clay aggregation	Mull, moder	Caricion firmae
Neoluvisol	Sol brun lessivé	Typic Hapludalf	OL-OF-A-E-BT	4.0–5.5	Clay leaching	Mull, moder	Nardion strictae
Alocrisol	Sol brun acide Sol brun ocreux	Typic Dystrochrept	OL-OF-A-S-C	4.0–5.5	Brunification Weak podzolization	Mull Moder	Oxytropido-Elynion Festucion variae
Podzosol	Podzol humo-ferrugineux	Humic Cryorthod	OL-OF-OH-A-E-BPh-BPs-C	3.0–5.0	Cheluviation	Moder, mor	Rhododendro-Vaccinion

5.3.2 Organic Matter—A Key Indicator

Even if each soil type, such as those presented in table 5.1, shows different rates of response to climate change, humus layer A will always be a key indicator of these changes because of its mixed organomineral constitution. In fact, the evolution of organic matter appears to provide an important means of understanding the changes that can affect the soil. Many essential soil properties depend on it: structure, water retention, mineral fertility, strength, and diversity of animal life and microflora. Soil organic matter is complex and a key regulator of global carbon fluxes. It also regulates the nitrogen cycle, in which the organic matter forms the principal store of this important nutritional element (except in the nival zone, where snow pack appears to be the main reservoir for nitrogen; Haselwandter et al. 1983; Körner 1989).

Soil processes involving organic matter, in addition to those concerning water retention, are the most revealing among the possible descriptors of rapid soil change (on the scale of a year to a century). Changes in soil organic matter can also indicate vegetation change, which can occur quickly because of climate change (Almendinger 1990). In contrast, the soil's mineral evolution (weathering of rocks, new formation of silicate minerals, removal of carbonates) is typically slower, even though Dambrine (1985) provided evidence that these changes were happening faster than expected.

Whereas the total carbon content is a good indicator of rapid changes in some young lowland soils (Fierz, Gobat, and Guenat 1995), it does not seem particularly helpful in the subalpine and alpine zones, where its inertial mass is large. Here, secondary mineralization is very weak. Some humus components may be several thousand years old, as Balesdent (1982) showed for the soils of the Haut-Jura (Jura range).

Tracking changes in subalpine and alpine soils may provide a useful means of qualitatively understanding the dynamics of organic matter. This knowledge may be important on a global scale when considering the terrestrial and atmospheric carbon budgets. The "missing sink" (about 0.4–4 Gt C/yr; Gifford 1994) needed to balance these two budgets may well be in soils.

5.3.3 Humification and Podzolization

Two promising processes have emerged from the research within the Ecocline framework: the physicochemical evolution of the litter and its transformation to humus, and the intensity of podzolization.

During humification (see Duchaufour 1983; Stevenson 1982; Schnitzer and Kahn 1989), litter undergoes various transformations between fresh debris and stable humine, the last stage in the evolution of soil organic matter. The three possible humification routes are inheritance (recovery of components present in the vegetation), polycondensation (aggregation of crenic, fulvic, and humic acids), and synthesis by bacteria (secretion of polysaccharides)

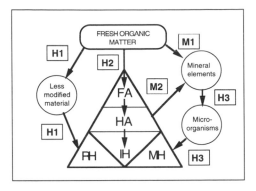

Explanations

A. Transformations
M1 = Primary mineralisation
M2 = Secondary mineralisation
H1 = Inheritance humification
H2 = Polycondensation humification
H3 = Microbial humification

B. Organic matter categories
FA = Fulvic acids
HA = Humic acids
RH = Residual humine
IH = Insolubilisation humine
MH = Microbial humine

Figure 5.3   Dynamics of organic matter in soil. (After Soltner 1992, simplified.)

(figure 5.3). The first stages are particularly interesting because they take place at rates that can be readily measured (Gallet and Lebreton 1995). These phases of humification have a direct functional link with podzolization.

Taking into account the organic material's degree of chemical evolution requires models sensitive to subtle changes in organic carbon. A more refined modeling approach is essential to predict reliably the transfer of carbon among the categories of organic material, each of which corresponds to a certain carbon retention period. Ågren and Wirkström (1993, 343) have demonstrated the difficulties in constructing such models of allocation: "In view of the criteria that should be satisfied by C-allocation models, we conclude that today there are no allocation models that satisfy all requirements," Perruchoud and Fischlin (1995) emphasized the same point. Bryant and Arnold (1994) and Powlson, Smith, and Smith (1996) evaluate some models (SOMM, CANDY, CENTURY, and others) that try to consider the different parts of soil organic matter and to integrate nitrogen.

Podzolization is one of the fastest soil-generating processes (Guillet 1972; Boudot, Bruckert, and Souchier 1981). It occurs only in acid soils and involves organometallic compounds (chelates) migrating from the O or A layers to the spodic layers BP, where they precipitate as organic molecules (BPh) and as metal oxides (BPs). Podzolization also requires a wet climate, so that even short periods of chemical reduction liberate iron. In the Alps, this happens particularly when the snow melts. Therefore, changes in the rate of

Sensitivity of Plant and Soil Ecosystems of the Alps to Climate Change

podzolization may be a useful indicator of environmental change. On steep slopes, however, solifluction may severely slow podzolization by constantly eroding the organic layers.

5.3.4 Summary

• Processes in soils operate at different rates, and many happen on a timescale of one year or tens of years.

• Organic matter, the key factor in soil processes, will be affected by climate change (directly or indirectly, qualitatively and quantitatively), changing main soil processes (humification, podzolization), and the nitrogen cycle.

• Podzolization, a fast soil-generating process on acid soils, is directly linked to the qualitative decomposition of organic matter (humification) in a wet climate.

5.4 ALPINE VEGETATION: THE DRIVING FACTORS

The vegetation of the European Alps is certainly the best known of those of all the high mountain ranges in the world (e.g., Ellenberg 1986, 1988, 1996; Favarger 1995a, 1995b; Mayer 1974; Mucina and Grabherr 1993; Mucina, Grabberr, and Ellmauer 1993; Mucina, Grabherr, and Wallnöfer 1993; Ozenda 1985, 1988). To understand how vegetation may respond to climatic changes, we review the relationships between climate and vegetation. "Vegetation" and "flora" are distinct terms. Flora is the set of all the plant taxa (species, subspecies, varieties) found in a given area. Vegetation is the set of all the plant communities resulting from ecological and competitive combinations between species. Flora and vegetation are closely linked, but two regions with almost the same flora may have different vegetation.

5.4.1 Complexity Levels in Vegetation

Vegetation is a key factor in ecosystem structure. Vegetation's physiognomy, one of its more obvious components, varies according to environmental conditions and human utilization. Vegetation's architecture determines many ecological parameters within a plant community, such as microclimate, energy budget, photosynthesis, water regime, surface runoff, snow gliding, and soil temperature (e.g., Cernusca 1976; Tappeiner and Cernusca 1991, 1994, 1996).

Vegetation can be classified according to its physiognomical characteristics into several types called *formations* (e.g., Ellenberg and Mueller-Dombois 1967; Mueller-Dombois and Ellenberg 1974). Within a formation, distinguishable variations in plant species composition result from the different ecological requirements, competitive abilities, and distribution of each species. Such variations can be used to distinguish plant communities or phytocoenoses within a formation.

A *plant community*, generally considered the fundamental organizational level of vegetation, is a more or less discrete entity. It is a part of an ecosystem (the microecosystem; Ellenberg 1973) and is defined by its plant combination (floristic structure) and ecological characteristics. There are several approaches for the study of formations and plant communities, including physiognomic, floristic, and ecosystemic approaches (e.g., Mueller-Dombois and Ellenberg 1974; Grabherr and Kojima 1993).

Because of their different life forms and ecological requirements, the species in a plant community do not occupy and use the habitat in the same manner. Therefore, a plant community may be divided into ecological compartments called *synusiae* (e.g., Gams 1918; Barkman 1980; Walter 1984, 1985; partial systems of Ellenberg 1973; microcoenoses). Synusiae constitute organizational levels at which the competition among the species within a single synusia is assumed stronger than among the species belonging to different synusiae. Because of their ecological specificities, synusiae may function independently. Hence, synusiae represent functional units for the microhabitats of plant communities, and they can be considered functional groups in the sense of Körner (1993b). The earlier concept of "guild" introduced by Root (1967) may be considered to be the equivalent of a synusia.

Landscape is the highest level of complexity. It integrates all the ecological factors and can provide strong variations in environmental gradients through its heterogeneity or patchiness (fine or coarse) and boundaries. In many ways, landscape can be considered the memory of the past in mountainous regions, preserving records of geological, geomorphological, and biological events, including human use. At the landscape level, plant communities are assembled into *vegetation complexes* (biogeocene complexes: Walter 1984, 1985) of several levels of complexity (Naveh and Liebermann 1984; Theurillat 1992a, 1992b, 1992c) relating to ecological gradients and geomorphology.

5.4.2 Climatic Factors

The vegetation ecosystems of the European Alps will doubtless change in response to changes in climate. However, as discussed in chapters 2 and 4, changes in climate are very likely to be complex, characterized by new combinations of temperature, precipitation, solar radiation, and other factors that will differ from one part of the Alps to another. Because mathematical complexity constrains present scientific understanding of the atmospheric system and many climate processes, it is impossible to predict changes in climate reliably at a spatial and temporal scale relevant to most plants. In the absence of this key knowledge, we limit our discussion of plant responses to climate change to the (in principle) potential effects of changes in temperature, precipitation, and the frequency of extreme events.

5.4.2.1 Temperature Temperature is a complex factor directly correlated to solar radiation, and both are key factors for plant growth. Given that climate change would not alter solar radiation intrinsically, the main effect of an elevation of temperature would be to extend the growing season length by sixteen to seventeen days per degree increase in mean annual air temperature. To some extent, this could enhance plant growth in general, particularly that of alpine species. However, as Körner (1994, 1995) indicated, photoperiod-sensitive alpine species might not benefit much from a longer growing season, as Prock and Körner (1996) showed for *Ranunculus glacialis* L. For other vegetation belts, plants—particularly trees—would be able to grow at higher altitudes because of a longer vegetation period only if all other requirements were fullfilled. For instance, the growing season of mountain pine (*Pinus mugo* subsp. *uncinata* (DC.) Domin) could be extended in autumn if the maximum temperature in August and September remained above a threshold level. Also, cold nights during September would influence bud and cambial initiation and thus influence growth in the following year (Rolland and Schueller 1995).

Elevation of temperature, would also stimulate mitochondrial respiration (dark respiration), an important physiological feature. If no acclimation occurred with the elevation of temperature, this catabolic process could lead to a negative carbon balance and a rapid exhaustion of the plant's carbohydrate reserves, eventually killing it. Larigauderie and Körner (1995) showed that some alpine plants such as *Cerastium uniflorum* Clairv., *Saxifraga muscoides* All., and *Saxifraga biflora* All. might have very low acclimation potentials, this may also be the case for plants from lower altitudes.

5.4.2.2 Precipitation Precipitation and cloudiness affect oceanicity and continentality, which in turn influence plant distribution (e.g., Holten and Carey 1992; Holten 1993; Pache, Michalet, and Aime 1996). For the two parameters, which represent opposite extremes of a common phenomenon, a hygric and a thermic aspect must be distinguished. The hygric aspect is the ratio of precipitation to altitude, and the thermic aspect is the amplitude of the variation among daily or seasonal temperatures. Elevated precipitations at a given altitude correspond to a high hygric oceanicity or to a low hygric continentality; an increase in temperature amplitude corresponds to a low thermic oceanicity and a high thermic continentality.

An increase in precipitation enhances hygric oceanicity. To some extent, this can slightly balance the increase in mean air temperature, because of water's high specific heat. However, if cloud cover also increases, this reduces solar radiation, resulting in a decrease in temperature and a reduction of temperature amplitude, which could lead to a higher thermic oceanicity. At the opposite extreme, a decrease in precipitation could increase solar radiation, with the increase in temperature increasing thermic continentality. A variation in hygric and thermic continentality or oceanity may shift the distribution of dominant tree species, particularly in transitional zones. According to Klötzli (1992, 1994), an increase in precipitation in the Alps to that of a peroceanic climate would enhance the development of cushion-like vegetation on alpine swards, and given simultaneous eutrophication, tall herbs and green alder (*Alnus viridis* (Chaix) DC.) shrubs would show greater development.

Changes in the timing of precipitation and in the variability in the depth of snow cover would affect vegetation and soil processes differentially. For instance, changes in snow cover would modify the contrasts between convexities or steep slopes and concavities or flat places. A decrease in snow cover would affect plant communities sensitive to frost, such as snowbed communities and some heaths. For example, blueberry (*Vaccinium myrtillus* L.) slowly loses its tolerance to cold during the winter through the diminution of soluble sugars as a result of respiration (Ögren 1996). Late frosts in spring could particularly affect dwarf shrubs (i.e., chamaephytes and nanophanerophytes). Snow cover allows microbial activity during winter by insulating the soil surface from the air temperature, maintaining a soil temperature slightly below 0°C. The resulting release of CO and N_2O account for more than 25 percent of the carbon and more than 10 percent of nitrogen fixed in annual aboveground production (Brooks, Schmidt, and Williams 1997). Changes in precipitation would also modify physical processes, especially periglacial phenomena for winter precipitation (see section 5.4.3.1).

5.4.2.3 Extreme Events Knowledge about potential changes in the frequency and duration of extreme events as a result of climate change is even more limited than that for temperature and precipitation. However, frequency and variability of extreme events exert a greater influence on the position of the timberline than do average values (Holtmeier 1994). If the frequency of extreme events like late frosts increases, because of warmer winters and earlier springs, it could slow the upward shift of the vegetation belts (Wigley 1985). In particular, trees are likely to be very sensitive in this respect, because air temperature is the most important factor regulating bud burst (e.g., Hänninen 1991). Moreover, late frosts could lead to a severe loss of needles by conifers, such as Arolla pine and Norway spruce. This could then start a positive feedback loop, with increasing sensitivity to winter cold because of reduced or even no summer growth due to a reduced photosynthetic capacity, even in a warmer climate. In return, this could make the supranival parts more sensitive to winter stress, as observed by Kullman (1996a) in the Swedish Scandes.

An increase in snow cover could increase the number of avalanches, snowslips, and snow creeping. This could counter the potential for subalpine trees and other biota to move upslope into the alpine zone where these phenomena occur, and, in some cases, could even lower the present timberline if these phenomena become more frequent and/or of greater magnitude. Snow accumulation may favor the development of parasitic snow fungi that can severely damage conifers. Stronger winds can also affect the elevation of the timberline, particularly in exposed zones.

Drought may not be a problem to native vegetation in the subalpine and alpine belts, as these ecosystems are not presently under water stress (Körner

1994; Körner et al. 1996). However, in lower elevations, it may affect the vegetation of dry places like the internal valleys (such as Vintschgau, Valais, Aosta, Maurienne, and Haute-Durance) and the Mediterranean part of the western Alps.

5.4.3 Edaphic and Orographic Factors

Where edaphic factors strongly influence the distribution of vegetation, the resulting vegetation forms specialized plant communities termed *edaphic climax communities* (pedobiomes; Walter 1984, 1985). On the opposite end of the spectrum, where climate is the main controlling factor, the resulting plant communities are termed *climatic climax communities*. Examples of edaphic climaxes are wetlands and vegetation on salty soils. When a climatic climax occurs in another climate different from the one where it is usually found because edaphic factors compensate for the general climate, such vegetation is called *extrazonal*. For instance, a plant community normally occurring in a dry climate as its climatic climax can also be found in a wetter climate as an edaphic climax in particular situations such as outcrops or well-drained soils (for example, oak woods with *Quercus pubescens* Willd., climatic climax in the Mediterranean or continental regions, and extrazonal, relict, edaphic climax only in very warm and dry places at low elevation in the suboceanic areas of western central Europe). The vegetation associated with cliffs is another example illustrating a compensation through edaphic factors. There, vegetation may be strongly influenced by geological substrate and slope, factors that influence water availability, and by aspect and slope, which influence temperature. As a consequence, the same type of cliff vegetation can be found in places with climates differing remarkably in their mean temperature and precipitation and at different elevations. Such vegetation is said to be *azonal*. Thus, relative to climate, many edaphic climaxes illustrate the Rubel's (1935) law of the replaceability of ecological factors, which is equivalent to Walter's law of the relative constancy of habitat (Walter and Walter 1953; Walter 1984, 1985).

5.4.3.1 Physical Processes Physical processes can mediate climatic change's impact on vegetation (e.g., Holten and Carey 1992). Their importance increases with elevation, and they typically dominate in the upper alpine belt.

From the alpine belt upward, the slopes become much steeper (figure 5.4). Beyond a slope of 40°, the environment becomes rocky, and downward transportation of altered material through erosion, runoff, snow creeping, solifluction, and landslips becomes prominent. These factors are of major importance in the development of soil. Active screes derived from cliffs may be another environment where factors other than the general climate more strongly influence vegetation. Because of their ability to free the plants growing in these habitats from the general climate, cliffs and screes are con-

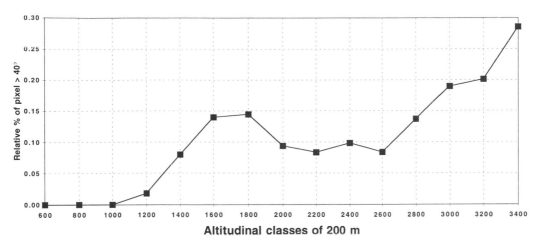

Figure 5.4 The increase in area of slope greater than 40° with elevation in the region of the Val d'Arpette (120 km²) and Belalp (98 km²) (Alps of the Valais). The relative percentage is calculated in both regions with the 25 m Digitized Elevation Model for Switzerland on the number of pixels greater than 40° in altitudinal sections of 200 m (i.e., 400–599 m a.s.l., 600–799 m, ..., 3,000–3,199 m), except for the last section, which comprises altitudes above 3200 m.

sidered to have played a key role in the persistence of the endemic Tertiary flora during the Pleistocene glaciations (e.g., Braun-Blanquet 1948; Rioux and Quézel 1949; Davis 1951; Pawlowski 1970; Snogerup 1971).

Geological factors play a role in the response of vegetation on limestone to climate change. The nature (soft marl vs. hard limestone), alteration processes, position, slope and orientation of the limestone layers may strongly influence the composition and distribution of plant communities (e.g. Béguin 1974; Béguin and Raemy 1995). These factors may greatly affect the altitudinal shift of vegetation zones, as they vary greatly with the sequence and thickness of the geological layers.

Periglacial phenomena, that is, those phenomena governed by frost and thaw in contact with glaciated areas, are important factors affecting vegetation and soils at higher elevations in the European Alps. Permafrost is potentially widely distributed (e.g., Keller 1994). In the upper alpine zone, persistence of permafrost influences the altitudinal shift of the vegetation of the lower alpine zone and the distribution of species. When permafrost melts, frozen slopes in the high elevations may become less stable, and this may generate more screes, stone falls and landslips, which could benefit pioneer vegetation.

Gelifluction (solifluction generated at thaw) and cryogenic processes (originated by frost) are other periglacial factors influencing vegetation that are closely linked to climate (e.g., Carbiener 1966; Veit and Höfner 1993). They occur in the upper subalpine zone. Their intensity and distribution increase in the alpine zone up to the transition to the nival zone. Gelifluction affects sediment transfer. Cryogenic processes (such as cryoturbation, the moving and sorting of material by frost) destroy closed vegetation such as alpine

Sensitivity of Plant and Soil Ecosystems of the Alps to Climate Change

swards. The intensity of gelifluction and cryoturbation depends mainly on snow cover and winter temperature. Mild winters with early snow cover result in shallow frost penetration in soil and little or no gelifluction. Severe winters with late or shallow snow cover result in intense gelifluction and cryogenic processes. Therefore, the shift of the alpine belt into the nival belt through global warming may depend on winter temperature and precipitation, particularly snow cover (depth, regime). Periglacial processes regulated by winter climate could either speed up alpine vegetation's colonization of the nival zone or slow it down.

Brockmann-Jerosch's mountain mass elevation effect (*Massenerhebungs-Effekt*) or Merriam effect (e.g., Brockmann-Jerosch 1919; Ellenberg 1986, 1988, 1996; Flenley 1995; Grubb 1971) is responsible for vegetation's altitudinal distribution. In the highest massifs and large aggregations of mountains, the vegetation belts extend higher than in the lower and isolated massifs. Gams' water continentality index expresses the mass effect (Gams 1931, 1932). Correlated to the mass effect, the elevation effect (van Steenis 1961; Backhuys 1968; Hengeveld 1985; Ellenberg 1986, 1988) corresponds to potential, more or less temporary abyssal and upper distributions outside the normal range, if the massifs are high enough for a permanent zone of establishment. According to Backhuys (1968), species growing in massifs in the Swiss Alps' that are higher than 2,000 meters are distributed, on average, 600–700 meters below this altitude, in part because of the elevation effect. However, although the elevation effect does exist, Backhuys' data cannot prove his explanation statistically (Hengeveld 1985).

5.4.3.2 Effect of Soil Variability in soils may limit vegetation's ability to respond to climate change. For example, a ranker (a poorly evolved, alpine soil on acidic rocks) could rapidly evolve into a podzosol (a well differentiated, evolved soil), progressing first to a cryptopodzolic ranker (or humic podzosol), then to a true podzol. Meadows whose soil undergoes such a transformation are very susceptible to colonization by acidifying heath plants and quickly disappear. On the other hand, a brown or ochre, acid soil (or alocrisol) resists transformation into a podzol longer, even if podzolization is already occurring in a diffuse form (Michalet and Bruckert 1986). Meadows with this type of soil may be less susceptible to colonization by heath vegetation. In addition, soils that formed under a vegetation different from that which they are currently supporting may accelerate recolonization by the initial vegetation, given the favorable pedological conditions. For example, in matgrass (*Nardus stricta* L.) pastures occurring on podzols formed under a heath, recolonization by heath will reinforce the initial soil conditions through a positive feedback through heath's acid litter. At present, however, pedological arguments alone cannot predict whether heath or meadow will prevail under a climatic change. One must also consider functional interactions among temperature, precipitation, carbon, and nitrogen (see section 5.6.5).

Complex scenarios involving the dynamics of soil-vegetation interactions must also consider the different inherent rates of change of the systems concerned. As a rule of thumb, on the phytocoenotic level, one can estimate that the vegetation has ten times less inertia than the soil as a whole. But at the lower organizational levels—synusia and soil layers—the vegetation strata and the pedological layers can have comparable responses.

It is a tenable hypothesis that the vegetation changes observed on limestone soils would essentially reflect climate changes, the internal changes in the soil being relatively insignificant because of their chemical buffering capacity. On acid soils, vegetation changes would occur through both climatic change and changes in soils induced by the vegetation itself.

As described in section 5.6.5.5, changes in soils' nitrogen supply may strongly affect vegetation. In synergy with climate change, this could accelerate or slow the vegetation's response to climate change.

5.4.4 Biological Factors

In addition to climatic and physical factors, biological factors also play a role in the distribution of species and plant communities. Interestingly, some biological factors are correlated with climatic factors, at least with temperature. Biological factors can act either in synergy with climatic and physical factors or against them. Factors not related to temperature, such as precipitation or wind, are not easily or reliably quantified at the relevant level of the microhabitat. Hence, because of their great variability, their effects are difficult to evaluate qualitatively, and they are too complex to model based on current available data.

Biological features such as competition, life strategies, and the ability of plants to adapt could also interact either positively or negatively with climate change. Competition among species limits their physiological possibilities. Therefore, the ecology of a species does not necessarily reflect its physiological optimum (that is, the conditions under which its best development would occur in the absence of competition with other species) or its physiological amplitude, and vice versa. Under new environmental conditions, competitive interactions among species change, possibly changing their ecology.

To some extent, clonal plants in heaths and swards are likely to be able to resist invading species under global warming, owing to characteristics such as their longevity, competitive ability, and rapid response to changed environmental conditions (Callaghan et al. 1992). Closed subalpine heaths like alpenrose (*Rhododendron ferrugineum* L.) heaths are a good example of resistance to trees' and shrubs' colonization. Their architectural characteristics, high biomass accumulation, and high humus accumulation present obstacles to seeds of other species and to their germination (Doche, Pommeyrol, and Peltier 1991; Pornon and Doche 1995b), as in the inhibition model of Connell and Slatyer (1977). Clonal, dominant species of alpine swards may also persist and withstand climate change, as evidenced for *Carex curvula* All. A clone of this sedge in the Swiss Alps was found to be around 2,000 years in age (Steinger, Körner, and Schmid 1996). Long-lived dominant species like trees may also not react immediately and therefore persist. Moreover, the physiological and morphological plasticities, as well as genetic adaptations, will increase ability to persist. Physiological factors, mainly plants' response to an increase in the atmospheric concentration of CO_2, may also modify climatic change's effect. These factors have been discussed by Woodward (1992), Woodward, Thompson, and McKee (1991), Körner (1993a), Schäppi and Körner (1996), and Körner et al. (1996). Körner (1994, 1995) reviewed atmospheric changes' potential ecophysiological effects on alpine vegetation (e.g., thermal acclimation, growth strategies, developmental processes, water relations, mineral nutrition, and carbon dioxide effects).

5.4.5 Phenology: A Link between Climate and Plant Development

The growth and development of plants are controlled by both internal factors (such as plant physiology, plant morphology) and external factors (such as nutrients and water availability, temperature; e.g., Odland 1995). Temperature is a key factor in plant development, influencing the timing of such key life history traits as bud burst, flowering time, growth, and length of vegetative period. "Phenology is generally described as the art of observing life cycle phases or activities of plants and animals in their temporal occurrence throughout the year" (Lieth 1974, 4). Phenological observations allow an assessment of many aspects of the development of plants and ecosystems. Many have published studies on the phenology of vascular plants (see, e.g., Dierschke 1990; Lieth 1974; Orshan 1989; Rathcke and Lacey 1985; Schnelle 1955).

Phenology also holds promise as a tool for assessing climate change's impacts on plant growth and development (e.g., Defila 1991; Diekmann 1996; Fitter et al. 1995; de Groot, Ketner, and Ovaa 1995; Hunter and Lechowicz 1992; Lechowicz and Koike 1995; Kramer 1994; Molau 1993, 1996; Moore 1995). Indeed, the existing correlations between phenophases (distinguishable phases in the life cycle such as flowering, bud burst) and climatic variables permit the use of phenological observations to derive climate data (e.g., Schreiber 1968; Schreiber et al. 1977; Bucher and Jeanneret 1994).

Grimme (1903) demonstrated that the sexual organs of bryophytes in central Germany may mature one to two months earlier than in Scandinavia, where Arnell (1875) studied the same species. Since these early accounts, few phenological observations have been published, and little information is available on bryophytes' reproductive adaptations to alpine environment. Climate affects formation and distribution of sex organs and development of the sporophyte generation (Longton and Schuster 1983), in contrast to vegetative propagation, which occurs throughout the growing season. Some species show distinct seasonal patterns of reproductive development, whereas others are more variable (Miles, Odu, and Longton 1989).

In the Ecocline project, phenological investigations in the subalpine and alpine belts with several species of bryophytes and vascular plants indicate how the decrease of temperature with elevation influences phenophases. These observations show also the close relationship between air temperature (or temperature at the surface of the vegetation canopy) and phenophases. For example, each of the flowering and fruit-setting stages of alpenrose (*Rhodendron ferrugineum* L.) can be correlated to a mean daily temperature of the air at two meters above ground level (figure 5.5; Schlüssel and Theurillat 1998). In the south-facing transect of the Val d'Arpette, there was a mean difference of 3.3 days in the developmental stages of plants for each 100 meters in elevation for the years 1993–96, which is in accordance with the observations of Puppi Branzi et al. (1994) on comparative phenology between grassland and heaths in the Apennines. In the east-facing transect of Belalp, there is a mean difference of 6.3 days for every 100 meters in elevation, in this case in accordance with the shortening of the growing season following thawing (six days per 100 meters). Contrary to some authors' opinions, phenology of alpine plants is not always more strongly determined by soil temperature than by air temperature. As Hegg (1977) has already observed, and according to our observations in the Ecocline project, soil temperature is not necessarily a determinant of phenology.

For bryophytes, the study of survival strategies (Lloret 1988; Longton 1994) of populations of *Pleurozium schreberi* (Brid.) Mitt. demonstrates that archegonia and antheridia (female and male sexual organs) are produced earlier at lower altitude, and that sporophytes occur less often at higher sites (Velluti and Geissler 1996).

Phenological observations permit an assessment of plants' habitat requirements in relation to temperature and other environmental parameters. Phenological data can also provide climatological information (Bucher and Jeanneret 1994). Global warming's first effect might be to lengthen the growing season. Not all plants may benefit, since many species commonly stop growing toward the end of summer (e.g., the dwarf Alpine willows *Salix herbacea* L., *S. reticulata* L., *S. retusa* L., and *S. serpyllifolia* Scop.). The same is true for plants in the arctic tundra (Bliss 1956). Wijk (1986) reported that August temperatures were not correlated to shoot elongation of the dwarf willow (*Salix herbacea* L.). For some species, phenology is determined more genetically than climatically (Körner 1994), and differential response to climatic parameters among populations of a single species can also be genetically determined (Rathcke and Lacey 1985). In bryophytes, phenological patterns can be determined both photoperiodically and climatically (Longton and Greene 1969b). The variety of phenological patterns observed plays a role in speciation and might be an expression of genetic differentiation (Stark 1985).

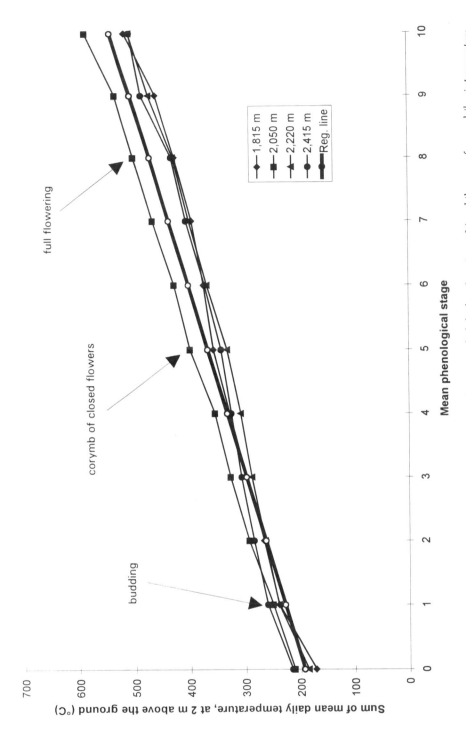

Figure 5.5 The relationship between mean phenological stages of alpenrose (*Rhododendron ferrugineum* L.) and the sum of mean daily air temperature (°C) at 2 m at four different altitudes (1,815 m a.s.l., 2,050 m a.s.l., 2,220 m a.s.l., 2,415 m a.s.l.) in the Val d'Arpette in 1995. Measurements were begun in each plot on the date of snow melt. The regression line over the four sites gives the overall correlation between temperature and phenophase. Phenological stages: 0: buds completely closed; 5: flowers closed and arranged in a corymb; 8: inflorescence with all flowers open.

5.4.6   Human Activities

Land use is a very important factor affecting ecosystems in the European Alps, where human activities have influenced every ecosystem for millenia, even the high alpine zone. To some extent, subalpine forests were transformed less, because of their protective value to human settlements. Man's transformations of the landscape enhanced the natural prehuman fragmentation of habitats in the Alps. Vegetation's response to climatic change may be very different when the effects of anthropic factors are included (e.g., Di Castri and Hansen 1992; Grime 1993). Very likely, the subalpine-alpine limit (ecocline) was one of the Alps' first ecosystems to be cleared and pastured during the Neolithic age, at the Atlantic period (6,000–5,000 yr BP). This is supported by the recent discovery of "Ötzi" in Tirol, on the Austrian–Italian border (e.g., Burga 1988; Bortenschlager 1993; Müller 1992; Pott 1993) and above all pedoanthracological investigations (that is, investigations of soil charcoal) in the alpine belt (Carcaillet, Barbero, and Talon 1996; Talon 1997). That clearing activities occurred above the present highest forest limit is also evidenced for the last thousand years by discoveries of macrofossils (trunks and branches) in alpine bogs (e.g., Tinner, Amman, and Germann 1996, for the Valais), and by toponymy (e.g., Mariétan 1929, for the Valais). Through the lowering of the upper limit of the forest by 200–400 meters and the development of heathlands and swards of replacement at the upper subalpine zone, the subalpine-alpine ecocline's climatic equilibrium has been modified (e.g., Holtmeier 1994).

Pollen analyses at the highest peat bog of the eastern Alps (2,760 meters) indicate that even the swards of the lower alpine belt may originate through pasturing, replacing low heaths (Bortenschlager 1993). Recent soil charcoal investigations in the southwestern Alps (Talon 1997) that show the regular presence of tree species like Arolla pine (*Pinus cembra* L.) and larch (*Larix decidua* Miller) up to 2,900 meters, 500 meters above the present tree line, support this hypothesis. These tree species, together with low heath species, were destroyed by fire at some point since the end of the Atlantic period (5,000 yr BP), very likely by human activities. Observations in the alpine belt in remnant undisturbed places—small, very infrequently grazed (or accessed) areas—currently support these findings, as does the ability of species at the tree line to tolerate climatic fluctuations by taking on a krummholz form (Kullman 1993, 1995; see section 5.5.3.1).

According to Wiersema (1989), the lowering of the forest's upper limit extended the habitat of the ibex (*Capra ibex* L.) downslope. Access to a greater forage supply very likely increased ibex populations, leading to greater grazing pressure that in turn may have helped extend the area of high alpine swards at the expense of low alpine heaths. Therefore, extensive grazing of alpine swards over millenia significantly shaped the distribution of plants and plant communities of the subalpine-alpine ecocline. Mass harvesting of medicinal plants probably also influenced this distribution, both locally and regionally. The harvest of the Celtic nard (*Valeriana celtica* subsp.

norica Vierh.) is exemplary. For centuries, the rhizomes (underground stems) of this small alpine plant were collected by the hundreds of kilos or even tons per year in some Austrian regions, where it was one of the major economic resources (Teppner cited in Jacquemoud 1989). No doubt this profoundly affected the vegetation where this plant grows.

Human activities can either assist or inhibit the upward shift of vegetation belts, in particular the elevation of tree line and the shift of subalpine forests into the alpine belt. Humans may try to accelerate this shift through, for example, afforestation to counter potential forest dieback, landslips or stone falls, or, at the tree line, by discontinuing grazing. On the other hand, humans may slow an upward shift of subalpine vegetation into the alpine belt by preventing the development of subalpine vegetation through more intensive pasturing of the alpine belt. Since the middle of this century, the alpine zone has been heavily altered in many places for skiing with the development of machine-graded ski runs. With global warming, recreational activities and settlement are very likely to increase in both the upper subalpine and the alpine zones. Hence, human activities may have a yet stronger impact, with more transformation of habitats and more habitat fragmentation.

5.4.7 Summary

• The structure of vegetation determines formations at the level of plant communities, and synusiae within plant communities.

• Mean air temperature is directly linked to growing season.

• Precipitation regime determines oceanicity and continentality, which in turn influence plant distribution.

• The frequency of extreme events (e.g., late frosts, avalanches) is an important factor determining timberline position.

• Winter snow cover is a key parameter for plants sensitive to frost, for winter soil microbial activity and resulting carbon and nitrogen biogeochemical dynamics, and for the activity of periglacial phenomena.

• Edaphic factors can compensate for climatic factors.

• Geological factors are important determinants of the vegetation on limestone.

• Periglacial phenomena (permafrost, gelifluction, cryoturbation) affect plant distribution.

• Soil may produce either a positive or a negative feedback loop in the response of plants to climate change.

• Clonal plants can resist invading species and may persist longer under global warming.

• Many aspects of plant development, especially growth and flowering, are directly linked to temperature and can provide climatological information by way of observations of the developmental phases (phenology).

• Humans lowered the tree line and forest line by 200–400 meters through fire and pasturing activities as early as the Atlantic period.

• Man can either accelerate or counteract vegetation's response to climate change through his activities, especially at the human-determined subalpine-alpine ecocline.

5.5 THE ROLE OF PLANT GENETIC DIVERSITY

Species' response to climate ultimately can be reduced to the populations' genetic structure and the mechanisms that determine it, especially reproductive type and genetic variation, both key factors of microevolution. Geographic variation in a species' genetic structure and the possibility of an ecotypic differentiation imply that climatic changes may affect populations' ability to survive or adapt to new environmental conditions differentially. For example, previous climate changes, such as glaciation, provoked successive fluctuations in species' range, which might have favored differentiation among populations by adaptation to different environmental conditions or by random genetic drift (i.e., the random change of the genetic constitution of a finite population through generations). Thus, it is important to consider climate change's genetical impact, which modifies all the processes of selection within and among populations.

5.5.1 Genetic Variability

Genetic variability may be investigated at different scales and is often partitioned within and among population components, showing both spatial and temporal patterns (e.g., Baur and Schmid 1996). Genetic variability depends on several factors such as life history traits, population size, and population history.

The reproductive type of plants is the key factor for genetic diversity. Sexuality promotes population variability, whereas asexual reproduction decreases it (Eriksson 1993). Moreover, a plant's breeding system (autogamy, allogamy, apomixis) may alter populations' genetic structure. Self-pollination (autogamy) promotes within-population uniformity and favors between-population differentiation when compared to cross-pollination (allogamy) (Hamrick 1989). The reproduction by seeds of the maternal genome (apomixis) induces genetical uniformity even if several clones may coexist in a single population (Loveless and Hamrick 1984). This is not always the case, however. Bayer (1989) found higher diversity indexes in the obligate gametophytic apomictic *Antennaria rosea* Greene than in its diploid relatives with sexual reproduction.

For within-population diversity, the size of the plant population is an important parameter. Small populations may show effects of inbreeding together with genetic drift. In extreme cases, genetic drift may lead to a bottleneck—the absence of variability. However, rare plants' genetic structure may vary considerably according to the species (Cosner and Crawford 1994), and small, isolated populations or fragmented populations may still maintain a significant level of genetic variability (e.g., Foré et al. 1992; Ballal, Foré, and Guttman 1994; Holderegger and Schneller 1994; Young, Boyle, and Brown 1996).

One way to approach the genetic diversity within a population is to measure the mean number of alleles (i.e., the different forms of the same gene) per gene locus (the site on the chromosome where a given gene is located; a site with several possible alleles is called a polymorphic locus). The greater the mean number of alleles per locus within a population of a species, the better its ability to adapt to a change.

Another aspect of the allelic diversity parallel to the mean number of alleles at a gene locus is the degree of heterozygosity, that is, the proportion of individuals carrying more than one allele at one locus, one on each of the two homologous chromosomes for a diploid species, contrasted with being homozygous (i.e., having the same allele on both homologous chromosomes). Usually, a population's gene diversity is evaluated in terms of the expected heterozygosity under a random mating system. A loss of heterozygosity can reduce individual fitness and population viability.

Overall, species vary widely, and generalization about genetic diversity should be made only with caution. Nevertheless, tree populations usually show a very high genetic variability in comparison with other vascular plants based on their allelic diversity, both in the number of alleles per locus and in the degree of heterozygosity (e.g., Hamrick 1989; Hattemer 1994; Müller-Stark 1994; Slatkin, Hindar, and Michalakis 1995), to which longevity and stress resistance capacity are related (Hattemer 1994).

5.5.2 Genetic Differentiation of Populations

Two opposing forces regulate genetic differentiation of plant populations: selection, which favors different genotypes in a heterogeneous environment, and gene flow (the movement of alleles between populations), which acts as a unifying factor. These two mechanisms are not exclusive, and genetic differentiation may occur in the presence of gene flow (Endler 1977; Caisse and Antonovics 1978).

Genetic novelties may arise through mutation that are then transmitted if they survive to selection. Nevertheless, the common process of adaptation is the selection of new combinations of genes issued from the recombination of parental genetic material during meiosis. These changes might affect either genetic variability as a whole or the variation at a single genetic marker. Global genetic variability may be interpreted as an evaluation of a plant species' evolutionary potential, and high genetic diversity is a prerequisite to adaptation to new environmental conditions. Two main approaches can be used to determine genetic differentiation among populations: the ecological approach and the genetic approach.

5.5.2.1 Ecological Approach The first ecological experiments to demonstrate plant's ecotypic differentiation were comparative morphological analyses of plants collected in different habitats and cultivated in experimental gardens under uniform conditions (Clausen, Keck, and Hiesey 1940; Turesson 1922). Ecotypic differentiation of a species implies that populations differ in their survival and fertility as a function of the environment. Genetic differentiation may be detected in a common environment if the reaction norms (the relationship between phenotypic variation and environmental variation) do not overlap under the given ecological conditions. Common-garden experiments are convenient for demonstrating genetic differentiation (Böcher 1949a, 1949b), but differences in the inverse direction may be observed when the reaction norms intersect. Consequently, a higher value of a character in one population than in another may not necessarily reflect the natural situation.

5.5.2.2 Genetic Approach The ecological approach uses morphological and phenological traits to measure genetic differentiation. These characters are often polygenic, however, and environmental factors may influence their expression. They are consequently not necessarily reliable for evaluating genetic diversity. Recently, genetic variation at the protein and DNA levels has been increasingly emphasized. Enzymes and DNA markers are used extensively for monitoring genetic variability (e.g., Gottlieb 1981; Hamrick 1989; Nybom 1993; Lynch and Milligan 1994). However, the degree and type of variation depend greatly on which part of the genome is being investigated. Also, patterns of genetic diversity depend on the set of markers used, and generalizations are difficult to make. Therefore, comparison of the measure of diversity with isozymes or DNA markers may yield similar results (Heun, Murphy, and Phillips 1994) or divergent ones (Zhang, Saghai Maroofand, and Kleinhofs 1993). Nevertheless, a greater differentiation of populations may be found with DNA markers than with allozymes.

5.5.3 Genetics and Environmental Change

Numerous descriptions of genetic polymorphism illustrate that genetic variability is not random. Relationships between ecological gradients and genetic clines have been found regularly. For instance, Lumaret (1984) showed that a clinal variation occurs along both altitudinal and latitudinal gradients in *Dactylis glomerata* L. However, an individual plant's response to ecological change at the genetic level has two important aspects: phenotypic plasticity and the selection of adapted genotypes.

5.5.3.1 Phenotypic Plasticity and Selection An organism's *phenotype* is its appearance according to its genetic constitution (genotype) interacting with the environment. In new environments, an individual may change its appearance, expressing new phenotypes (phenotypic plasticity). Modifications may be adaptative (e.g., increase of survival or of reproductive success) or nonadaptative (e.g., Scheiner 1993). Phenotypic plasticity appears, however, to have a genetic determinism. Therefore, one can distinguish between "passive plasticity," that is, plasticity not regulated by the organism but resulting in the direct effect of environmental factors on cellular metabolism, and "active plasticity," that is, plasticity regulated on complex genetic-developmental mechanisms (Pigliucci 1996). An example of phenotypic adaptation to the environment is the occurrence of trees in high mountain regions. At the upper altitudinal limit of spruce (*Picea abies* (L.) Karsten) in the Scandes Mountains in Sweden, trees 4,700–4,800 years old survived climatic changes during the Holocene by taking on the krummholz form when temperature decreased (Kullman 1995). Similarly, Kullman (1993, 769) reaches the same conclusions about birches: "Possibly, birches with dead main stems may survive as copses of low growing sprouts more or less indefinitely."

Strong selection may create very rapid genetic differentiation within plant populations. Rapid genetic changes in plants are known from human activities that create many sharp disturbances of the environment. For instance, tolerance to heavy metals (Antonovics 1971) or to herbicides (Warwick 1991) provides several examples. Although they do not relate directly to climate changes, these examples suggest that climate change might induce measurable genetic changes. Another example is in the grass species *Anthoxanthum odoratum* L., where genetic differentiation in terms of plant height, yield, and susceptibility to disease was observed between adjacent meadows subjected to various fertilizer and liming treatments over a span of fifty years and over distances of thirty meters (Snaydon and Davies 1972). Many changes were already detectable within six years (Snaydon and Davies 1982). With the same species, differences in salt tolerances were observed in less than thirty years between a population located at the edge of a highway and those of the surrounding pastures (Kiang 1982).

5.5.3.2 Ecogeographical Gradients Gradual genetic variation along an ecological gradient can result from selection. For instance, selection was considered a possible explanation for genetic differentiation among populations of *Pinus ponderosa* Lawson along an altitudinal transect (Mitton, Sturgeon, and Davis 1980). Another convincing argument for selection is the study of acid phosphatase in *Picea abies* (L.) Karsten along a latitudinal gradient in Finland and an altitudinal transect in the Austrian Alps and in different populations in the Swiss Alps. In the three groups of populations, enzymic variation was consistent according to the climate gradients (Bergmann 1978). In the two sites of the Ecocline project, *Anthoxanthum alpinum* A. & D. Löve, a widespread grass growing in the subalpine and alpine belts, was tested for six enzyme systems including ten loci. When the populations were pooled, three enzyme systems, glutamate oxaloacetate transaminases (GOT), peroxidases (PX) and malate dehydrogenases (MDHB) showed a significant corre-

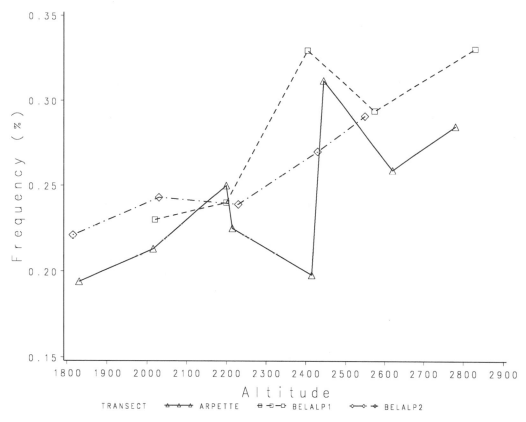

Figure 5.6 Variation of allele MDHB (malate dehydrogenase) in *Anthoxanthum alpinum* A. & D. Löve in relation to altitude in three transects at two sites in the Valais.

lation between one allele frequency and altitude. Figure 5.6 shows the increase of allele MDHB as a function of altitude.

On the other hand, interpretation of geographical clines is not straight-forward because they may be caused either by the neutral diffusion of an allele during migration or by the action of selection on ecological gradients. Evidence of a selective effect may nevertheless be obtained by the demonstration of similar correlations at different locations or at different scales. In *Abies alba* Miller, geographical clines were explained by historical factors such as postglacial migration (Breitenbach-Dorfer et al. 1992; Hussendörfer and Müller-Starck 1994), and in *Pinus muricata* D. Don by recent contact with formerly isolated populations (Millar 1983). Enzymic variation in *Fagus sylvatica* L. was related either to climate or to geography, depending on the marker investigated (Felber and Thiébaut 1984; Cuguen et al. 1985).

5.5.3.3 Physiological Properties of Genetic Markers
A parallel issue to the relationship between enzymes and environment is the variation in enzymes' physiological properties. It has been demonstrated repeatedly that alleles coding for enzymes may differ according to environmental conditions.

For example, differences in specific enzymes have been found in comparing related species occupying somewhat different thermal niches in the same area (Liu, Sharitz, and Smith 1978) or in comparing two clonal populations (Simons, Potuin, and Blanchard 1983; Simons, Charest, and Peloquin 1986).

Differences in function for the same enzyme have also been described in several widespread plant species at the population level (McNaughton 1974). Kinetic parameters of an enzyme can be associated with the climate the organisms experience, in that enzymes appear to function with the greatest efficiency or regulatory ability under the conditions an organism normally encounters.

Relatively small temperature differences among habitats may be associated with enzyme differences at the interspecific and intraspecific levels. Lumaret's (1984) study of polymorphism of the glutamate oxaloacetate transaminase (GOT1) of *Dactylis glomerata* L. found that the enzyme's relative temperature activity was consistent with the expected temperature in the clines of the natural populations based on the altitudinal and latitudinal gradients. Another example is that of isocitrate dehydrogenase (IDH) in European silver fir, *Abies alba* Miller. A latitudinal frequency cline was observed for the two forms of that system. Thermostability measurements demonstrated that the most frequent form in the South was more adapted to high temperature (Bergmann and Gregorius 1993).

5.5.3.4 Genetic Response to Climate Change—An Outlook Investigating genetic diversity in the context of projected climate change would present a number of opportunities. For example

1. Research could focus more strongly on assessing how plants react to natural or induced environmental changes. Phenotypic plasticity is a complementary approach to genetic studies. Phenotypic plasticity would determine species' short-term ecological response to climatic changes and would indicate how populations react to these according to elevation, and, in particular, if marginal populations differed from central ones. In addition, in some cases phenotypic plasticity might buffer the effect of climatic change.

2. Under the hypothesis that spatial scale mimics timescale, the variation of a single genetic marker might be investigated along environmental gradients paralleling the forecasted climate change. Correlations might be found between genetic markers and environmental variables, and their response to selection might be tested. Tools might be thus developed for monitoring a species' genetic response to climate change. However, correlation is not causality, and it would be important to test the response of markers to selection. If a genetic marker were directly subject to selection, physiological differentiation should be found in the different expressions of the genes (alleles). In that context, the activity of alleles could then be tested according to the suspected ecological parameter. Yet to be used, a genetic marker should not necessarily itself be submitted to selection but could just be tightly linked

to the selected gene. If recombination between the selected gene and the observed genetic marker were sparse, there would be no restriction on using such a marker. In that case, selection could be tested by reciprocal transplant experiments. However, an impressive number of genetic markers is available, and many dead ends might be encountered that could not be forecast. Therefore, more species differing in life history traits and distribution patterns would have to be analysed, and more genetic markers would have to be tested before generalizing.

5.5.4 Summary

• Genetic diversity is a prerequisite to adaptation to new environmental conditions; in general, big populations are less prone to genetic uniformity than small ones.

• Two opposing factors regulate genetic differentiation: selection (differentiating) and gene flow (unifying).

• Phenotypic plasticity is an important factor in the response to ecological changes.

• Adapted genotypes can be selected within a few years following an ecological change.

• Relationships between ecological gradients like climatic gradients and genetic clines have been regularly found.

• Genetic clines along ecological gradients may be correlated with physiological differences at the enzyme level(s).

5.6 THE RESPONSE OF THE SUBALPINE-ALPINE VEGETATION AND SOILS TO CLIMATE CHANGE: INTEGRATING ECOSYSTEM COMPONENTS ACROSS MULTIPLE SPATIAL SCALES

In the face of climate change, plants have three options: adapt in situ to the new conditions, migrate to more suitable habitats, or become extinct (Bazzaz 1996). Many contributions have considered the possible response of vegetation and ecosystems to the impact of climate change due to the doubling of the atmospheric CO_2 concentration and the increase of other greenhouse gases in the atmosphere (e.g., Bazzaz 1996; Bolin et al. 1989; Cramer and Leemans 1993; Gates 1993; Huntley 1991; Markham, Dudley, and Stolton 1993; Peters and Lovejoy 1992; Tegart, Sheldon, and Griffiths 1990; and Woodward 1992). The reaction of mountain and alpine plants and vegetation to climatic change has already been discussed several times, and various hypotheses and predictions have been proposed (e.g., Gottfried, Pauli, and Grabherr 1994; Grabherr, Gottfried, and Pauli 1994; Grabherr et al. 1995; Guisan, Theurillat, and Spichiger 1995; Halpin 1994a, 1994b; Hofer 1992; Holten and Carey 1992; Klötzli 1992, 1994; Körner 1992, 1993a, 1994, 1995; Markham, Dudley, and Stolton 1993; Ozenda and Borel 1991, 1994;

Theurillat 1995). Two main hypotheses have been put forward repeatedly concerning the migration of species:

1. Species will respond individually.

2. Species' response will not keep pace with the forecasted changes.

Thus, a climatic change might induce shifts in the abundance and distribution of species, extinctions, or adaptations to new ecological conditions (Holt 1990).

However, species also interact at different levels of complexity such as synusiae, plant communities, vegetation complexes, and landscapes. Because of their complex organization and structure and their relationship with abiotic factors, climate change can affect these levels differentially, and they may affect in return species' individual response. The following section attempts to evaluate such a differential response according to levels of complexity, using evidence from changes in the vegetation during the Holocene.

5.6.1 Synusiae

Climate change will directly affect synusiae of the dominant layers, like tree synusiae of forests, if they belong to climatic climax plant communities, although the impact may be gradual (see the example with *Fagus grandifolia* Ehr. and *Tsuga canadensis* (L.) Carrière below). In mixed forests, the dominance hierarchies may change if the species have different physiological amplitudes to temperature, for instance for bud burst, which initiates the growth period. These synusiae may be less affected if they belong to an azonal or extrazonal edaphic climax, where the habitat conditions could compensate for the new climate. Simulated distributions in response to climate change have been modeled for several important European tree species (e.g., Huntley et al. 1995; Sykes, Prentice, and Cramer 1996).

In the subalpine-alpine belts, some heath synusiae, determined by a few dominant clonal species, appear to be finely tuned to climatic factors, whereas others have a thermal inertia of at least 3 K based on their altitudinal amplitude. Investigations in the Ecocline project (Schlüssel and Theurillat 1996) show that the thermophilous ecosystem, with *Arctostaphylos uva-ursi* (L.) Sprengel, is composed of two systems of synusiae that intergrade into each other gradually. With a temperature increase the lower system would push the higher system upward, but without any change in the species composition, because the two systems' dominant synusiae have the same floristic composition. The synusial structure of the mesophilous ecosystem, with *Rhododendron ferrugineum* L. and *Empetrum hermaphroditum* subsp. *hermaphroditum* (Hagerup) Böcher, presents a great altitudinal uniformity for the dominant synusiae. Therefore, an increase in temperature alone would not modify the general structure of this ecosystem, whose driving factor is snow cover protecting species to sensitive late frost. However, a reduction in the length of snow cover could lead to an important change both in the synusial structure and in the species composition through frost damage.

Synusiae of the lower layers, like small herb synusiae or bryophyte synusiae of forests or shrub communities, may not react directly to climate change, but only indirectly through a change in the dominant synusiae of the phytocoenose, which in turn affects the conditions of the microhabitats such as microclimate, water and nutrient availability, and quantity and quality of light (e.g., Knapp, Smith, and Young 1989; Bahn et al. 1994; Svensson, Floderus, and Callaghan 1994). In particular, bryophyte synusiae may behave independently from synusiae dominated by vascular plants, implying that it may be useful to establish the present relationship of bryophyte synusiae to vascular plant vegetation. To our knowledge, no such analysis has been conducted in the European Alps since Frey 1922. Factors to consider may include changes in diversity, productivity, reproductive pattern, or interaction with other plants.

In the future, we could imagine a greater development of the cover of a dominant synusia with big leaves; such a development might cause the synchronal lower synusiae to disappear (Körner 1991, 1993a). Alternatively, lower synusiae or a part of them may well persist in new evolving phytocoenoses if the conditions of their microhabitats are left substantially unchanged. This would be the case for the lower synusiae developing before the others in early spring, in the absence of competition for light and nutrients at the time they grow, although these species might have a lower level of photosynthetic activity in an enhanced CO_2 environment (e.g., *Primula elatior* (L.) L.) compared to dominant species (Körner 1991, 1993a).

A historical example of an understory species' persistence may be the present distribution of the Compositae *Aposeris foetida* (L.) Less. at the foothills of the northeastern Alps (Bavaria, Austria). Its present distribution exactly matches the former distribution of the Norway spruce (*Picea abies* (L.) Karst.) before beech (*Fagus sylvatica* L.) replaced spruce a few thousand years ago in the region (Küster 1990).

Global warming would also modify competitive relationships among plant functional types, such as life forms (Körner 1992, 1993a). For example, deciduous dwarf shrubs (chamaephytes) such as bilberry (*Vaccinium myrtillus* L.) and bog bilberry (*V. uliginosum* L.) are unlikely to prevent low nanophanerophytes such as alpenrose (*Rhododendron ferrugineum* L.) from growing in mesophilous conditions with enough snow cover in winter, or dwarf juniper (*Juniperus communis* subsp. *alpina* (Suter) Celak) from growing in meso-thermophilous conditions if these nanophanerophytes can colonize open microsites. Alpenrose heaths are likely to be invaded by green alder (*Alnus viridis* (Chaix) DC.), however, if mean air temperature increases (i.e., there is a longer growing season). But alpenrose may not be completely outcompeted under the cover of the green alder or associated tall herbs unless precipitation and/or eutrophication increase strongly, because it can grow as an understory plant.

Another important effect of climate change at the synusial level would be that the exotic species would be likely to invade particular synusiae of plant communities, at least in the lowest vegetation belts (colline, montane; see section 5.6.2).

5.6.2 Plant Communities

Climatic climax and edaphic climax plant communities may respond differently to climate change. Climatic climax communities respond by searching for a novel equilibrium with the available flora (e.g., Graham and Grimm 1990; Tallis 1991). This is considered to have been the case for the changes in vegetation in general since the last glaciation. For example, it has been shown for the beech-hemlock forest (*Fagus grandifolia* Ehrh.—*Tsuga canadensis* (L.) Carrière) in the Great Lakes region of eastern North America that the ranges of the two species have been fully coincident only during the past 500 years, whereas they have overlapped since around 6,000 years BP (Graham and Grimm 1990; Moore 1990).

A climatic community may persist in places with suitable edaphic conditions and so withstand the new climate as an extrazonal, edaphic climax (Walter 1984, 1985). For example, the pioneer alluvial pinewoods with Scots pine (*Pinus sylvestris* L.), buckthorn (*Hippophae rhamnoides* L.) and juniper (*Juniperus communis* L.) are presently edaphic plant communities, but were climatic climax at the Alleröd period (Pott 1993). The absence of competing, better-adapted species can even result in the persistence of the vegetation, even vegetation that does not correspond to the climax, given that the species can endure the new conditions physiologically or adapt genetically. The climatic climax of the thermocolline belt in the Insubrian climate at the Alps' southern border corresponds to forests with laurophyllous species. Because this climax vegetation could not recolonize the area from their refugia during the Holocene, the belt is now occupied by oak woodlands with chestnut (Gianoni, Carraro, and Klötzli 1988; see also Klötzli 1988).

For edaphic climaxes, plant communities can sustain a climatic change providing their limiting factors are not modified ("buffering"; Jonasson 1993). However, their range may change in relation to the positive or negative synergy of the climate change on such limiting factors. Thus, we can recognize a persistence (Pimm 1984) of the edaphic plant communities at three levels: persistence in time (several thousand years), persistence in space (several thousand kilometers), and persistence in altitude (several hundred meters). The examples in table 5.2, taken from the vegetation classes of Europe, particularly of the Alps, illustrate these three types of persistence during the Holocene. Cliffs and rock cavities offer examples illustrating the persistence of vegetation in time. For instance, carbonate rock cavities with dripping water in the northwestern Mediterranean area (mountains of Spain, Maritime Alps, Apennines) are colonized by maidenhair fern communities (*Adiantum capillus-veneris* L.) with several ancient endemic species of butter-

Table 5.2 Persistence of vegetation

	Time persistence	Spatial persistence	Altitudinal persistence
Asplenietea trichomanis[a]	×[1,2,3]	×	×
Montio-Cardaminetea[b]	×[4]	×	×[4]
Scheuchzerio-Caricetea fuscae[c]	×[5]	×	×
Isoëto-Nanojuncetea[d]	×	×[6]	
Adiantetea[e]	×[7]	×[7]	
Salicetea herbaceae[f]	×	×[1,8]	
Elyno-Seslerietea[g]	×[3]	×	
Thlaspietea rotundifolii[h]	×[1,3]	(×)	
Koelerio-Corynephoretea	×	(×)	
Caricetea curvulae[i]	(×)	(×)	
Festuco-Brometea[j]	×		
Oxytropido-Elynetea[k]	×		
Potametea[l]		×	
Phragmiti-Magnocaricetea[l]		×	
Oxycocco-Sphagnetea[m]		×	
Loiseleurio-Vaccinietea[n]		×[7]	
Vaccinio-Piceetea[o]		×[7]	

Note: ( ) partial persistence.

[a] cliffs; [b] springs; [c] swamps; [d] temporarily wet, bare habitats; [e] calcareous rock cavities with dripping water; [f] snowbeds; [g] cryophilous alpine swards; [h] screes; [i] some acidophilous alpine swards; [j] some dry grasslands; [k] some calcicolous alpine swards; [l] aquatic habitats; [m] peat bogs; [n] subalpine and alpine heaths; [o] boreo-montane forests. [1] Braun-Blanquet 1948; [2] Rechinger 1965; [3] Pawlowski 1970; [4] Zechmeister and Mucina 1994; [5] Braun-Blanquet 1967; [6] Grabherr and Kojima 1993; [7] Deil 1989, 1994; [8] Braun-Blanquet 1964.

worts *(Pinguicula* ser. *Longifoliae)* since the Miocene (Deil 1989, 1994). Cliffs of the southern Alps also offer examples of vegetation persistence in time, especially plant communities on limestone in the Maritime Alps with their ancient Tertiary relict species of the genera *Saxifraga* L. and *Moehringia* L. (Rioux and Quézel 1949).

New plant communities are likely to develop and partially replace the present ones, most likely of an intermediate composition compared to present communities. However, provided that there is no immigration of new species, they will still belong to the higher phytosociological classification units (alliances, orders, classes) known for the European Alps, because of their similarity to present species composition, and because of the limits for rapid dispersion of new invaders, particularly in the alpine belt. However, climate change may favor cultivated exotic species' invasion of natural plant communities. For instance, in the oak-chestnut woodlands of the thermocolline belt under the Insubrian climate at the southern border of the Alps, a whole array of cultivated, exotic laurophyllous species are invading the understory in the warmest places, where they now successfully reproduce ("laurophyllisation"; (Gianoni, Carraro, and Klötzli 1988; Klötzli et al. 1996).

In the most favorable sites, they are already part of the tree layer (Walther 1996; Carraro et al. in press). A plausible explanation for this expansion is the unprecedented sere of mild, nearly frostless winters which have occurred for several years.

5.6.3 Landscape

The natural landscape is constantly changing with time, independent of climatic variations. Changes in the landscape may be either slow and progressive or rapid and abrupt. The landscape can change slowly and inconspicuously toward an equilibrium with climatic conditions, as in the case of the beech-hemlock forest of eastern North America discussed above, or it can change rapidly, as in the case of natural catastrophes (e.g., volcanic eruptions, rock falls, floods, windstorms, fires, or climatic anomalies such as long droughts; e.g., Stine 1994; Street-Perrott 1994).

The cultural landscape (*Kulturlandschaft*) is changing very rapidly through human activities. Human impact is often more significant than other factors. The ecosystems and landscapes that respond to climate change may be very different than those found today (Di Castri 1992).

Landscape may respond very noticeably and differentially to climate change as it integrates all ecological and historical factors. Several examples during the Holocene illustrate a differential response of the vegetation, either lagging or expanding very rapidly within a landscape.

As an example of differential response within the same region, Küster (1990) showed, for the Auerberg Mont (1055 m a.s.l.) in the foothills of the northern Alps in southern Bavaria (Germany), significant differences in the development of past vegetation among three sites separated by only three kilometers with no physical barrier between them. Pollen profiles indicate that Norway spruce (*Picea abies* (L.) Karsten) arrived as much as 700 years later at the western site compared to the southern and the eastern sites. Also, the local extension of the species differed significantly among the sites. Around 7,000 years BP, the southern site supported less than 10 percent *Picea*, whereas pollen profiles for the western and the eastern site suggested 24 percent and 30 percent, respectively. These kinds of differences among the sites also occurred later during the expansion of European silver fir (*Abies alba* Miller) and beech (*Fagus sylvatica* L.).

The change in vegetation after the Wisconsin glaciation in the Grand Canyon (Arizona) is another example of differential response to general climate. There, changes in vegetation lagged changes in climate by 1,000–3,000 years (Cole 1985; Lewin 1985).

An example of a rapid expansion over a larger area is the expansion of Norway spruce (*Picea abies* (L.) Karsten) in the Scandes Mountains in Sweden. There, the presence of spruce has been recently confirmed more than 5,000 years prior to the date assumed using inferences from pollen data (Kullman 1995, 1996b). Spruce appears to have expanded its range very early after deglaciation to specific microhabitats up to some 1,000 km away from the nearest source area. According to Kullman (1996b), this demonstrates a climate-plant equilibrium prevailing throughout the Holocene.

5.6.4 Assessments

5.6.4.1 Impact of Global Change on Bryophytes Although bryophytes possess mechanisms (poikilohydry) allowing them to withstand drought, their broad altitudinal distribution is nevertheless related to general climate. Therefore, climate change might threaten the survival of stenoicious species restricted to a specialized habitat. For example, species living in the coldest habitats might disappear, like the bryophyte community *Dermatocarpetum rivularis* from seasonal high altitude, cold springs of melting water of permafrost block screes, with *Hydrogrimmia mollis* (B., S. & G.) Loeske, the moss species characteristic of this habitat, which has hardly ever been found with sporophytes in Europe (see Geissler 1976). Another example might be the coprophilous *Voitia nivalis* Hornsch. growing on chamois (*Rupicapra rupicapra* L.) excrements in *Carex curvula* All. swards between 2,500 and 2,700 meters a.s.l. in the eastern Alps. Generally speaking, species with a wider ecological amplitude than stenoicious species should not be threatened, especially if they show a high genetic variability and phenotypic plasticity, although fragmentation of their populations might occur.

Increasing atmospheric CO_2 concentration and nitrogen emissions, temperature, and humidity in a projected climate change would stimulate bryophyte growth (Sveinbjörnsson and Oechel 1992). Jauhiainen, Vasande, and Silvola (1994) showed through laboratory experiments with *Sphagnum fuscum* (Schimp.) Klinggr. that elevated CO_2 increase reduces length increment but has no effect on biomass production, and that supraoptimal nitrogen supply limits growth (see also Lee, Baxter, and Emes 1990). Nitrogen then becomes available to the higher plants, thereby changing the competitive relationships between mire species. Thus, although both productivity and the frequency of sexual reproduction might increase under predicted climate change, abundance might be reduced because of competition with more successful higher plants (Herben 1994). For the special habitat of northern peat lands, Gignac and Vitt (1994) predicted a decrease in bryophyte cover as a consequence of elevated evapotranspiration with global warming.

Wet acidic depositions would mainly affect neutrophilous taxa, with less effect on acidophilous species, which have natural resistance to low pH, and on calcicolous species, which have a naturally higher buffer capacity (Farmer, Bates, and Bell 1992). Many rare alpine bryophyte species belong to this latter ecological group. However, below pH 3.0, growth and reproduction decrease significantly, even with acidophilous species (Raeymakers and Glime 1986).

5.6.4.2 Broad-Scale Impact on Vascular Plants Because of the different distribution patterns of orophytes and endemics, many aspects are involved in assessing how climate change might affect their respective distributions that relate to other important factors instrumental in determining their distribution (such as history, topography, climate, and physiology).

According to Scharfetter (1938), glaciation was more destructive than the warmest interglacial periods. However, the warmest interglacial periods enabled forests to climb higher toward the summits of low mountains (1,800–2,300 m), thereby reducing many orophyte populations. This would be the case in global warming for many isolated endemics and orophytes living presently in refugia, such as tops of low mountains, mainly in the eastern Alps, because they would have almost no possibility to migrate higher (in the present nival belt), either because they could not move there rapidly enough, or because the nival zone is already absent (Grabherr, Gottfried, and Pauli 1994; Grabherr et al. 1995; Gottfried, Pauli, and Grabherr 1994). In addition to competition and changes of habitat conditions, one of the most important factors for orophytes might be their potential to acclimate their respiration at night (dark respiration) to a higher temperature (see Larigauderie and Körner 1995). The present altitudinal range, population size, conservation status (endangered/threatened or not), ecology (cool, wet or dry, warm habitats), and genetic and phenotypic diversity are parameters to be taken into consideration when assessing the general response of both a species and a particular population. However, a linear response to a temperature increase would be very unlikely. An increase of 1–2 K would still likely be in the range of tolerance of most alpine and nival species, but a greater increase (3–4 K) might not be (Theurillat 1995) (see also sections 5.6.4.4 and 5.6.4.5).

Endemics and orophytes with extant distributions are not expected to disappear in the climate change predicted. Likewise, species having a great potential for adaptative responses through genetic diversity, phenotypic plasticity, high abundance, or significant dispersal capacities are at the least risk of extinction (Holt 1990). However, the European Alps do not constitute a uniform orobiome (Walter 1984, 1985), and climate change might not have the same impact everywhere. Ranges of species showing a disjunction (north-south, east-west) or fragmented distribution might become even more fragmented, with regional disappearances, if they could not persist or adapt. Many plant populations interconnected today along a continuous alpine zone could become separated into isolated populations on the tops of some mountains (habitat fragmentation).

Plants from cold habitats, and from mesophilous habitats to some extent, would have to migrate upward and northward to find habitats with suitable conditions, unless they could persist locally because of favorable edaphic conditions. Isolated arctic, stenoicious, relict species that are pioneers in wet habitats might disappear, since these habitats are very limited and the implantation of artificial lakes for hydroelectric plants has already destroyed many. On the other hand, an increase in temperature should not affect endemics and orophytes inhabiting rock fissures in the montane and subalpine zones of the southern Alps because they are already adapted to extreme conditions, particularly if they grow on limestone. This flora appears to have survived abrupt and rapid warming (7 K) during the glacial-interglacial period (Zahn 1994).

5.6.4.3 Responses of Forests For the climatic climax or zonal vegetation, a quantitative change directly proportional to global warming could be expected, as Brzeziecki, Kienast, and Wildi (1995) showed for the forests of Switzerland. In proposed climate scenarios, warming is expected to be greater in the northern part of the Alps, and precipitation greater in the southern part. Because an increase in precipitation should mitigate the effects of the temperature increase, changes in the vegetation are less likely to occur under the humid Insubrian climate (e.g., Brzeziecki, Kienast, and Wildi 1995). For a 1–1.4 K increase in temperature, 30–55 percent of the forested pixels of Switzerland show a change of their classification types (phytosociological suballiances), according to a Geographic Information System–assisted sensitivity assessment (Kienast, Brzeziecki, and Wildi 1995, 1996). Classification changes are predicted for 55–89 percent of the forested pixels for an increase of 2–2.8 K. Analysis of the adaptation potential for the dominating tree species with a diameter at breast height (DBH) of at least 12 cm shows an increase of poorly adapted pixels from 25–30 percent for today's situation to 35–60 percent for a 2–2.8 K increase. This situation is reduced to 8–15 percent for the 2–2.8 K scenario when the analysis includes trees with a DBH of less than 12 cm.

However, forests' responses to climate change, and in particular tree species' responses, cannot be assessed based only on static modeling, without taking into account the dynamic aspect of transient stages. But even with dynamic models (see chapter 6), not every aspect can be considered, such as chilling temperatures, and the effect of late frosts, or the adaptation potential of tree species through their high genetic diversity, regional variability, and the maintenance of that genetic diversity. For instance, populations of Norway spruce (*Picea abies* (L.) Karsten) in northern Italy show a lower genetic variability than those in Switzerland and south Germany (mean of 1.8 alleles versus 2.4–2.6 per locus; 16.2 percent heterozygosity versus 22.2–25.2) (Müller-Starck 1994).

5.6.4.4 Responses of the Subalpine-Alpine Ecocline One widespread hypothesis (e.g., Ozenda and Borel 1991, 1994) is that global warming would shift vegetation belts upward in a more or less regular pattern. For example, the subalpine belt would shift upward into the area presently occupied by the alpine belt, which itself would shift into the present nival belt (figure 5.7). In contrast, Halpin (1994a, 1994b) hypothesized that a global, linear, upward shift in vegetation would not take place. Instead, he proposed that the vegetation zones' size and composition would change dramatically, though not in a regularly ordered pattern. Nevertheless, the difference

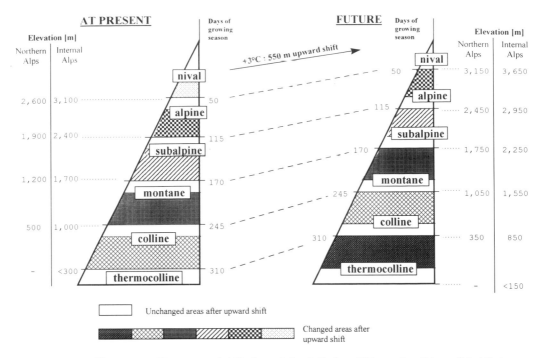

Figure 5.7 Linear upward shift of vegetation belts for a 3°C warming. It is predicted that more than 75 percent of each of the present belts would be occupied by the vegetation belt currently below; that is, less than 25 percent would stay unchanged. Elevations are given for the upper limits of each vegetation belt at present and for the future.

between these two viewpoints seems mainly conceptual (bioclimatic zone vs. ecoclimatic zone). In reality, the potential climatic climax is the same in both cases. Although the hypothesis of an upward shift of the present vegetation is tenable at a broad scale, the impact of climatic change will have to be appreciated at finer, regional, ecological scales. Therefore, regional climate scenarios are essential.

At the tree line, changes in vegetation would be complex. There, extreme events, edaphic and topographic factors already prevail, as in the alpine vegetation belt (see sections 5.4.2.3 and 5.4.3). Harsh topoclimates influence the establishment and growth of trees (e.g., Tranquillini 1979; Slatyer and Noble 1992; Holtmeier 1994). For phanerophytes, the *kampfzone* represents a temperature-related threshold whose inertia compensates for positive and negative variations in climate, preventing a linear variation of the forest limit (figure 5.8). For the tree line to expand upslope, even where it was artificially lowered by man, would require a more favorable climate for at least 100 years (Holtmeier 1994). Photogrammetric observations in Glacier National Park in northwest Montana during the last twenty to seventy years confirm this hypothesis, because tree lines under natural conditions have not yet moved, despite a 0.5 K warming during this century (Butler et al. 1994). Principally from palynological and macrofossil investigations, it was deter-

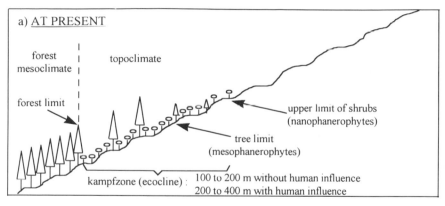

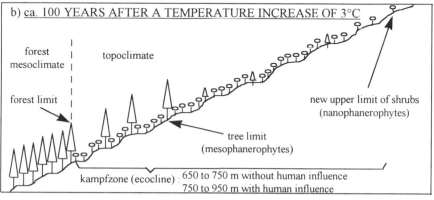

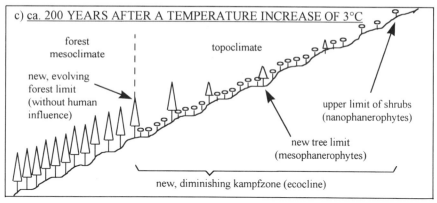

Figure 5.8 The subalpine-alpine ecocline (a) at present and its predicted evolution after a 3 K warming (b) for 100 years and (c) for 200 years.

mined that the forest limit did not extend upward more than 100–300 meters during the warmest periods of the Boreal and Atlantic periods of the Holocene (e.g., Burga 1988, 1993, 1995; Bortenschlager 1993; Lang 1993). Based on the general agreement that the mean annual temperature was 2–4 K greater than today during these periods, the forest limit would have been expected to extend from 350 to more than 700 meters higher for the 0.55 K/ 100 meter gradient, reaching the high alpine belt, in obvious nonconformity

Sensitivity of Plant and Soil Ecosystems of the Alps to Climate Change

with palynological observations. Interestingly enough, however, these elevations match recent pedoanthracological observations (i.e., of soil charcoal) (Carcaillet, Barbero, and Talon 1996; Talon 1997). The apparent contradiction between the two methods could be elucidated by taking into consideration that charcoal of trees from the alpine belt is very likely that of *krummholz* (nano- and microphanerophyte) plants, which rarely or never produce pollen under the conditions of the alpine belt, explaining to some extent their absence in alpine pollen profiles.

Therefore, we can put forward the hypothesis that on one hand, an increase in mean annual temperature of 1–2 K might not shift the present forest limit upward by much more than 100–200 meters. Yet the tree line and the subalpine-alpine ecocline in general might show a greater and quicker response than the forest limit, with low trees occupying favorable places in the low alpine belt. On the other hand, it is inconceivable that the inertia of the temperature-related threshold would withstand a 3–4 K increase, equal to the temperature amplitude of an entire vegetation belt. With such an increase, the *kampfzone* would very likely invade the alpine belt, with a later shift in the forest limit into the low alpine. Indeed, heaths (chamaephytes) and to a lesser extent shrubs (nanophanerophytes) would likely react faster than forest ecosystems to a warming, invading alpine meadows within a few decades. Hence, the subalpine-alpine ecocline might extend upward with expansion of heaths following a warming, whereas the upslope expansion of forest would likely be much slower (panel (b) of figure 5.8). If a temperature increase of more than 2 K persisted for several centuries, forests could develop at higher elevations than at any time since the last glaciation. For a 1–2 K increase, human use might play a substantial role either in slowing down the upward shift of the subalpine vegetation through grazing or accelerating it through the abandonment of agricultural practices and afforestation.

Given the predicted climatic scenarios, those plants most apt to colonize the *kampfzone* and the low alpine zone would be those that combine properties such as climatic optimum in the subalpine level, moderate insensitivity to the present pH of the soil, efficient use of nitrogen, ability to have a positive feedback on the soil, resistance to soil movements, and a strong competitive ability, both in the short term (*r* characteristics) and in the long term (*K* characteristics). In the northern Alps, green alder (*Alnus viridis* (Chaix) DC.) appears to be a suitable candidate, at least under a suboceanic climate. In the southwestern Alps and in the dry intra-alpine valleys, larch (*Larix decidua* Miller), a pioneer species, would very likely extend its distribution upward.

The situation of the subalpine-alpine ecocline would provide a wealth of possibilities for phytoindicators of climate change through (1) the modification of species and the dominant life forms in heaths, heath swards, and swards; (2) the balance between heaths and meadows; (3) the invasion of subalpine shrubs like green alder or prostrate mountain pine and trees like larch and arolla pine; (4) the persistence or disappearance of vegetation types; and (5) the positive or negative effects of soils, orography, and geology. However, the movement of the tree line itself, and even more of the timberline, would be unlikely to be used to monitor the effects of climatic change because of the slow rate of upward spread and possible downward movements through extreme events (Slatyer and Noble 1992).

5.6.4.5 Responses of the Alpine Vegetation Because it occurs at the top of high mountains, alpine vegetation would potentially be the most endangered by climatic change, even if a nival belt lay above the alpine belt. Indeed, edaphic factors (relief, aspect, substrate, soils, periglacial processes) would play a key role in the new distribution patterns of alpine vegetation in the present nival belt. Increasing steepness at the upper alpine and nival belt (slope greater than 40°; see figure 5.4) would totally modify the present ratio in the cover of plant communities. Very likely, the present middle and high alpine plant communities on slopes less than 40° (for example, snow beds with *Salix herbacea* L., alpine swards with *Carex curvula* All.) would shrink and even disappear in some places. However, partial persistence would be likely, at least at higher elevations and cooler orientations where cryogenic processes and permafrost would limit the altitudinal shift of more thermophilic species. Steep slopes and active, unstable screes (particularly of carbonate rocks) would react much more slowly. Vegetation colonization might happen extremely slowly, as shown by observations on permanent plots on machine-graded ski runs (Delarze 1994). Nevertheless, we should not forget that tree species occupied the alpine belt at very high elevations prior to the neolithic period, according to pedoanthracological research (Carcaillet, Barbero, and Talon 1996; Talon 1997).

As for the subalpine-alpine ecocline, we can put forward the hypothesis that on the one hand, an increase in mean annual temperature of 1–2 K might affect only the lower alpine belt. On the other hand, a 3–4 K increase, which equals the temperature amplitude of an entire vegetation belt, would be likely to change the vegetation of the present alpine belt substantially. Today's alpine belt might show, after warming, a mosaic of subalpine elements (isolated arolla pine, fragments heaths and swards) in the warmest and most stable places and alpine elements in the coldest and least stable ones (open cryophilous swards with *Elyna myosuroides* (Villars) Fritsch, *Juncus trifidus* L., *Festuca halleri* All., *Festuca quadriflora* Honck., *Carex firma* Host on ridges and ledges, cushion plant communities with *Silene exscapa* All., *S. acaulis* (L.) Jacq., *Minuartia sedoides* (L.) Hiern, *Saxifraga bryoides* L. on superficial mineral soil with solifluction). The present high alpine vegetation would shift into the nival belt, where it already exists, but its extension might depend more strongly on edaphic factors and less on the general climate.

To illustrate the possible evolution of the upper subalpine and alpine vegetation in the near future according to a mean plausible climatic scenario, table 5.3 provides an assessment for the two regions investigated in the Valais in the Ecocline project.

Table 5.3 Predicted evolution under environmental changes of the plant communities of the subalpine-alpine ecocline in Belalp and Val d'Arpette at the end of next century

Vegetation units	Belalp		Val d'Arpette	
	Present	Predicted	Present	Predicted
Subalpine *Picea abies* forest	−	?	+++	>
Subalpine *Pinus uncinata* forest	−	−	+++	>
Thermophilous *Pinus cembra* forest	−	−	+++	>
Mesophilous *Pinus cembra* forest	−	−	++	<
Prostrated *Pinus uncinata* scrub	−	−	+++	>
Alnus viridis scrub	++	≫	+	= / <
Subalpine mesophilous *Rhododendron ferrugineum* heaths	+++	<	+	<
Subalpine meso-thermophilous *Rhododendron* and dwarf *Juniperus* heaths	++	≫	++	= / <
Subalpine meso-thermophilous *Calluna vulgaris* heath	+++	<	−	−
Thermophilous *Arctostaphylos uva-ursi* and dwarf *Juniperus* heaths	+	>	+++	> / =
Cryophilous *Empetrum hermaphroditum* and *Loiseleuria procumbens* low heaths	+++	<	+	<
Meso-thermophilous *Vaccinium* spp. and dwarf *Juniperus* heaths	−	≫	+++	>
Subalpine *Nardus stricta* swards	+++	≪	−	−
Subalpine-alpine *Carex sempervirens* and *Nardus stricta* swards	+++	< / =	++	< / =
Subalpine-alpine *Carex sempervirens* and *Festuca puccinellii* swards	++	< / =	+++	<
Mesophilous subalpine-alpine *Agrostis schraderana* meadows	+++	<	++	<
Subalpine thermophilous *Festuca varia s.l.* swards	+	>	++	=
Subalpine mesophilous *Calamagrostis villosa* meadows	−	>	++	=
Alpine *Festuca varia s.l.* (heaths-) swards	−	−	+++	=
Alpine *Festuca halleri* swards	++	<	+	=
Alpine *Carex curvula* swards	+++	<	−	−
Alpine *Elyna myosuroides* swards	++	= / <	−	−
Alpine *Salix herbacea* snow beds	++	<	−	−
Eriophorum angustifolium neutral swamps	++	= / <	−	−
Trichophorum cespitosum neutral swamps	+++	= / <	+	<
Neutral swamps with *Carex* spp.	+++	=	−	−
Neutral swamps with *Carex nigra*	++	=	−	−
Molinia caerulea wet meadows	−	—	+	<
Springs sensu lato	++	=	+	<
Rock fissures	+	=	+++	=
Screes	++	=	+++	=

Note: Predictions are based on the following environmental changes: increase of mean annual air temperature of 3 K, increase of precipitation by 20%, increase of atmospheric CO_2 concentration to 700 $\mu g \cdot 1^{-1}$, nitrogen deposition of 5–30 $kg \cdot ha^{-1} \cdot year$, ongoing acidification, and no change in human use. Present cover of plant communities: −: absent; +: 1–100 m²; ++: 100 m²–1ha; +++: > 1ha. Predicted evolution: >: increase; ≫: strong increase; <: decrease; ≪: strong decrease; =: no change; ?: uncertain. Both sites located on siliceous rocks. The Belalp site (3 km², east oriented) is larger than the Val d'Arpette site (2 km², south oriented).

5.6.5 Assessments of the Response of the Subalpine-Alpine Soils

Climate change might have direct and indirect effects on subalpine and alpine soil ecosystems. The most important effects would concern the evolution of permafrost, the processes of material transfer, changes in the organic matter, and changes in nitrogen supplies. Johnson (1993) discussed several issues about the potential impact of climate change on the carbon cycle.

5.6.5.1 Differences between Carbonate and Noncarbonate Soils Climate change would have different effects on carbonate and noncarbonate soils. Because limestone soils have large chemical buffering capacity (Duchaufour 1983), their internal processes are less sensitive to changes than those of acid soils. Carbonate soils are characterized by an iron deficiency, large phosphate and HCO_3^- supplies, and the presence of nitrates instead of ammonium (e.g., Gigon 1971, 1983, 1987).

The weak physicochemical resistance of acid soils (except for the most evolved podzols) rapidly leads to changes in the soil's internal functioning. For example, the clay-humus complex may break up, acidity may increase, or aluminium may dissolve at toxic levels because of its solubility at low pH. Breymeyer and Uba (1989) arrived at similar conclusions by observing the decomposition of litter, which appeared to be faster on granitic soils.

5.6.5.2 Changes in Permafrost Cryosoils are closely associated with the distribution of permafrost. These soils have a permanently frozen layer at a certain depth below ground level and reduced biological activity. They can be practically bare of humus in the most extreme conditions where vegetation has hardly taken root, or they can show signs of organomineral evolution (humus-containing cryosoils). Study of the two types of surface, including colonization by vegetation and the evolution of organic residues, yields useful information about changes in permafrost. This pedological approach is a useful aid to studying permafrost, the thawing of which could be a major and rapid consequence of global warming. Depending on its type and thickness, the thaw of a permafrost body might take decades to centuries after an increase in the mean annual air temperature (Häberli 1994, 1995; Keller 1994).

5.6.5.3 Changes in the Processes of Material Transfer Soil supports numerous upward and downward movements of material. In an alpine climate, downward movement generally dominates upward because of a positive balance of precipitation over evapotranspiration. However, certain mesoclimates may be exceptions to this rule: the southern Alps, influenced by the Mediterranean climate with a dry summer season, and the inner Alpine valleys, with a continental tendency up to fairly high altitudes (e.g., at Zermatt, Valais). Thus, scenarios for material transfer under climate change must take mesoclimate into account.

Sensitivity of Plant and Soil Ecosystems of the Alps to Climate Change

The balance between precipitation and evapotranspiration determines the intensities of the three main downward transport mechanisms: *leaching* of dissolved ions, *flushing* of clay and *cheluviation* of organometallics. Any scenario under climate change must therefore take into account the distribution of precipitation over the year of the observed changes, especially in the spring (plants begin to grow, snow melts, podzols develop further) and in summer (maximum temperatures, maximum productivity).

In a more general context, Post et al. (1982) suggested that the pool of organic matter increases with precipitation, decreases with temperature (at constant precipitation), and increases with the ratio of precipitation over evapotranspiration. Johnson (1993) experimentally studied the effect of temperature increase's effect on the release of organic acids in podzolized subalpine soils. Their main conclusions: "Warming will likely lead to a decrease in organic acid leaching and to an increase in carbonic and/or nitric acid leaching, depending on soil nitrogen status ... Thus, a significant increase in soil temperature in cold-region soils may well lead to a shift from organic to carbonic acid dominance, with attendant changes in soil solution pH (increase) and iron and aluminium transport by cheluviation (decrease)." (See also Johnson et al. 1977, 359).

Leaching of basic cations could also be affected. Dambrine (1985) demonstrated the utility of the gravity flow fraction of the major elements in the soil solution for detecting short-term processes, and Keller (1991) investigated the case of heavy metals. Although these results appear to be clearcut as regards temperature, they say nothing about the combined effects of changes in precipitation, evapotranspiration, and temperature. It is well known in ecology that the single-factor behavior of a process can be quite different from its multifactorial behavior.

5.6.5.4 Changes in Organic Matter

How organic carbon in the soil would change as a result of climate change is unclear. Although experiments suggest that soil carbon would increase with rising temperature (Van Cleve, Oechel, and Hom 1990; Anon 1992; Gifford 1994), global models provide contrary patterns (Olson 1963; Melillo, Aber, and Muratore 1982). Studies on local mountainsides indicate that it is premature to draw conclusions because of the intervention of other ecological factors (e.g., links between organic matter and clays) and because of the diversity of organic matter in the soil. For example, Johnson (1993, 357) stated, "Thus it would seem unlikely that soil organic matter will follow the same idealised decay curves as fresh litter does."

Besides the stocks of organic matter, another aspect of soil organic matter that climate change could modify is its rate of decomposition, which is itself linked to the vegetation (Tinker and Ineson 1990). Higher temperatures would increase animal and microflora activity, which might enhance decomposition rates of organic matter in the subalpine and alpine soils. However, the quality of organic matter and litter prevails in the processes of decomposition, in particular, nitrogen content and the relative concentration of lignin and acid-soluble carbohydrates (Hobbie 1996). Thus, other effects might result from climate change.

For instance, subalpine ericaceous heaths might be able to colonize alpine meadows as the mean temperature increases. The quality of the litter might change with this change in flora. The more resistant, acidifying residues of the heath plants might dominate the more easily decomposable debris of the herbaceous meadow plants. Woody structures and thick cuticles in the litter are more resistant to physical degradation, and their chemical decomposition releases products like phenolic acids, which form complexes with metals, or like tannins, which are toxic to bacteria. A multiple effect could then result: increase in the carbon stock in the holorganic layers, decrease in biological activity, and increase in chelates.

Another important aspect to be taken into account is the increase of atmospheric concentration of CO_2, which could enhance primary productivity and thereby modify the carbon/nitrogen (C/N) ratio and, consequently, the chemical quality of litter (e.g. Bazzaz 1990; Woodward, Thompson, and McKee 1991; Coûteaux et al. 1991; Anderson 1992; Lambers 1993; Körner et al. 1996). Qualitatively, the litter and its rate of decomposition might change without any change in the flora's composition. However, not all species would follow this principle, as shown by the in situ quality of litter obtained under high CO_2 in temperate and tropical environments, which "was not significantly different from that produced in similar communities maintained at current ambient CO_2" (Hirschel, Körner, and Arnone 1997, 387). Although there was an effective small decrease in litter quality of the alpine sedge *Carex curvula* All. in an experiment of CO_2 enrichment conducted in an alpine sward in the Swiss Alps, Hirschel, Körner, and Arnone (1997) conclude that the decomposition rate might remain the same over the long-term.

Tracking changes in soil organic matter is important for understanding soil-vegetation dynamics and the carbon budget, but in mathematical ecosystem models organic matter is still too often regarded as merely a simple, temporary stock of carbon. To be plausible, any explanation of how the carbon stock evolves in the soil would have to include some detailed understanding of the dynamics of organic material, and in particular of the role that nitrogen content plays in the litter and of the conditions for releasing complexing acids (see section 5.6.5.5). Presently, no standard method exists for physicochemical analysis of the organic material, and so it is difficult to compare results from different studies (Johnson 1993).

5.6.5.5 Changes in Nitrogen Supply Global warming might enhance decomposition rates of soil organic matter, with consequent changes in the nitrogen cycle, an increase in nitrogen mineralization, lowering of the C/N ratio, and change of pH. Global warming might also affect the nitrogen content in subalpine soils by modifying the form of nitrogen in the spring. With low temperatures, mineral nitrogen is essentially in the ammonium form (NH_4^+). Given a 2–4 K increase of soil temperature, it would be transformed into nitrate (NO_3^-) (Pornon and Doche 1994).

An increase in nitrogen content of soils would have direct consequences for vegetation. In oligotrophic ecosystems like subalpine and alpine ones, slow-growing plants with low nitrogen requirements, low productivity, and low nutrients cannot use additional nutrients (e.g., Chapin and Shaver 1989; Steubing 1993) but faster-growing alpine plants can (Schäppi and Körner 1996). As a result of increased productivity and a more rapid turnover of soil nutrients, oligotrophic plant communities might disappear (Berendse 1993; Vitousek 1994). On the other hand, nitrogen availability to plants could decrease if an increase of carbohydrates released by nonmycorrhizal plants stimulated the abundance and activity of soil microorganisms (Díaz et al. 1993; Bazzaz 1996). A negative feedback loop could also operate through modification of the organic matter's quality, which would then benefit the oligotrophic vegetation. For a typical alpine meadow with Carex curvula All., Körner et al. (1996) showed that with elevated CO_2, only small amounts of additional nonstructural carbohydrates accumulate, both in above and underground parts. Thus, they suggested that a part of the additional net assimilation may be sequestered in the soil as a faster turnover of carbohydrates in roots or root exudation.

A change in the form of nitrogen available to plants, that is, a modification of the nitrate/ammonium ratio, might also have important ecological consequences. Pornon and Doche (1994, 1995a) observed that the alpenrose (Rhododendron ferrugineum L.) performed better for growth and germination where the nitrate/ammonium ratio was nearer to 0 than to 1. Troelstra, Wagenaar, and Smant (1995a, 1995b) demonstrated that heathland species adapted to ammonium-based nutrition can grow with nitrate as a source of nitrogen, but that more carbon is then allocated to root growth, changing the shoot/root ratio. However, transformation of ammonium into nitrates through nitrification is an acidifying process, with the leaching of basic cations and mobilization of aluminium. At the subalpine-alpine belts, where precipitation exceeds evapotranspiration, a permanent acidification can be expected. Under an oligotrophic vegetation, nitrates produced may also be leached if none of the species present can absorb nitrate selectively over ammonium, as may be the case under a closed heath. Then, a negative feedback loop could operate, leading to nutrient depletion.

Predicting changes in soil nitrogen is a complex question that cannot be dissociated from those of soil organic carbon (e.g., Körner 1989; Körner and Arnone 1992; Bowler and Press 1993; van de Geijn and van Veen 1993; Gifford 1994; Coûteaux, Bottner, and Berg 1995; Bazzaz 1996). Both positive and negative feedback mechanisms can occur in this relationship and affect the C/N ratio. Coûteaux, Bottner, and Berg (1995) concluded that increased atmospheric concentrations of CO_2 could decrease the nitrogen content of litter, and Körner et al. (1996) have confirmed this for some forbs in an alpine meadow. In contrast, Gifford (1994) suggested that the nitrogen cycle might track the carbon cycle in the long term through biological nitrogen fixation and, therefore, the nitrogen concentration of soils might increase. In addition to potential changes in nitrogen supply because of warming, nitrogen supply in subalpine and alpine soils is already increasing through atmospheric deposition by 5–30 kg/ha per year (Körner 1989).

5.6.6 Summary

• In highly structured vegetation, synusiae of the understory are unlikely to react directly to climate change but rather only indirectly to a change in their microhabitat conditions.

• Present climatic climax plant communities are likely to persist in climate change as edaphic climaxes.

• Present edaphic climaxes are likely to persist temporally and spatially in climate change according to their buffering capacities, that is, so far as their limiting factors are not changed.

• Natural landscape is likely to change at a differential rate, according to its component elements (lagging vs. rapid change).

• The cultural landscape that will undergo climate change may be very different from today's.

• The response of species, plant communities and landscape to a temperature increase is not likely to be linear, and there may be an inertia of 1–2 K.

• A temperature increase of 3–4 K would very likely have a profound effect at every level of complexity, since it equals the temperature amplitude of an entire vegetation belt.

• Fragmentation, diminution of populations, and selective extinctions of high alpine and arctic stenoicious relict plant populations of bryophytes and vascular plants are likely over the entire Alps if the temperature increases by 3–4 K.

• Alpine endemics restricted to tops of low mountains (i.e., those lacking nival belts, mainly in the eastern Alps) are likely to be severely endangered of disappearance/extinction if the temperature increases by 3–4 K.

• Phenotypic plasticity and genetic adaptation might buffer the effect of climatic change in some cases for vascular plants, and perhaps more frequently for bryophytes.

• Changes in forest types might occur in 30–55 percent of the forested area for an increase of 1–1.4 K, and in up to 55–89 percent for a 2–2.8 K increase, according to static modeling of Swiss forests (Kienast, Brzeziecki, and Wildi 1995, 1996).

• Adaptation of forests and tree species to climate change cannot be suitably established without taking into account the typically high genetic diversity of tree populations and the maintainance of that diversity.

• For an increase of 1–2 K in the mean annual temperature, the present upper subalpine forest limit is not likely to shift upward much more than 100–200 meters, because of its temperature-related inertia, but the *kampfzone* might move into the low alpine belt in favorable places.

• For an increase of 3–4 K in the mean annual temperature, the *kumpfzone* would be very likely to invade the alpine belt, with an upward shift of the forest limit into the low alpine belt.

• With an increase of 3–4 K in the mean annual temperature, subalpine elements would be likely to invade the entire alpine vegetation belt, and new plant communities would be likely to replace in part the present communities.

• For an increase of 1–2 K in the mean annual temperature, human use might slow down the subalpine vegetation's upward shift, in particular that of heaths, shrubs, and trees, through an intensification of pasturing. In contrast, humans could also accelerate the shift through abandoning agricultural activities and afforestion of the upper subalpine belt.

• The upward shift of the present alpine vegetation into the nival belt, where it exists, might depend more on edaphic factors than on climatic ones because of the upper alpine and nival belts increasing steepness. Therefore, it is likely that present alpine plant communities on gentle slopes would disappear.

• Carbonate soils would likely be less prone to react to climate change than noncarbonate soils, thus acting nonsynergistically on vegetation change.

• Processes in soils operate at different rates, and many happen on the time-scale of one year or tens of years. The accepted idea that soils evolve on a scale of centuries or millenia must be abandoned in considerations of climate change.

• Organic matter in soils is the key factor that would be affected, directly or indirectly, qualitatively and quantitatively, by climate change, but its evolution cannot be dissociated from that of soil nitrogen content.

• Nitrogen content, form, and cycle would be affected, and cannot be dissociated from changes in organic matter, as, for example, an increase in humification.

5.7   CONCLUSION

The plant and soil ecosystems of the European Alps are important for many biological, socioeconomic and cultural reasons. Because of their great ecological diversification, the Alps offer numerous habitats, which explain their high biodiversity.

On the basis of the information presented in this chapter, it appears that mountain ecosystems and soils would present a contradictory response to predict climate change, apparently being simultaneously characterized by sensitivity and resistance to it. Many uncertainties remain as to climate change's possible effects on subalpine and alpine ecosystems and soils in the

Alps, and it seems impossible to propose an ecosystem evolution valid for the entire Alpine chain simply because of the diverse mesoclimates, with all possible gradations between oceanic, continental, and Mediterranean influences. For this reason, and because of the interdependence of ecological and biological processes, their multifactorial, nonlinear effects, and the uncertainties about climate changes themselves, it is impossible to predict reliably soil and vegetation ecosystems' response to climate change at the regional level. Only general assessments can be made, such as those summarized in section 5.6.6.

Vegetation's apparently contradictary response to climate change comes from differential reactions, both temporally and spatially, of climatic and edaphic plant communities and ecosystems, because the climatic ecosystems are inclined to react, whereas the edaphic ones are not. This phenomenon grows in importance as ecological diversity increases. Biological inertia may also act synergistically with the edaphic factors, dependent on the ecosystem structure and the strategy of the species as well as their physiological and genetic characteristics.

As far as the biota is concerned, high mountains seem able to counterbalance the impact of climatic change to a certain extent. Indeed, mountains offer options for potential adjustment to the climatic requirement of the biota mainly through the possibility of altitudinal migration or shifting from more south-facing slopes to more north-facing ones. Thus, high mountains act as sanctuaries for preserving biodiversity; the presence of very old species in many mountain systems bears witness to this. However, such counterbalancing has limits. Glaciation did breach this effect, but certainly as much through mechanical destruction and the physical barriers of glaciers as through the very cold climate itself. With warming, no such destruction and obstructions would occur. Nevertheless, we do not know to what extent the Alps' protective capacities could stand a rapid increase in temperature. We are inclined to agree with some authors (e.g., Körner 1995) that an increase of 1–2 K would not affect the diversity of the Alpine chain very much. Paleoecological data showing the small variation (about 100–300 m) in the tree line during the Holocene period support this viewpoint. Within this temperature range, changes in the flora would occur on the ecological level (micro- and macrohabitats), and these would have limited implications on the phytogeographical level. An increase of temperature greater than 3 K and lasting for centuries would have serious implications on both the ecological and phytogeographical levels, because such an increase would be on the order of or greater than the temperature range of a vegetation belt (3.8 K).

To understand and assess the impact of climate change on plant distributions in high mountain areas and to produce models, we must consider the differential response between climatic and edaphic ecosystems, along with biological characteristics, from the structure of ecosystems to genetics, within the paradigm of the individual reaction of species, for which the following principles can be proposed:

1. Proceed regionally, to work as much as possible under a uniform meso-climate.

2. Determine the ecological inertia of the main ecosystems and soils with respect to climate, in particular relative to edaphic factors.

3. Evaluate the biological inertia of ecosystems with respect to climate, in particular relative to their structural complexity, and of species relative to their phenotypic plasticity and their genetic diversity.

4. Conceive studies at related spatial scales, from population to landscape and vice versa (strategic cyclical scaling; Root and Schneider 1995), and on a temporal basis, to avoid pseudopredictability (Gosz 1991, 1992),

5. Combine present data and palaeodata, field experiments, modeling, and monitoring, along with regional climate models.

ACKNOWLEDGMENTS

The assistance of the following persons is kindly acknowledged: Dr. D. Aeschimann, Geneva; Dr. C. Béguin, Fribourg; Dr. J.-L. Borel, Grenoble; A. Bushnell, Sierre; Prof. U. Deil, Freiburg; Prof. P. J. Edwards, Zürich; Prof. C. Favarger, Neuchâtel; Prof. G. Grabherr and colleagues, Vienna; Dr. J. Holten, Trondheim; Dr. F. Jacquemoud, Geneva; Dr. F. Keller, Samedan; Prof. F. Klötzli, Zürich; Prof. C. Körner, Basel; Dr. L. Kohorn, Durham; Dr. R. E. Longton, Reading; Prof. E. Martini; Genova, and Dr. J. Stöcklin, Basel.

GLOSSARY

Definitions are mostly taken or adapted from: Allaby, M. 1985. *The Oxford Dictionary of Natural History*. Oxford: Oxford University Press. Art, H. W. 1993. The *Dictionary of Ecology and Environmental Science*. New York: Henry Holt & Co. Foth, H. D. 1990. *Fundamentals of Soil Science*. New York: John Wiley & Sons.

Acclimation: Response of a plant to a change in its environment through modification of its physiological mechanisms to support the new conditions.

A-horizon: See *horizon*.

allele: Common shortening of the term "allelomorph." One of the several forms of a gene arising by mutation, hence differing in DNA sequence, and occupying the same relative position (locus) on homologous chromosomes.

  allele frequency: The commonness of an allele in a population.

allogamy: Mating system consisting of cross-fertilization among different, individuals belonging to the same species that is, in plants, fertilization of a flower with pollen produced by another individual.

allozyme: Common shortening of the term "alloenzyme." One of the isozymes coded for by different alleles at the same locus.

alpine: In the phytogeographical sense, the zone in high mountains above the tree line and below the snow line. By this meaning, the term "alpine" does not necessarily apply to the Alps. See also vegetation belt.

alpine belt: See *vegetation belt.*

alpine zone: See *vegetation belt.*

ammonium: The ion NH_4^+.

antheridium (plural antheridia): Male sex organ or gametangium, within which male gametes (spermatozoids) are formed in algae, fungi, bryophytes, and ferns. In bryophytes, it is a multicellular globose to broadly cylindric, stalked structure.

apomixis: Asexual reproduction in plants without fertilization or meiosis, consisting of the development of an individual from an unfertilized egg (reproductive cell) or a somatic cell (any other nonreproductive cells).

archegonium (plural archegonia): Female sex organ or gametangium, containing the single egg cell in algae, fungi, bryophytes and ferns. In bryophytes, it is a multicellular, flask-shaped structure consisting of a stalk, a swollen base or venter, and a slender neck.

Arcto-Tertiary: Term indicating the subtropical flora occupying the Northern Hemisphere during the Tertiary, from which orophytes originated in the high mountains of the Northern Hemisphere.

asexual reproduction: Reproduction without the sexual processes of gametes (sex cells), for example, apomixis.

autogamy: Mating system consisting of the self-fertilization of an individual, that is, in plants, fertilization of a flower with its own pollen. Also extended in the genetic sense to fertilization of a flower with the pollen produced by other flowers of the same individual.

azonal vegetation: Vegetation that is independent of the climate (as opposed to zonal vegetation, that is, orobiomes and zonobiomes). For example, the vegetation of cliffs, screes, and wet places is azonal.

belt: See *vegetation belt.*

biogeochemical turnover: The movement of chemical elements from organism (biological part) to physical environment to organism in a more or less circular pathway. The form and the quantity of elements varies through the cycle.

biomass: The total weight of the living components in an ecosystem at any moment, usually expressed as dry weight per unit area.

biome: Biological subdivision of the earth's surface that reflects the ecological and physiognomic character of the vegetation. Biomes are the largest geographical biotic communities.

  orobiome: Biome resulting from the altitudinal succession of particular vegetation belts in high mountains because of the regular diminution of temperature with increasing elevation.

  pedobiome: Biome resulting from particular edaphic (soil) conditions instead of climatic ones.

  zonobiome: Biome resulting from a climatic region. On a broad level, zonobiomes are arranged latitudinally, for example, tropical rainforest biome, desert biome, temperate biome, tundra biome.

biosphere: The part of the earth's environment in which living organisms are found.

BP: See *horizon.*

BPh: See *horizon.*

Bps: See *horizon.*

bryophyte: A division of plants, that includes the mosses, liverworts, and hornworts. They are nonvascular, green plants whose gametophytic generation is conspicuous and dominant and whose sporophytic generation is ephemeral.

buffer capacity: A measure of the ability of a buffer to neutralize. Loosely, the ability of a solution to absorb acids or bases without changing its pH.

C/N ratio: The ratio of the mass of organic carbon to the mass of organic nitrogen in soil or in biomass.

canopy: Usually, the upper level of woody communities where tree branches and leaves are most dense and the distinctive microenvironment which they form. The term may be applied to any terrestrial plant community in which a distinctive habitat is formed in the upper, denser regions of the taller plants.

carbohydrate: A class of organic molecules produced by plants through photosynthesis, including sugars and starches, such as glucose and cellulose.

carbon balance: The physiological ratio in plants between the production of carbohydrates and their use as a source of energy through catabolism. A negative carbon balance leads to the death of the plant once all reserves have been used, since energy demand exceeds energy production.

carbonic acid: The molecule H_2CO_3.

catabolism: The part of cellular metabolism encompassing the reactions that yield energy through the degradation of food molecules.

cation: Ion carrying a positive electrical charge.
  basic cations: Ions that can neutralize acids, such as Ca^{++}, Mg^{++}, K^+, Na^+.

chamaephyte: A plant in which surviving organs (perennating buds) lie very close (less than 50 cm) to the ground. (One life form of Raunkier s classification system.)

chelates: Organo-metallic molecules that form a ring structure secured by multiple chemical bonds.

cheluviation: The removal of organo-metallic chelates from the soil, for example, from the upper layers during podzolization.

clay: (a) A soil component consisting of particles smaller than 0.002 millimeters in equivalent diameter.
(b) an aluminium silicate mineral having a layered structure (phyllosilicate) and the ability to take up and lose water readily, such as kaolinite and vermiculite.

climax community: The final stage of a plant succession, in which vegetation reaches a state of equilibrium with the environment. The region's climate and soil largely determine its nature.
  climatic climax: The final stage of plant succession that climate mainly determines.
  edaphic climax: The final stage of plant succession that edaphic (soil) processes mainly control.

cline: Gradual change in gene (allele) frequencies or character states within a species across its geographic distribution.

clonal plant: A plant that, through vegetative growth, forms more or less extensive patches constituted of modular parts without actual detachment. When separated, each modular part can grow independently and propagate the plant.

colline belt: See *vegetation belt*.

colluvial: Adjective from *colluvium*. A general term applied to deposits on a slope or at the foot of a slope or cliff that were moved there chiefly by gravity.

complexing acids: The organic acids that have the property of forming a chelate that can bind metallic cations such as iron or aluminium.

continentality: A measure of how a place's remoteness from the oceans and oceanic air affects its climate.

coprophilous: Growing on or in dung.

crenic acids: Small organic acids, precursors of the fulvic acids.

cryogenic: In soils, processes produced by freeze-and-thaw cycles.

cryophilous: Preferring to grow at low temperatures.

cryosoil: Soil characterized by a temporarily or perennially frozen soil horizon (permafrost).

cryoturbation: General term for the disturbance of ground due to freezing and thawing, such as upward movement of the ground surface or individual particles due to the formation of ice in the soil.

cryptopodzol: See *podzol*.

cuticule: (a) Thin, waxy, protective layer covering the surface of the leaves and stems of plants. (b) The outer layer of arthropods (e.g., insects) formed by proteins and chitin.

Devonian: A period of the Paleozoic era (the first era of "obvious" life) that began 410 million years ago and lasted approximately 65 million years.

dioecious: Possessing male and female reproductive organs on separate, unisexual, individual plants.

diploid: An organism with two chromosome sets in each cell (excluding the reproductive cells), one from each parent.

DNA (deoxyribonucleic acid): A complex organic molecule found in all animals and plants cells that contains the genetic information passed from one generation to the next.

ecocline: Gradual change in community composition or ecosystem type along a major environmental gradient.

ecophysiology: The study of the physiology of organisms in relation to their environment.

ecosystem: A functioning unit of nature that combines communities of living organisms and the environmental parameters with which they interact.

ecotone: Narrow and fairly sharply defined transition zone between two or more different communities.

ecotype: A locally adapted genetic variant within a species. Ecotypes retain their physiological and morphological differences when transplanted to a single location.

ecotypic differentiation: Different selection pressures of different environments result in the development of different ecotypes within a single species.

edaphic: Of or influenced by the soil.

edaphic climax: See *climax*

endemic: Indigenous to and restricted to a particular geographic region.

subendemic: An organism extending slightly outside a particular geographic region to which it is mainly restricted.

enzyme: The proteins of the cell that catalyze chemical reactions (see also *allozyme, isozyme*).

Eocene: Tertiary epoch that began approximately 55 million years ago and ended 17 million years later, during which mammals underwent much expansion.

eutrophication: Process of nutrient enrichment, especially of nitrogen and phosporus in aquatic ecosystems, and by extension in all ecosystems. Human activity greatly speeds up this otherwise slow process (over geological time). Contrast with *oligotrophic*.

evapotranspiration: Term for water loss from a given area during a specified period of time by evaporation from the soil surface and transpiration from plants combined.

extrazonal vegetation: Vegetation found outside the range where it is zonal (i.e., determined by climate) because edaphic factors compensate for the difference in climate. The vegetation which normally occurs in a dry, warm climate is an extrazonal vegetation when it is found in a wet, temperate climate.

flora: All the plant species that make up the vegetation of a given area.

flushing: The removal of clay minerals from the soil.

formation: In vegetation description and analysis, a classificatory unit that implies a distinctive physiognomy rather than a distinctive species composition.

fulvic acids: The yellow mixture of complex organic molecules that remains in alkaline solution of organic matter after acidification removes humic acid.

functional type: Grouping of species according to some of their functional traits (e.g., legumes fixing atmospheric nitrogen through bacteria in their root nodules). Functional types allow a comparison among ecosystems differing in their species composition.

gametophyte: The haploid phase (the phase with one set of chromosomes) of a plant during which gametes are produced by mitosis.

gelifluction: The downslope movement of water-saturated layers of soil or rock debris over a frozen ground. Gelifluction may occur on slopes of even as little as 1°.

gene flow: The movement of genes within a population and from one population to another by the dispersal of pollen, seeds, or individuals.

genetic drift: Random fluctuations in gene frequencies occurring in an isolated population from generation to generation. Genetic drift results from chance combinations of different characteristics and is independent of the forces of selection. Although drift occurs in all populations, its effects are most marked in very small isolated populations, in which it gives rise to the random fixation of alternative alleles.

genetic marker: A marker at the genetic level (such as a specific enzyme).

genetic polymorphism: See *polymorphism.*

genotype: The genetic composition, latent or expressed, of an organism, as opposed to its phenotype; often, the particular combination of alleles at one locus or at a set of loci.

geophyte: A land plant that survives unfavorable periods by means of underground food storage organs, such as bulbs, tubers, and rhizomes. Buds arise from these to produce new aerial shoots when favorable growth conditions return. (One life form of Raunkier's classification system.)

habitat: Living place of an organism or community, characterized by its physical or biotic properties.

microhabitat: The portion of a more general habitat actually frequented by a specific organism, group of organisms within a community, or small, specialized plant community, and often characterized by a distinct microclimate.

holorganic: Strongly dominated by organic material (e.g., top soil layers like O-horizon).

homeohydry: Stage of plants that are able to regulate their rate of water loss. Homeohydric plants (like ferns or flowering plants) are opposed to poikilohydric plants.

horizon: In pedology, a relatively uniform soil layer that lies, at any depth in the soil profile, parallel, or nearly so, with the soil surface, and that is differentiated from genetically related horizons above and below by contrasts in physical, chemical, and biological properties or characteristics.

A-horizon: Organo-mineral horizon formed at the surface and characterized by an accumulation of humified organic matter intimately mixed with the mineral fraction.

BP: A soil horizon in which chelates carried from the overlaying layers has been precipitated from solution (**BPh**—with humus accumulation; **BPs**—with iron and aluminium accumulation).

O-horizon: The litter layer with its subhorizons of decomposition (fragmentation, fermentation, humification).

humic acids: A dark mixture of complex organic substances that can be extracted from soil by various alkaline agents and is precipitated by acidification to pH 1 or 2.

humification: Development of humus from dead organic material. Also, the process whereby the carbonic molecules of organic residues are transformed and converted to humic substances through biochemical and/or chemical processes.

humine: The most humified part of organic matter, very resistant to alteration.

humus: An organic soil material so thoroughly decayed that the identity of the biologic source cannot be recognized, that is, all of the organic compounds in soil exclusive of undecayed plant and animal tissues, their partial decomposition products, and the soil biomass. Resistant to further alteration.

intraspecific: Occurring among individuals within a species.

isozyme: Common shortening of the term "Isoenzyme." Each form of an enzyme that catalyzes a given chemical reaction.

kampfzone (German word): Transition zone between the subalpine and the alpine belts of stunted and often prostrate trees (designated as Krummholz) found between the upper limit of tall, erect trees growing in forest densities (the timber line) and the extreme upper limit of tree growth (the tree line); that is, the limit of the microphanerophytic life form.

kinetic: Describes in particular the speed at which enzymes catalyze chemical reactions.

krummholz zone (German word): See *kampfzone*.

layer (in soils): See *horizon*.

leaching: The removal of materials (especially cations and anions) in solution from the soil by the action of percolating liquid.

life form: An organism's structure, form, habits, and life history. In plants, especially, characteristic life forms, in particular morphological features, are associated with different environments. This observation has formed the basis of life form schemes for the classification of vegetation, such as Raunkier's system. (See, e.g., *chamaephyte, geophyte, phanerophyte, therophyte*.)

lignin: Complex aromatic substance, resulting from the polymerization of phenyl propanoid subunits, present in many plant cell walls, especially the wood of vascular plants, not yet detected in bryophytes. Its function appears to be to cement and anchor cellulose fibers and to stiffen the cell wall.

liverwort (hepatic): Common name for plants of the division Hepatophyta (bryophytes).

locus (plural loci): Specific place on a chromosome where a given gene is located. At each locus there is one gene; if that gene can take several forms (alleles), only one will be present at a given locus. Homologous chromosomes contain identical sets of loci in the same linear order.

loess: Unconsolidated material transported and deposited by wind, consisting of predominantly silt-sized particles (0.002–0.05 millimeters) and showing little or no stratification.

meiosis: The form of cell division in which the number of chromosomes is reduced by half, occuring in reproductive cells of sexually reproducing organisms. Meiosis involves two consecutive cell divisions with only one replication of the chromosome set, thereby producing four sexual cells (gametes) characterized by one-half the parental genetic information.

mesoclimate: A local variation in climate found only in a restricted area, such as a small valley.

metabolism: All the chemical and physical activities that sustain an organism. Metabolism involves the breakdown of organic compounds to create energy (catabolism) that is used to create complex compounds (anabolism) such as proteins from simpler compounds. Plant metabolism involves the creation of organic compounds through photosynthesis as well as their use to create energy.

microclimate: A very small-scale variation from the overall climate pattern, usually caused by local physical conditions such as topography (topoclimate) or the structure of a plant community (e.g., the microclimate at the ground level, at the canopy, etc.). Microclimates are related to microhabitats.

microhabitat:: See *habitat*.

microphanerophyte: See *phanerophyte*.

mineralization: The breakdown of organic matter into its inorganic chemical components as a result of faunal and microbial activity in soils.

Miocene: Tertiary epoch that began approximately 24 million years ago and ended 5 million years ago.

mitochondria: Specialized, capsule-shaped or threadlike functional structures (organelles) in the cytoplasm of cells that contain genetic material and many enzymes. Mitochondria are the site of processes (primary reaction of respiration) that provide energy for cell metabolism. They are also involved in the synthesis of proteins.

montane belt: See *vegetation belt*.

moss: Common name for a plant belonging to the division of bryophytes.

mutation: (a) Process by which a gene or a chromosome set undergoes a structural, inheritable change. (b) Gene or chromosome set that has undergone a structural change at random. The majority of mutations are changes within individual genes (e.g., of the DNA sequence at some point of the chromosome), but some are gross structural changes of chromosomes or changes in the number of chromosomes.

mycorrhizae: Literally "fungus root." The association, usually symbiotic, of specific fungi with the roots or other structures of higher plants. The fungal mycelia help the plant absorb minerals and in return absorb energy compounds the plant produces.

nanophanerophyte: See *phanerophyte*.

nitrate: The ion NO_3^-. Nitrates are important as concentrated sources of nitrogen. They are very water soluble and easily leached from the soil.

nitric acid: HNO_3, a very corrosive strong acid.

nitrification: Biological oxidation of ammonium to nitrite and nitrate, or a biologically induced increase in the oxidation state of nitrogen. The various stages of nitrification are carried out by a group of bacteria collectively called nitrobacteria or nitrifying bacteria.

nival belt: See *vegetation belt*.

nival zone: See *vegetation belt*.

norm of reaction: The relation between phenotype and environment (see also *phenotypic plasticity*).

O-horizon: See *horizon*.

oceanicity: Maritime influences' effects on a climate.
  peroceanic: Showing a high oceanicity.

oligotrophic: Of low nutrient content, especially with regard to nitrogen, and therefore with low productivity. Contrast with *eutrophic*.

organic acids: Acids composed of organic molecules and the functional group –COOH.

orobiome: See *biome*.

orophyte (also *oreophyte*): Plant growing at high altitudes (usually from the subalpine belt) and showing specialized adaptations to cold conditions.

pedobiome: See *biome*.

pedoanthracology: The study of charcoal in soils.

pedosphere: The part of the biosphere that contains the soil layer.

periglacial: A term referring to an environment in close proximity to a glacier, more generally applied to any environment in which freeze-and-thaw cycles are dominant surface processes.

peroceanic: See *oceanic*.

pH: A measure of the relative concentration of hydrogen ions in a solution. It is calculated as the negative logarithm to the base ten of the hydrogen ion concentration in moles per liter. The lower the pH, the more acidic is the medium; the higher the pH, the more alkaline.

phanerophyte: A woody plant (tree or shrub) in which surviving organs (perennating buds) are located more than 50 cm above the soil level. (One life form of Raunkier's classification system.) Phanerophytes may be subdivided further according to their height as nano- (less than 2 m), micro- (2–8 m), meso- (8–30 m), and megaphanerophytes (more than 30 m). Another subdivision distinguishes between evergreen and deciduous.

microphanerophyte: Young tree or shrub 2–8 meters in height.

nanophanerophyte: Young tree or shrub 0.5–2 meters in height.

phenolic acids: The phenols with an acidic functional group –COOH.

phenology: The study of the relationship between climate and periodic natural phenomena such as bud bursting or flowering of plants.

phenophase: A distinguishable phase in the life cycle such as flowering or bud burst that is used in phenological studies.

phenotype: The observable manifestation of a specific genotype; those properties of an organism produced by the genotype in conjunction with the environment that are observable. Organisms with the same genotype may have different phenotypes because of the effects of the environment and of the gene interaction.

phenotypic plasticity: Variation in the phenotypic expression of a given genotype; a measure of the amount of variation in the observable aspects of a quantitative genetic character among individuals.

photosynthesis: The series of chemical reactions by which plant cells transform light energy into chemical energy by producing simple sugars (or other energy compounds) and oxygen from carbon dioxide (CO_2) and water.

phytocoenose: A plant community or the vegetative components of a biotic community (biocoenose).

phytogeography: The study of plant distribution throughout the world, with a focus on how environmental factors have influenced evolution and distribution.

phytosociology: The branch of plant ecology that studies the description and classification of plant communities based primarily on floristic rather than life form or other considerations, and the interrelations among the populations of various plant species. Also called "plant sociology."

plant community: A group of plants living together in the same area, usually interacting or depending on each other for existence.

Pleistocene: The first epoch of the Quaternary subera of geologic time. The Pleistocene lasted from approximately 2 to 0.1 million years ago.

podzol: A mineral soil type formed at an advanced stage of leaching by podzolization and identified by an eluviated and bleached horizon and an iron-colored BP horizon (spodic horizon). Also called "spodosol."

cryptopodzol: A soil type lacking the visual characteristics of a podzol but in which podzolization nevertheless occurs.

podzolization: One of the major processes of soil evolution consisting of the process of removal of iron and aluminium, humus, and clay minerals, with a very strong acidification, from the surface of soil horizons by cheluviation.

poikilohydry: Stage of plants that are not able to regulate their rate of water loss and therefore contain roughly the same amount of moisture as their immediate environment. Poikilohydric plants (like bryophytes or lichens) are resistant to prolonged drought periods.

polygenic character: A trait whose variation is determined by several genes.

polymorphism: The existence of phenotypic variation in a population, whether or not genetically based.

genetic polymorphism: The existence of genetically based variation in a population.

polyploid: An individual having more than two sets of homologous chromosomes, such as a tetraploid, which contains four sets of homologous chromosomes.

polysaccharide: A biological polymer made up of many sugar molecules. Polysaccharides are complex carbohydrates including starch and cellulose. They can be broken down into their simple sugar components and used by organisms as a source of energy.

pseudovicariance: See *vicariance*.

ranker: A poorly evolved soil on acidic rocks, with an A horizon directly laying on the bedrock.

recombination: The process occurring at meiosis through which offspring can display new genetic combination of the DNA producing different gene associations than either parents, resulting in new combinations of traits.

reduction: The opposite of oxidation; a chemical reaction in which a compound undergoes a decrease in oxidation state by gaining electrons, for example the reduction from Fe^{3+} to Fe^{2+}. Reduction of one compound is always accompanied by oxidation of another compound.

relict: Applied to organisms that have survived while other related ones have become extinct.

relict community: Community that formerly had a much wider distribution but that now occurs only very locally. Such contraction can be caused by various factors, including climatic change.

runoff: The portion of precipitation that flows (freely) over the soil surface into streams.

sexual reproduction: Reproduction involving the fusion of gametes (sex cells).

solifluction: The gradual downslope movement of soil saturated with water.

specific heat: The quantity of heat in calories required to raise the temperature of one gram of a substance by 1°C.

spodic: A term applied to a soil horizon characterized by illuvial accumulations of amorphous materials composed of aluminium and of organic carbon with or without iron. A spodic horizon is the diagnostic feature of a podzol (or spodosol).

sporophyte: The spore-producing diploid generation in the life cycle of plants. In higher plants, the sporophyte is the dominant generation, forming the conspicuous plant. In lower plants such as bryophytes, the gametophyte is the dominant and conspicuous generation, and the sporophyte remains attached to the gametophyte and is partially dependent on it.

stenoicious: Applied to an organism that can live only in a restricted range of habitats.

subendemic: See *endemic*.

subalpine belt: See *vegetation belt*.

synusia (plural **synusiae**): A functional unit within a distinct layer of a plant community composed of plants of similar life forms, phenology, and ecological requirements.

tannins: A group of acidic, aromatic compounds occurring widely in plants, such as oak bark or tea leaves, very resistant during the decay of the organic matter. Tannins are toxic substances with astringent properties whose principal function appears to be to render plant tissues unpalatable to herbivores.

Tertiary: Period in the earth's history that began 65 million years ago and ended some 2 million years ago, followed by the Pleistocene. During the Tertiary, mountains such as the European Alps, Andes, and Rockies were formed, especially during the last 40 million years of the epoch. Most of the orophytes separated from the surrounding flora of the warm lowlands in the late Tertiary (Pliocene).

texture: In pedology, the relative proportions of sand, silt, and clay in the fine earth of a soil sample, which give a distinctive feel to the soil when handled and which are defined by classes of soil texture.

thermocolline belt: See *vegetation belt*.

therophyte: A plant that completes its life cycle rapidly during periods when conditions are favorable and survives unfavorable conditions (cold, heat) as a seed; it is thus an annual or ephemeral plant. (One life form of Raunkier's classification system.)

tiller: In grasses and sedges, a lateral shoot arising at ground level.

timberline: (also **forest line**, **forest limit**, *Waldgrenze*):   Line that marks the altitudinal limit of trees that grow erect and tall (more than 8 m; mesophanerophytes) and form a forest.

topoclimate: See *microclimate*.

toponymy: The study of place names.

tree line: (Also **tree limit**, *Baumgrenze*) A line marking the limit of a zone of stunted tree growth (2–8 m; microphanerophytes), as is often found above the elevations at which trees grow uniformly erect and form a forest (timberline).

vascular plants: A division of plants that have vascular tissues, that is, conducting tissues through which water and nutrients are transported. The division comprises the ferns and fernlike plants (pteridophytes), the conifers and conifer-like plants (gymnosperms), and the flowering plants (angiosperms).

vegetation belt: Altitudinal section of a mountain characterized by a specific climate determining the growing period and vegetation. A vegetation belt forms a bioclimatic zone. In the European Alps, a vegetation belt has an altitudinal extension of approximately 700 meters, corresponding to a temperature range of 3.8 K, and a variation in growing period that ranges across 65 days.

    thermocolline belt: The lowest vegetation belt in the temperate regions, characterized by a growing season of more than 310 days and zonal hop-hornbeam forests.

    colline belt: The vegetation belt in the temperate regions occurring above the thermocolline belt, characterized by a growing season of 245 to 310 days and oak and hornbeam forests, suitable for the cultivation of vine, corn, and chestnut.

    montane belt: The vegetation belt in the temperate regions occurring above the colline belt, characterized by a growing season of 180 to 245 days and beech, silver fir, and scots pine forests, suitable for the culture of cereals.

    subalpine belt: The vegetation belt in the temperate regions, occurring above the montane belt, characterized by a growing season of 115 to 180 days and coniferous forests, suitable for the mowing of meadows. The subalpine belt is the upper forest belt.

    alpine belt: The vegetation belt in the temperate regions occurring above the subalpine belt, characterized by a growing season of 50 to 115 days and low heaths and swards, suitable only for pasturing.

    nival belt: The highest vegetation belt in the temperate regions, occurring above the alpine belt, characterized by a growing season of less than 50 days, with snow remaining on flat ground throughout the summer and only a loose, sparse vegetation of vascular plants.

vegetation complex: A particular, spatial association of plant communities, often mosaic-like, that is related to a specific combination of topographical and edaphic factors and that forms a relatively homogenous part of the landscape.

vicariance: In general, the replacement of a species inhabiting a territory by one or several neighboring species in other areas (geographical vicariance). The replacement according to ecological conditions (e.g., siliceous soils and calcareous soils) is called the ecological vicariance.

    false vicariance: Applies to two unrelated species having the same level of ploidy (as opposed to true vicariance).

    pseudovicariance: Applies to a species where two different stages of polyploidy (e.g., diploid and tetraploid) show continuous and complementary areas of distribution.

    true vicariance: Applies to two species having the same level of ploidy that differentiated from a common ancestor (as opposed to false vicariance).

zonal vegetation: Vegetation determined by the general climate in orobiomes and zonobiomes.

zonobiome: See *biome*.

REFERENCES

Ågren, G. I., and J. F. Wirkström. 1993. Modelling carbon allocation: A review. *New Zealand Journal of Forestry Science* 23:343–53.

Almendinger, J. C. 1990. The decline of soil organic matter, total N and available water capacity following the late Holocene establishment of Jack pine in sandy mollisols, north-central Minnesota. *Soil Science* 150:680–94.

Anchisi, E. 1985. Quatrième contribution à l'étude de la flore valaisanne. *Bulletin de la Murithienne* 102:115–26.

Anderson, J. M. 1992. Responses of soils to climate change. *Advances in Ecological Research* 22:163–210.

Anon, N. 1992. Soil-warming experiments in global change research. In *Woods Hole, Massachussetts*, ed. National Science Foundation. Washington, D.C.: National Science Foundation.

Antonovics, J. 1971. The effect of heterogeneous environment on the genetics of natural populations. *American Naturalist* 59:593–9.

Arnell, H. W. 1875. *De Skandinaviska Löfmossernas Kalendarium*. Uppsala, Sweden: Esaias Edquist.

Arnold, R. W., Szabolcs, I., Tasgulian, V. O. (eds.) 1990. *Global Soil Change*. International Soil Science Society.

Atkin, O. K., B. Botman, and H. Lambers. 1996. The causes of inherently slow growth in Alpine plants: An analysis based on the underlying carbon economies of Alpine and lowland Poa species. *Functional Ecology* 10:698–707.

Backhuys, W. 1968. Der Elevations-Effekt bei einigen Alpenpflanzen der Schweiz. *Blumea* 16:273–320.

Bahn, M., A. Cernusca, U. Tappeiner, and E. Tasser. 1994. Wachstum krautiger Arten auf einer Mähwiese und einer Almbrache. *Verhandlungen der Gesellschaft für Ökologie* 23:23–30.

Baize, D., and M.-C. Girard. 1992. Référential pédologique. Principaux Sols d'Europe. Paris: Institut National de la Recherche Agronomique.

Balesdent, J. 1982. Etude de la dynamique de l'humification de sols de prairies d'altitude (Haut-Jura) au moyen des datations ^{14}C des matières organiques. Thèse Doct. Ing., Nancy I., France

Ballal, S. R., S. A. Foré, and I. Guttman. 1994. Apparent gene flow and genetic structure of Acer saccharum subpopulations in forest fragments. *Canadian Journal of Botany* 72:1311–5.

Barkman, J. J. 1980. Synusial approaches to classification. In *Classification of Plant Communities*, 2d ed., ed. R. H. Whittaker, 111–65. Den Haag: Junk.

Bartoli, F., and G. Burtin. 1979. Etude de quatre séquences sol-végétation à l'étage alpin. *Documents de Cartographie Ecologique, Université Scientifique et Médicale de Grenoble* 21:79–93.

Baur, B., and B. Schmid. 1996. Spatial and temporal patterns of genetic diversity within species. In *Diversity: A Biology of Numbers and Difference*, ed. K. J. Gaston, 169–201. London: Blackwell Science.

Bayer, R. J. 1989. Patterns of isozyme variation in the Antennaria rosea (Asteraceae:Inuleae) polyploidagamic complex. *Systematic Botany* 14:389–97.

Bazzaz, F. A. 1996. *Plants in a changing environment: Linking physiological, population, and community ecology*. Cambridge: Cambridge University Press.

Bazzaz, F. A. 1990. The response of natural ecosystems to the rising global CO_2 level. *Annual Review of Ecology and Systematics* 21:167–96.

Béguin, C. 1974. Contribution à l'étude phytosociologique et écologique du Haut-Jura. *Beiträge zur geobotanischen Landesaufnahme der Schweiz* 54:1–190.

Béguin, C., and C. Raemy. 1995. Clairières naturelles sur le versant sud du premier anticlinal jurassien. *Ukpik. Cahiers de l'Institut de Géographie de Fribourg* 10:7–23.

Berendse, F. 1993. Ecosystem stability, competition, and nutrient cycling. In *Biodiversity and ecosystem function*, Ecological Studies 99, ed. E.-D. Schulze and H. A. Mooney, 409–31. Heidelberg: Springer.

Bergmann, F. 1978. The allelic distribution at an acid phosphatase locus in Norway spruce (Picea abies) along similar climatic gradients. *Theoretical and Applied Genetics* 52:57–64.

Bergmann, F., and H. R. Gregorius. 1993. Ecogeographical distribution and thermostability of Isocitrate Dehydrogenase (IDH) alloenzymes in European Silver fir (Abies alba). *Biochemical Systematics and Ecology* 21:597–605.

Bliss, L. C. 1956. A comparison of plant development in microenvironments of Arctic and Alpine tundras. *Ecological Monographs* 26:303–37.

Böcher, T. W. 1949a. Racial divergences in Prunella vulgaris in relation to genetic and environmental factors. *Dansk Botanisk Arkiv. Udgivet af Dansk Botanisk Forening* 11:1–20.

Böcher, T. W. 1949b. Racial divergences in Prunella vulgaris in relation to genetic and environmental factors. *New Phytologist* 48:285–14.

Bolin, B., B. R. Döös, J. Jäger, and R. A. Warrick, eds. 1989. *The greenhouse effect, climatic change, and ecosystems*. Scope, vol. 29. 1986. Reprint, New York: Wiley & Sons.

Bortenschlager, S. 1993. Das höchst gelegene Moor der Ostalpen "Moor am Rofenberg" 2760 m. *Dissertationes Botanicae* 196:329–34.

Bottner, P. 1972. Evolution des sols en milieu carbonaté. Thèse. Faculté des Sciences, Montpellier, France.

Boudot, J.-P., S. Bruckert, and B. Souchier. 1981. Végétation et sols climax sur les Grauwackes de la série du Markstein (Hautes-Vosges). *Annales des Sciences Forestières* 38:87–106.

Bowler, J. M., and M. C. Press. 1993. Growth responses of two contrasting upland grass species to elevated CO_2 and nitrogen concentration. *New Phytologist* 124:515–22.

Braun-Blanquet, J. 1967. Vegetationsskizzen aus dem Baskenland mit Ausblicken auf das weitere Ibero-Atlantikum. II. Teil. *Vegetatio* 14:1–126.

Braun-Blanquet, J. 1964. *Pflanzensoziologie*. 3d ed. Vienna: Springer.

Braun-Blanquet, J. 1948. La végétation alpine des Pyrénées orientales. *Communication de la Station Internationale de Géobotanique Méditerranéenne et Alpine, Montpellier* 98:1–306.

Breitenbach-Doefer, M., W. Pinsker, R. Hacker, and F. Müller. 1992. Clone identification and clinal allozyme variation in populations of Abies alba from the Eastern Alps (Austria). *Plant Systematics and Evolution* 181:109–20.

Breymeyer, A., and L. Uba. 1989. Organic production and decomposition in mountain transects: Their response to climate change. In *European Conference on Landscape-Ecological Impact of Climate Change in Lunteren, The Netherlands*, ed. M. M. Boer and R. de Groot, 119. Amsterdam: IOS Press.

Brockmann-Jerosch, H. 1919. Baumgrenze und Klimacharakter. *Beiträge zur geobotanischen Landesaufnahme der Schweiz* 6:1–255.

Brooks, K. N., P. F. Folliott, H. M. Gregersen, and J. L. Thames. 1991. *Hydrology and Management of Watersheds*. Ames: University of Iowa Press.

Sensitivity of Plant and Soil Ecosystems of the Alps to Climate Change

Brooks, P. D., S. K. Schmidt, and M. W. Williams. 1997. Winter production of CO_2 and N_2O from alpine tundra: Environmental controls and relationship to inter-system C and N fluxes. *Oecologia, Berlin* 110:403–13.

Bryant, R. B., and R. W. Arnold, eds. 1994. *Quantitative Modeling of Soil Forming Processes.* Madison, WI: Soil Science Society of America

Brzeziecki, B., F. Kienast, and O. Wildi. 1995. Modelling potential impacts of climate change on the spatial distribution of zonal forest communities in Switzerland. *Journal of Vegetation Science* 6:257–8.

Bucher, F., and F. Jeanneret. 1994. Phenology as a tool in topoclimatology. In *Mountain Environments in Changing Climates*, ed. M. Beniston, 270–80. London: Routledge Publishing Co.

Burga, C. A. 1995. Végétation et paléoclimatologie de l'Holocène moyen d'une ancienne tourbière située au front du Glacier du Rutor, 2510 m (Vallée d'Aoste, Italie). *Revue de Géographie Alpine* 1:9–16.

Burga, C. A. 1993. Das mittelholozäne Klimaoptimum Europas: Palynologische Untersuchungen an einem ehemaligen hochgelegenen Moor am Rutor-Gletscher, 2510 m (Aosta-Tal, Italien). *Dissertationes Botanicae* 196:335–46.

Burga, C. A. 1988. Swiss vegetation history during the last 18,000 years. *New Phytologist* 110:581–602.

Busby, J. R., L. C. Bliss, and C. D. Hamilton. 1978. Microclimate control of growth rates and habitats of the boreal forest mosses, Tomenthypnum nitens and Hylocomium splendens. *Ecological Monographs* 48:95–110.

Butler, D. R., C. Hill, G. P. Malanson, and D. M. Cairns. 1994. Stability of alpine treeline in Glacier National Park, Montana, U.S.A. *Phytocoenologia* 22:485–500.

Caisse, M., and J. Antonovics. 1978. Evolution in closely adjacent populations IX: Evolution of reproductive isolation in clinal populations. *Heredity* 40:371–84.

Callaghan, T. V., B. A. Carlsson, I. Jónsdóttir, B. M. Svensson, and S. Jonasson. 1992. Clonal plants and environmental change: Introduction to the proceedings and summary. *Oikos* 63:341–7.

Carbiener, R. 1966. Relations entre cryoturbation, solifluxion et groupements végétaux dans les Hautes-Vosges (France). *Oecologia Plantarum* 1:335–67.

Carcaillet, C., M. Barbero, and B. Talon. 1996. Arbres et incendies au cours de l'holocène 300 m au-dessus de la limite actuelle des arbres dans les Alpes du nord-ouest. Colloque Végétation et sols de Montagne. Diversité, fonctionnement et évolution, Grenoble, 8.-13.7. Laboratoire des Ecosystèmes alpins, Université Joseph Fourier, Grenoble.

Carraro, G., G. Gianone, R. Mossi, F. Klötzli, and G.-R. Walther. In press. Studio sulla percezione dei cambiamenti della vegetazione in relazione con il riscaldamento dell' atmosfera. Zürich: ZDE.

Cernusca, A. 1976. Bestandesstruktur, Bioklima und Energiehaushalt von alpinen Zwergstrauchbeständen. *Oecologia Plantarum* 11:71–102.

Chapin III, F. S., W. C. Oechel, K. Van Cleve, and W. Lawrence. 1987. The role of mosses in the phosphorus cycling of an Alaskan black spruce forest. *Oecologia* 74:310–5.

Chapin, F. S. III, and G. R. Shaver. 1989. Differences in growth and nutrient use among arctic plant growth forms. *Functional Ecology* 3:73–80.

Clausen, J., D. D. Keck, and W. M. Hiesey. 1940. Experimental studies on the nature of species I: The effect of varied environments on Western North American plants. *Publication of the Carnegie Institution of Washington* 520:1–422.

Cole, K. 1985. Past rates of change, species richness, and a model of vegetational inertia in the Grand Canyon, Arizona. *The American Naturalist* 125:289–303.

Connell, J. H., and R. O. Slatyer. 1977. Mechanisms of succession in natural communities and their role in community stability and organization. *The American Naturalist* 111:1119–44.

Corley, M. F. V., and A. C. Crundwell. 1991. Additions and amendments to the mosses of Europe and the Azores. *Journal of Bryology* 16:337–56.

Cosner, M. E., and D. J. Crawford. 1994. Comparisons of isozyme diversity in three rare species of Coreopsis (Asteraceae). *Systematic Botany* 19:350–8.

Coûteaux, M.-M., P. Bottner, and B. Berg. 1995. Litter decomposition, climate and litter quality. *Trends in Ecology and Evolution* 10:63–6.

Coûteaux, M.-M., M. Mousseau, M.-L. Celkerier, and P. Bottner. 1991. Increased atmospheric CO_2 and litter quality decomposition of sweet chestnut litter with animal food webs of different complexities. *Oikos* 61:54–64.

Cramer, W. P., and R. Leemans. 1993. Assessing impacts of climate change on vegetation using climate classification systems. In *Vegetation Dynamics and Global Change*, ed. A. M. Solomon and H. H. Shugart, 190–217. London: Chapman & Hall.

Cuguen, J., B. Thiébaut, F. Ntsiba, and G. Barrière. 1985. Enzymatic variability of beechstands (Fagus sylvatica L.) on three scales in Europe: Evolutionary mechanism. In *Genetic Differentiation and dispersal in Plants*, NATO ASI Series, vol. G5, ed. P. Jaquard, G. Heim and J. Antonovics: 17–39. Berlin: Springer.

Dambrine, E. 1985. Contribution à l'étude de la répartition et du fonctionnement des sols de haute montagne (massif des Aiguilles Rouges et du Mont-Blanc). Thèse Doctorat d'Etat, Paris VII.

Davis, P. H. 1951. Cliff vegetation in the eastern Mediterranean. *Journal of Ecology* 39:63–93.

Defila, C. 1991. Pflanzenphänologie der Schweiz. *Veröffentlichungen der Schweizerischen Meteorologischen Anstalt* 50:1–235.

de Groot, R. S., P. Ketner, and A. H. Ovaa. 1995. Selection and use of bio-indicators to assess the possible effects of climate change in Europe. *Journal of Biogeography* 22:935–43.

Deil, U. 1994. Klassifizierung mit supraspezifischen Taxa und symphylogenetische Ansätze in der Vegetationskunde. *Phytocoenologia* 24:677–94.

Deil, U. 1989. Adiantetea-Gesellschaften auf der Arabischen Halbinsel, Coenosyntaxa in dieser Klasse sowie allgemeine Überlegungen zur Phylogenie von Pflanzengesellschaften. *Flora* 182:247–64.

Delarze, R. 1994. Dynamique de la végétation sur les pistes ensemencées de Crans-Montana (Valais, Suisse): Effets de l'altitude. *Botanica Helvetica* 104:3–16.

Di Castri, F. 1992. Ecosystem evolution and global change. In *Perspectives on Biological Complexity*, Monograph Series vol. 6, International Union of Biological Sciences, ed. O. T. Solbrig and G. Nicolis, 189–217. Paris: International Union of Biological Sciences.

Di Castri, F., and J. A. Hansen. 1992. The environment and development crises as determinants of landscape dynamics. In *Landscape Boundaries: Consequences for Biotic Diversity and Ecological Flows*, ed. A. J. Hansen and F. di Castri, 3–18. Heidelberg: Springer.

Díaz, S., J. P. Grime, J. Harris, and E. McPherson. 1993. Evidence of a feedback mechanism limiting plant response to elevated carbon dioxide. *Nature* 364:616–7.

Diekmann, M. 1996. Relationship between flowering phenology of perennial herbs and meteorological data in deciduous forests of Sweden. *Canadian Journal of Botany* 74:528–37.

Diemer, M. 1996. The incidence of herbivory in high-elevation populations of Ranunculus gla-cialis: a re-evaluation of stress-tolerance in alpine environments. *Oikos* 75:486–492.

Dierschke, H. 1990. Bibliographia symphaenologica. Excerpta botanica. Sectio B. *Sociologica* 28:49–87.

Dixon, R. K., S. Brown, R. A. Houghton, A. M. Solomon, M. C. Trexler, and J. Wisniewski. 1994. Carbon pools and flux of global forest ecosystems. *Science* 263:185–90.

Dixon, R. K. and D. P. Turner. 1991. The global carbon cycle and climate change: Responses and feedbacks from below-ground systems. *Environmental Pollution* 73:245–62.

Doche, B., V. Pommeyrol, and J. P. Peltier. 1991. Les landes à éricacées (callunaies, rhodoraies) et les vitessses de transformation des paysages végétaux en montagne (Massif Central et Alpes). *Bulletin d'Ecologie* 22:221–6.

Duchaufour, P. 1983. *Pédologie I. Pédogenèse et classification.* 2d ed. Paris: Masson.

Duchaufour, P. 1977. *Pédologie. I. Pédogenèse et Classification.* Paris: Masson.

During, H., and B. F. van Tooren. 1987. Recent developments in bryophyte population ecology. *Trends in Ecology and Evolution* 2:89–93.

Ellenberg, H. 1996. *Vegetation Mitteleuropas mit den Alpen in ökologischer, dynamischer und histo-rischer Sicht.* 5th ed. Stuttgart: Ulmer.

Ellenberg, H. 1988. *Vegetation ecology of Central Europe.* 4th ed. Cambridge: Cambridge University Press.

Ellenberg, H. 1986. *Vegetation Mitteleuropas mit den Alpen in ökologischer Sicht.* 4d ed. Stuttgart: Ulmer.

Ellenberg, H. 1973. Versuch einer Klassifikation der Ökosysteme nach funktionalen Gesichts-punkten. In *Ökosystemforschung,* ed. H. Ellenberg. Heidelberg: Springer.

Ellenberg, H., and D. Mueller-Dombois. 1967. A key to Raunkier plant life forms with revised subdivisions. *Berichte des Geobotanischen Institutes der Eidg. Techn. Hochschule Stiftung Rübel* 37:56–73.

Endler, J. A. 1977. Geographic variation, speciation and clines. Princeton, NJ: Princeton University Press.

Eriksson, O. 1993. Dynamics of genets in clonal plants. *Trends in Ecology and Evolution* 8:313–16.

Farmer, A. M., J. W. Bates, and J. N. Bell. 1992. Ecophysiological effects of acid rain on bryo-phytes and lichens. In *Bryophytes and Lichens in a changing environment,* ed. J. W. Bates and A. M. Farmer, 284–313. Oxford: Clarendon Press.

Favarger, C. 1995a. *Flore et végétation des Alpes. 1. Etage alpin.* 3d ed. Lausanne, Switzerland: Delachaux & Niestlé.

Favarger, C. 1995b. *Flore et végétation des Alpes. 2. Etage subalpin.* 3rd ed. Lausanne, Switzerland: Delachaux & Niestlé.

Favarger, C. 1989. La flore. La végétation. In *Guide du naturaliste dans les Alpes,* 2d ed., ed. J.-P. Schaer, P. Veyret, C. Favarger, G. Du Chatenet, R. Hainard, and O. Paccaud, 131–257. Neu-châtel, Switzerland: Delachaux & Niestlé.

Favarger, C. 1972. Endemism in the montane floras of Europe. In *Taxonomy, Phytogeography and Evolution,* ed. D. H. Valentine, 191–204. London: Academic Press.

Favarger, C. 1962. Contribution de la biosystématique à l'étude des flores alpine et jurassienne. *Revue de Cytologie et de Biologie Végétales* 25:397–410.

Felber, F., and B. Thiébaut. 1984. Etude préliminaire sur le polymorphisme enzymatique du hêtre, Fagus sylvatica L.: Variation génétique de deux systèmes de peroxydases en relation avec les conditions écologiques. *Acta Oecologia Plantarum* 5:133—50.

Fierz, M., J.-M. Gobat, and C. Guenat. 1995. Quantification et caractérisation de la matière organique de sols alluviaux au cours de l'évolution de la végétation. *Annales des Sciences Forestières* 52:547—59.

Fitter, A. H., R. S. R. Fitter, I. T. B. Harris, and M. H. Williamson. 1995. Relationships between first flowering date and temperature in the flora of a locality in central England. *Functional Ecology* 9:55—60.

Flenley, J. R. 1995. Cloud forest, the Massenerhebung effect, and ultraviolet insolation. In *Tropical Cloud Montane Forests*, Ecological Studies 110, ed. L. S. Hamilton, J. O. Juvik, and F. N. Scatena, 150—5. Heidelberg: Springer.

Foré, S. A., R. J. Hickey, J. L. Vankat, S. I. Guttman, and R. L. Schaefer. 1992. Genetic structure after forest fragmentation: A landscape ecology perspective on Acer saccharum. *Canadian Journal of Botany* 70:1659—68.

Frey, E. 1922. Die Vegetationsverhältnisse der Grimselgegend im Gebiet der zukünftigen Stauseen. *Mittheilungen der Naturforschenden Gesellschaft Bern* 6:1—195.

Friend, A. D., and F. I. Woodward. 1990. Evolutionary and ecophysiological responses of mountain plants to the growing season environment. *Advances in Ecological Research* 20:59—124.

Frisvoll, A. 1988. A taxonomic revision of the Racomitrium heterostichum group (Bryophyta, Grimmiales) in N. and C. America, N. Africa, Europe and Asia. *Gunneria* 59:1—289.

Gallet, C., and P. Lebreton. 1995. Evolution of phenolic patterns in plants and associated litters and humus of a mountain forest ecosystem. *Soil Biology and Biochemistry* 27:157—65.

Gams, H. 1932. Die klimatische Begrenzung von Pflanzenarealen und die Verteilung der hygrischen Kontinentalität in den Alpen. II. Teil. *Zeitschrift der Gesellschaft für Erdkunde zu Berlin* 1932:52—68, 178—98.

Gams, H. 1931. Die klimatische Begrenzung von Pflanzenarealen und die Verteilung der hygrischen Kontinentalität in den Alpen. *Zeitschrift der Gesellschaft für Erdkunde zu Berlin* 1931:321—46.

Gams, H. 1918. Prinzipienfragen der Vegetationsforschung. *Vierteljahrsschrift der Naturforschenden Gesellschaft in Zürich* 63:293—403.

Gates, D. M. 1993. *Climate Change and its Biological Consequences*. Sunderland, U.K.: Sinauer.

Geissler, P. 1976. Zur Vegetation alpiner Fliessgewässer. *Beiträge zur Kryptogamenflora der Schweiz* 14(2):1—52.

Gianoni, G., G. Carraro, and F. Klötzli. 1988. Thermophile, an laurophyllen Pflanzenarten reiche Waldgesellschaften im hyperinsubrischen Seenbereich des Tessins. *Berichte des Geobotanischen Instituts der Eidg. Techn. Hochschule Stiftung Rübel* 54:164—80.

Gifford, R. M. 1994. The global carbon cycle: A viewpoint on the missing link. *Australian Journal of Plant Physiology* 21:1—15.

Gignac, L. D., and D. H. Vitt. 1994. Responses of northern peatlands to climate change: Effects on bryophytes. *Journal of the Hattori Botanical Laboratory* 75:119—32.

Gigon, A. 1987. A hierarchic approach in causal ecosystem analysis: The calcifuge-calcicole problem in alpine grasslands. In *Potentials and limitations of ecosystems analysis*, ed. E.-D. Schulze and H. Zwölfer, 228—244. Heidelberg: Springer.

Gigon, A. 1983. Welches ist der wichtigste Standortsfaktor für die floristischen Unterschiede zwischen benachbarten Pflanzengesellschaften? *Verhandlungen der Gesellschaft für Ökologie* 11: 145–59.

Gigon, A. 1971. Vergleich alpiner Rasen auf Silikat- und auf Karbonatboden. *Veröffentlichungen des geobotanischen Institutes der ETH, Stiftung Rübel, Zürich* 48:1–159.

Gosz, J. R., 1992. Ecological functions in a biome transition zone: Translating local responses to broad-scale dynamics. In *Landscape boundaries: Consequences for biotic diversity and ecological flows*, Ecological Studies 92, ed. A. J. Hansen and F. Di Castri, 55–75. Heidelberg: Springer.

Gosz, J. R. 1991. Fundamental ecological characteristics of landscape boundaries. In *Ecotones: The role of landscape boundaries in the management and restoration of changing environments*, ed. M. M. Holland, P. G. Risser, and R. J. Naiman, 55–75. London: Chapman & Hall.

Gottfried, M., H. Pauli, and G. Grabherr. 1994. Die Alpen im "Treibhaus": Nachweis für das erwärmungsbedingte Höhersteigen der alpinen und nivalen Vegetation. *Jahrbuch des Vereins zum Schutz der Bergwelt* 59:13–27.

Gottlieb, L. D. 1981. Electrophoretic evidence and plant populations. *Progress in Phytochemistry* 7:1–46.

Grabherr, G., M. Gottfried, A. Gruber, and H. Pauli. 1995. Patterns and current changes in alpine plant diversity. In *Arctic and alpine biodiversity: patterns, causes and ecosystem consequences*, Ecological Studies 113, ed. F. S. Chapin III and C. Körner, 167–81. Heidelberg: Springer.

Grabherr, G., M. Gottfried, and H. Pauli. 1994. Climate effects on mountain plants. *Nature* 369:448.

Grabherr, G., and S. Kojima. 1993. Vegetation diversity and classification systems. In *Vegetation Dynamics and Global Change*, ed. A. M. Solomon and H. H. Shugart, 218–32. London: Chapman & Hall.

Grabherr, G., E. Mähr, and H. Reisigl. 1978. Nettoprimärproduktion und Reproduktion in einem Krummseggenrasen (Caricetum curvulae) der Ötztaler Alpen, Tirol. *Oecologia Plantarum* 13:227–51.

Graham, R. W., and E. C. Grimm. 1990. Effects of global climate change on the patterns of terrestrial biological communities. *Trends in Ecology and Evolution* 5:289–92.

Grime, J. P. 1993. Vegetation functional classification systems as approaches to predicting and quantifying global vegetation change. In *Vegetation Dynamics and Global Change*, ed. A. M. Solomon and H. H. Shugart, 293–305. London: Chapman & Hall.

Grime, J. P. 1979. *Plant Strategies and Vegetation Processes*. Chichester: Wiley & Sons.

Grimme, A. 1903. Über die Blüthezeit deutscher Laubmoose und die Entwickelungsdauer ihrer Sporogone. *Hedwigia* 42:1–75.

Grolle, R. 1983. Hepatics of Europe including the Azores: An annotated list of species, with synonyms from the recent literature. *Journal of Bryology* 12:403–59.

Grubb, P. J. 1971. Interpretation of the "Massenerhebung" effect on tropical mountains. *Nature* 229:44–5.

Guillet, B. 1972. Relations entre l'histoire de la végétation et la podzolisation dans les Vosges. Thèse Doctorat d'Etat, Nancy, France.

Guisan, A., J.-P. Theurillat, and R. Spichiger, 1995. Effects of climate change on alpine plant diversity and distribution: the modelling and monitoring perspectives. In *Potential ecological impacts of climate change in the Alps and Fennoscandian mountains*, ed. A. Guisan, J. I. Holten, R. Spichiger, and L. Tessier, 129–35. Geneva: Conservatoire et Jardin botaniques.

Häberli, W. 1995. Glaciers and permafrost in the Alps. In *Potential ecological impacts of climate change in the Alps and Fennoscandian mountains*, ed. A. Guisan, J. I. Holten, R. Spichiger, and L. Tessier, 113–20. Geneva: Conservatoire et Jardin botaniques.

Häberli, W. 1994. Accelerated glacier and permafrost changes in the Alps. In *Mountain environments in changing climates*, ed. M. Beniston, 91–107. London: Routledge Publishing Co.

Halpin, P. N. 1994a. GIS analysis of the potential impacts of climate change on mountain ecosystems and protected areas. In *Mountain environments and Geographic Information Systems*, ed. M. F. Price and D. I. Heywood, 281–301. London: Taylor and Francis Ltd.

Halpin, P. N. 1994b. Latitudinal variation in the potential response of mountain ecosystems to climatic change. In *Mountain environments in changing climates*, ed. M. Beniston, 180–203. London: Routledge Publishing Co.

Hamrick, J. L. 1989. Isozymes and the analysis of genetic structure in plant populations. In *Isozymes in Plant Biology*, ed. D. E. Soltis and P. S. Soltis, 87–105. Portland, Or: Dioscorides Press.

Hänninen, H. 1991. Does climatic warming increase the risk of frost damage in northern trees? *Plant, Cell and Environment* 14:449–54.

Haselwandter, K., A. Hofmann, H.-P. Holzmann, and D. J. Read. 1983. Availability of nitrogen and phosphorus in the nival zone of the Alps. *Oecologia* 57:266–9.

Hattemer, H. H. 1994. Die genetische Variation und ihre Bedeutung für Wald und Waldblume. *Schweizerische Zeitschrift für Forstwesen* 145:953–75.

Hegg, O. 1977. Mikroklimatische Wirkung der Besonnung auf die phänologische Entwicklung und auf die Vegetation in der alpinen Stufe der Alpen. *Bericht über das Internationale Symposium der Internationalen Vereinigung für Vegetationskunde* 1975:249–70.

Hengeveld, R. 1985. On the explanation of the elevation effect by a dynamic interpretation of species distribution along altitudinal gradients. *Blumea* 30:353–61.

Herben, T. 1994. The role of reproduction for persistence of bryophyte populations in transient and stable habitats. *Journal of the Hattori Botanical Laboratory* 76:115–26.

Heun, M., J. P. Murphy, and T. D. Phillips. 1994. A comparison of RAPD and isozyme analysis for determining the genetic relationships among Avena sterilis L. accessions. *Theoretical and Applied Genetics* 87:589–696.

Hirschel, G., C. Körner, and J. A. Arnone III. 1997. Will rising atmospheric CO_2 affect leaf litter quality and in situ decomposition rates in native plant communities? *Oecologia, Berlin* 110:387–92.

Hobbie, S. E. 1996. Temperature and plant species control over litter decomposition in Alaskan tundra. *Ecological Monographs* 66:503–22.

Hofer, H. R. 1992. Veränderungen in der Vegetation von 14 Gipfeln des Berninagebietes zwischen 1905–1985. *Berichte des Geobotanischen Instituts der Eidg. Techn. Hochschule Stiftung Rübel* 58:39–54.

Holderegger, R., and J. J. S. Schneller. 1994. Are small isolated populations of Asplenium septentrionale variable? *Biological Journal of the Linnean Society* 51:377–85.

Holt, R. D. 1990. The microevolutionary consequences of climate changes. *Trends in Ecology and Evolution* 5:311–5.

Holten, J. I. 1993. Potential effects of climatic change on distribution of plant species, with emphasis on Norway. In *Impacts of climatic change on natural ecosystems, with emphasis on boreal and arctic/alpine areas*, ed. J. I. Holten, G. Paulsen, and W. C. Oechel, 84–104. Trondheim: Norwegian Institute for Nature Research.

Holten, J. I., and P. D. Carey. 1992. *Responses of climate change on natural terrestrial ecosystems in Norway.* Trondheim, Norway: Norsk Inst. Naturforskning.

Holtmeier, F.-K. 1994. Ecological aspects of climatically caused timberline fluctuations. In *Mountain environments in changing climates,* ed. M. Beniston, 220–233. London: Routledge Publishing Co.

Holtmeier, F.-K. 1989. Ökologie und Geographie der oberen Waldgrenze. *Tuexenia* 1:15–45.

Hunter, A. F., and M. J. Lechowicz. 1992. Predicting the timing of budburst in temperate trees. *Journal of Applied Ecology* 29:597–604.

Huntley, B. 1991. How plants respond to climate change: Migration rates, individualism and the consequences for plant communities. *Annals of Botany* 67 (suppl. 1):15–22.

Huntley, B., P. M. Berry, W. Cramer, and A. P. McDonald. 1995. Modelling present and potential future ranges of some European higher plants using climate response surfaces. *Journal of Biogeography* 22:967–1001.

Hussendörfer, E., and G. Müller-Starck. 1994. Genetische Inventuren in Beständen der Weisstanne (Abies alba Mill.)—Aspekte der nacheiszeitlichen Wanderungsgeschichte. *Schweizerische Zeitschrift für Forstwesen* 145:1021–9.

Jacquemoud, F. 1989. Excursion de la Société botanique de Genève dans les Alpes autrichiennes (10–19 juillet 1988): compte-rendu floristique. *Saussurea* 20:45–69.

Jauhiainen, J., H. Vasande, and J. Silvola. 1994. Response of Sphagnum fuscum to N deposition and increased CO_2. *Journal of Bryology* 18:83–95.

Jerosch, M. C. 1903. *Geschichte und Herkunft der schweizerischen Alpenflora.* Leipzig, Germany: Engelmann.

Johnson, D. W. 1993. Carbon in forests soils—Research needs. *New Zealand Journal of Forestry Science* 23:354–66.

Johnson, D. W., D. W. Cole, S. P. Gessel, M. J. Singer, and R. V. Minden. 1977. Carbonic acid leaching in a tropical, temperate, subalpine, and northern forest soil. *Arctic and Alpine Research* 9:329–43.

Jonasson, S. 1993. Buffering of arctic plant responses to a changing climate. In *Global change and arctic terrestrial ecosystems: an International Conference 21–26 August 1993, Oppdal, Norway. Abstracts,* ed. T. Gilmanov, J. I. Holten, B. Maxwell, W. C. Oechel, and B. Sveinbjörnsson, 19. Trondheim: Norwegian Institute for Nature Research.

Keller, C. 1992. Le Mont-Blanc: de la Montagne Maudite à l'observatoire pour l'environnement. *Bulletin de la Société Vaudoise des Sciences Naturelles* 81:181–97.

Keller, C. 1991. Etude du cycle biogéochimique du cuivre et du cadmium dans deux écosystèmes forestiers. Thèse de doctorat, Ecole Polytechnique de Lausanne, Switzerland.

Keller, F. 1994. Interaktionen zwischen Schnee und Permafrost. *Mitteilungen der Versuchsanstalt für Wasserbau, Hydrologie und Glaziologie der Eidgenössischen Technischen Hochschule Zürich* 127:1–145.

Kiang, Y. T. 1982. Local differentiation of Anthoxanthum odoratum L. populations on roadsides. *American Midland Naturalist* 107:340–50.

Kienast, F., B. Brzeziecki, and O. Wildi. 1996. Long-term adaptation potential of Central European mountain forests to climate change: a GIS-assisted sensitivity assessment. *Forest Ecology and Management* 80:133–53.

Kienast, F., B. Brzeziecki, and O. Wildi. 1995. Simulierte Auswirkungen von postulierten Klimaveränderungen auf die Waldvegetation im Alpenraum. *Angewandte Landschaftsökologie* 4:83–101.

King, G. A., and R. P. Neilson. 1992. The transient response of vegetation to climate change: a potential source of CO_2 to the atmosphere. *Water, Air, and Soil Pollution* 64:365–83.

Klötzli, F. 1994. *Stabilität und Diversität in alpinen Ökosystemen unter der Wirkung veränderter Umweltbedingungen.* AlpenForum'94, Disentis. Poster.

Klötzli, F. 1992. Alpine Vegetation: Stabil und natürlich? In *Die Alpen—ein sicherer Lebensraum?* Ed. J. P. Müller and B. Gilgen, 70–83. Publikation der Schweizerischen Akademie der Naturwissenschaften, vol. 5, Disentis.

Klötzli, F. 1988. On the global position of the evergreen broad-leaved (non-ombrophilous) forest in the subtropical and temperate zones. *Veröffentlichungen des geobotanischen Institutes der ETH, Stiftung Rübel, Zürich* 98:169–96.

Klötzli, F., G.-F. Walther, G. Carraro, and A. Grundmann. 1996. Anlaufender Biomwandel in Insubrien. *Verhandlungen der Gesellschaft für Ökologie* 26:537–50.

Knapp, A. K., W. K. Smith, and D. R. Young. 1989. Importance of intermittent shade to the ecophysiology of subalpine herbs. *Functional Ecology* 3:753–8.

Körner, C. 1995. Impact of atmospheric changes on alpine vegetation: the ecophysiological perspective. In *Potential ecological impacts of climate change in the Alps and Fennoscandian mountains,* ed. A. Guisan, J. I. Holten, R. Spichiger, and L. Tessier, 113–20. Geneva: Conservatoire et Jardin botaniques.

Körner, C. 1994. Impact of atmospheric changes on high mountain vegetation. In *Mountain environments in changing climates,* ed. M. Beniston, 155–66. London: Routledge Publishing Co.

Körner, C. 1993a. CO_2 fertilization: The great uncertainty in future vegetation development. In *Vegetation dynamics and global change,* ed. A. M. Solomon and H. H. Shugart, 53–70. London: Chapman & Hall.

Körner, C. 1993b. Scaling from species to vegetation: the usefulness of functional groups. In *Biodiversity and ecosystem function,* Ecological Studies, vol. 99, ed. E.-D. Schulze and H. A. Mooney, 117–40. Heidelberg: Springer.

Körner, C. 1992. Response of alpine vegetation to global climate change. *Catena* 22 (suppl.):85–96.

Körner, C. 1991. Some often overlooked plant characteristics as determinants of plant growth: a reconsideration. *Functional Ecology* 5:162–73.

Körner, C. 1989. The nutritional status of plants from high altitudes. A worldwide comparison. *Oecologia* 81:379–91.

Körner, C., and J. A. Arnone III. 1992. Responses to elevated carbon dioxide in artificial tropical ecosystems. *Science* 257:1672–5.

Körner, C., and M. Diemer. 1987. In situ photosynthetic responses to light, temperature and carbon dioxide in herbaceous plants from low and high altitude. *Functional Ecology* 1:179–94.

Körner, C., M. Diemer, B. Schäppi, and L. Zimmermann. 1996. Response of alpine vegetation to elevated CO_2. In *Carbon dioxide and terrestrial ecosystems,* ed. G. W. Koch and H. A. Mooney, 177–196. New York: Academic Press.

Körner, C., and W. Larcher. 1988. Plant life in cold climates. In *Plants and temperature,* ed. S. F. Long and F. I. Woodward, 25–57. Company Biol. Ldt, Symp. Soc. Exper. Biol., vol. 42, Cambridge.

Körner, C., M. Neumayer, S. Pelaez Menendez-Riedl, and A. Smeets-Scheel. 1989. Functional morphology of mountain plants. *Flora* 182:353–83.

Kramer, K. 1994. Selecting a model to predict the onset of growth of Fagus sylvatica. *Journal of Applied Ecology* 31:172–81.

Kullman, L. 1996a. Recent cooling and recession of Norway spruce (Picea abies (L.) Karst.) in the forest-alpine tundra ecotone of the Swedish Scandes. *Journal of Biogeography* 23:843–54.

Kullman, L. 1996b. Norway spruce present in the Scandes Mountains, Sweden at 8,000 BP: new light on Holocene tree spread. *Global Ecology and Biogeography Letters* 5:94–101.

Kullman, L. 1995. New and firm evidence for Mid-Holocene appearance of Picea abies in the Scandes Mountains, Sweden. *Journal of Ecology* 83:439–47.

Kullman, L. 1993. Tree limit dynamics of Betula pubescens ssp. tortuosa in relation to climate variability: evidence from central Sweden. *Journal of Vegetation Science* 4:765–72.

Kuster, H. 1990. Gedanken zur Entstehung von Waldtypen in Süddeutschland. *Bericht der Reinhold-Tüxen-Gesellschaft* 2:25–43.

Labroue, L., and C. Tosca. 1977. Dynamique de la matière organique dans les sols alpins. *Bulletin d' Ecologie* 8:289–98.

Lambers, H. 1993. Rising CO_2, secondary plant metabolism, plant-herbivore interactions and litter decomposition. *Vegetatio* 104–105:263–71.

Landolt, E. 1992. *Unsere Alpenflora*. 6th ed. Stuttgart: Fischer.

Landolt, E., and D. Aeschimann. 1986. *Notre flore alpine*. Bern: Club Alpin Suisse.

Lang, G. 1993. Holozäne Veränderungen der Waldgrenze in den Schweizer Alpen—Methodische Ansätze und gegenwärtiger Kenntnisstand. *Dissertationes Botanicae* 196:317–327.

Larcher, W. 1994. Hochgebirge: An den Grenzen des Wachstums. In *Ökologische Grundwerte in Österreich—Modell für Europa?* Ed. W. Morawetz, 304–43. Vienna: Österreichische Akademie der Wissenschaften.

Larcher, W. 1980. *Physiological plant ecology*. Heidelberg: Springer.

Larigauderie, A., and C. Körner. 1995. Acclimation of leaf dark respiration to temperature in alpine and lowland plant species. *Annals of Botany* 76:245–52.

Lean, J., and D. A. Warrilow. 1989. Simulation of the regional climatic impact of Amazon deforestation. *Nature* 342:411–3.

Lechowicz, M. J., and T. Koike. 1995. Phenology and seasonality of woody plants: An unappreciated element in global change research? *Canadian Journal of Botany* 73:147–8.

Lee, J. A., R. Baxter, and M. J. Emes. 1990. Responses of Sphagnum species to atmospheric nitrogen and sulphur deposition. *Botanical Journal of the Linnean Society* 104:255–65.

Legros, J.-P., and Y. Cabidoche. 1977. Les types de sols et leur répartition dans les Alpes et les Pyrénées cristallines. *Documents de Cartographie Ecologique*, 19. Université Scientifique et Médicale de Grenoble, 1–19.

Lewin, R. 1985. Plant communities resist climatic change. *Science* 228:165–6.

Lieth, H., ed. 1974. *Phenology and seasonality modeling*. Ecological Studies, vol. 8. Heidelberg: Springer.

Liu, E. H., R. R. Sharitz, and M. H. Smith. 1978. Thermal sensitivities of malate dehydrogenase isozymes in Typha. *American Journal of Botany* 65:214–20.

Lloret, F. 1988. Estrategías de vida y formas de vida en briófitos del Pirineo oriental (España). *Cryptogamie, Bryologie-Lichénologie* 9:189–217.

Longton, R. E. 1994. Reproductive biology in bryophytes. In *Bryophytes and Lichens in a changing environment*, ed. J. W. Bates and A. M. Farmer, 32–76. Oxford: Clarendon Press.

Longton, R. E. 1992. The role of bryophytes and lichens in terrestrial ecosystems. The challenge and the opportunities. *Journal of the Hattori Botanical Laboratory* 76:159–72.

Longton, R. E., ed. 1988. *Biology of polar bryophytes and lichens*. Cambridge: Cambridge University Press.

Longton, R. E. 1984. The role of bryophytes in terrestrial ecosystems. *Journal of the Hattori Botanical Laboratory* 55:147–63.

Longton, R. E. 1980. Physiological ecology of mosses. In *Mosses of North America*, ed. R. J. Taylor and A. E. Leviton, 77–113. San Francisco: Pacific Division American Associstion for the Advancement of Science.

Longton, R. E. 1974. Genecological differentiation in bryophytes. *Journal of the Hattori Botanical Laboratory* 38:49–65.

Longton, R. E., and S. W. Greene. 1969a. Relationship between sex distribution and sporophyte production in Pleurozium schreberi (Brid.). Mitt. *Annals of Botany* 33:107–26.

Longton, R. E., and S. W. Greene. 1969b. The growth and reproductive cycle of Pleurozium schreberi (Brid.). Mitt. *Annals of Botany* 33:83–105.

Longton, R. E., and R. M. Schuster. 1983. Reproductive biology. In *New manual of bryology*, vol. 1, ed. R. M. Schuster, 386–462. Nichinan, Japan: The Hattori Botanical Laboratory.

Loveless, M. D., and J. L. Hamrick. 1984. Ecological determinants of genetic structure in plant populations. *Annual Review of Ecology and Systematics* 15:65–95.

Lumaret, R. 1984. The role of polyploidy in the adaptative significance of polymorphism at the GOT 1 locus in the Dactylis glomerata complex. *Heredity* 52:153–69.

Lynch, M., and B. G. Milligan. 1994. Analysis of population genetic structure with RAPD markers. *Molecular Ecology* 3:91–9.

MacArthur, R. 1972. *Geographical Ecology*. New York: Harper & Row.

Maier, E., and P. Geissler. 1995. Grimmia in Mitteleuropa. *Herzogia* 11:1–80.

Mariétan I. 1929. Notes floristiques sur la partie supérieure de la vallée de Bagnes (Fionney). *Bulletin de la Murithienne, Société Valaisanne des Sciences Naturelles* 46:32–51.

Markham, A., N. Dudley, and S. Stolton. 1993. *Some like it hot*. Gland, Switzerland: World Wide Fund International.

Mayer, H. 1974. *Wälder des Ostalpenraumes*. Stuttgart: Fischer.

McNaughton, S. J. 1974. Natural selection at the enzyme level. *American Naturalist* 108:616–24.

Melillo, J. M., J. Aber, and J. F. Muratore. 1982. Nitrogen and lignin control of hardwood leaf litter decomposition dynamics. *Ecology* 63:621–6.

Michalet, R., and S. Bruckert. 1986. La podzolisation sur calcaire du subalpin du Jura. *Science du Sol* 24:363–76.

Miles, C. J., E. A. Odu, and R. E. Longton. 1989. Phenological studies on British mosses. *Journal of Bryology* 15:607–21.

Millar, C. I. 1983. A steep cline in Pinus muricata. *Evolution* 37:311–9.

Miller, N. G. 1983. Tertiary and quaternary fossils. In *New manual of bryology*, vol. 2, ed. R. M. Schuster, 1194–232. Nichinan, Japan: The Hattori Botanical Laboratory.

Mitton, J. B., K. B. Sturgeon, and M. L. Davis. 1980. Genetic diferentiation in ponderosa pine along a steep elevational transect. *Silvae Genetica* 29:100–3.

Molau, U. 1996. Phenology and reproductive success in arctic plants: susceptibility to climate change. In *Global change and arctic terrestrial ecosystems*, ed. W. C. Oechel, T. Callaghan, T. Gilmanov, J. I. Holten, B. Maswell, U. Molau, and B. Sveinbjörnsson, 153–170. Heidelberg: Springer.

Molau, U. 1993. Relationships between flowering phenology and life history strategies in tundra plants. *Arctic and Alpine Research* 25:391–402.

Moore, D. M. 1995. Opening time by degrees. *Nature* 375:186–7.

Moore, P. D. 1990. Vegetation's place in history. *Nature* 347:710.

Mucina, L., and G. Grabherr, eds. 1993. *Die Pflanzengesellschaften Österreichs. Teil 2. Natürliche waldfreie Vegetation*. Stuttgart: Fischer.

Mucina, L., G. Grabherr, and T. Ellmauer, eds. 1993. *Die Pflanzengesellschaften Österreichs. Teil 1. Anthropogene Vegetation*. Stuttgart: Fischer.

Mucina, L., G. Grabherr, and S. Wallnöfer, eds. 1993. *Die Pflanzengesellschaften Österreichs. Teil 3. Wälder und Gebüsche*. Stuttgart: Fischer.

Mueller-Dombois, D., and H. Ellenberg. 1974. *Aims and methods of vegetation ecology*. London: Wiley.

Müller, U. A. 1992. Geschichte der alpinen Forschung. In *Die Alpen—ein sicherer Lebensraum?* Ed. J. P. Müller and B. Gilgen, 7–45. Publikation der Schweizerischen Akademie der Naturwissenschaften, vol. 5, Disentis.

Müller-Starck, G. 1994. Die Bedeutung der genetischen Variation für die Anpassung gegenüber Umweltstress. *Schweizerische Zeitschrift für Forstwesen* 145:977–97.

Naveh, Z., and A. S. Lieberman. 1984. *Landscape ecology*. Heidelberg: Springer.

Neilson, R. P. 1993. Vegetation redistribution: a possible biosphere source of CO_2 during climate change. *Water, Air & Soil Pollution* 70:659–73.

Nybom, H. 1993. Applications of DNA fingerprinting in plant population studies. In *DNA fingerprinting: state of the science*, ed. S. D. J. Pena, R. Charkraborty, J. T. Epplen, and A. J. Jeffreys, 293–309. Basel, Switzerland: Birkhäuser.

Odland, A. 1995. Frond development and phenology of Thelypteris limbosperma, Athyrium distentifolium, and Matteucia struthiopteris in Western Norway. *Nordic Journal of Botany* 15: 225–36.

Oechel, W. C., and B. Sveinbjörnsson. 1978. Primary production processes in arctic bryophytes at Barrow, Alaska. In *Vegetation and production ecology of an Alaskan arctic tundra*, Ecological Studies 29, ed. L. L. Tieszen, 269–298. Heidelberg: Springer.

Oechel, W. C., and K. Van Cleve. 1986. The role of bryophytes in nutrient cycling in the taiga. In *Forest ecosystems in the Alaskan taiga*, Ecological Studies, vol, 57, ed. K. Van Cleve, F. S. Chapin III, P. W. Flanagan, L. A. Viereck, and C. T. Dyrness, 121–37. Heidelberg: Springer.

Ögren, E. 1996. Premature dehardening in Vaccinium myrtillus during a mild winter: a cause for winter dieback? *Functional Ecology* 10:724–32.

Olson, J. E. 1963. Energy storage and the balance of producers and decomposers in ecological systems. *Ecology* 44: 322–31.

Orshan, G., ed. 1989. Plant pheno-morphological studies in Mediterranean type ecosystems. *Geobotany*, vol. 12, Dordrecht, the Netherlands: Kluwer.

Ozenda, P. 1994. *Végétation du continent européen*. Lausanne, Switzerland: Delachaux & Niestlé.

Ozenda, P. 1988. *Die Vegetation der Alpen im europäischen Gebirgsraum*. Stuttgart: Fischer.

Ozenda, P. 1985. *La végétation de la chaîne alpine dans l'espace montagnard européen*. Paris: Masson.

Ozenda, P., and J.-L. Borel. 1994. Potential effects of a future global climatic change on the terrestrial ecosystems of the Alps. In *The sensitivity of Alpine ecosystems to potential climate change.*

Final Report, ed. L. Bourjot, 81–98. Centre International pour l'Environnement Alpin (ICALPE), Le Bourget-du-Lac, France.

Ozenda, P., and J.-L. Borel. 1991. *Les conséquences écologiques possibles des changements climatiques dans l'Arc alpin*. Centre International pour l'Environnement Alpin (ICALPE), Rapport FUTURALP 1, Chambéry, France.

Pache, G., R. Michalet, and S. Aimé. 1996. A seasonal application of the Gams (1932) method, modified Michalet (1991): the example of the distribution of some important forest species in the Alps. *Dissertationes Botanicae* 258:31–54.

Paternoster, M. 1981. Colonisation par la végétation et pédogenèse initiale sur les moraines latérales historiques du Glacier d'Aletsch. Thèse spécialisation agro-éco-pédologie, Nancy I. France.

Pawlowski, B. 1970. Remarque sur l'endémisme dans la flore des Alpes et des Carpathes. *Vegetatio* 21:181–243.

Perruchoud, D., and A. Fischlin. 1995. The response of the carbon cycle in undisturbed forest ecosystems to climate change: A review of plant-soil models. *Journal of Biogeography* 22:2603–18.

Peters, R. L., and T. E. Lovejoy. 1992. *Global warming and biological diversity*. London: Yale University Press.

Pielke, R. A., T. J. Lee, T. G. F. Kittel, T. N. Chase, J. M. Cram, and J. S. Baron. 1994. Effects of mesoscale vegetation distributions in mountainous terrain on local climate. In *Mountain environments in changing climates*, ed. M. Beniston, 121–36. London: Routledge Publishing Co.

Pigliucci, M. 1996. How organisms respond to environmental changes: From phenotypes to molecules (and vice versa). *Trends in Ecology and Evolution* 11:168–73.

Pimm, S. L. 1984. The complexity and stability of ecosystems. *Nature* 307:321–6.

Pornon, A., and B. Doche. 1995a. Minéralisation et nitrification de l'azote dans différents stades de colonisation des landes subalpines à Rhododendron ferrugineum L. (Alpes du Nord; France). *Comptes Rendus de l'Académie des Sciences, Paris. Série 3. Sciences de la Vie* 318:887–95.

Pornon, A., and B. Doche. 1995b. Influence des populations de Rhododendron ferrugineum L. sur la végétation subalpine (Alpes du Nord-France). *Feddes Repertorium* 29:179–91.

Pornon, A., and B. Doche. 1994. Dynamics and functioning of Rhododendron ferrugineum subalpine heathlands (northern Alps, France). In *Mountain environments in changing climates*, ed. M. Beniston, 244–58. London: Routledge Publishing Co.

Post, W. M., W. R. Emanuel, P. J. Zinke, and A. G. Stangenberger. 1982. Soil carbon pools and world life zones. *Nature* 298:156–9.

Pott, R. 1993. *Farbatlas Waldlandschaften*. Stuttgart: Ulmer.

Powlson, D. S., P. Smith, and J. U. Smith, eds. 1996. *Evaluation of Soil Organic Matter Models*. Heidelberg: Springer.

Prock, S., and C. Körner. 1996. A cross-continental comparison of phenology, leaf dynamics and dry matter allocation in arctic and temperate zone herbaceous plants from contrasting altitudes. *Ecological Bulletins* 45:93–103.

Proctor, M. C. F. 1982. Physiological ecology: water relations, light and temperature responses, carbon balance. In *Bryophyte ecology*, ed. A. J. E. Smith, 333–81. London: Chapman & Hall.

Puppi Branzi, G., A. L. Zanotti, and M. Speranza. 1994. Phenological studies on Vaccinium and Nardus communities. *Fitosociologia* 26:63–79.

Raeymakers, G., and J. Glime. 1986. Effects of simulated acidic rain and lead interaction on the phenology and chlorophyll content of Pleurozium schreberi (Brid.). *Mitt. Journal of the Hattori Botanical Laboratory* 61:525–41.

Rathcke, B., and E. P. Lacey. 1985. Phenological patterns of terrestrial plants. *Annual Review of Ecology and Systematics* 16:179–214.

Rechinger, K. H. 1965. Der Endemismus in der grieschischen Flora. *Revue Roumaine de Biologie, Série de Botanique* 10:135–8.

Rioux, J., and P. Quézel. 1949. Contribution à l'étude des groupements rupicoles endémiques des Alpes-Maritimes. *Vegetatio* 2:1–13.

Robert, M., Y. Cabidoche, and J. Berrier. 1980. Pédogenèse et minéralogie des sols de hautes montagnes cristallines (étages Alpin et Subalpin)—Alpes-Pyrénées. *Science du Sol* 18:313–36.

Robinson, A. L., D. H. Vitt, and K. P. Timoney. 1989. Patterns of community structure and morphology of bryophytes and lichens relative to edaphic gradient in the subarctic forest-tundra of Northwestern Canada. *Bryologist* 92:495–512.

Rolland, C., and J. F. Schueller. 1995. Relationships between mountain pine and climate in the French Pyrenees (Font-Romeu) studied using the radiodensitometrical method. *Pirineos* 143–144: 55–70.

Root, R. B. 1967. The niche exploitation pattern of the blue-gray gnatcatcher. *Ecological Monographs* 37:317–50.

Root, T. L. and S. H. Schneider. 1995. Ecology and climate: Research strategies and implications. *Science* 269:334–41.

Rübel, E. 1935. The replaceability of ecological factors and the law of the minimum. *Ecology* 16:336–341.

Running, S. W., and R. R. Nemani. 1991. Regional hydrologic and carbon balance responses of forests resulting from potential climatic change. *Climatic Change* 19:342–68.

Schäppi, B., and C. Körner. 1996. Growth responses of an alpine grassland to elevated CO_2. *Oecologia, Berlin* 105:43–52.

Scharfetter, R. 1938. *Das Pflanzenleben der Ostalpen*. Vienna: Deuticke.

Scheiner, S. M. 1993. Genetics and evolution of phenotypic plasticity. *Annual Review of Ecology and Systematics* 24:35–68.

Schlüssel, A., and J.-P. Theurillat. 1998. Phenology of Rhododendron ferrugineum L. in the Valaisian Alps (Switzerland). *Ecologie* 29:135–139.

Schlüssel, A., and J.-P. Theurillat. 1996. Synusial structure of heathlands at the subalpine/alpine ecocline in Valais (Switzerland). *Revue Suisse de Zoologie* 163:795–800.

Schnelle, F. 1955. *Pflanzen-Phänologie*. Leipzig, Germany: Geest & Portig.

Schnitzer, M., and S. U. Kahn, eds. 1989. *Soil Organic Matter*. 4th ed. Amsterdam: Elsevier.

Schreiber, K.-F. 1968. Les conditions thermiques du canton de Vaud. *Beiträge zur geobotanischen Landesaufnahme der Schweiz* 49:1–31.

Schreiber, K.-F., N. Kuhn, C. Hug, R. Häberli, C. Schreiber, W. Zeh, and S. Lautenschlager. 1977. *Wärmegliederung der Schweiz*. Bern, Switzerland: Eidgenössisches Justiz- und Polizeidepartement.

Schröter, C. 1923–1926. *Das Pflanzenleben der Alpen*. 2nd ed. Zürich: Raustein.

Simons, J. P., C. Charest, and M. J. Peloquin. 1986. Adaptation and acclimatation of higher plants at the enzyme level: Kinetic properties of NAD malate dehydrogenase in three species of Viola. *Journal of Ecology* 74:19–32.

Simons, J. P., C. Potuin, and M. H. Blanchard. 1983. Thermal adaptation and acclimatation of higher plants at the enzyme level: Kinetic properties of NAD malate dehydrogenase and glutamate oxaloacetate transaminase in two genotypes of Arabidopsis thaliana (Brassicaceae). *Oecologia* 60:143–8.

Slack, N. G. 1988. The ecological importance of lichens and bryophytes. In *Lichens, bryophytes and air quality*, Bibliotheca Lichenologica, vol. 30, ed. T. H. Nash and V. Wirth, 23–53. Vaduz, Liechtenstein: Cramer.

Slatkin, M., K. Hindar, and Y. Michalakis. 1995. Processess of genetic diversification. In *Global biodiversity assessment*, ed. V. H. Heywood, 213–25. Cambridge: Cambridge University Press.

Slatyer, R. O., and I. R. Noble. 1992. Dynamics of montane treelines. In *Landscape boundaries: consequences for biotic diversity and ecological flows*, Ecological Studies, vol. 92, ed. A. J. Hansen and F. Di Castri, 346–59. Heidelberg: Springer.

Smith, T. M., and H. H. Shugart. 1993. The transient response of terrestrial carbon storage to a perturbed climate. *Nature* 361:523–6.

Snaydon, R. W., and M. S. Davies. 1972. Rapid population differentiation in a mosaic environment II: Morphological variation in Anthoxanthum odoratum. *Evolution* 26:390–405.

Snaydon, R. W., and T. M. Davies. 1982. Rapid divergence of plant populations in response to recent changes in soil conditions. *Evolution* 36:289–97.

Snogerup, S. 1971. Evolutionary and plant geographical aspects of chasmophytic communities. In *Plant life of South-West Asia*, ed. P. H. Davis, P. C. Harper, and I. C. Hedge, 157–170. Edinburgh: The Botanical Society of Edinburgh.

Soltner, D. 1992. *Les bases de la production végétale* vol. 1 *Le Sol*. Sainte-Gemmes-sur-Loire, France: Sciences et techniques agricoles.

Spaltenstein, H. 1984. Pédogenèse sur calcaire dur dans les Hautes Alpes calcaires. Thèse de doctorat, Ecole Polytechnique de Lausanne, Switzerland.

Stark, L. 1985. Phenology and species concepts: a case study. *The Bryologist* 88:190–8.

Steinger, T., C. Körner, and B. Schmid. 1996. Long-term persistence in a changing climate: DNA analysis suggests very old ages of clones of alpine Carex curvula. *Oecologia* (Berlin) 105:94–9.

Steubing, L. 1993. Der Eintrag von Schad- und Nährstoffen und deren Wirkung auf die Vergrasung der Heide. *Berichte der Reinhold-Tüxen-Gesellschaft* 5:113–33.

Stevenson, F. J. 1982. *Humus chemistry. Genesis, composition, reactions*. New York: Wiley & Sons.

Stine, S. 1994. Extreme and persistent drought in California and Patagonia during mediaeval time. *Nature* 369:546–9.

Street-Perrott, F. A. 1994. Drowned trees record dry spells. *Nature* 369:518.

Sveinbjörnsson, B., and W. C. Oechel. 1992. Controls on growth and productivity of bryophytes: environmental limitations under current and anticipated conditions. In *Bryophytes and Lichens in a changing environment*, ed. J. W. Bates and A. M. Farmer, 77–102. Oxford: Clarendon Press.

Svensson, B. M., B. Floderus, and T. V. Callaghan. 1994. Lycopodium annotinum and light quality: Growth response under canopies of two Vaccinium species. *Folia geobotanica et phytotaxonomica* 29:159–66.

Sykes, M. T., I. C. Prentice, and W. Cramer. 1996. A bioclimatic model for the potential distributions of north European tree species under present and future climates. *Journal of Biogeography* 23:203–33.

Tallis, J. H. 1991. *Plant community history*. London: Chapman & Hall.

Talon, B. 1997. *Evolution des zones supra-forestières des Alpes sud-occidentales françaises au cours de l'Holocène. Analyse pédoanthracologique.* Thèse, Université de Droit, d'Economie et des Sciences d'Aix-Marseille III, Institut Méditerranéen d'Ecologie et de Paléoécologie. Facultés des Sciences et Techniques de Saint-Jérôme, Marseille, France.

Tappeiner, U., and A. Cernusca. 1996. Microclimate and fluxes of water vapour, sensible heat and carbon dioxide in structurally differing subalpine plant communities in the Central Caucasus. *Plant, Cell and Environment* 19:403–17.

Tappeiner, U., and A. Cernusca. 1994. Bestandestruktur, Energiehaushalt und Bodenatmung einer Mähwiese, einer Almweide und einer Almbrache. *Verhandlungen der Gesellschaft für Ökologie* 23:49–56.

Tappeiner, U., and A. Cernusca. 1991. The combination of measurements and mathematical modelling for assessing canopy structure effects. In *Modern ecology: Basic and applied aspects*, ed. G. Esser and D. Overdieck, 161–93. Amsterdam: Elsevier.

Tegart, W. J. McG., G. W. Sheldon, and D. C. Griffiths. 1990. *Climate change. The IPCC impacts assessment.* Canberra: Australian Government Publication Service.

Theurillat, J.-P. 1995. Climate change and the alpine flora: Some perspectives. In *Potential ecological impacts of climate change in the Alps and Fennoscandian mountains*, ed. A. Guisan, J. I. Holten, R. Spichiger, and L. Tessier, 121–7. Geneva: Conservatoire et Jardin botaniques.

Theurillat, J.-P. 1992a. L'analyse du paysage végétal en symphytocoenologie: ses niveaux et leurs domaines spatiaux. *Bulletin d'Ecologie* 23:83–92.

Theurillat, J.-P. 1992b. Etude et cartographie du paysage végétal (symphytocoenologie) dans la région d'Aletsch (Valais, Suisse). Développement historique et conceptuel de la symphytocoenologie, niveaux de perception, méthodologie, applications. *Beiträge zur geobotanischen Landesaufnahme der Schweiz* 68:1–384

Theurillat, J.-P. 1992c. Abgrenzungen von Vegetationskomplexen bei komplizierten Reliefverhältnissen, gezeigt an Beispielen aus dem Aletschgebiet (Wallis, Schweiz). *Berichte der Reinhold-Tüxen-Gesellschaft* 4:147–66.

Theurillat, J.-P., 1991. Les étages de végétation dans les Alpes centrales occidentales. *Saussurea* 22:103–47.

Theurillat, J.-P., D. Aeschimann, P. Küpfer, and R. Spichiger. 1994. Habitats et régions naturelles des Alpes. *Colloques Phytosociologiques* 22:15–30.

Thornwaite, C. W. 1954. Topoclimatology. In *Toronto Meteorological Conference, 1953*, 227–232. London: Royal Meteorological Society.

Tinker, P. B., and P. Ineson. 1990. Soil organic matter and biology in relation to climate change. In *Soils on a warmer earth.* Amsterdam: Elsevier.

Tinner, W., B. Amman, and P. Germann. 1996. Treeline fluctuations recorded for 12,500 years by soil profiles, pollen, and plant macrofossils in the central Swiss Alps. *Arctic and Alpine Research* 28:131–47.

Tranquillini, W. 1979. *Physiological ecology of the alpine timberline.* Ecological Studies, vol. 31. Heidelberg: Springer.

Troelstra, S. R., R. Wagenaar, and W. Smant. 1995a. Nitrogen utilization by plant species from acid heathland soils I. Comparison between nitrate and ammonium nutrition at constant low pH. *Journal of Experimental Botany* 46:1103–12.

Troelstra, S. R., R. Wagenaar, and W. Smant. 1995b. Nitrogen utilization by plant species from acid heathland soils II. Growth and shoot/root partioning of No_3^- assimilation at constant low pH and varying No_3^-/Nh_4^+ ratio. *Journal of Experimental Botany* 46:1113–21.

Turesson, G. 1922. The genotypical response of the plant species to the habitat. *Hereditas* 3:211–350.

Urmi, E., I. Bisang, P. Geissler, H. Hürlimann, L. Lienhard, N. Müller, I. Schmid Grob, N. Schnyder, and L. Thöni. 1992. *Rote Liste. Die gefährdeten Moose der Schweiz.* Bern, Switzerland: Eidgenössische Druchsachen- und Materialzentrale.

Vaccari, L. 1913. Contributo alla briologia della Valle d'Aosta. *Nuovo Giornale Botanico Italiano* 20:417–96.

Vaccari, L. 1911. La Flora nivale del Monte Rosa. *Bulletin. Société de la Flore Valdôtaine* 7:129–35.

Van Cleve, K., W. D. Oechel, and J. L. Hom. 1990. Response of black spruce (Picea mariana) ecosystems to soil temperature modifications in interior Alaska. *Canadian Journal of Forest Research* 20:291–302.

van de Geijn, S. C., and J. A. van Veen. 1993. Implications of increased carbon dioxide levels for carbon input and turnover in soils. *Vegetatio* 104–105:283–92.

van Steenis, C. G. G. 1961. An attempt towards an explanation of the effect of mountain mass elevation. *Proceedings, Koninklijke Nederlandse Akademie van Wetenschappen. Series C, Biological and Medical Sciences* 64:435–42.

Veit, H., and T. Höfner 1993. Permafrost, gelifluction and fluvial sediment transfer in the alpine/subnival ecotone, central Alps, Austria: present, past and future. *Zeitschrift für Geomorphologie* N.F. 92 (suppl.):71–84.

Velluti, C., and P. Geissler. 1996. Preliminary results on the phenology of alpine bryophytes: Pleurozium schreberi (Brid.). Mitt. *Colloques Phytosociologiques* 24:771–77.

Vitousek, P. M. 1994. Beyond global warming: ecology and global change. *Ecology* 75:1861–76.

Walter, H. 1985. *Vegetation of the earth and ecological systems of geo-biosphere.* 3d ed. Heidelberg: Springer.

Walter, H. 1984. *Vegetation und Klimazonen.* Stuttgart: Ulmer.

Walter, H. 1931. *Die Hydratur der Pflanze und ihre physiologisch-ökologische Bedeutung.* Jena, Germany: Fischer.

Walter, H., and S.-W. Breckle. 1991. *Ökologischer Grundlagen in globaler Sicht.* 2d ed. *Ökologie der Erde,* vol. 1. Stuttgart: Fischer.

Walter, H., and S.-W. Breckle. 1985. *Ecological systems of the geobiosphere. 1. Ecological principles in global perspective.* Heidelberg: Springer.

Walter, H., and E. Walter. 1953. Einige allgemeine Ergebnisse unserer Forschungsreise nach Südwestafrika 1952/53: Das Gesetz der relativen Standortskonstanz; das Wesen der Pflanzenge-meinschaften. *Berichte der Deutschen botanischen Gesellschaft* 66:228–36.

Walther, G.-R. 1996. Distribution and limits of evergreen broad-leaved plants in the southern Ticino. *Bulletin of the Geobotanical Institute* ETH 62:115–6.

Wardle, J. 1983. Causes of alpine timberline: a review of the hypotheses. In *Forest development in cold climates,* ed. J. Alden, J. L. Mastrantonio, and S. Odum, 89–103. London: Plenum Press.

Warwick, S. I. 1991. Herbicide resistance in weedy plants—Physiology and population biology. *Annual Review of Ecology and Systematics* 22:95–114.

Watson, R. T, M. C. Einjowero, R. H. Moss, and D. J. Dotten (eds.) 1996. Climate change 1995. Impacts, adoptations and mitigation of climate change: scientific-technical analyses. Cambridge: Cambridge University Press.

Wiersema, G. 1989. Climate change and vegetation characteristics of ibex habitats in the European Alps. *Mountain Research and Development* 9:119–28.

Wigley, T. M. L. 1985. Impact of extreme events. *Nature* 316:106–7.

Wijk, S. 1986. Influence of climate and age on annual shoot increment in Salix herbacea. *Journal of Ecology* 74:685–92.

Wildi, B., and C. Lütz. 1996. Antioxydant composition of selected high alpine plant species from different altitudes. *Plant, Cell and Environment* 19:138–46.

Wisniewski, J., R. K. Dixon, J. D. Kinsman, R. N. Sampson, and A. Lugo. 1993. Carbon dioxide sequestration in terrestrial ecosystems. *Climate Research* 3:1–5.

Woodward, F. I., ed. 1992. *The ecological consequences of global climate change.* Advances in Ecological Research, vol. 22. London: Academic Press.

Woodward, F. I, ed. 1987. *Climate and Plant Distribution.* Cambridge: Cambridge University Press.

Woodward, F. I., G. B. Thompson, and I. F. McKee. 1991. The effects of elevated concentrations of carbon dioxide on individual plants, populations, communities and ecosystems. *Annals of Botany* 67 (suppl. 1):23–38.

Wyatt, R. A. 1994. Population genetics of bryophytes in relation to their reproductive biology. *Journal of the Hattori Botanical Laboratory* 76:147–57.

Wyatt, R. A., A. Stoneburner, and I. J. Odrzykosski. 1989. Bryophyte isozymes: systematics and evolutionary implications. In *Isozymes in plant biology*, ed. D. E. Soltis and P. S. Soltis, 221–40. Portland: Dioscorides Press, Portland.

Yaalon, D. 1990. The relevance of soils and paleosols in interpreting past and ongoing climatic changes. *Palaeogeography, Palaeoclimatology, Palaeoecology* 82:63–64.

Young, A., T. Boyle, and T. Brown. 1996. The population genetic consequences of habitat fragmentation for plants. *Trends in Ecology and Evolution* 11:413–8.

Zahn, R. 1994. Fast flickers in the tropics. *Nature* 372:621–2.

Zechmeister, H., and L. Mucina. 1994. Vegetation of European springs: High-rank syntaxa of the Montio-Cardaminetea. *Journal of Vegetation Science* 5:385–402.

Zhang, Q., M. A. Saghai Maroofand, and A. Kleinhofs. 1993. Comparative diversity of RFLPs and isozymes within and among populations of Hordeum vulgare ssp. spontaneum. *Genetics* 134:909–16.

6 Vegetation Responses to Climate Change in the Alps: Modeling Studies

Heike Lischke, Antoine Guisan, Andreas Fischlin,
Jann Williams, and Harald Bugmann

6.1 INTRODUCTION

High altitude plants are well adapted to their present environment (Körner 1992), but their ability to adapt to the changes in climate expected over the next century remains uncertain (Guisan, Holten, Tessier et al. 1995). Chapter 3 presented a few glimpses of plants and ecosystems' reactions to changing climates in the past. Chapter 5 reviewed the present knowledge about montane ecosystems and hypotheses for their possible responses to climate change in the Alps, based on experiments and field observations. In this chapter, we present a modeling approach for assessing climate change's impact on the vegetation of the Alps.

Our current understanding of the systems and processes that characterize natural vegetation offers a wide array of answers to the question of how vegetation will respond to expected climate change (Idso 1980a, 1980b; Körner and Arnone 1992a, 1992b). For example, the spectrum of predictions for forests ranges from severe diebacks (Neilson 1993) to a general increase in productivity due to elevated temperatures, precipitation, and atmospheric CO_2 concentrations (Blum 1991; Graybill and Idso 1993; Idso and Idso 1994; Kimball and Idso 1983). For montane forests, invasions of trees into areas above present timberline are predicted (Bugmann 1994; Kienast 1989, 1991; Kräuchi and Kienast 1993). For herbaceous or dwarf shrub species of the alpine zone (above tree line), predictions range from global vegetation shifts and species invasions or extinctions (Ozenda and Borel 1990) to local variations in species' abundances or community composition regulated by microhabitat distribution. However, except for the most extreme climate change scenarios, natural mosaic-like microtopography might provide refuges for species that would otherwise be threatened as a result of climate change (Körner 1995).

Different studies diverge in their assessments of climate change's effects because multiple processes can influence ecosystem dynamics in conflicting directions; for example, increased atmospheric concentrations of CO_2 may enhance potential productivity (Eamus and Jarvis 1989; Morison 1987; Robinson 1994), whereas superoptimal temperatures and drought may reduce it. Moreover, these processes and their typically nonlinear responses to climate work at different temporal, spatial, and organizational scales, also impeding a direct assessment.

One way to integrate our current knowledge of the ecosystem processes' effects to evaluate climate change impacts is to use simulation models of vegetation occurrence or dynamics (Guisan, Holten, Spichiger, et al. 1995). Not only do such models help improve our understanding of the processes taking place at the interface between climate and ecosystems, but they also allow a preliminary assessment of climate change's potential impacts on vegetation by comparing simulation results under present and future climate.

In addition to conceptual models (Romme and Turner 1990), numerous computer-based simulation models have been developed for assessing terrestrial ecosystems' response to anticipated climate change. They differ in their spatial scale and resolution, in the level of detail at which they work, and also in whether and how they treat vegetation's variability and temporal development. Solomon and Leemans (1989) Walker (1990) and Kirschbaum and Fischlin (1996) review these models, which assess environmental change's ecological impact, and detail their advantages, drawbacks, spatiotemporal-scale requirements, limitations, and applications.

Dynamic models describe how vegetation changes with time, even when input values are constant. Process rates in these models usually depend explicitly or implicitly on climate.

Dynamic ecophysiological process models (Bossel 1987, 1991; Bossel et al. 1991) describe in detail many ecophysiological processes contributing to plant growth, birth, and death. These models' complexity confines them to the local scale and to relatively short time windows.

Succession models, such as the forest gap or patch dynamics models like JABOWA (Botkin, Janak, and Wallis 1970, 1972a, 1972b), FORET (Shugart and West 1977) or FORSKA (Leemans and Prentice 1989), are less detailed and offer the advantage of mimicking vegetation dynamics' long-term characteristics. They work at the local to regional scale and at the level of individuals. The forest patch models FORECE (Kienast 1989, 1991; Kienast and Kräuchi 1989; Kienast and Kuhn 1989), FORSUM (Kräuchi 1993; Kräuchi and Kienast 1993) and FORCLIM (Bugman 1994, 1996; Fischlin, Bugmann, and Gyalistras 1995) all apply in the Alpine region, because their species sets contain most of the dominant tree species found in this region. Recently, several gap models have been simplified to more efficient structured population models (Fulton 1991; Kohyama and Shigesada 1995; Lischke, Löffler, and Fischlin 1998).

Also many nondynamic models or modeling procedures are used to assess possible impacts of climate change. Global biogeography models (e.g., BIOME (Prentice et al. 1992), TVM (Leemans and van-den-Born 1994), BIOME2 (Haxeltine, Prentice, and Cresswell 1997), MAPPS (Neilson 1995), and DOLY (Woodward and Smith 1994; Woodward, Smith, and Emanuel 1995)) are static equilibrium vegetation models that can be driven by transient input variables and are sometimes coupled with dynamic nutrient cycling models. Vegetation is considered in terms of plant functional types, vegetation complexes, or biomes. Potential vegetation composition is determined empirically or causally from ecophysiological constraints, such as yearly day-degree sum, and from dominance tables or maximum net ecosystem production (NEP) or the leaf area index that can be reached under the given moisture and nutrient conditions.

Many models of vegetation or single species are based on statistical analyses of interactions between species and their ecological environment. Such models are more descriptive and often noncausal, although they increasingly include ecophysiologically meaningful variables. Busby (1988, 1991) applied his BIOCLIM approach (a fitted, species-specific, p-dimensional environmental envelope) to alpine vegetation of southeastern Australia using 0.1 degree latitude-longitude grid cells. As an improvement, Carpenter, Gillison, and Winter (1993) developed the DOMAIN model to map potential distributions of species. It is based instead on a point-to-point similarity metric (measure of multivariate distances) and has been proven more suitable to applications where available records are limited.

As an alternative, a large range of regression methods (least square, nonlinear regressions, regression trees, generalized linear or additive models (GLMs, GAMs); Nicholls 1989; Yee and Mitchell 1991; see also section 6.2) have been developed for modeling species' distributions and ecological tolerances (realized niches). Numerous such studies have been successfully applied to climate change ecological impact assessment (Austin 1992; Brzeziecki, Kienast, and Wildi 1994; de Swart et al. 1994; Hill 1991; Huntley et al. 1995).

Thus, overall, there are many distinct ecological models or approaches, and each has its particular advantages and drawbacks depending on the particular application's goals. However, for our purpose not all of these models are adequate. In our studies, we wanted to focus on the level of individual plant species. This excluded the use of biogeography models, which aggregate species to plant functional types or biomes. We were also interested in assessing climate change impacts at a scale between local and regional (1–1,000 km), however, with the fine resolution (1–100 m) required for dealing with the rugged microtopography typical of the Alps as well as the associated high ecological complexity. This excluded models either restricted to single locations because of too great computing time and input value demands, such as detailed ecophysiological models, or those working at too large a resolution, such as the biogeography models already mentioned.

We were also interested in two different aspects of vegetation responses: We wanted to predict the potential future distribution of alpine species in large areas, particularly those of rare species, for reasons such as conservation or biodiversity management. At the same time, we were interested in the dynamics of dominant forest species during the next century to assess local impacts of climate change, such as that on the regulation of water runoff, or to explore montane forests' potential role in the future global carbon cycle (see chapter 5). Therefore, we chose two different modeling approaches, each suitable for its specific application and working on comparable spatial and hierarchical scales.

The first model aims to assess the potential future distribution of alpine plant species (sensu stricto, those above tree line). To include many species, including rare ones, this approach has to be simple, because detailed species-specific information about ecological processes is not available for all species. It is therefore a static model similar to those Brzeziecki, Kienast, and Wildi (1993, 1994) and Kienast, Brzeziecki, and Wildi (1994) developed for forest communities but focusing on herbaceous and dwarf shrub species above forest limits. It is empirical and comparative, in that it uses statistical analyses to relate present plant distributions to environmental covariates (e.g., climate) and predicts probabilities of plant species occurrence in geographic space (section 6.2).

The second model focuses on the temporal development of a limited number of dominant species, in this case forest trees. A static approach is not appropriate and also not required, because rather detailed information is available about forest tree species' ecological processes. Therefore, we chose the mechanistic (i.e., more causal) and stochastic forest patch dynamics approach, which explicitly simulates birth, growth, and death of individual trees and is suitable for site-specific, realistic simulations of the mid- and long-term temporal development of tree biomass and species compositions for the coming century (section 6.3).

Our two models differ from those few existing models that combine both static and dynamic aspects (Solomon and Leemans 1989). The static model used here is clearly spatially based, working at a regional scale with a fine resolution and depending on the resolution of the available digital elevation model, yet it lacks a time dimension. The dynamic model, in contrast, is temporally explicit but deficient regarding the spatial dimension, that is, for each site a separate simulation has to be run. Both models, together with their development and validation, are discussed in terms of their respective properties, advantages, and drawbacks, particularly with respect to their application to climate change (Brzeziecki, Kienast, and Wildi 1995; Fischlin 1995; Guisan, Theurillat, and Spichiger 1995).

6.2 THE STATIC ASPECT: MODELING THE POTENTIAL HABITAT OF ALPINE PLANTS

6.2.1 Static Plant Modeling

Static modeling procedures are not new. In fact, as long as ecologists have tried to relate vegetation or plant distributions directly to the physical environment, they have made static analyses based on assumptions of pseudo-equilibrium. Static modeling is now being reconsidered in studies of global climate change as one possible method for obtaining rapid primary impact assessments over large areas.

Since the 1980s, many bioclimatological studies have considered the distribution of plant species in both environmental and geographical spaces (Hill 1991). A species' environmental space corresponds to its realized ecological niche (the combination, or envelope, of ecological conditions that a species can tolerate in a multidimensional environmental space); its geographical space is its actual geographical distribution (dependent, for instance, on historical factors and human influences). Both spaces were judged to be necessary for assessing climate change impacts on plant species distributions (Hill 1991). Many examples of such bioclimatic studies exist, but almost all are concerned with low elevation areas (Guisan et al. 1998). Few bioclimatic studies have been conducted in high-altitude areas (alpine and snow belts), whose rugged topography requires a much higher spatial resolution to obtain reliable results. Fischer (1990) used topographic factors, radiation, land use, precipitation, and snow cover (which integrates temperature, precipitation and microrelief in time and space) to predict the distribution of plant communities, including alpine communities, in the region of Davos (Switzerland) and achieved a rather high (70 percent) correspondance with actual vegetation maps. Brzeziecki, Kienast and Wildi (1993) applied another vegetation model to all of Switzerland; the model included the seventy-one forest community types described by Ellenberg and Kiötzli (1972). More recently, Zimmermann and Kienast (in press) improved the model to include alpine plant communities. They performed quite successful modeling experiments in the Swiss alpine region of Grindelwald and recently generalized their model to the whole country (see figure 6.3b). However, no model has yet focused on specific alpine plant species distributions in Switzerland.

Such a static approach is primarily based on statistical methods (such as multiple regression, decision trees) and focuses mainly on determining the potential present and future distribution of plant or animal species or communities. The range of possible applications includes biodiversity and endangered flora management as well as primary, short-term assessments of climate change impacts over specific areas. In particular, such an approach could enable researchers to identify species or communities that might be particularly threatened by a change in climate and those that might be favored. It assumes, within the resolution and time frames of interest, that current vegetation outside of areas of intense human impact is in a quasi-stationary equilibrium. This postulate is considered true for alpine areas, where any kind of modification to the local climate could break such a fragile ecological equilibrium and modify the composition and structure of ecosystems (Galland 1982).

At the scale of the whole Alpine arc, the distribution of plant species can be described, at best, by indicating presence or absence within broad biogeographic units (Welten and Sutter 1982), within political entities (see

Aeschimann et al. in press) or within relatively large grid divisions (Hartel et al. 1992). Such macroscale information indicates clearly the overall biogeographical and historical distribution trends but does not allow for an accurate ecological description of species' specific habitats (Theurillat 1995).

Modeling at a finer resolution requires knowledge of a species' ecological requirements at the microscale level (for instance, along a mountainside or within a small alpine catchment). For Switzerland, knowledge about alpine species' environmental requirements is summarized by ecological indices for species (Landolt 1977). However, in practice, such ecological values are difficult to include in phytogeographical models because of their semi-quantitative nature, and because their reliability is too limited geographically. Our study thus required us to undertake our own intensive field sampling to obtain more reliable data on the ecology of species.

Most models and results presented in this section are related to the ALPLANDI project, which is part of the wider Ecocline coordinated project (Theurillat et al. 1997) developed in the framework of the Priority Programme Environment of the Swiss National Science Foundation (Guisan, Theurillat, and Spichiger 1995; Guisan 1997; Guisan et al. 1998).

6.2.2 Model Construction and Calibration

The modeling of alpine plant distributions involves many successive steps and also requires many different analytical techniques and associated tools. As an illustration, the methodology we are following requires the use of a triangulated constellation consisting of a geographical information system (GIS), a statistical package, and a database package (see figure 6.1). The basic kernel of all our modeling procedures is a 25-meter resolution Digital Elevation Model (DEM) covering the whole study area (see figure 6.1).

The following sections describe some important practical and theoretical aspects involved in constructing a static alpine plant distribution model (see also Buckland and Elston 1993 for a more general review). They are illustrated with concrete examples from the ALPLANDI project.

6.2.2.1 Data Sources The data set used for calibrating the model was sampled in the field during a three-year summer campaign (1993–95). Classical Braun-Blanquet *relevés* (measure of abundance-dominance of species) were made at each point of the sampling design in a four-square-meter plot. The first two years (1993–94) were devoted to sampling points to be used for calibrating the model. An additional summer (1995) allowed us to sample independent *relevés* for validating the model.

Spatial variations in environmental factors within the area were determined by modeling procedures on the DEM (for example, solar radiation, annual mean temperature, permafrost; following Hutchinson and Bishof 1983 and Brown 1994) by digitizing existing maps (geology, hypdrology, lithology) and through derivation from aerial photographs or satellite scenes

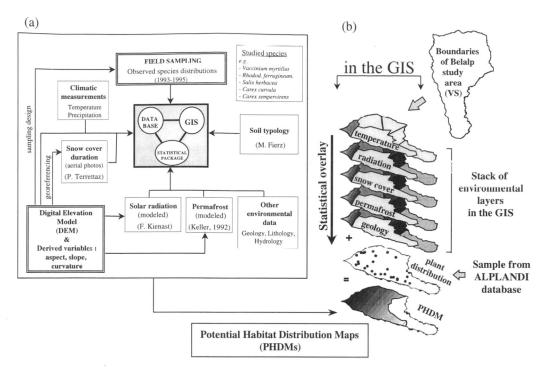

(a)

(b)

Figure 6.1 Structure and organization of the data and tools used in the ALPLANDI methodology, clearly showing the central role of the 25-meter digital elevation model (DEM) regarding sampling design, variable derivation, and data georeferrencing. Shading on the Gis maps is for illustrative purposes only.

(snow cover, vegetation cover, unmapped hydrology). These variables were stored as environmental layers within the Geographical Information System (see figure 6.1).

6.2.2.2   Spatial Scale and Autocorrelation   In formulating a statistical model, selecting the spatial scale is a crucial decision, because of the scale dependency of most statistical relationships in ecological studies (Greig-Smith 1983; Jongman, Ter Braak, and van Tongeren 1987). In particular, if spatial autocorrelation (AC) is observed for the dependent variable (that is, at a measured spatial distance, pairs of observations take more (positive AC) or less (negative AC) similar correlation values than would be expected if pairs were randomly associated), then correlations and estimation of model parameters can change with scale (Anselin and Getis 1992). From a purely statistical point of view, the optimum scale for modeling should preferably be "just small enough" to avoid spatial AC. However, when observed at the level of species distributions, spatial AC can also be considered an indicator of spatial processes such as dispersal and be used to incorporate dispersal functions into classical analyses of the distribution of species abundance (Malanson 1985). At the scale of this study, that is, one point sampled every 250 meters, no autocorrelation was observed in the data.

Vegetation Responses to Climate Change in the Alps: Modeling Studies

6.2.2.3 Ideal versus Reasonable Sampling Strategy In a single spatially limited study area, it is very difficult to sample the broad range of ecological conditions that a species can tolerate. A sampling limited to a part of the actual range of conditions the species experiences can make the species' observed response to an environmental variable appear different from what it actually is, thus "truncating" its environmental profile (see section 6.2.2.5).

Efficient sampling strategies therefore aim to give complete and accurate information on species' responses along an identified environmental gradient. However, the problem becomes much more complex when, as in our case, numerous species must be sampled using the same sampling strategy (in a global survey), because the position of their maximum occurrence may differ along the main gradient (in our case, elevation), and because other environmental features also determine their distributions. In fact, setting up an efficient random-stratified sampling for more than one species is, in practice, very difficult. For all the species in a region, it becomes unrealistic.

Given these limitations, a grid-sampling scheme was considered the best alternative. Our set of calibration points was obtained by sampling all points (208) of a 250 m × 250 m DEM (the only one available at that time) covering the study area. The set of 100 validation points was sampled randomly at a later date among the points of the newly aquired 25 m × 25 m DEM.

6.2.2.4 Selection of Environmental Variables The next step in model construction entails retaining, from a broad set of environmental variables, those most highly correlated with the species distribution pattern. To keep the model statistically valid, the potential problem of multicollinearity between these variables should first be managed. That is, each variable or combination of variable should ideally be independent of the others. This can be achieved by selecting them on the basis of the degree of correlation between them or, in some cases, by combining them linearly to obtain artificial orthogonal factors explaining most of the variance (Franklin et al. 1995). With the latter operation, however, ecological interpretation of the new principal components remains problematic unless they are strongly correlated to simple combinations of the basic variables.

Furthermore, Austin, Cunningham, and Good (1983) demonstrated that many of the so-called environmental variables (or combinations of variables) used in vegetation modeling had no direct physiological impact on plants. Because static plant distribution models' basic aim is clearly not to analyze cause-and-effect relationships but rather to obtain reliable final predictions validated by actual field data, nonexplicit environmental variables can be used in principle. However, in such a case, no physiological interpretation should be derived from the results. Therefore, if possible, physiologically meaningful environmental variables should be preferred to noncausal variables (for example temperature values rather than altitude).

In particular, solar radiation and snow cover are important variables for alpine landscape modeling (Brown 1994; Fischer 1990). They can be derived simply for wide areas from a DEM and satellite images (e.g., Parlow and Scherer 1991). As synthetic factors, they integrate several single variables (e.g., slope, aspect, microtopography and altitude), thus limiting the number of variables included in the model. Moreover, they are physiologically important for most high-altitude plants (solar radiation budget is directly related to photosynthesis and snow cover to the duration of the growing season). Snow cover presents the additional advantage of being directly related to climate. This is particularly interesting if climate change scenarios must be derived, as in our case. Other important variables in alpine landscapes are geology (the nature of bedrock, acid versus basic influences), rocky cover (form of substrate: e.g., moving or fixed screes, cliffs), hydrology (proximity of mountain streams, marshes, or springs), permanently frozen soils (permafrost), microclimate, and human-induced or natural disturbances (e.g., grazing, fire). Soil is a more problematic component, as most soil types result primarily from the underlying geology and are probably secondary, as much influenced by vegetation as an influence upon it. Moreover, in alpine regions, their typology remains too understudied to include them in the present ecological models (see chapter 5).

The final set of quantitative environmental variables retained were: solar radiation index (obtained by extracting the first axis of a principal component analysis on nineteen individual days of solar radiation), snow cover index (derived from aerial photographs), slope and curvature (a scale going from concave to convex) of the sample point, and mean annual temperature (derived from elevation using field measurements). In addition, three classes of nominal variables that proved, from exploratory analyses, to have an important power for discriminating species distributions were also retained: two classes of lithology (screes and cliffs) and a class of geology (moraines). Because it is not a physiological variable and because it is highly correlated with solar radiation index, slope aspect was not included in the set of input variables for the specific models to be developed. Precipitation was not included either because, to our knowledge, no reliable spatial extrapolation of precipitation could yet be made in an alpine landscape and because precipitation information is already partly included in snow cover information. A second step used permafrost (Keller 1992), and a water accumulation index (derived from the DEM and the hydrological layer in the GIS) as filtering factors of the preliminary maps (i.e., species once observed or not in the corresponding factor class).

6.2.2.5 Ecological Responses of Species As a next step, a species' physiological response to a given environmental variable should ideally be identified and integrated into the model (Huisman, olff, and Fresco 1993). However, a species' actual response to an environmental variable (that is, its ecological response curve) may not follow its physiological response obtained from laboratory experiments under controlled conditions, because inter- and intraspecific competition and other biotic factors influence physiological responses in the natural environment. Thus, the response cannot be considered solely physiological but, rather, at a higher level of ecological complexity. This exemplifies the overall problem of incorporating results from laboratory experiments into such static ecological models.

In plant ecology, the debate about the form of plant response curves remains partly unresolved. Whittaker (1956) proposed that most response curves approach a normal distribution (i.e., Gaussian: bell-shaped and symmetric), and most plant ecologists accepted this as typical of species' response to ecological variables (Brown 1984; Ter Braak and Gremmen 1987; Ter Braak and Looman 1986). However, the evidence remains equivocal. There is no a priori reason to assume that such response curves need to be symmetrical or indeed that any ideal or ubiquitous response curve exists (Austin, Cunningham, and Good 1983; Austin and Gaywood 1994; Huisman, Olff, and Fresco 1993; Jongman, Ter Braak, and van Tongeren 1987). Austin (1979) showed that bimodal and even more complex responses of species to a single environmental variable are common. He felt that many of the statistical techniques commonly used in vegetation modeling were inappropriate, given such complex responses. Austin, Nicholls, and Margules (1990) discussed observations of skewed responses to changes in variables such as temperature, organic matter content, or total nitrogen in soils. More recently, Austin and Gaywood (1994), using beta functions (measures of shape: skewness and kurtosis), tested and confirmed two hypotheses: Species response curves differ significantly from Gaussian-shaped curves, and the direction of skew is a function of the species' position along the environmental gradient. They showed that a great variety of response types exist depending on the organisms and variables considered. Such knowledge should now be better integrated into static model construction.

As discussed in section 6.2.2.3, the type of response also depends strongly on how the sampling is undertaken along the environmental gradient (Austin 1987; Green 1979; Mohler 1983). For example, curved responses may appear linear or close to linear if only one side of the mode is sampled (Jongman, Ter Braak, and Tongeren van 1987).

As an example, figure 6.2 shows different species' empirical response curves along a temperature gradient. In this case, response curves were drawn using techniques for smoothing histograms (with a large smoothing parameter) as an exploratory method for approaching the response's shape. Unimodal species' responses were observed (bell shaped, skewed, or even more complex ones, as for *Salix herbacea*), suggesting the need for at least a second-order term in the model. The use of simple responses, including an additional second-order term, was first investigated by running exploratory univariate GLMs including various-order terms of the same variable $(x, x^2, x^3, \ldots)$. Cubic terms were not retained because they always generate sinusoidal curves, which are not subject to simple ecological interpretation. This approach enabled us to develop satisfactory models for some species (see section 6.2.3). More complex responses (e.g., using beta functions) were

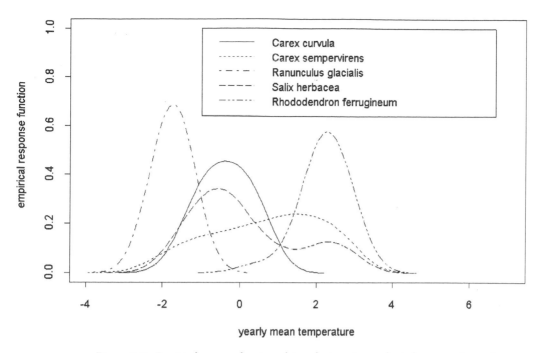

Figure 6.2 Empirical curves of various alpine plant species, ecological response to yearly mean temperature. The aim here is merely to explore possible shapes of response curves, here obtained by techniques for smoothing histograms. The abscissa's scale is in density function units (the curve's surface area must be unity).

considered for species for which model results were not satisfying (e.g., *Carex sempervirens*).

6.2.2.6 Statistical Models As discussed previously, curves of plant response to environmental regimes are not generally Gaussian. In this case, ordinary least-squares linear regression models are unlikely to have Gaussian residuals, and so they cannot be applied. Alternative methods of regression are thus more likely to produce representative results. The use of GLMs (see McCullagh and Nelder 1989) is a possible alternative when the response function's distribution family is known, because GLMs allow nonnormal response distributions to be modeled by transforming them to linearity.

GLMs rely on the following principles. If y is the dependent variable (i.e., response variable), with a known (but not necessarily Gaussian) frequency distribution, and $x_1, \ldots, x_n$ are the n explanatory variables, it is assumed that in a GLM, these variables can influence the distribution of y only through a single linear function called the *linear predictor*. The mean of y is thus a smooth, inversible function of the linear predictor, and its inverse function is called the *link function*. Hence, a multiple regression is made between the linear predictor (LP) and the explanatory part of the equation; for example,

$$LP = a + b_1 x_1 + \cdots + b_n x_n,$$

Vegetation Responses to Climate Change in the Alps: Modeling Studies where a, b_1, $\ldots$, b_n are the different regression coefficients to estimate, using, for example, the maximum likelihood principle. The probability of species occurrence, in response to a single environmental variable x, can then be expressed simply as a function of the linear predictor. More complex models including quadratic and interaction terms can similarly be considered. Common examples of GLM families include Gaussian, binomial, Poisson, inverse Gaussian and gamma response distributions and their most current associated link functions: identity, logit, log, 1/2, and inverse (see McCullagh and Nelder 1989 for details). Such GLMs have been successfully used in many recent ecological studies (Austin 1992; Austin et al. 1994; Austin, Nicholls, and Margules 1990; Brown 1994; Ferrer-Castán et al. 1995; Yee and Mitchell 1991) and are also applied in our study.

6.2.2.7 GIS: A Necessary Tool for Modeling Potential Habitat
The analyses of geographical data in this study differ somewhat from those in other ecological studies where the exact geographical location of observations in not important (e.g., when focusing on a particular ecological factor and species without aiming to make predictions from the results). When the aim is to predict the spatial distribution of species' potential habitat, that is, when a large amount of geographical data has to be handled simultaneously, a GIS provides both an analytical and a cartographic tool (Haslett 1990). Examples of GIS approaches applied to plant or animal species are numerous (Davis and Delain 1986; Jensen et al. 1992; Lancia, Adams, and Lunk 1986; Leninhan 1993; Lyon et al. 1987; Mille, Stuart, and Howell 1989; Ormsby and Lunetta 1987; Pereira and Itami 1991). These systems have been conceived in such a way that they are able to handle data of different geographical types (vector versus raster, empirical (field) versus derived data, digitized maps versus satellite scenes or aerial photographs). This is particularly useful when studying relationships between plant species and ecological factors, which often need an approach combining both field data and existing data (from herbaria and from the literature; Rhoads and Thompson 1992). Every specific set of data constitutes a single, monovariate layer in the system (see section 6.2.2.1).

In addition to producing a nice cartographic output to classical statistical analyses, GIS also enables particular analyses of geographical data to be performed that could not be performed in another way (neighboring analysis, for instance). Such GIS-specific handling of data can allow the derivation of new layers (synthetic factor) by modifying elements (e.g., slope, aspect, or curvature from the DEM), by combining layers (e.g., Gams index; see chapter 5) or by modeling new ones (e.g., permafrost, temperature, solar radiation). Once the model is developed and potential maps are calculated and stored within the GIS, geographical information from the newly generated map (for model evaluation, for example) can be accessed quickly.

6.2.3 Model Results: Potential Habitat Distribution Maps

Once the species' multiple response (i.e., its global ecological profile) is derived by statistical and geographical modeling, its associated potential distribution within the modeled area can be determined. As previously stated, modeling plant species' potential distribution is equivalent to modeling their potential habitat (in the sense of Whittaker, Levin, and Root 1973) and it is relevant to speak about Potential Habitat Distribution Maps (PHDMs; d'Oleire-Oltmanns 1995; Schuster 1990). PHDMs correspond to probability maps (figure 6.3) and are cartographic representations, for all points of the DEM, of the probability of finding the species. Figure 6.3 provides examples of PHDMs for *Carex curvula*, *Carex sempervirens*, and *Rhododendron ferrugineum*. An example of a similar static potential distribution modelling is given by Zimmermann (1996), who predicted the distribution of most alpine plant communities under present climatic conditions (see, e.g., fig. 6.3d).

6.2.4 Model Evaluation

The procedure's next stage is testing the model's quality. The study should be initiated (started) with two distinct data sets, one for building the model and another for its evaluation. In addition to the primary visual comparison of predicted and actual distributions in the study area, a more reliable model evaluation is conducted by checking the predictions over the set of evaluation points.

The quality or the model's robustness, also varies according to the species modeled. As an example, more than 66 percent of the predictions made for *Carex curvula* or for *Rhododendron ferrugineum* (over the set of 100 validation points) fell within a 10 percent interval around the actual values; for modeling in plant ecology, this is considered a rather good result. The models for *Carex sempervirens* reached only 25 percent agreement. However, as the choice of this interval is subjective, considering a slightly larger interval would increase by far the model's quality. Given the present results, the methodology was considered successful for the prediction of some alpine plant distributions. Including data on moisture potential, however, would greatly improve it, because this variable has proven its importance as a latent environmental gradient in exploratory analyses of the species data (correspondence analysis) and could help determine distributions for less successfully modeled species.

As a rule, static presence-absence or abundancy models, especially at high spatial resolution, rarely explain more than 60–70 percent of the variance in the distribution of plant species, communities (Brzeziecki, Kienast, and Wildi 1993; Zimmermann 1996) or vegetation boundaries (Brown 1994). Indeed, explaining 50 percent of the variation is sometimes considered a good result (e.g., Brown 1994). However, these models are sufficiently valid with regard to the purpose for which they were constructed: to permit comparative studies.

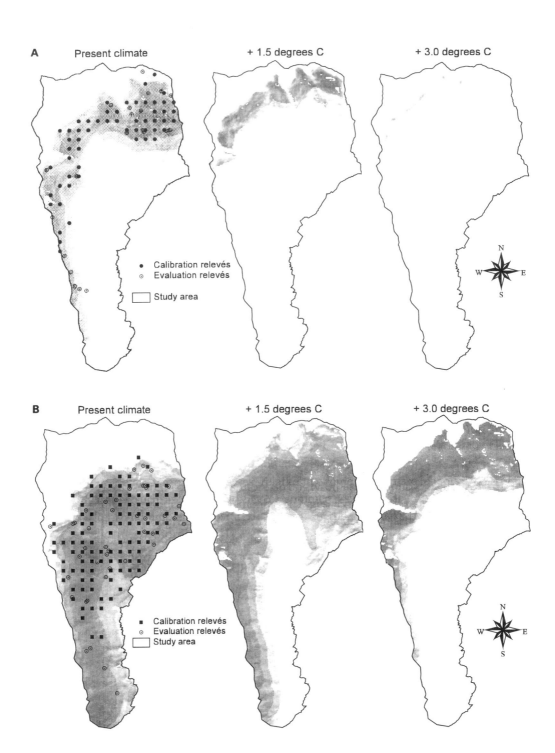

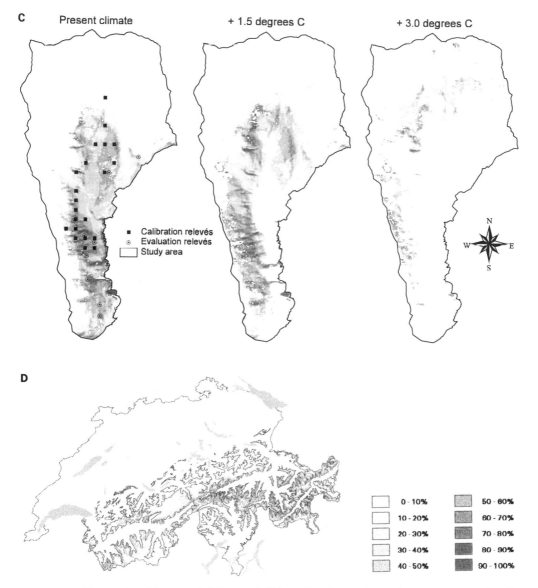

Figure 6.3 Three-paneled Potential Habitat Distribution Maps for (a) *Carex curvula* All., (b) *C. sempervirens* Villars and (c) *Rhododendron ferrugineum*. L., under present climatic conditions and with a + 1.5 and a + 3.0 degree increase in yearly mean temperature (ymt). Gray scale: probability of presence of the species. Black: probability = 1 (100%). White: probability = 0 (d) Potential distribution maps (kindly provided by Klaus Zimmermann (1996)) of the alpine plant community *Caricion curvulae* (characterized, among others species, by a high density of the sedge *Carex curvulae*).

The numerous uncertainties involved in this kind of modeling explain why the models show a relatively low level of correspondence with the data. First, it is difficult to gather complete environmental information at a high resolution and to follow an adequate multispecies sampling scheme (see section 6.2.2.3). Second spatial studies superimposing external and GIS-derived environmental layers in a GIS have a high risk of error propagation. The error accumulation probably starts during data collection, when species can be misobserved, grazed, not yet in flower, or simply not present at the time the *relevé* is made. Third, the possibility of coexistence of species with similar niches, and thus the existence of processes of competition and nonuniform seed dispersal (Shmida and Ellner 1984), adds another error component to species distribution.

However, we still believe that increasing the spatial resolution to one meter and using the most powerful global positioning systems (GPS), now reaching an accuracy of one centimeter, to locate the sampling points in the study area should greatly enhance the models—one aim of the second stage of our project.

6.2.5 Deriving Climate Change Impact Scenarios

Deriving accurate climate change impact scenarios for alpine regions requires two main products: a good climate change scenario, reliable at the local scales used for modeling, and the ecological model itself, including the same driving variables (which should be physiologically meaningful for the species) as predicted by the climate change model. However, good climate models at the local scale are not yet available. Predictions for some locations are derived by downscaling procedures (Gyalistras et al. 1994; Wanner and Beniston 1995), but their results cannot yet be extrapolated to the local spatial scale. Thus, provisional climate change impact scenarios for this study were simply built by including mean annual temperature in the model as a function of elevation and calibrated from actual field measurements. The two climate change scenarios considered (CC1 and CC2) (obtained from Gyalistras and Fischlin 1995, chap. 4) did not include any precipitation change, likewise because of lack of availability and accuracy of precipitation change scenarios and the difficulty of introducing such changes into the snow index used in the model (instead of precipitation). They were

- *CC1:* a mean annual temperature change of +1.5°C; downscaled climate under ECHAM1/LSG 2 × CO_2 (344 to 720 ppmv; Cubasch et al. 1992; see chapter 4), and

- *CC2:* a mean annual temperature change of +3.0°C; downscaled climate under CCC GCMII 2 × CO_2 (Boer, McFarlane, and Lazare 1992).

The predicted probabilities of finding the species were then divided into five categories of prediction: A (absent), I (improbable), P (probable), V (very probable) and S (sure). Table 6.1 gives the number of sites falling into each category (over a total of 32,000 sites).

Table 6.1 Results of climate change simulation on the distribution of three alpine species

Counts	Categories				
	A	I	P	V	S
Present					
cc	22,000	2,800	2,900	2,800	1,500
cs	82,00	3,100	4,800	12,800	3,100
rf	21,900	5,400	2,700	1,400	600
CC1					
cc	26,400	1,600	1,700	1,700	600
cs	10,200	4,200	5,200	9,200	3,200
rf	23,250	4,800	2,100	1,400	450
CC2					
cc	30,500	600	400	400	100
cs	19,200	2,400	2,000	4,800	3,600
rf	28,500	1,800	700	700	300
Relative differences CC1	(percentages)				
cc	+20.00	−42.86	−41.38	−75.00	−60.00
cs	+24.39	+35.48	+8.33	−28.13	+3.23
rf	+6.16	−11.11	−22.22	0	−25.00
CC2					
cc	+38.64	−78.57	−86.21	−85.71	−93.33
cs	+134.15	−22.58	−58.33	−62.50	+16.13
rf	+30.14	−66.67	−74.07	−50.00	−50.00

Note: cc—*Carex curvula*; cs—*Carex sempervirens*; rf—*Rhododendron ferrugineum*. The five categories of prediction are: A (absent), I (improbable), P (probable), V (very probable), and S (sure). PRESENT: present climate; CC1: +1.5°C warming; CC2: +3°C warming. Counts are number of pixels with associated probability class, estimated from a histogram of the probability maps in the GIS. The table's lower section gives precentages of changes from current levels under the two climate change scenarios and for each species.

The results (figure 6.3) clearly show that strong decreases in occurrence are to be expected for two of three species studied (excluding *Carex sempervirens*). Under both scenarios, the number of sites where *Carex curvula* is predicted to be absent increases, and those where the species is predicted to be present decreases, with a stronger decrease observed for the most probable categories. The pattern is similar for *Rhododendron ferrugineum*, although the distribution of predicted decreases is not the same within categories. For all three species, distribution declines are predicted (i.e., more sites where the species will be absent). For *C. curvula*, 20% disappearence is predicted under CC1 and 38.64% under CC2; for *C. sempervirens*, 24.39% (CC1) and 134.15% (CC2); and for *R. ferrugineum*, 6.16% (CC1) and 30.14% (CC2).

6.2.6   Discussion

Specific models were built for alpine species that assessed more particularly the species' bioclimatological space to allow the derivation of climate change impact scenarios. Driving variables were mean annual temperature, solar radiation, snow index, slope, and a class of lithology (screes). The water accumulation index and permafrost were then used as posterior filtering factors. The preliminary results from these models indicate that under both low (CC1) and middle (CC2) IPCC climate change scenarios, tremendous change could be observed in the potential distributions of species. In particular, the present range of strictly high-alpine species (see figure 6.3(a) for *Carex curvula*) could become more suitable for lower-elevation alpine species (see figure 6.3(b) for *Carex sempervirens*). Because their competitive abilities would probably be lowered, and because higher elevation becomes decreasingly suitable for vegetation establishment and growth (as greater surface area is covered by sterile screes, permanent snowy area, and permafrost), the range of high-alpine species could diminish.

However, at this stage of the discussion, we should mention some important limitations. First, the results from static modeling indicate changes in species' *potential* distribution. These potential distributions have been successfully evaluated for present situations, but no evaluation is possible for future scenarios. Because of the model's nondynamic nature, results must be considered with care. In particular, soils' reaction to climate change is very uncertain (see chapter 5). Because soils are very closely related to the establishment of new species and soils in turn developed through plant establishment, the time for which the responses are predicted could well be delayed (ecological inertia). Second, new situations could also alter competitive relationships, potentially in favor of the indigenous high-altitued species. These species sometimes form dense underground networks of roots (as observed with populations of *Carex curvula*) and might not allow invader species to establish. Third, many species in the high-alpine zone are not in a climatic climax but rather in a site-specific climax (see chapter 5 for a definition of both types of climax). This could greatly postpone their reaction to climate change. Moreover, in such sites where species can be independent from the general climate, only primary successional species can develop, thus limiting again the number of invader species that can establish themselves. Fourth, the fact that a precipitation change was, in this first step, not included in the scenario, owing to the lack of reliability of climate scenarios in alpine regions and to the difficulty in obtaining spatial distributions in such complex landscapes, also limits the results. Precipitation change should be factored in as soon as climatologists provide reliable downscaled scenarios for precipitation change in alpine landscapes.

6.3   THE DYNAMIC APPROACH: EVALUATING THE FATE OF FORESTS

Even in comparison with alpine plants (chapter 5, section 6.2), forest trees are characterized by long generation times and life spans on the order of several decades to millennia. As a result, forest ecosystems respond slowly to changing environmental conditions before reaching a new equilibrium, if they reach it at all. Thus, a static, that is, an equilibrium-based approach is inappropriate for predicting forest responses to continuously changing climatic conditions in the coming centuries. An approach is required that can simulate forest ecosystems' temporally changing (transient) behavior. In addition, forest dynamics' large timescales hinder an empirical approach; experiments studying the responses of entire forests are exorbitantly expensive and limited in scope.

Dynamic forest models are appropriate tools to overcome these problems. They can integrate knowledge about forest ecosystems and simulate both transient and equilibrium behavior of the forests.

We made use of one such dynamic forest model, FORCLIM (Bugmann 1994, 1996; Fischlin, Bugmann, and Gyalistras 1995), which we briefly present here. Methods and problems of model validation, as well as case studies of quantitative assessments of possible forest responses to climate change scenarios, are demonstrated by means of this model.

6.3.1   The Forest Model FORCLIM

The model FORCLIM focuses on trees, neglecting other biota found in forests. Trees dominate the ecosystem "forest" in two respects. First, trees (in general, not species specific) strongly influence the other biota, for example, by shading and microclimatological effects, whereas the feedback from other biota to the trees is much smaller. Second, forest trees are more important for carbon sequestration (or fixing), both at a local and global level (see chapter 5), than nontree components of the ecosystem.

FORCLIM is a patch model, that is, it simulates the fate of individual age classes of trees, called cohorts, (panels (a) and (b) of figure 6.4) within small (1/12 ha) patches by mimicking establishment, growth, and death processes. The growth submodel is deterministic; the growth rate depends nonlinearly on environmental conditions (panel (a) of figure 6.4), such as the yearly day-degree sum; nitrogen; light, which is controlled by the shading of the trees; and drought, which is affected by monthly temperatures and precipitation. Climate input fluctuates stochastically around mean values. Establishment and death are formulated as stochastic processes, with mortality influenced by the same abiotic factors as growth, and with establishment depending on minimum winter temperature, yearly day-degree sum and light availability on the forest floor. Other abiotic forcing factors such as forest fires or flooding, which in the Alpine region play only a minor role, are implicitly included

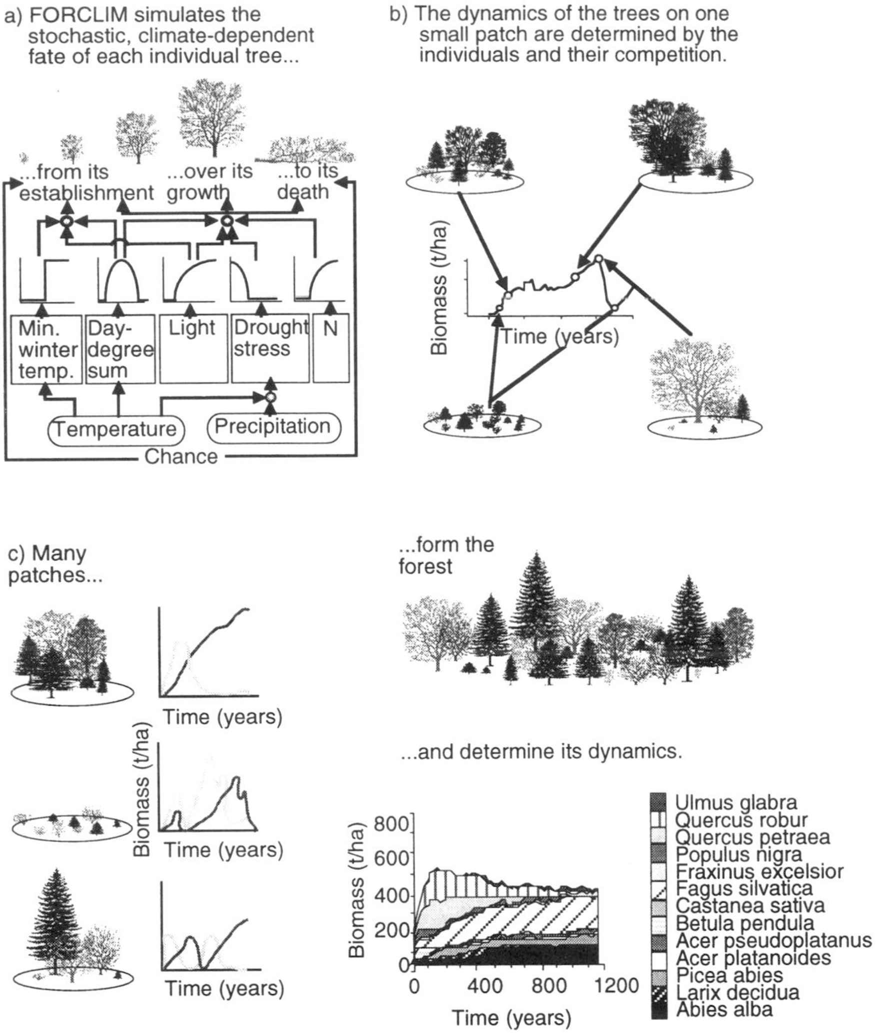

a) FORCLIM simulates the stochastic, climate-dependent fate of each individual tree...

...from its establishment ...over its growth ...to its death

Min. winter temp. | Day-degree sum | Light | Drought stress | N

Temperature   Precipitation
— Chance —

b) The dynamics of the trees on one small patch are determined by the individuals and their competition.

Biomass (t/ha)
Time (years)

c) Many patches...

Biomass (t/ha)
Time (years)
Time (years)
Time (years)

...form the forest

...and determine its dynamics.

Biomass (t/ha)
800
600
400
200
0
0   400   800   1200
Time (years)

Ulmus glabra
Quercus robur
Quercus petraea
Populus nigra
Fraxinus excelsior
Fagus silvatica
Castanea sativa
Betula pendula
Acer pseudoplatanus
Acer platanoides
Picea abies
Larix decidua
Abies alba

Figure 6.4 Principal functioning of the forest model FORCLIM (Bugmann 1994; Bugmann 1996; Fischlin, Bugmann, and Gyalistras 1995): (a) Climatic parameters drive a weather generator, which is used to determine values of bioclimatic variables. Species-specific functions of response to environmental factors (only qualitative shape is shown) influence process rates of individual trees living on a small patch (1/12 ha). (b) Because the model is stochastic, its behavior over 1,200 years needs to be sampled repeatedly (Monte Carlo simulation). (c, left) In this study, we always sampled 200 patches and used mean abundances to describe the changes in species compositions (c, right).

in the stochastic mortality rate. Seeds of all tree species are assumed always to be present. Genetic adaptation is not taken into account, partly because of trees' long generation times, which render an adaptation by mutation and selection improbable, and partly because information about intraspecific genotypic variability is lacking. Given the site characteristics such as climate parameters and field capacity, FORCLIM is typically used to estimate the average temporal evolution of a forest (panel (c) of figure 6.4) by simulating 200 forest patches for 1,200 years using a Monte Carlo simulation. Typically, such simulation averages reach a quasi-stationary state (equilibrium) after about 1,000 years (Bugmann and Fischlin 1992). As the input data for forest patch models describe the characteristics of a single specific site, these models act at the local scale. Covering larger heterogeneous areas requires a large number of simulations.

6.3.2 Validation in the Present and in the Past

Prior to its application, a model as complex as FORCLIM should be evaluated (figure 6.5) for its ability to predict a set of observations independent of those used for structuring the model and estimating its parameters. Two sources of independent data are required: first, a record of the input data the model requires (e.g., temperature and rainfall parameters), and second, measurements of those variables that the model calculates (e.g., plant species composition). Then the measured variables can be compared quantitatively with the calculated ones.

Model validation, or at least plausibility tests, should be performed for a range of conditions (for example, for the climatic input) similar to the range of conditions expected for the planned model application to test not only the model's precision but also its general applicability. For a climate-driven forest model like FORCLIM, such a range of conditions may be defined by climate gradients in space (validation at different locations in the present, panel (a) of figure 6.5) and time (paleoecological validation, panel (b) of figure 6.5).

Data used to evaluate dynamic forest models may come from many sources. There exists a potential wealth of observations on past and current forests, such as yield tables, forest inventories, long-term data from permanent plots, tree ring chronologies, pollen records, remote sensing data, and phytosociological *relevés*. However, most of these data either do not cover a long time span (e.g., forest inventory and remote sensing data), lack the temporal aspect (e.g., phytosociological descriptions), or are available at only a small number of sites and do not adequately cover climate gradients (e.g., permanent plots, tree ring chronologies, and pollen records).

Therefore, the validation must rely on a combination of several data sources. Two, phytosociological descriptions and pollen data, are used below to illustrate the potential benefits and pitfalls of validating a complex ecosystem model such as FORCLIM.

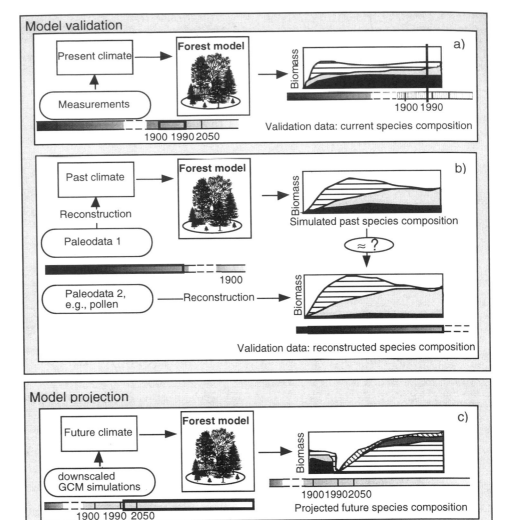

Figure 6.5 (a) Validation with present data: The forest model is validated with today's climate as input by comparing the equilibrium species distribution with that of modern forests. (b) Paleoecological validation: The model's transient behavior must be validated with past data. Here both climate input and comparison data are reconstructed from paleodata, such as ice cores and pollen data. (c) Model application: The model aims to project the impact of a future climate change as derived from GCM simulations. The black frames and the black vertical line in (a) indicate the time window of data and simulation.

6.3.2.1 Data from the Present FORCLIM's behavior was tested along transects through the European Alps and through eastern North America. The required input data were monthly expected values, standard deviations, and cross correlations of temperature and rainfall for climate parameters, and species-specific parameter values (e.g., maximum height or shading tolerance) of the most abundant tree species in the respective subcontinents. In most cases, the simulated equilibrium species compositions and the total aboveground biomass along the transects corresponded to the observed near-natural forests (Bugmann 1994; Bugmann and Solomon 1995).

However, this kind of model validation presents several difficulties. In the first place, the simulations do not include natural disturbances such as fire and windstorms, although these effects introduce additional stochastic variability that can strongly shape natural vegetation. Second, most of the forests in the simulated regions are at best near-natural; that is, they have been more or less intensively managed during the last 1,000 years. Thus, the simulated natural species' composition rarely mirrors that observed in today's landscape. Third, it is not sufficient that the simulated species' biomass and composition correspond with observed data in an equilibrium state. To study the impacts of transient climate change, it is also necessary to assess the model's transient (i.e. time-dependent) behavior.

6.3.2.2 Paleoecological Data Validating a dynamic forest model's transient behavior requires long time series of input data and independent validation data from natural forests. Both can be obtained from paleoecological proxy data sources (figure 6.5b). However, these data must meet certain requirements that can be problematic to fulfill.

First, no direct data source exists for the required temperature and precipitation variables. Instead, these data have to be reconstructed (e.g., from ice core or lake level data; see chapter 3), usually by using another model. Such data reconstruction models may contain errors of their own and have limited precision (Bradley 1991). Moreover, the reconstruction's temporal resolution is often restricted to annual or seasonal values, and it is therefore not possible to reconstruct the full annual cycle (Guiot, Harrison, and Prentice 1993). Hence, to arrive at a resolution of monthly intervals, it is necessary to make additional assumptions, which add to the uncertainties of the reconstruction itself.

Second, forest patch models simulate the abundance or biomass of tree species. Information about such variables in the distant past must be derived from proxy data (e.g., pollen found in lake or peat bog sediments). Pollen data usually are converted into relative biomass per tree species using the Iversen factors (Faegri and Iversen 1975; Lotter and Kienast 1992), which only estimate roughly the amount of pollen deposited per unit tree biomass.

Third, the reconstructed temporal sequence of data must be mapped onto the ecosystem model's time axis. However, the dating of proxy records may be inaccurate or biased because of coarse temporal resolution in the record as well as uneven sedimentation rates. Thus, differences between proxy data and model output are difficult to interpret, because they may stem from a system-intrinsic time lag or a dating error.

Finally, the pollen accumulated in lake sediments represents the plants of at least a whole catchment area. However, catchments are not homogenous. Therefore, a given pollen record represents the differential contributions of different areas within the catchment, each with their distinct site characteristics (in, for example, soil water-holding capacity, slope, aspect, and climate) and hence forest types. A forest model like FORCLIM, however, simulates a forest only at one particular site with a well-defined set of specific characteristics. Therefore, the site characteristics used in the simulation are implicitly assumed to represent a mixture of all the sites within the whole pollen source area.

In general, at least three models with all their inherent uncertainties are involved in paleoecological validation of an ecosystem model: the ecosystem model itself and two auxiliary models to convert proxy data into the form needed for input and comparison data. Together with the uncertainty of dating and of the pollen source area, these issues complicate the validation of a model using paleoecological data.

6.3.2.3 Pollen Data from Soppensee To compare tree species succession as simulated by the forest patch model FORECE with documented pollen sequences, Lotter and Kienast (1992) used an annually layered pollen stratigraphy from Soppensee (Swiss Plateau, elevation 596 m) covering about 4,000 years of the early Holocene (Lotter 1989; panel (a) of figure 6.6). We repeated this experiment by using the model FORCLIM to illustrate the potential and the limitations associated with such a study (Bonan and Hayden 1990; Solomon et al. 1980; Solomon and Tharp 1985; Solomon, West, and Solomon 1981).

An optimal scenario to be used as climate input in such a study should yield reliable information about the transient climate in a high temporal and spatial resolution (e.g., seasonal values at the specific study site). Additionally, it should be as much as possible independent from pollen data. However, since no data were available to fulfill all these requirements one at a time, we had to rely on a compromise consisting of transient, nearly pollen-independent, seasonal temperature and precipitation anomalies in small to medium temporal and spatial resolution.

The scenario used was based on the current climate at Huttwil (SMA 1901–90) near Soppensee, assuming that anomalies in annual precipitation and seasonal temperatures changed over the entire simulation period (table 6.2). It was constructed from temperature anomalies obtained by GCM simulations for the Holocene (Huntley and Prentice 1993) with the Community Climate Model (Kutzbach and Guetter 1988) and from precipitation anomalies reconstructed by Guiot, Harrison, and Prentice (1993) based on Holocene pollen and lake level data. Furthermore, we assumed a mesic soil with a field capacity of 30 cm water (Bugmann 1994). The simulation was run for 4,000 years, and the pollen record was assumed to start around 10,000 BP.

Although the GCM simulations used must be interpreted with caution and their spatial and temporal resolution was coarse (about 500 km and 3,000 years), we got an indication of transient seasonal temperature anomalies, which were obtained independently from pollen data. The precipitation scenario in turn was originally derived from pollen data but corrected with data about Holocene lake levels and can therefore be considered more or less in-

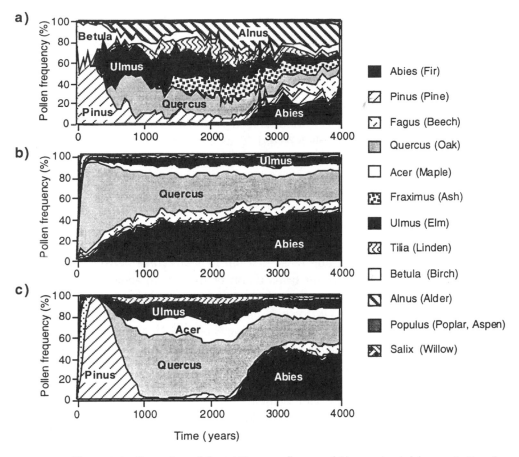

Figure 6.6 Comparison of the 4,000-year pollen record (a) reconstructed from a stratigraphy from Soppensee and (b and c) redrawn from (Lotter and Kienast 1992) with the vegetation simulated by FORCLIM (Bugmann 1994, 1996; Fischlin, Bugmann, and Gyalistras 1995) during 4,000 years. Climate input scenario (table 6.2): Seasonally varying and transient temperature anomalies and negative precipitation anomalies. (b) All species assumed present from start of simulation. (c) All species assumed to be present when reaching more than 1 percent in the pollen record. In all cases, the pollen sum includes only arboreal pollen without *Corylus*. The biomass values simulated by FORCLIM were converted to pollen frequencies using inverted pollen representation factors (Faegri and Iversen 1975) and summed to genera.

Table 6.2 Paleoclimate scenario used for paleovalidation of FORCLIM

Time (years BP)	Temperature anomaly (°C)		Precipitation anomaly (cm/month)
	Winter	Summer	
10,000	−1.17	1	−3.33
9,000	0	2	−3.33
6,000	−2	1	−0.83

Note: The scenario is obtained by interpolation between the given anomaly values. After 6,000 BP, climate was assumed to remain constant. Temperature anomalies are from Huntley and Prentice (1993) and Kutzbach and Guetter (1998), and precipitation anomalies from Guiot, Harrison, and Prentice (1993).

dependent from pollen. Futhermore it was available at a spatial resolution of one degree. The simulated species-specific biomass values were converted to pollen percentages using inverted Iversen factors of pollen representation (Lotter and Kienast 1992) and summed to genera.

The first simulation studies based on the above experimental setup (panel (b) of figure 6.6) yielded a forest dominated by species of the genera *Quercus* (oak) and *Abies* (fir) toward the end of the first 1,000 years, sharply different from the known pollen record. Thus, the model cannot explain the pattern found in the pollen record under the given paleoclimate scenario.

In the following, we tried to evaluate the influence of another boundary condition. To a certain degree, the pollen record reflects the immigration history of tree species into the Swiss Plateau after the last Pleistocene glaciation; for example, *Abies* appears rather late in the pollen record. Such a late arrival of tree species at the pollen source area might have resulted from insufficient migration rates, or from long distances between glacial refugia and available sites for colonization or from geographic or climatic barriers between them, such as high mountain ranges. Therefore, the subsequent simulation assumed that the complete absence of a genus in the pollen diagram was due not to competition, but to delayed immigration. This was simulated by allowing the tree species to establish only after they had attained more than 1 percent in the pollen record.

The results of these simulations (panel (c) of figure 6.6) show a more satisfactory agreement with the pollen record (panel (a) of figure 6.6). The model simulates the transition from the early *Betula-Pinus* birch-pine forest (years 0–500) to a *Quercus-Ulmus* (oak-elm) forest (years 500–2,700) followed by a mixed deciduous forest where *Abies* becomes important as well as *Fagus* (beech) (years 2,700–4,000). Major quantitative discrepancies are an over-representation of the pollen percentage of *Quercus* and *Acer* (maple) and an underestimation of the pollen percentage of *Tilia* (linden), *Fraxinus* (ash), and *Alnus* (alder).

The fact that in the simulation the region was regarded as uniform with regard to field capacity, where a mean value was used throughout the region, which favors species adapted to this mean field capacity, may explain the remaining differences between model results and observations from the pollen record. From present conditions we must assume that the pollen source area consisted of various soil types, including very wet soils, where flood-resistant species such as *Alnus* and *Fraxinus* probably had a competitive advantage. Furthermore, the simple empirical model for pollen-to-biomass conversion can lead to deviations, because a relatively small change in the pollen representation factors can dramatically change the relative values of the pollen record.

Inferring the time of species immigration from the pollen record violates the requirement that the input data be independent of data used to evaluate model behavior. However, at least for *Abies* and *Fagus*, we can assess from the pollen maps of Huntley and Birks (1983) that in actuality their absence is probably due to delayed immigration. For example, presumably the high elevations of the central Alps formed both a topographic and a climatic barrier for the immigration of *Abies* from its glacial refuge in northern Italy. Finally, it would be desirable to estimate the points in time when species immigrated by the explicit modeling of tree migration. By taking into account the climatic conditions on the migrational path, the specific reasons for delayed immigration could be evaluated.

For other species, such as *Quercus*, the pollen maps do not support the hypothesis of delayed immigration; their premature appearance in the first simulation can be attributed to uncertainties in the climate input scenario, which was derived from GCM simulations on a coarse scale, or in the dating of the pollen record as well as to uncertainties in the model.

Therefore, the validation of the model's transient behavior can be judged neither successful nor unsuccessful unless the uncertainties in the input scenario or the boundary conditions can be narrowed. One approach for further validation could be to determine a plausible set of input scenarios and boundary conditions with which the simulation matches pollen proxy data at one specific site, then to validate the model with data from other sites, using this specific set.

6.3.2.4 Potential and Pitfalls of Model Validation

Evaluating the performance of dynamic forest models requires a combination of approaches that test models' transient and equilibrium behavior under a range of conditions. From the case study of model validation using the Soppensee pollen proxy data, we conclude that paleoecological validation of forest ecosystem models has potential but also limitations.

Paleoecological proxy data offer time series of adequate length of unmanaged forest dynamics and mirror forest changes under a climate that was fluctuating at various rates, thereby increasing our confidence that the model behaves realistically under scenarios of future climate change. However, a paleoecological model validation has several limitations. First, it requires a high quality, highly resolved and independent record of past climatic input conditions. Such data will potentially be available soon (cf. chapter 3). Second, a number of uncertainties are introduced in addition to the potential errors in the ecosystem model. To reduce these uncertainties, further ecological models are needed: a process-oriented model of pollen production, transport, and sedimentation is required that could replace the transfer functions often used to reconstruct past vegetation compositions. In addition, models of tree migration are needed, because migration potentially plays a key role in transient forest responses to climate change, especially during phases of extreme climate change such as at the end of the last glacial period, and as is projected for the next century.

6.3.3 Responses to Future Climates: Surprises and Inertia

The Soppensee case study indicates that a final validation of the model's transient behavior is still problematic. However, even with this remaining uncertainty, simulations with a dynamic model such as FORCLIM are still useful for projecting transient forest responses under climate change (if one keeps in mind this uncertainty), in particular because such models are the only ones available for this purpose. FORCLIM makes it possible to study forests' sensitivity at various sites to an identical climate change scenario and to examine the the variability's effect within climate change scenarios.

6.3.3.1 Different Responses at Different Sites For several sites at a range of altitudes, we used the site-specific monthly temperature and precipitation anomalies obtained from semiempirical downscaling of transient GCM simulations (cf. chapter 4, Gyalistras et al. 1994) as input data for simulations with FORCLIM. These scenarios were derived from the same GCM simulations (Cubasch et al. 1992) and based on the IPCC "Business-as-Usual" CO_2 scenario A (Houghton, Jenkins, and Ephraums 1990). The simulations were run for 1,000 years with present climatic conditions to allow the simulated forests to reach an equilibrium state. The climate then was assumed to change instantaneously in 2080 and held constant afterward. The simulated forests were again allowed to adapt to the new climatic conditions for 1,000 years to assess the new equilibrium vegetation (Bugmann and Fischlin 1994).

These step response simulations demonstrated that a large range of possible responses can be expected from the same projected global climate change depending on the climatic characteristics of the site and the forests simulated for present conditions (figure 6.7). At midaltitudes, represented by the simulations for Bern, on the periphery of the Alps, the forest appears to be well buffered against the projected climatic changes, because the model predicts no extreme changes in species composition (panel (b) of figure 6.7). However, at high altitudes, such as at Bever in the central Alps, the model predicts that the species composition responds strongly (panel (a) of figure 6.7). A forest similar to those currently found in the montane zone would replace the subalpine forest simulated for the present, thereby reflecting an altitudinal shift of the forest belts. These predicted changes are drastic, and the associated transient forest diebacks might cause problems ranging from soil erosion to slope destabilization in vulnerable areas. Finally, the model predicts surprisingly strong responses at the low altitude site of Sion (panel (c) of figure 6.7) in a central part of the Alps, where none of the tree species present in the model would survive. Here the projected warming in combination with a relatively small change in the precipitation regime so increase the water pressure deficit in the simulation that an enduring drought stress prevented the survival of all trees (Fischlin, Bugmann, and Gyalistras 1995).

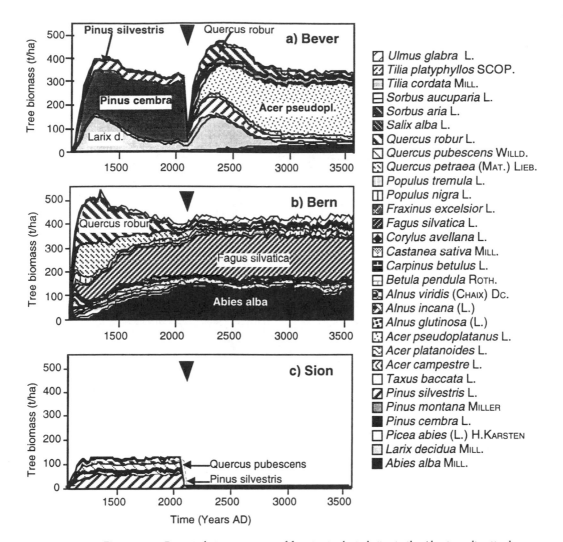

Figure 6.7 Projected step responses of forests at selected sites in the Alps to a climatic change as downscaled separately for each month from a transient IPCC Business as Usual GCM run (Gyalistras and Fischlin 1995, Gyalistras et al. 1994) for the year 2080. All simulations were made with the forest model FORCLIM (Bugmann 1994, 1996; Fischlin, Bugmann, and Gyalistras 1995) assuming the climate remains constant after the year of climate change (step at arrow). (For common English names of genera, see figure 6.6.)

6.3.3.2 Sensitivity of FORCLIM to Different Climate Scenarios

The chain of assumptions, methods, and models used to obtain regionally differentiated climate change scenarios contains many uncertainties. Principal sources of uncertainty range from CO_2 emission scenarios over GCM simulations to the downscaling of their results to the regional scale.

We tested the effect of these uncertainties on the forest model's behavior by simulating step responses for a set of climate scenarios at Bever (figure 6.8). All simulations were run with ECHAM-GCM results (Cubasch et al.

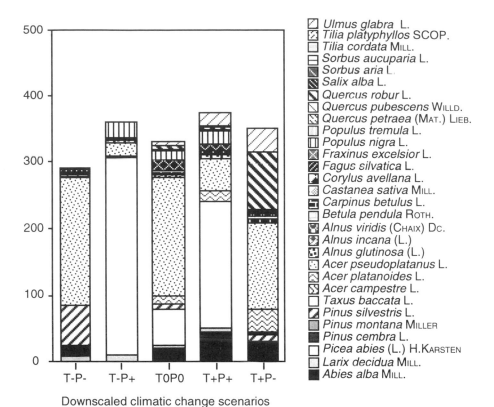

500 —
400 —
300 —
200 —
100 —
0 —

T-P-   T-P+   T0P0   T+P+   T+P-

Downscaled climatic change scenarios

Ulmus glabra L.
Tilia platyphyllos SCOP.
Tilia cordata MILL.
Sorbus aucuparia L.
Sorbus aria L.
Salix alba L.
Quercus robur L.
Quercus pubescens WILLD.
Quercus petraea (MAT.) LIEB.
Populus tremula L.
Populus nigra L.
Fraxinus excelsior L.
Fagus silvatica L.
Corylus avellana L.
Castanea sativa MILL.
Carpinus betulus L.
Betula pendula ROTH.
Alnus viridis (CHAIX) DC.
Alnus incana (L.)
Alnus glutinosa (L.)
Acer pseudoplatanus L.
Acer platanoides L.
Acer campestre L.
Taxus baccata L.
Pinus silvestris L.
Pinus montana MILLER
Pinus cembra L.
Picea abies (L.) H.KARSTEN
Larix decidua MILL.
Abies alba MILL.

Figure 6.8 Simulated sensitivity of species compositions at equilibrium to uncertainties within the climatic change scenarios. All climate anomalies wree downscaled (Gyalistras et al. 1994, Gyalistras and Fischlin 1995) separately for winter and summer from the ECHAM-GCM (Cubasch et al. 1992) simulations based on the transient IPCC Business as Usual global CO_2-emission scenario A (Houghton, Jenkins, and Ephraums 1990) for 2080. Simulations were made with the ecosystem model FORCLIM (Bugmann 1994, 1996; Fischlin, Bugmann, and Gyalistras 1995). T—Temperature; P—Precipitation; TøPø—Best estimate; T−P−, T+P+—respectively, the minimal and maximal estimated changes (95 percent confidence intervals). (T—Temperature, P—Precipitation; TøPø—Best estimate, T−P−, T+P+—respectively the minimal and maximal estimated changes, *T−P+ and T+P− the combinations of minimal temperature and maximal precipitation change and vice versa* (95% confidence intervals). (For common English names of genera, see figure. 6.6.)

1992) which were based on the IPCC "Business-as-Usual" global CO_2-emission scenario (Houghton, Jenkins, and Ephraums 1990) and downscaled separately for winter and summer half year to the Bever site (Gyalistras et al. 1994). We show only the simulated species composition at equilibrium. The center represents the best estimate TøPø obtained by the downscaling procedure for the changes in temperature (T) and precipitation (P). The left and the right sides represent, respectively, lower estimates (T−P−, T− P+) and higher estimates (T+P+, T+P−) for temperature, where T+ and P+ represent the upper estimated limits of the 95 percent confidence interval and T− and P− the lower limits. This study reveals that forests in the Alps, such as the subalpine forest at Bever, may show strongly differing responses depending on which climate change scenario is assumed.

6.4 CONCLUSIONS

Ecosystems are highly complex (see chapter 5), which renders a simple, direct extrapolation of their behavior to future climatic conditions impossible, even if deterministic predictions of climate were possible. For the same reason, and because of the slow reaction times and high inertia of forest and alpine ecosystems, it is likewise difficult to learn experimentally about their responses. In this situation models are indispensable, despite all their drawbacks, especially when we attempt to assess the ecosystems' future behavior under a changing climate. Two models have been presented: a static, equilibrium-based distribution model for high-alpine plant species and a dynamic succession model for montane forests.

Because no model can capture the full complexity of the system it simulates, both models focus on a limited number of key components and features of the ecosystems at the expense of others. The forest model simulates the dynamics of dominant tree species at single sites, neglecting other forest species. The climate input variables of temperature and precipitation drive the forest model, which describes the processes determining the dynamics of forests such as climate-dependent stochastic tree growth or death. In turn, whereas the static, equilibrium-based, alpine plant species model can take into account many species, including rare ones, in an entire area, and include more abiotic factors, it does not include information about specific processes.

The major difference between these models is in the way they treat the temporal development of the ecosystems or biological entities they model. The static alpine species distribution model considers only equilibrium or climax states of vegetation. The alpine ecosystems modeled are assumed to be currently in equilibrium. Moreover, future potential species distributions are predicted as though ecosystems could rapidly reach a new equilibrium. Although the first assumption seems justified for high-alpine meadows (i.e., "it is estimated that, without climate change, the present state of the vegetation would not evolve in a significant manner" (Galland 1982)), the second is certainly debatable, because the alpine species' dynamics tend to be gradual in any case (see, e.g., chapter 5), notwithstanding time scales for soil dynamics on the order of millennia (see section 5.2). Thus, alpine vegetation will likely not be able to reach a new equilibrium as long as climate is changing.

Therefore, such static models do not provide a specific answer to the question of how climate change would affect future species' distributions, and they must not be taken as predictions. Rather, they attempt to explore possible ranges of vegetation change and thus offer a basis for discussion about, for example, species' potential to reach new habitats. Moreover, they constitute the basic layers for future spatiotemporal vegetation models (Halpin 1994; Solomon and Leemans 1989).

The time constants of forest dynamics are so large, probably exceeding those of alpine vegetation, that here the transient behavior clearly must be simulated to assess consistently the response to a transiently changing climate in the coming centuries. The forest model's dynamic approach accomplishes this.

However, this model's dynamic nature results in comparably long simulation times, which its stochastic nature further increases. The stochastic simulations enable us to capture forests' intrinsic variability. Yet many replicates of the stochastic process must be calculated to obtain reliable results (Bugmann, Fischlin, and Kienast 1996). These long simulation times impede spatial applications at a fine resolution, which would require a prohibitive number of simulation runs.

The application of all models, particularly those developed in a particular region, to a combination of conditions outside the range used to obtain the parameter values (by model calibration, as in the static model, or from empirical and experimental studies of single processes, as in the forest model) can cause problems. The mechanistic forest model offers some hope that the formulation of its process functions will still be valid under changed conditions, although a higher uncertainty due to its larger number of parameters might outweigh this. Therefore, models should not only be calibrated but also thoroughly validated, that is, tested by comparing their results to vegetation data under a broad range of conditions of an amplitude at least similar to that expected for the next century's putative climate change.

The evaluation of the plant distribution model is mainly limited by data requirements, that is, the gathering of supplementary species abundance data and associated environmental covariates. Therefore evaluation has been possible only in a small alpine area. Because the results were satisfying within the usual range of uncertainties encountered in such studies, we anticipate extending this model to other areas. Ideally, if input data about species, climate, and other abiotic factors were available at the same fine resolution, simulations could be conducted for all alpine regions of Switzerland or even at the scale of the entire Alps and results compared to vegetation data.

Both the equilibrium and the transient behavior of the dynamic forest model need to be further tested, which requires data from gradients in space as well as in time. Paleoecological data offer promising opportunities for model validation, but at the same time they introduce new problems, such as the many incertitudes in their interpretation. In particular, problems associated with tree species migration pose an unpleasant obstacle to our process of model testing, but they also point clearly to tree species migration's potential importance to future ecosystem adjustment to climate change.

Given a thorough validation under a larger range of climatic conditions as was performed in part for the forest model and planned for the alpine plant distribution model, both models should succeed in generating more valuable scenarios of vegetation responses to a changing climate. However, the two models presented already allow preliminary glances at future potential vegetation and at the associated uncertainties. Based on both simulations, we can draw some general and tentative conclusions for the potential responses of montane forests and alpine herbaceous species' distributions to future changes in climate.

Although some forests might profit slightly from the envisaged climate changes or would be completely unaffected, some might suffer drastically. No uniform, simple response of the mountain forests to a climatic change could be expected. Moreover, the mountain forests might prove sensitive to climatic features that remain very uncertain in the regional climate change projections. The major sensitivities were found in subalpine conditions and in areas that border on having a continental climate already, as is typical for some of the large Alpine valleys. When we contrast these findings with what we have learned about the peripheral forests at lower altitudes, we conclude that mountain forests are especially sensitive to climate change.

Both a low and a medium IPCC climate change scenario predict tremendous changes in the potential distribution of species of herbaceous and dwarf shrub alpine plants. However, these results must be interpreted very carefully, because great uncertainties remain associated with the postulate of the future equilibrium of alpine ecosystems: uncertainties in alpine soils' potential reaction to climate change (see chapter 5), in the future competitive relationships among plant species, and in the interaction of these factors. Hence, the prediction of plant species distributions mainly indicates that a changing climate would displace present locations of ecological optima and tolerance limits of plants geographically. It does not say whether plants will be able to reach and colonize new habitats or to adapt to such modifications of their present environment.

A sound assessment of alpine ecosystems' potential vulnerability beyond these first steps demands, first of all, full validation of both models. Toward this goal, the alpine plant distribution model should be calibrated using data from additional, ecologically distinct regions. Validating the forest model's transient behavior requires paleoclimatic scenarios that were reconstructed independently from past vegetation (see chapter 3).

In both models species' ability to migrate under altered climatic conditions turned out to be crucial. Many factors affect the facility of migration, such as seed dispersal capacities, climate-dependent vegetation dynamics along the path of migration, natural and human barriers, and suitability of substrates. Explicitly modelling migration as a combination of seed dispersal and climate-dependent community dynamics along the migrational path could reduce the uncertainty concerning species migration.

Another uncertainty remains concerning the plants' potential to adapt to changing conditions in ways other than migration, that is, through natural selection or through physiological and phenological adjustments or other kinds of phenotypic plasticity. These traits are currently not captured by the presented models and would require further experimental and modeling efforts.

The responses of both alpine herbaceous or shrub vegetation and forest ecosystems for the entire Alpine region could be systematically evaluated either by covering the entire Alpine arc with a grid of point simulations or by subdividing the Alps into "representative" regions, as defined by the most frequent combinations of abiotic factors, and performing simulations for each of these regional classes only. This would require spatially highly resolved data for abiotic input variables, in particular local climatic change scenarios, such as changes in monthly temperatures, monthly precipitation, annual distribution of precipitation, or snow cover duration, such as those that other researchers within the CLEAR project are currently developing (Gyalistras et al. 1994; Wanner and Beniston 1995).

In the long term, both ecosystem modeling approaches might converge to spatially highly resolved, species-based, dynamic vegetation models. Such models would provide the necessary basis for people within the Alpine region, in particular decision makers, to assess the possible consequences of policy decisions made today, either by them or by people from other parts of the globe. Despite all their deficiencies and drawbacks, vegetation models offer indispensible means of glimpsing the possible future plant covers of the Alpine region. Of course, local as well as global environmental changes, not all covered to the same extent by a particular model, might affect vegetation. But global changes like those anticipated for climate pose a particular challenge and call strongly for means of projecting the future as precisely as possible, especially because of the intrinsic, lengthy time delays between the decisions made today and their consequences. These decisions' impact on the vegetation may lie as far as centuries into the future. A manifestation of just such "vegetational pictures" begins to emerge from efforts like those presented here.

ACKNOWLEDGMENTS

We would like to thank Felix Kienast, Felix Bucher, Niolas Wyler, and particularly Klaus Zimmermann for their valuable scientific inputs and comments. We are very grateful to Loraine Kohorn for her constructive and efficient help concerning the English. Furthermore, we would like to thank Jean Paul Theurillat for the first efforts in editing the chapter.

REFERENCES

Aeschimann, D., P. Kuepfer, and R. Spichiger. (in press). Projet pour une flore des Alpes. In P. Kuepfer, and P. Galland, (eds.) *Actes Congr. Int. Ecol. Biogeogr. Alpines*, La Thuile, Italy (2–6 Sept. 1991).

Anselin, L., and A. Getis. 1992. Spatial statistical analysis and geographic information systems. *Annals of Regional Science* 26:19–33.

Austin, M. P. 1992. Modelling the environmental niche of plants: Implications for plant community response to elevated CO_2 levels. *Australian Journal of Botany* 40:615–30.

Austin, M. P. 1987. Models for the analysis of species' response to environmental gradients. *Vegetatio* 69:35–45.

Austin, M. P. 1979. Current approaches to the nonlinearity problem in vegetation analysis. In *Satellite Program in Statistical Ecology*, ed. G. P. Patil and M. Rosenzweig. Fairland: International Co-operative Publishing House, pp. 187–210.

Austin, M. P., R. Cunningham, and R. B. Good. 1983. Altitudinal distribution of several eucalypt species in relation to other environmental factors in southern New South Wales. *Australian Journal of Ecology* 8:169–80.

Austin, M. P., and M. J. Gaywood. 1994. Current problems of environmental gradients and species response curves in relation to continuum theory. *Journal of Vegetation Science* 5:473–82.

Austin, M. P., A. O. Nicholls, M. D. Doherty, and J. A. Meyers. 1994. Determining species response functions to an environmental gradient by means of a Beta-function. *Journal of Vegetation Science* 5:215–28.

Austin, M. P., A. O. Nicholls, and C. R. Margules. 1990. Measurement of the realized qualitative niche: Environmental niches of five Eucalyptus species. *Ecological Monographs* 60:161–77.

Blum, H. 1991. Treibhauseffekt und Pflanzen wachstum. *Vierteljahresschrift der Naturforschenden Gesellschaft in Zürich* 136(4):189–206.

Boer, G. J., N. A. McFarlane, and M. Lazare. 1992. Greenhouse gas–induced climate change simulated with the CCC second-generation general circulation model. *Journal of Climate* 5:1045–77.

Bonan, G. B., and B. P. Hayden. 1990. Using a forest stand simulation model to examine the ecological and climatic significance of the Late-Quaternary pine-spruce pollen zone in eastern Virginia, U.S.A. *Quaternary Research* 33:204–18.

Bossel, H. 1991. Modelling forest dynamics: Moving from description to explanation. *Forest Ecology and Management* 42:129–42.

Bossel, H. 1987. A dynamic simulation model of tree development under pollution stress. *Erwin-Riesch-Workshop, Advances in Systems Analysis of Biological processes. 2nd Ebernburger Working Conference.* Vieweg, Braunschweig, 46–58.

Bossel, H., H. Krieger, H. Schäfer, and N. Trost. 1991. Simulation of forest stand dynamics, using real-structure process models. *Forest Ecology and Management* 42:3–21.

Botkin, D. B., J. F. Janak, and J. R. Wallis. 1972a. Rationale, limitations and assumptions of a Northeastern forest growth simulator. *IBM Journal of Research and Development* 16:101–16.

Botkin, D. B., J. F. Janak, and J. R. Wallis. 1972b. Some ecological consequences of a computer model of forest growth. *Journal of Animal Ecology* 60:849–73.

Botkin, D. B., J. F. Janak, and J. R. Wallis. 1970. A simulator for Northeastern forest growth: A contribution of the Hubbard Brook ecosystem study and IBM research: IBM Research Report no. 3140. Yorktown Heights, N.Y.

Bradley, R. 1991. *Global Changes of the Past*. Boulder, Colorado: Office for Interdisciplinary Earth Studies, University Corporation for Atmospheric Research.

Brown, D. G. 1994. Predicting vegetation types at treeline using topography and biophysical disturbance variables. *Journal of Vegetation Science* 5:641–56.

Brown, J. H. 1984. On the relationship between abundance and distribution of species. *The American Naturalist* 124:255–79.

Brzeziecki, B., F. Kienast, and O. Wildi. 1995. Modelling potential impacts of climate change on the spatial distribution of zonal forest communities in Switzerland. *Journal of Vegetation Science* 6:257–8.

Brzeziecki, B., F. Kienast, and O. Wildi. 1994. Potential impacts of a changing climate on the vegetation cover of Switzerland: A simulation experiment using GIS technology. In *Mountain Environments and Geographic Information Systems*, ed. M. F. Price and D. I. Heywood. London: Taylor and Francis Ltd., pp. 263–279.

Brzeziecki, B., F. Kienast, and O. Wildi. 1993. A simulated map of the potential natural forest vegetation of Switzerland. *Journal of Vegetation Science* 4:499–508.

Buckland, S. T., and A. Elston. 1993. Empirical models for the spatial distribution of wildlife. *Journal of Applied Ecology* 30:478–95.

Bugmann, H. 1996. A simplified forest model to study species composition along climate gradients. *Ecology* 77(7):2055–74.

Bugmann, H. 1994. On the ecology of mountainous forests in a changing climate: A simulation study. Diss. ETH, Nr. 10638 Swiss Federal Institute of Technology, Zürich.

Bugmann, H., and A. Fischlin. 1994. Comparing the behaviour of mountainous forest succession models in a changing climate. In *Mountain Environments in Changing Climantes*, ed. M. Beniston. London: Routledge Publishing Co., pp. 237–255.

Bugmann, H., A. Fischlin, and F. Kienast. 1996. Model convergence and state variable update in forest gap models. *Ecological Modelling* 89:197–208.

Bugmann, H., and A. M. Solomon. 1995. The use of a European forest model in North America: A study of ecosystem response to climate gradients. *Journal of Biogeography*:477–84.

Busby, J. R. 1991. BIOCLIM—A bioclimate analysis and prediction system. In *Nature Conservation: Cost Effective Biological Surveys and Data Analysis*, ed. C. R. Margules and M. P. Austin. Melbourne: Commonwealth Scientific and Industrial Research Organization, pp. 64–68.

Busby, J. R. 1988. Potential impacts of climate change on Australia's flora and fauna. In *Greenhouse: Planning For Climate Change*. Melbourne: Commonwealth Scientific and Industrial Research Organization.

Carpenter, G., G. Gillison, and J. Winter. 1993. DOMAIN: A flexible modelling procedure for mapping potential distributions of plants and animals. *Biodiversity and Conservation* 2:667–80.

Cubasch, U., K. Hasselmann, H. Höck, E. Maier-Reimer, U. Mikolajewicz, B. Santer, and R. Sausen. 1992. Time-dependent greenhouse warming computations with a coupled ocean-atmosphere model. *Climate Dynamics* 8(55):55–69.

Davis, L. S., and L. I. Delain. 1986. Linking wildlife-habitat analysis to forest planning with ECOSYM. In *Wildlife 2000*, ed. J. Verner, M. L. Morrison, and C. J. Ralph. Madison: University of Wisconsin Press, pp. 361–370.

de Swart, E. O. A. M., A. G. van der Valk, K. J. Koehler, and A. Barendregt. 1994. Experimental evaluation of realized niche models for predicting response of plant species to a change in environmental conditions. *Journal of Vegetation Science* 5:541–52.

d'Oleiro-Oltmanns, W., T. Mingozzi, and U. Brendel. 1995. Effects of climate change on bird populations. In *Potential Ecological Impacts of Climate Change in the Alps and Fennoscandian Mountains*, ed. A. Guisan, J. I. Holten, R. Spichiger, and L. Tessier, 173–5. Geneva: Conservatoire et Jardin Botaniques.

Eamus, D., and P. G. Jarvis. 1989. The direct effects of increase in the global atmospheric CO_2 concentraion on natural and commercial temperate trees and forests. *Advances in Ecological Research* 19:1–55.

Ellenberg, H., and F. Klötzli. 1972. Waldgesell schaften und Waldstandorte der Schweiz. *Mitteilungen der Eidgenössischen Anstalt für forstliches Versuchswesen* 48:589–930.

Faegri, K., and J. Iversen. 1975. *Textbook of Pollen Analysis.* Copenhagen, Denmark: Munksgaard.

Ferrer-Castán, D., J. F. Calvo, M. A. Esteve-Selma, A. Torres-Martínez, and L. Ramírez-Díaz. 1995. On the use of three performance measures for fitting species response curves. *Journal of Vegetation Science* 6:57–62.

Fischer, H. S. 1990. Simulating the distribution of plant communities in an alpine landscape. *Coenoses* 5:37–43.

Fischlin, A. 1995. Assessing sensitivities of forests to climate change: Experiences from modelling case studies. In *Potential Ecological Impacts of Climate Change in the Alps and Fennoscandian mountains,* ed. A. Guisan, J. I. Holten, R. Spichiger, and L. Tessier, 145–147. Geneva: Conservatoire et Jardin Botaniques

Fischlin, A., H. Bugmann, and D. Gyalistras. 1995. Sensitivity of a forest ecosystem model to climate parametrization schemes. *Environmental Pollution* 87(3):267–82.

Franklin, S. B., D. J. Gibson, A. Robertson, J. T. Pohlmann, and J. S. Fralish. 1995. Parallel Analysis: A method for determining significant principal components. *Journal of Vegetation Science* 6:99–106.

Fulton, M. R. 1991. A computationally efficient forest succession model: Design and initial tests. *Forest Ecology and Management* 42:23–34.

Galland, P. 1982. *Etudes de la végétation des pelouses alpines au parc national suisse.* Faculté des Sciences, diss., Université de Neuchâtel.

Graybill, D. A., and S. B. Idso. 1993. Detecting the aerial fertilization effect of atmospheric CO_2 enrichment in tree-ring chronologies. *Global Biogeochemical Cycles* 7:81–95.

Green, R. H. 1979. *Sampling Design and Statistical Methods for Environmental Biologists.* New York: John Wiley.

Greig-Smith, P. 1983. *Quantitative Plant ecology.* Vol. 9: *Studies in Ecology.* Oxford: Blackwell Scientific Publications.

Guiot, J., S. P. Harrison, and I. C. Prentice. 1993. Reconstruction of Holocene precipitation patterns in Europe using pollen and lake-level data. *Quaternary Research* 40:139–49.

Guisan, A., J. I. Holten, R. Spichiger, and L. Tessier, eds. 1995. *Potential Ecological Impacts of Climate Change in the Alps and Fennoscandian Mountains. An annex to the Intergovernmental Panel on Climate Change (IPCC), Second Assessment Report, Working Group II-C (Impacts of Climate Change on Mountain Regions).* Geneva: Conservatoire et Jardin Botaniques.

Guisan, A., J. I. Holten, L. Tessier, W. Haeberli, and M. Baumgartner. 1995. Understanding the impact of climate change on mountain ecosystems: An overview. In *Potential Ecological Impacts of Climate Change in the Alps and Fennoscandian Mountains,* ed. A. Gusian, J. I. Holten, R. Spichiger, and L. Tessier, 15–37. Geneva: Conservatoire et Jardin Botaniques.

Guisan, A. 1997. ALPLANDI: évalues la réponse des plantes alpines aux changements climatiques à travers la modelisation des distributions actuelles et futures de leur habitat potentiel. *Bulletin de la Murithienne,* Societé Valaisanne de Sciences Naturelles 114:187–196.

Guisan, A., J.-P. Theurillat, and F. Kienast. 1998. Predicting the potential distribution of plant species in an alpine environment. *Journal of Vegetation Science* 9:65–74.

Guisan, A., J.-P. Theurillat, and R. Spichiger. 1995. Effects of climate change on alpine plant diversity and distribution: The modelling and monitoring perspectives. In *Potential Ecological Impacts of Climate Change in the Alps and Fennoscandian Mountains*, ed. A. Guisan, J. I. Holten, R. Spichiger, and L. Tessier, 129–35. Geneva: Conservatoire et Jardin Botaniques.

Gyalistras, D., and A. Fischlin. 1995. Derivation of climate change scenarios for mountainous ecosystems: a GCM-based method and the case study of Valais, Switzerland, Systems Ecology Report 22, Systems Ecology, Swiss Federal Institute of Technology ETHZ, Zurich, Switzerland.

Gyalistras, D., H. von-Storch, A. Fischlin, and M. Beniston. 1994. Linking GCM-simulated climatic changes to ecosystem models: Case studies of statistical downscaling in the Alps. *Climate Research* 4(3):167–89.

Halpin, P. N. 1994. GIS analysis of the potential impacts of climate change on mountain ecosystems and protected areas. In *Mountain Environments and Geographic Information Systems*, ed. M. F. Price and D. I. Heywood, 281–301. London: Taylor and Francis Ltd.

Hartel, H., G. Kniely, G. H. Leute, H. Niklfeld, and M. Perko. 1992. *Verbreitungsatlas der Farn- und Blütenpflanzen Kärntens*. Klagenfurt, Austria: Naturwissenschftlicher Verein für Kärnten.

Haslett, J. R. 1990. Geographic informatic systems: A new approach to habitat definition and the study of distributions. *Trends in Ecology and Evoluton* 5:214–8.

Haxeltine, A., I. C. Prentice, and I. D. Cresswell. 1997. A coupled carbon and water flux model to predict vegetation structure. *Journal of Vegetation Science* (in press).

Hill, M. O. 1991. Patterns of species distribution in Britain elucidated by canonical correspondence analysis. *Journal of Biogeography* 18:247–55.

Houghton, J. T., G. J. Jenkins, and J. J. Ephraums, eds. 1990. *Climate change—The IPCC scientific assessment*. Cambridge: Cambridge Univiversity Press.

Huisman, J., H. Olff, and L. F. M. Fresco. 1993. A hierarchical set of models for species response analysis. *Journal of Vegetation Science* 4:37–46.

Huntley, B., P. M. Berry, W. Cramer, and A. P. McDonald. 1995. Modelling present and potential future ranges of some European higher plants using climate response surfaces. *Journal of Biogeography* 22:967–1001.

Huntley, B., and H. J. B. Birks. 1983. *An Atlas of Past and Present Pollen Maps for Europe: 0–13,000 years Ago*. Cambridge: Cambridge University Press.

Huntley, B., and I. C. Prentice. 1993. Holocene vegetation and climates of Europe. In *Global Climates Since the Last Glacial Maximum*. H. E. J. Wright, J. E. Kutzbach, T. Webb III, W. F. Ruddiman, F. A. Street-Perrott, and P. J. Bartlein (eds.). Minneapolis, London: University of Minnesota Press, pp. 136–168.

Hutchinson, M. F., and R. J. Bishof. 1983. A new method for estimating the spatial distribution of mean seasonal and annual rainfall applied to Hunter valley. *Australian Meteorological Magazine* 31:179–84.

Idso, S. B. 1980a. The climatological significance of a doubling of earth's atmospheric carbon dioxide concentration. *Science* 207:1462–3.

Idso, S. B. 1980b. Answers to: The climatological significance of a doubling of earth's atmospheric carbon dioxide concentration. *Science* 210:6–8.

Idso, K. B., and S. B. Idso. 1994. Plant responses to atmospheric CO_2 enrichment in the face of environmental constraints: A review of the past 10 years' research. *Agricultural and Forest Meteorology* 69:153–203.

Jensen, J. R., S. Narumalani, O. Watherbee, K. S. Morris Jr., and H. E. Mackey Jr. 1992. Predictive modelling of cattail and waterlily distribution in a South Carolina reservoir using GIS. *Photogrammetric Engineering & Remote Sensing* 58:1561–8.

Jongman, R. H. G., C. J. T. Ter Braak, and O. F. R. Tongeren van. 1987. *Data Analysis in Community and Landscape Ecology.* Wageningen, Netherlands: Pudoc.

Keller, F. 1992. Automated mapping of mountain permafrost using the program PERMAKART within the geographical Information System ARC/INFO. *Permafrost and Periglacial Processes* 3(2):133–138.

Kienast, F. 1991. Simulated effects of increasing atmospheric CO_2 and changing climate on the successional characteristics of Alpine forest ecosystems. *Landscape Ecology* 5(4):225–38.

Kienast, F. 1989. Simulated effects of increasing CO_2 on the successional characteristics of Alpine forest ecosystems. Paper presented at conference: Landscape Ecological Impact of Climatic Change on Alpine Regions, with Emphasis on the Alps, at Lunteren, The Netherlands, Dec. 3–7, 1989.

Kienast, F., B. Brzeziecki, and O. Wildi. 1994. Computergestützte Simulation der räumlichen Verbreitung naturnaher Waldgesellschaften in der Schweiz. *Schweizerische Zeitschrift für das Forstwesen* 145:293–309.

Kienast, F., and N. Kräuchi. 1989. Simulated response of alpine forest ecosystems to various environmental changes: application of a forest succession model. In *Forests of the World: Diversity and Dynamics,* ed. E. Sjögren. Uppsala: University of Uppsala, 1989, pp. 139–140.

Kienast, F., and N. Kuhn. 1989. Simulating forest succession along ecological gradients in southern central Europe. *Vegetatio* 79:7–20.

Kimball, B. A., and S. B. Idso. 1983. Increasing atmospheric CO_2: Effects on crop yield, water use and climate. *Agricultural Water Management* 7:55–72.

Kirschbaum, M., and A. Fischlin. 1996. Climate change impacts on forests. In *Climate Change 1995—Impacts, Adaptations and Mitigation of Climate Change: Scientific-Technical Analyses: Contribution of Working Group II to the Second Assessment Report of the InterGovernmental Panel on Climate Change,* R. Watson, ed. Cambridge: Cambridge University Press, pp. 95–131.

Kohyama, T., and N. Shigesada. 1995. A size-distribution–based model of forest dynamics along a latitudinal environmental gradient. *Vegetatio* 121:117–26.

Körner, C. 1995. Impact of atmospheric changes on alpine vegetation: The ecophysiological perspectives. In *Potential Ecological Impacts of Climate Change in the Alps and Fennoscandian Mountains,* ed. A. Guisan, J. I. Holten, R. Spichiger, and L. Tessier, 113–20. Genera: Conservatoire et Jardin Botaniques.

Körner, C. 1992. Response of alpine vegetation to global climate change. *Catena* 22 (Supplement):85–96.

Körner, C., and J. A. Arnone III. 1992a. Recent correspondence related to the paper: Responses to elevated carbon dioxide in artificial tropical ecosystems. Science 257:1672–5.

Körner, C., and J. A. Arnone III. 1992b. Responses to elevated carbon dioxide in artificial tropical ecosystems. *Science* 257:1672–5.

Kräuchi, N. 1993. Potential impacts of a climate change on forest ecosystems. *European Journal of Forest Pathology* 23:28–50.

Kräuchi, N., and F. Kienast. 1993. Modelling subalpine forest dynamics as influenced by a changing environment. *Water, Air, & Soil Pollution* 68:185–97.

Kutzbach, J. E., and P. J. Guetter, 1988. The influence of changing orbital parameters and surface boundary conditions on climate simulations for the past 18,000 years. *Journal of Atmospheric Sciences* 43(16):1726–59.

Lancia, R. A., D. A. Adams, and E. M. Lunk. 1986. Temporal and spatial aspects of species-habitat models. In *Wildlife 2000,* ed. J. Verner, M. L. Morrison, and C. J. Ralph. Madison: University of Wisconsin Press, pp. 177–182.

Landolt, E. 1977. Ökologische Zeigerwerte zur Schweizer Flora. *Veröffentlichungen des Geobotanischen Institutes der ETH Zürich* 64:5–208.

Leemans, R., and I. C. Prentice. 1989. FORSKA: A general forest succession model. Uppsala, Sweden: Institute of Ecological Botany.

Leemans, R., and G. J. van-den-Born. 1994. Determining the potential distribution of vegetation, crops and agricultural productivity. *Water, Air, & Soil Pollution* 76:133–61.

Leninhan, J. M. 1993. Ecological response surface for North American boreal tree species and their use in forest classification. *Journal of Vegetation Science* 4:667–80.

Lischke, H., T. J. Löffler, and A. Fischlin. 1998. Aggregation of individual trees and patches in forest succession models: Capturing variability with height structured, random, spatial distributions. *Theor. Popul. Biol.* (in press).

Lotter, A. 1989. Evidence of annual layering in Holocene sediments of Soppensee, Switzerland. *Aquatic Sciences* 51(1):19–30.

Lotter, A., and F. Kienast. 1992. Validation of a forest succession model by means of annually laminated sediments. Paper read at Conference: Laminated Sediments, June 4–6, 1990, Lammi Biological Station, Finland.

Lyon, J. G., J. T. Heinen, R. A. Mead, and N. E. G. Roller. 1987. Spatial data for modelling wildlife habitat. *Journal of Surveying Engineering* 113:88–100.

Malanson, G. P. 1985. Spatial autocorrelation and distributions of plant species on environmental gradients. *Oikos* 45(2):278–280.

McCullagh, P., and J. A. Nelder. 1989. *Generalized Linear Models.* 2d ed. London: Chapman and Hall.

Mille, R. I., S. N. Stuart, and K. M. Howell. 1989. A methodology for analysing rare species distribution pattern utilizing GIS technology: The rare birds of Tanzania. *Landscape Ecology* 2:173–89.

Mohler, C. L. 1983. Effect of sampling pattern on estimation of species distributions along gradients. *Vegetatio* 54:97–102.

Morison, J. I. L. 1987. Intercellular CO_2 concentration and stomatal responses to CO_2. In *Stomatal Function,* ed. E. Zeiger, G. D. Farquhar, and I. R. Cowan. Stanford, CA: Stanford University Press, pp. 229–251.

Neilson, R. P. 1995. A model for predicting continental-scale vegetation distribution and water balance. *Ecological Applications* 5(2):362–385.

Neilson, R. P. 1993. Vegetation redistribution: A possible biosphere source of CO_2 during climate change. *Water, Air, & Soil Pollution* 70:659–73.

Nicholls, A. O. 1989. How to make biological surveys go further with generalized linear models. *Biological Conservation* 50:51–75.

Ormsby, J. P., and R. S. Lunetta. 1987. Whitetail deer food availability maps from Thematic Mapper data. *Photogrammetric Engineering & Remote Sensing* 53:1081–5.

Ozenda, P., and J.-L. Borel. 1990. The possible responses of vegetation to a global climatic change. Scenarios for western Europe, with special reference to the Alps. Paper read at conference: Landscape-Ecological Impact of Climatic Change, Lunteren, The Netherlands, Dec. 3–7, 1989.

Parlow, E., and D. Scherer. 1991. Auswirkungen von Vegetationsänderungen auf den Strahlungshausalt: Eine methodische Studie aus Schwedisch-Lappland. *Regio Basiliensis* 32:33–42.

Pereira, J. M. C., and R. M. Itami. 1991. GIS-based habitat modeling using logistic multiple regression: A study of the Mt. Graham red squirrel. *Photogrammetric Engineering & Remote Sensing* 57:1475–86.

Prentice, I. C, W. Cramer, S. P. Harrison, R. Leemans, R. A. Monserud, and A. M. Solomon. 1992. A global biome model based on plant physiology and dominance, soil properties and climate. *Journal of Biogeography* 19:117–134.

Rhoads, A. F., and L. Thompson. 1992. Integrating herbarium data into a geographic information system: Requirements for spatial analysis. *Taxon* 41:43–49.

Robinson, J. M. 1994. Atmospheric CO_2 and plants. *Nature (London)* 368:105–6.

Romme, W. H., and M. G. Turner. 1990. Implications of global climate change for biogeographic patterns in the greater Yellowstone ecosystem. *Conservation Biology* 5(3):373–86.

Schuster, A. 1990. Ornithologische Forschung unter Anwendung eines Geographischen Informationssystems. *Salzburger Geographische Materialien* 15:115–23.

Shmida, A., and S. Ellner. 1984. Coexistence of species with similar niches. *Vegetatio* 58:29–55.

Shugart, H. H., and D. C. West. 1977. Development of an Appalachian deciduous forest succession model and its application to assessment of the impact of the chestnut blight. *Journal of Environmental Management* 5:161–79.

SMA (Schweizerische Meteorologische Anstalt). 1901–1990. *Annalen der Schweizerischen Meteorologischen Anstalt.* Zurich, Switzerland: Schweizerische Meteorologische Anstalt.

Solomon, A. M., H. R. Delcourt, D. C. West, and T. J. Blasing. 1980. Testing a simulation model for reconstruction of prehistoric forest-stand dynamics. *Quaternary Research* 14:275–93.

Solomon, A. M., and R. Leemans. 1989. Climatic change and landscape-ecological response: Issues and analysis. Paper read at Landscape-Ecological impact of climatic change, Proceedings of a European conference, at Lunteren, The Netherlands, Dec. 3–7, 1989.

Solomon, A. M., M. L. and Tharp. 1985. Simulation experiments with late quaternary carbon storage in midlatitude forest communities. In *The Carbon Cycle and Atmospheric CO_2: Natural Variations Archean to Present*, E. T. Sundquist and W. S. Broecker (eds.). Washington, D.C.: American Geophysical Union, pp. 235–250.

Solomon, A. M., D. C. West and J. A. Solomon. 1981. Simulating the role of climate change and species immigration in forest succession. In *Forest Succession: concepts and application*, D. C. West, H. H. Shugart, and D. B. Botkin (eds.). New York: Springer, pp. 154–177.

Ter Braak, C. J., and N. J. M. Gremmen. 1987. Ecological amplitudes of plant species and the internal consistency of Ellenberg's indicator values for moisture. *Vegetatio* 69:79–87.

Ter Braak, C. J., and C. W. M. Looman. 1986. Weighted averaging, logistic regression and the Gaussian response model. *Vegetatio* 65:3–11.

Theurillat, J.-P. 1995. Climate change and the alpine flora: Some perspectives. In *Potential Ecological Impacts of Climate Change in the Alps and Fennoscandian Mountains*, ed. A. Guisan, J. I. Holten, R. Spichiger, and L. Tessier, 121–27. Genera: Conservatoire et Jardin Botaniques.

Theurillat, J.-P., F. Felbes, P. Geissler, A. Guisan, and J.-M. Gobat. 1997. Le projet "Ecocline" et Programme Prioritaire "Environnement." *Bulletin de la Murithienne*, Societé Valaisanne de Sciences Naturelles 114:151–162.

Walker, B. H. 1990. A framework for modelling the effects of climate and atmospheric change on terrestrial ecosystems. Paper read at Global Change IGBP Report, Woods Hole, Masschusetts, April 1989; pp. 1–22.

Wanner, H., and M. Beniston. 1995. Approaches to the establishment of future climate scenarios for the alpine region. *Potential Ecological Impacts of Climate Change in the Alps and Fennoscandian Mountains*, ed. A. Guisan, J. I. Holten, R. Spichiger, and L. Tessier, 87–95. Geneva: Conservatoire et Jardin Botaniques.

Welten, M., and R. Sutter. 1982. *Verbreitungsatlas der Farn- und Blütenpflanzen der Schweiz*. 2 vols. Basel, Switzerland: Birkhäuser.

Whittaker, R. H. 1956. Vegetation of the Great Smoky Mountains. *Ecological Monographs* 26:1–80.

Whittaker, R. H., S. A. Levin, and R. B. Root. 1973. Niche, habitat, and ecotope. *American Naturalist* 107:321–38.

Woodward, F. I., and T. M. Smith. 1994. Predictions and measurements of the maximum photosynthetic rate at the global scale. In *Ecophysiology of Photosynthesis*, edited by E. D. Schulze and M. M. Caldwell. New York: Springer.

Woodward, F. I., T. M. Smith, and W. R. Emanuel. 1995. A global land primary productivity and phytogeography model. *Global Biogeochemical Cycles* 9(4):471–90.

Yee, T. W., and N. D. Mitchell. 1991. Generalized linear models in plant ecology. *Journal of Vegetation Science* 2:587–602.

Zimmermann, K. 1996. Ein klimasensitives, räumliches Vegetationsmodell für die alpine Stufe der Schweiz. Ph.D. diss., Bern University.

Zimmermann, K., and F. Kienast. (In press.) Predictive mapping of alpine grasslands in Switzerland: species versus community approach. *Journal of Vegetation Science*.

7 Innovative Social Responses in the Face of Global Climate Change

Bernhard Truffer, Peter Cebon, Gregor Dürrenberger,
Carlo C. Jaeger, Roman Rudel, and Silvia Rothen

7.1 INNOVATION-ORIENTED POLICIES IN THE FACE OF GLOBAL CLIMATE RISKS

7.1.1 The Need for an Innovation-Oriented Policy Approach

A further increase of GHG concentrations in the atmosphere due to human activities implies substantial risks. This assessment, already expounded in previous chapters of this book, has meanwhile found wide acceptance in global climate politics. In view of those risks, a preventive approach to climate policy, that is, a stabilization or even reduction of GHG emissions, has been accepted in the United Nations Framework Convention on Climate Change (UNFCCC) as a basis for future negotiations. The UNFCCC was prepared for the Rio de Janeiro conference in 1992 and was signed by 154 nations (O'Riordan and Cameron 1995; Mintzer and Leonard 1994). By May 1997, more than 160 governments had ratified the convention.

However, the impressive support expressed at the start of the framework convention has not yet translated into a consensus on specific strategies at a national level (Lee 1995). A substantial reduction of GHG emissions would require a considerable transformation of production and consumption patterns in many countries. Hence, it comes as no surprise that blocking coalitions are forming to oppose any substantial regulation that could come out of the UNFCCC (Sebenius 1995; Boehmer-Christiansen 1994a, 1994b). Many leading countries simply deny the need to take any immediate actions. Therefore, even in countries whose governments have endorsed the preventive approach to climate policy, actual climate policy appears to resemble a wait-and-adapt strategy (Victor and Salt 1995). A global consensus on the climate problem is apparently not a sufficient condition for mitigating the anticipated risks of global climate change. Much more attention must therefore be paid to the options for and barriers to reducing GHG emissions at the national, regional, and individual levels (Rayner 1993).

In this chapter, we argue that analyzing sociotechnical innovation processes is crucial for assessing GHG reduction potentials at the national level. These innovation processes often have an essentially regional (or mesoscale) dimension. Especially when radical transformations of products and technologies are at stake, regional networks of social actors gain in importance. Their input may not be triggered automatically by "global" incentives like taxes, information, or national regulations.[1]

The focus on these mesoscale processes is especially necessary, we argue, because the theoretical framework that currently informs large parts of climate policy tends to abstract from these processes and thus leads to a misrepresentation of the transformation potential of GHG emission patterns. This bias is a major reason for the reserved attitude of many economic and political actors and may, therefore, be decisive in limiting the realization of the FCCC.

We, therefore try to identify the points where the prevailing theoretical approaches regularly neglect fundamental constituents of sociotechnical innovation processes. We cannot develop a wholly consistent theoretical counterproposal. Rather, we try to sketch major points where an innovation-oriented regional approach to global climate change politics could make a real difference. In the first section, we discuss how conventional economic theory informs climate policy. Economics assumes an essentially mechanistic relationship between technical and social dimensions of those production and consumption patterns that cause GHG emissions (Brown 1992).[2] Such a mechanistic view is mainly due to treating the dynamics of technological change and preference formation as exogenous to the market. It is often found in computer models meant to help solve policy problems as well as in market-based policy instruments. In criticizing this view, we argue that it is a consequence of economic theory's axiomatic assumptions.

We then outline a theoretical framework that makes sociotechnical innovation processes a focus of attention. We point out four dimensions that are essential components of a dynamic analysis of technologies and preferences and that conventional theory does not treat appropriately: (1) the role that economies of scale at the level of the market play in stabilizing specific technologies and preferences, (2) production and consumption's irreducibly social aspects (or social construction), (3) social contact networks, central importance in developing and diffusing innovations, and (4) institutions beyond these networks that are able to supply or withhold resources and legitimacy for particular changes of technologies or preferences. These dimensions imply, at the actor level, a focus on processes of "social learning," the interplay between innovation processes at the individual level and their embeddedness into the social environment. At the market level, social learning may involve what could be called "multiple equilibria." We emphasize the inappropriateness of the metaphor of a single optimal equilibrium, a metaphor that underlies many current discussions about policy options.

A dynamic concept of technologies and preferences has major implications for the role of policymakers confronting the risks of climate change. One general consequence is that attempts to support robust development paths based on processes of social learning should replace the search for an "optimal policy strategy." We argue that such development paths often originate in regional contexts and that regional experiments may become an important innovation-oriented policy tool. This, however, does not rule out the application of conventional policy instruments such as regulations, taxes, and information campaigns. However, regional experiments and the associated development paths may enhance their potential considerably.

To illustrate the usefulness of an innovation-oriented regional approach, we present two case studies, both taken from the transportation sector and indicating a substantial reduction potential for GHG emissions in the Alpine region as well as globally. The first case study reports on a secular infrastructure investment project in the Swiss Alps—the NEAT (Neue Eisenbahn Alpentransversale). This project aims to switch transalpine freight traffic from road to rail through the construction of two fifty-kilometer baseline tunnels crossing the Alps. We argue that if the actors of the freight sector (transport industry, manufacturers, and the national railway companies) and national policymakers are unable to mobilize mesoscale processes that lead to a fundamentally new understanding of freight transport, the exorbitant infrastructure investment NEAT involves will achieve neither the stated environmental goals nor profitability. The second case study reports on the development, manufacturing, and market conditions of a technical alternative to conventional gasoline cars—lightweight vehicles (LWVs). A large-scale substitution of LWVs for cars would considerably reduce GHG emissions due to individual mobility. We analyze how economies of scale, the social construction of technologies and preferences, regional contact networks, and external institutions have to be taken into account for understanding the barriers to and potentials of this innovation.

7.1.2 The Theoretical Basis of Economic Advice for Climate Policy

Most economics-based policy advice referring to problems like global climate change draws on the theoretical apparatus of neoclassical economic theory. This theoretical approach treats the economic actors' choice, namely what and how much to produce and to consume, as a rational decision of atomistic actors. In particular, conventional analysis involves the following four claims:

1. Economic actors are either firms or consumers. The theory treats both kinds of actors as essentially isolated individuals who make their production and consumption decisions to maximize their respective individual profits or utilities. Producers and consumers communicate exclusively by adjusting prices and observing quantities demanded and supplied.

2. Firms make decisions based on the structure of the technologies they have at their disposition. Consumers decide on the basis of their preferences. It is characteristic for the neo-classical argument that both means (especially technologies) and goals (especially preferences) usually are exogenous parameters of a given choice situation.

3. To guarantee the existence of a unique supply-and-demand equilibrium, economists usually rule out increasing returns to scale and restrict the analysis to simple preference patterns that can be ascribed to a representative consumer.

4. Furthermore, as a benchmark for policy advice, the theory assumes that there is full competition among actors and that all relevant effects of consumption and production alternatives are priced; that is, no externalities exist. Public goods problems are then analyzed against this background.

If all four assumptions are fulfilled, then price signals are a sufficient means to achieve an equilibrium that coordinates quantities of goods offered and demanded in a socially optimal way.[3] This approach has proven its adequacy and analytical strength for a wide range of situations. However, when long-term dynamics of industrial society are at stake, as in the case of global climate change, major difficulties appear with all four assumptions. In general, they lead to a static and overly reductionist account of economic behavior. As Nelson (1993, 9) has noted: "Virtually all the models of technical innovation ... treat technical change as a process that can be predicted and planned in advance. They ignore the fundamental uncertainty that surrounds any effort to make a significant move forward in technology." Given innovations' cardinal role in modern societies, many analysts no longer consider this theoretical gap tolerable. Therefore, the search for analytical frames that can account for empirically observed change processes has intensified in economics in particular and in the social sciences in general. Such a search begins with a criticism of the four assumptions above.

Many social science studies of technical innovations and the history of technology have criticized the first assumption, that economic actors may be represented as isolated individuals maximizing their profit and utility. Several authors have repeatedly pointed to the importance of social processes and social contact networks in the development of new technologies (Edge 1995; Bimber 1994; Bijker and Law 1992; Dierkes and Hoffman 1992; Farmer and Matthews 1991; Bijker, Hughes, and Pinch 1987). One essential idea of these studies is that technological development cannot be explained by referring exclusively to technological artifacts' material properties. Instead, the evolution of new technologies can be explained only by focusing on the nexus between technical artifacts and social processes.[4] A central explanatory variable in many empirical studies is the actors' considerable "interpretative flexibility" in describing and valuating technologies and commodities (Bijker 1993). However, these interpretations depend on the technologies' material and social conditions. The degree to which specific technological solutions are subjected to shared interpretation and use in a society is one of the central dimensions of study.

The second assumption, that technologies and preferences can be treated as exogenous to the market, has been criticized most extensively from within economics, particularly by evolutionary economists (Metcalfe 1995; Nelson 1995; Eliasson 1992; Hodgson and Screptani 1991; Witt 1991; Burton 1989;

Dosi et al. 1988; Nelson and Winter 1982). Evolutionary approaches aim at giving an empirically informed approach to economic theory. For this reason, they start with a behavioral account of microeconomic activity. As in neoclassical economics, economic actors are assumed to have differing assets and capabilities of achieving their production and consumption goals. However, in evolutionary economics this diversity is not seen as a kind of suboptimal friction but forms the very basis of economic creativity and growth (Metcalfe 1995). An evolutionary mechanism is postulated to work at the level of the routines that guide the behavior of firms and consumers. Variation and selection of routines occurs through social and political processes in response to individual actors' experiences using these routines. As a consequence, technological development often follows specific paths that are stabilized by various cognitive, social, and political mechanisms. The resulting technological trajectories may, however, not be reduced to a sequence of short-term equilibria (Nelson 1995).

Studies on lock-in phenomena based on economies of scale at the level of the market have criticized most competently the third assumption, that production technology exhibits nonincreasing returns to scale and that preferences can be aggregated via a representative consumer. Technological alternatives do—in an early phase of development—often compete for dominance in a market. Economies of scale imply that one technology's superiority over another depends on the relative pace of expansion of its market share. A technology able to expand its market share more quickly than others may outpace its rivals, independently of whether it actually is superior (Arthur 1989). Although increasing returns to scale pose a major problem to conventional economic theory, for most manufactured commodities they are reality (Utterback 1994; Krugman 1991).

Lock-in phenomena may be important for understanding the spatial differentiation of economies. The institutional environment of economic activity, including its geographical dimension, influences economic growth. With regard to the relationship between international trade and technological change, Krugman (1995) argues that different starting conditions may influence the growth paths of national economies. In his opinion, economics accounts inappropriately for issues like development and international competitiveness. Other studies have repeatedly emphasized the importance of institutional structures for the competitiveness and innovativeness of national economies (Porter 1990; Lundvall 1992).[5]

Combining these criticisms, we are led to question the fourth assumption, namely, whether markets should be treated as though in equilibrium and whether this equilibrium should be treated as unique. Gradually, current research has eroded this assumption. A number of approaches in economics start from an explicit disequilibrium conception to describe the state of modern economies (Amendola and Gaffard 1995; Baumol and Wolff 1995; Thomsen 1992; Amendola and Gaffard 1988). Process and product innovations, when understood as feedback mechanisms in economic growth, may continually push an economic system away from a state of equilibrium. The achievement of equilibrium would, in these accounts, be equivalent to the economic system's stagnation. Recent advances in modeling complex systems and stochastic processes with mathematical tools and computer simulations have enabled economists to analyze disequilibrium states. (See, for instance, Nelson 1995; Cowan, Dines, and Meltzer 1994; Chiaromonte, Dosi, and Orsenigo 1993; Waldorp 1992; Stein and Nadel 1990; Anderson, Arrow, and Pines 1988; and Nelson and Winter 1982.)

The above-cited theoretical approaches are not mutually exclusive. Many areas of overlap may be identified. All criticize conventional economic theory by demanding that a theory of technological development and the behavior of markets over time be more realistic. For policy making, it seems most important to note the following four elements of this criticism:

1. increasing economies of scale at the market level

2. the significant extent to which meaning and existence of particular technologies and peoples' preferences in the market are socially constructed

3. the importance of social contact networks and the regional dimension in innovation processes

4. the institutional environment's role in economic activity.

If networks of interacting agents and social construction of technologies and preferences are important, then we cannot assume that actors can be treated as essentially isolated individuals. If technologies lead to increasing returns to scale, or if resource availability is institutionally mediated, the existence of a unique optimal equilibrium cannot be guaranteed. If technologies and preferences are socially constructed, we cannot assume that price signals are sufficient to determine action. If all these factors combine to yield technologies and preferences constructed in a way endogenous to the market, then we cannot assume that technologies and preferences evolve in exogenous processes.

Taken together, these four criticisms imply that local conditions defining the innovating actors' decision context may play an essential role in an innovation's actual fate. The result is not independent of the path taken in the beginning. Regional contact networks often deliver vital inputs for these processes. Later, specific alternatives may be locked in because of increasing returns to scale and become fixed as solutions for a long time thereafter. In such situations, global incentives for action are not sufficient instruments to guarantee an optimal innovation choice. The regional conditions for the emergence and diffusion of technologies and lifestyles must be analyzed much more carefully.

In line with this analysis, we argue that the kind of advice that conventional economic theory offers to the policymaker does not appropriately account for important dimensions of economic and social reality. This may be linked to the fact that there are at least two fundamentally differing judgments about modern societies' technological transformation potential that have not yet been reconciled at a theoretical level.

7.1.3 Economics Applied to Climate Policy

In the economic debate on climate policy, two important strands of literature assess quite differently the relative costs associated with policy alternatives. These assessments are often discussed as the "top-down" and the "bottom-up" approaches respectively (OECD 1994). Top-down approaches analyze climate change's economy-wide effects and the corresponding policies (Dean 1994). In general, they attempt to assess the costs of specific policies compared to those of an unhindered economy. They often conclude that little action is called for because impacts are highly uncertain and may be discounted over long time horizons, whereas the resources needed to reduce GHG emissions will lead to high costs today. Bottom-up approaches, in contrast, start with an analysis of currently known best practices. They often identify considerable emission reduction opportunities at no cost or even at a net benefit[6] (Johansson and Swisher 1994; von Weizsäcker, Lovins, and Lovins 1995; Goldemberg 1987; Lovins 1977).

No one has yet managed to reconcile these two approaches. Top-down modeling has the advantage of being firmly grounded in conventional economic theory. However, it is criticized for its misleading assumption that the market is in an equilibrium before the policy measure is introduced. Endemic market failures are hard to ignore in markets related to environmental pollution and energy efficiency (Flavin and Lenssen 1995; Sanstad and Howarth 1994; Cebon 1992; Okken, Swart, and Zwerver 1989; Shipper et al. 1992; Geller 1991). Bottom-up adherents, on the other hand, often have to retire to common sense arguments not easily integrated into the economic discipline's prevailing theoretical framework. As a consequence, bottom-up approaches are often criticized for making misleading claims about innovation potentials. Either these technological alternatives are actually superior and utility- and profit-maximizing actors will therefore automatically choose them, or else their nonacceptance is due to the existence of hidden costs that make them an inferior alternative for the decisionmakers in the marketplace (Cline 1994).

These two approaches need to be reconciled (OECD 1994). Our analysis suggests that very likely, conventional economic theory systematically underestimates real innovation potentials. Because reducing GHG emissions requires ongoing technological and social change, this might be crucial. One should, on the contrary, take innovation potentials identified by bottom-up approaches very seriously. But the analysis will have to incorporate hidden costs, that is, the barriers to social transformation. Against this perspective, we now turn to policies regarding climate change. We criticize the advice conventional economics provides and elaborate alternatives based on social science in a broader sense.

Conventional economics enters the climate policy debate in two ways. First, economists construct assessment models that help determine the seriousness of the climate change problem. Second, they construct and propose policy instruments intended to remedy the problems that have been identified. These two roles may be independent. It is possible to use an economics-based technique to determine the extent of the problem but develop a non–economics-based policy instrument to solve it. Similarly, it is possible to construct economics-based policy instruments to remedy any assessed damage, independently of the way that damage was assessed.

It is not possible to determine how well assessment models perform their purported function. They are, after all, mere projections into an unknown future. We can not test the future until we get there. But if we examine the way models are constructed (which we do in section 7.1.3.1), we find at least that computational and data limits aggravate theoretical simplifications, among them the treatment of technological change as exogenous to the market. It is possible, on the other hand, to examine how well policy instruments perform. There are a number of at least quasi-experimental and theoretical tests of their efficacy. Their results suggest that for an understanding of practical problems with policy instruments, it is important to recognize the incorrectness of the above-mentioned axiomatic assumptions of conventional economic theory.

7.1.3.1 Economics-Based Assessment Models To construct optimal economics-based instruments, it is necessary to quantify the benefits and costs of measures responding to climate change, so this is generally the aim of economics-based assessment efforts. Such methods achieve quantification by calculating the monetary costs society will incur through the impacts of climate change or through the resources that must be spent to prevent corresponding damages. This way of framing the problem allows us to rank policy alternatives in terms of the prevention they achieve and the costs to society. On the basis of such a ranking, it then becomes sensible to ask which of these policies should be given priority. The most elaborate assessments are constructed with computer models. Such economic climate models have been developed as prevention models, impact models, or cost-benefit models.

Models focusing on prevention try to estimate the cost a society faces when it implements technological alternatives meant to achieve lower GHG emissions. These models are formalized either as optimization models or as computable general equilibrium (CGE) models. Optimization models evaluate the least-cost technological system able to fulfill an exogenous constraint, for instance, a limit for CO_2 emissions, by assuming idealized market conditions. CGE models are more market-oriented. A market equilibrium for a good is reached when its equilibrium price is found, that is, a price at which the quantities offered and demanded are equal.[7]

Impact studies, on the other hand, have tried either to feed results directly from climate models into economic models (e.g., the McKenzie Basin impact study; see Cohen 1993) or to infer, from historical reference cases, the potential damages of future climate change (as in the MINK study; see Rosenberg et al. 1993). Apart from these integrated assessment studies, several sectoral-impact models exist, especially for agriculture (see Tobey, Reilly, and Kane 1992). For the alpine region, sectoral-impact studies have been carried out for tourism, energy production, and agriculture (Abegg and Froesch 1994; Breiling and Charamza 1994).

Cost-benefit models synthesize prevention and impact models and are designed to allow a selection among alternative policies. Cost-benefit models determine the socially optimal level of GHG emissions by balancing costs of impacts against costs of prevention measures (Cline 1994; Nordhaus 1994; Nordhaus 1991).[8] They interpret benefits as prevented damage from greenhouse warming and estimate costs through the goods and services withdrawn from consumption or investment to finance preventive policy measures.

Two kinds of criticisms have been leveled at the economic modeling approach (Jacobs 1994; Orr 1993; Brown 1991; Rothman and Chapman 1991): First, the economic models are essentially deterministic. Uncertainties related to the emergence of new technological options and qualitative shifts in consumer preferences cannot be treated appropriately. Second, these models tend to be blind to potential damages that have not yet been or which cannot be monetized. Therefore, they often underestimate the costs of wait-and-adapt strategies and overestimate the costs of mitigation strategies. We consider these points in more detail.

Economic models largely treat technological innovations and changes in preferences as exogenous to the market; that is, innovations are not seen as the result of decisions of economic actors (Goulder and Schneider 1995). The core of an optimization model, for instance, is a set of predefined technologies. External constraints, such as a limitation on CO_2 emissions, induce substitutions between them. New technologies, not yet available may be included in principle if their exact technical characteristics and costs are known in advance. CGE models, on the other hand, contain no explicit technologies. The supply side is modeled by means of production functions. Innovations are operationalized as a time-dependent (i.e., price-independent) parameter for the efficiency improvement of one single input or, alternatively, for the production function as a whole (Parson 1994).

These models operationalize improvements as occurring in an autonomous way, that is, independently of any policy intervention. As Goulder and Schneider (1995) suggest, this operationalization of innovations is one of the most sensitive aspects with respect to the implications of models for policies. At the same time, it is an area of research in which additional efforts promise important insights. Although the representation of induced technical change Goulder and Schneider (1995) chose is far from satisfying (and its authors are working on more elaborate versions), it has the considerable merit of highlighting the even greater weaknesses of models that ignore induced technical change altogether.

On an empirical level, Oravetz and Dowlatabadi (1995) tried to estimate the magnitude of an "autonomous energy efficiency improvement" trend in the U.S. economy from 1958 to 1990. Most macroeconomic models take a positively valued trend for granted. However, the authors could not identify such a trend in the period considered. The respective data suggest even a negative value for 1960–75. Only between 1975 and 1990 could the authors identify a positive trend. The explicit consideration of strong political signals may trigger energy efficiency as a design criterion for new products and technologies after the first oil shock seems to have had an effect even as real energy prices were falling later on. This suggests that strong political signals may trigger energy efficiency improvements but that no automatic improvement is guaranteed, especially not over long time horizons. Much further research is needed to clarify these issues.

Let us now compare treatment of technological innovation in prevention models to that of conventional economic theory. The latter assumes that new technologies diffuse in response to price signals, but that factors exogenous to the market determine the specific technical characteristics. The relative preferences for price-independent aspects of technologies are also determined in a way exogenous to the market. As noted above, there are many reasons to believe this is too simplistic an assumption for an effective climate policy. However, the prevention models are even more simplistic than neoclassical economics. In particular, shifts in the technology mix are generally not even price sensitive.

Cost-benefit models, because they incorporate prevention models, share a fortiori their shortcomings. In addition, they include weaknesses associated with assessing environmental change's future costs (Rohner and Edenhofer 1996; Rothman and Chapman 1991). Whereas the costs of prevention are felt directly within the economic system, damages occur mostly outside it. The economy is affected only indirectly through damages caused to natural systems. Impact assessment has mostly concentrated on the effects directly measurable in monetary terms, for instance, in the form of crop losses. It essentially ignores the apparently substantial costs of replacing the unvalued and unpaid services that flow constantly from the environment into the economic system.[9]

As we can see, these models tend to bias the assessment of both costs and benefits of specific policies against immediate intervention (Dowlatabadi 1995). Some commentators have, therefore, concluded that economic modeling is immature and an uncertain basis of advice for climate policy (Wallace 1995). A similar judgment is conferred upon the second form of policy advice based on economic theory, namely the evaluation and construction of policy instruments designed to influence GHG emission patterns.

7.1.3.2 Economics-Based Policy Instruments   Most economic advisors argue that the optimal policy approach consists of an appropriate mix of instruments (Baumol and Oates 1988; OECD 1994). Nevertheless, market-based instruments such as environmental taxes and emission permits are given prominence over command and control policies (Stram 1995; Tietenberg and Victor 1994; Pillet et al. 1993; Dower and Zimmerman 1992). Representatives of business (Schmidheiny and BCSD 1992) and politics (Gore 1992) arrive at a similar judgment,[10] since, under standard assumptions, market instruments achieve policy goals at minimal overall cost to society (OECD 1993b; Baumol and Oates 1988).

In these debates, the key markets in society are usually assumed to tend toward a single equilibrium. Taxes, on the other hand, aim at moving prices and quantities of the produced and demanded commodities to a socially optimal level. Hence environmental taxes are supposed to move the point of equilibrium away from its original position to attain some desired quality in a specific dimension, for instance, energy efficiency or reduced environmental disruption. The more the point of equilibrium has to be brought away from its original position, the costlier this endeavor is (normally expressed in lower-than-possible-GNP figures) (Jaeger and Kasemir 1996; Dean 1994). This cost may be justified only if the tax leads to a corresponding reduction of those costs to society not reflected by existing prices, that is, if there is a so-called "externality." On the basis of this conceptual framework, economic experts have been particularly preoccupied with assessing the optimal level of taxes (e.g., Dornbusch and Poterba 1992).

However, this theoretical analysis is based on a rather static conception of the way environmental regulation influences economic development. Porter and van der Linde (1995) show that actual regulations have often induced innovations that lead to either generally enhanced product quality or a reduction in production or disposal costs. Therefore, environmental regulations may actually reduce costs to society and to the firm if innovations are appropriately supported.

On the other side, actual experience with market-based instruments in industrialized countries indicates that so far they have not induced a substantial reduction of GHG emissions. In part, this is because for complex decision contexts like radical innovations, the actual working mechanisms of price incentives are still poorly understood. In these cases, environmental taxes might easily have no—or even adverse—effects (Cebon 1995). Furthermore, on a practical level, political resistance to new taxes has not been qualitatively different to resistance to traditional regulations. Not only does that mean that they are hard to implement at the national level, it also gives rise to doubts about their general appropriateness as a sound strategy. Many economic experts state, for instance, that taxes will be efficient only if they are introduced at least OECD-wide. Yet prospects for this to happen are judged as rather low (OECD 1994). The combination of resistance at home plus the necessity of collective action for the strategy to be effective has induced most states to drop CO_2 taxes or energy taxes from the policy agenda.[11]

The above criticisms identify shortcomings of economic models and market-based instruments that are principally due to the analysts' incomplete treatment of social and technical transformation processes. Model results depend heavily on the way these processes are represented (or not represented) in the model. The effectiveness or ineffectiveness of taxes and permits depends on their influence on technological development and preference dynamics. Climate policy, in particular, may be only partly effective if it is informed exclusively by conventional economic theory. In the following, we therefore expound how a nonmechanistic theoretical approach, which incorporates technological change and preference dynamics, could be of advantage for issues of climate policy.

7.1.4 Elements of an Innovation-Oriented Theoretical Approach

Before we sketch elements of an innovation-oriented climate policy, we show how the dynamics of technologies and preferences can be analyzed according to the criticisms of the traditional approach discussed above. On the basis of these arguments, we define the emergence and dynamics of technologies and preferences as "processes of social learning." Processes of social learning are to be understood in contrast to the reactive optimizing behavior that, in conventional economic theory, underlies the decisions of producers and consumers. A focus on processes of social learning affects the way societal options for action are framed. We conclude that the metaphor of a single optimum may be quite misleading as an orientation mark for an innovation-oriented policy. Instead, a conception of multiple equilibria is better suited for judging consumption and production alternatives that may have a lasting impact on GHG emissions.

7.1.4.1 Economies of Scale A technical but nonetheless important insight that has major implications for developing an innovation-oriented theoretical approach refers to "economies of scale" in the production and consumption of goods. Economies of scale are said to exist if average costs of production fall with the expansion of production quantities.

Although economies of scale were a central concern in the works of Adam Smith, most of his successors have excluded them from theoretical analysis.[12] This may be in large part because with economies of scale, consumers' and producers' production and consumption plans may give rise to nonunique equilibria (Varian 1992). Theoretically, nonunique equilibria are difficult to deal with because market processes are then not a sufficient determinant for achieving a socially optimal solution. As a consequence, it took a long time for economics to consider economies of scale a legitimate topic of theoretical analysis rather than an exotic anomaly (Arthur 1994; Krugman 1991).

As an empirical reality, economies of scale are, however, quite pervasive in industrial societies, especially through mass production, consumption, and diffusion of industrial commodities. The best-known examples are automobiles, refrigerators, and television sets. For these, economies of scale came from several sources. Besides innovations in manufacturing techniques like the production chain, applying new managerial concepts (Taylorism) as well as redefining the state's role in supporting economic development generated substantial productivity gains (Piore and Sabel 1984; Aglietta 1976). Although neglecting economies of scale may have been appropriate for analyzing rural societies, their existence can no longer be ignored in industrialized societies (Mokyr 1990).

At the level of the firm, unused capacity that can be used cheaply creates economies of scale. More relevant for our purposes, however, economies of scale at the level of the market may result from an expansion of the market that makes innovation in products and processes cost-effective. These innovations reduce product cost, lead to product differentiation in the market, or increase utility in some dimension, thus making a product attractive for a larger portion of the population. Economies of scale may, for instance, result from a reorganization of labor by specialization of specific tasks. They may also be associated with network externalities due to large infrastructure investments, technological standards and regulations (Hughes 1987; Nelson and Winter 1982). Finally, economies of scale may result when institutional factors give rise to de facto standards for specific technologies, as in the case of personal computers, video recorders, and the like (Powell 1991).

Preference formation through learning by using may lock in certain specific products and technologies on the consumer side as well (Teubal 1979). One standard examples is the history of the QWERTY keyboard. This layout of the keys has become the standard for virtually every typewriter and computer keyboard, even if other designs could be more efficient and comfortable (David 1985; Knie 1992). Users would have to invest a considerable amount of time in learning to use a new layout. Switching to a different layout would therefore annul the investment in the capabilities of myriads of people who have been trained on the QWERTY keyboard.

Economies of scale may lead to rigidities that hinder the development or adoption of radically different technologies even if they would offer a net benefit. In other words, economies of scale can fix a sociotechnical system on a specific development trajectory (Rosenberg 1994) even if many other equally feasible trajectories would be possible at largely comparable or even inferior overall costs. This assessment has consequences for the way sociotechnical equilibria should be understood as well as for the kind of policies that should be followed. We return to these points later.

7.1.4.2 The Social Construction of Technologies and Preferences If technologies are created and preferences are formed exogenous to the market, then technological and social factors can be analyzed separately. However, as many studies in the history and sociology of science and technology have demonstrated, an unambiguous separation of the two spheres is rarely feasible (Collins 1992; Granovetter 1991; Bijker, Hughes, and Pinch 1987; Callon 1987). Various social and cultural influences affect the way people perceive and value the relevant characteristics of technologies and consumer products.

Social and cultural factors affect decisions about what technological projects should be pursued or abandoned. Furthermore, once a project is initiated, both the product's conception and the design practices used are socially and culturally mediated. Some authors have called these design practices "technological paradigms" (Dosi 1988). Such paradigms may orient corporate decisions over long time spans of technological development (Knie 1994).

Consider, for example, the emergence of the gasoline-powered automobile. Early in this century, steam, electric, and gasoline vehicles coexisted. None of these alternatives provided unambiguously superior transportation. As we know, the gasoline car eventually outpaced its rivals, largely because a technological paradigm formed among the constructors of the early gasoline engine (Canzler and Knie 1994). No similar paradigm emerged for either electric or steam vehicles.

As long as a technological paradigm prevails, fierce competition among existing firms leads to incremental improvement of the technologies and products involved (Tushman and Anderson 1986). However, the same technological paradigm may also blind potential innovators to alternatives that would be available even at a net benefit. Empirical analyses of innovation histories have shown that radical technical changes often come from outsiders who question long-established rules of product construction and conception (Utterback 1994; Tushman and Anderson 1986).

Preferences are subjected to the influence of social and cultural factors in an even more obvious way than technologies. In fact, the formation of preferences can hardly be understood in terms of the exclusively individualistic standpoint, the assumption from which conventional economic theory starts. Here, consumers are hypothesized as having perfect knowledge about the consumption alternatives and having preferences either for commodities directly or for some of their material characteristics (Lancaster 1991). Consumer demand is then formed through the aggregation of myriad individual optimization operations that match product characteristics of available commodities with the individuals' preferences.

Marketing research maintains the concept of a largely individualistic consumer choice, at least in its early studies, but drops the idea of a perfectly informed individual. It also drops the optimization paradigm, assuming instead that consumers are uncertain about the optimality of their choices most of the time. As a consequence, they can be influenced to change their decisions (Frenzen, Hirsch, and Zerillo 1994). However, despite researchers' and practitioners' ongoing efforts, marketing has not yet been able to define a global disciplinary theoretical framework (Frenzen, Hirsch, and Zerillo 1994, 413).

Recently, interest in consumption has increased in the social sciences, as may be observed in many disciplines, from economics and psychology to anthropology and geography (Miller 1995). In respective studies, consumption no longer is seen as the mere opposite of production but is understood as a central forum of societal coordination in modern societies. Such a reconceptualization requires a shift in focus from the individual consumer to the consumer in a social context.

These developments have had major impacts on marketing as a social science. More socialized and integrated models of the consumer have challenged the individualistic approach (Belk 1995). The consumer is now seen as a social being, both expressing solidarity with particular social groups and conveying an image of himself/herself through consumption. He/she does so by projecting symbols, extracted from the cultural environment, back into that environment (Featherstone 1991; Bourdieu 1979). Thus, perception, interpretation, and valuation of product characteristics—and hence formation and dynamics of preferences—must not be attributed exclusively to subconscious psychological states but must be understood as socially constructed (McCracken 1988).

Therefore, emergence and development of technologies and preferences cannot be reduced to objective material characteristics of decision alternatives. The perception of characteristics of technologies and commodities, the identification of services that may arise as a benefit of using these technologies and commodities, and the value attributed to these services are all subject to social and cultural processes.

7.1.4.3 The Importance of Social Networks and the Regional Dimension
The above considerations are based on the assumption that the social construction of technologies and preferences does not happen in an abstract space of a "cultural fluidum" but rather within social contact networks. Often, although by no means always, these contact networks have a regional dimension. A closer look at the role of regional contact networks in the development of sociotechnical innovations may therefore be valuable. In general, regions often form the territorial space in which social contact networks are located. In particular, specific regions have remained competitive over quite long periods in economic history.

Conventional economic theory does not consider this theme very important, because it is based on an individualistic conception of economic actors, be they producers or consumers. Producers interact with each other and with consumers either through markets or through hierarchical intrafirm relationships (Williamson 1975). Consumers, on the other hand, interact with firms through markets—or more indirectly through their voting behavior—but do not interact with each other in any meaningful sense (Granovetter 1985).

This theoretical conception of individual actors finds a parallel in understanding recent historical tendencies as leading to ever more centralized and globalized decision structures (Harvey 1989). It follows from this view that technological development and economic competence are concentrated in large corporations with hierarchical decision structures. These corporations are often multinational firms that transcend the regulatory power of individual nation-states and are therefore an independent force in the shaping of global society. Along with this concentration of production goes a globalization of consumer goods and lifestyles. At the end of this process, the world will have become a single undifferentiated market. Mechanistic models of markets will then apply because regional diversity, perceived mainly as hindering the market's proper functioning, will have been eradicated.

Some evidence runs counter to the tendency toward an increase in firm sizes. Several studies have pointed to the importance of regional production networks, which are competitive on the world market and able to act very flexibly in volatile product markets (Scott 1988; Storper and Walker 1989; Piore and Sabel 1984). Large multinational corporations, on the other hand, have decentralized their decision structures and flattened their highly differentiated internal hierarchies, with significant decreases in size associated with a reorganization of the logic of production. The form of reorganization has varied according to the task and institutional context involved. Important models include "lean production" (Clark and Fujimoto 1991; Womack, Jones, and Roos 1990; Appelbaum and Batt 1994) and self-managed teams (Appelbaum and Batt 1994).

One of the major forces driving decentralization has been a change in the relative importance and acquisition cost of information needed to respond flexibly to rapidly changing circumstances. Because radical innovations often challenge taken-for-granted routines and world views, trust is needed to stabilize the social environment so that search and learning can occur. Regional networks may derive their strength from their ability to mobilize and stabilize high-trust relations in situations of considerable uncertainty (Smelser and Swedberg 1994; Hansen 1992; Lorenz 1992; Sabel 1989; Granovetter 1985). Large corporations, in contrast, often lack this kind of capacity to compensate for uncertainties and may therefore be at an inherent disadvantage in developing radical innovations, compared with more informally organized flexible networks.

Globalizing tendencies are not antithetical to regional production systems. Rather the contrary is true. Regional networks can often combine local sources of competence and know-how with the ability to operate in global markets. Some authors have called these particular regional contact networks "milieux innovateurs" (Maillat et al. 1995; Crevoisier 1991; Aydalot 1986) and have identified them in various parts of the world. Many of these milieus are situated in the Alpine region, such as the department of "Rhône-Alpes" in France, the Land of Baden-Württemberg in southern Germany, and the Jura region in Switzerland (Ernste and Jaeger 1989).[13]

Innovative networks may include users and consumers as well as producers. Von Hippel (1988) noted, for instance, that innovation often occurs outside the laboratory walls of the industry sector with the greatest experience regarding a particular product. Users of a technology often contribute significantly to the development of a new technology or product. To bring about a specific innovation, a whole network of rather diverse actors must be mobilized (Maillat et al. 1995; Lundvall 1992; Sørensen and Levold 1992; McKelvey 1991). These networks may even include citizens' movements engaged in the creation of new technological solutions (Irwin, George, and Vergragt 1994).

Like technological development, preference formation and dynamics often depend on processes within contact networks. The literature on the diffusion of innovations (Rogers 1983) emphasizes the lawlike way innovations spread through contact networks. Whereas early works on innovation diffusion assumed that innovations spread through networks by communication of their virtues, recent work has concentrated on another motor for diffusion. Competition among structurally equivalent people with ample information, that is, people with comparable positions in a social structure, may also be a relevant factor (Burt 1987). Given this, it seems important to focus, at an early stage in the diffusion process, on the cultural codes attached to innovative lifestyles (McCracken 1988).[14]

Networks of consumers may not only be relevant to consumers' consumption acts, but they often also strongly influence consumers' political votes, since here opinions are formed and political initiatives take shape. So networks have become increasingly important in the framing of policies, especially with regard to environmental problems (Elzinga and Jamison 1995). The rising importance of nongovernmental organizations (NGOs) in environmental policy is a key example (Commission on Global Governance 1995, 32).

Social contact networks are, therefore, a suitable unit for analyzing innovations in the fields of both technologies and preferences. As we have seen, contact networks may span the entire globe, and information and telecommunication technologies are clearly increasing a respective tendency. Such developments do not negate the regional dimension's importance, though. Specific regional settings may still play a central role in stabilizing trust relations and constructing interfaces between different kinds of actors. We hypothesize, therefore, that an innovation-oriented theoretical approach could gain a lot from a regional focus of analysis.

7.1.4.4 Institutional Impacts on the Flow of Resources and Legitimacy The innovative network does not exist in a social vacuum. Rather, it depends on a large number of local institutions that provide it with both resources (Pfeffer and Salancik 1978) and legitimacy (Powell and DiMaggio 1991; Scott 1992). Conventional economic theory implies that an action's legitimacy is not an important part of social life. Rather, resources can simply be acquired at the market rate. In reality, however, a person's position in society correlates strongly with his/her access to capital, skilled labor, government favors, or myriad other resources needed to bring about technical change. And whereas conventional economic theory does not necessarily assume a continuous supply of critical resources, economic theory as it is applied to policy problems generally does. That is, most analyses assume that a given resource is readily available at the market price, and that, should it be unavailable, a substitute can easily be procured.

We see in the case studies below that many successes and failures of innovative initiatives seem to depend on the availability of specific resources in the region that essentially lack satisfactory substitutes. For example (among many other aspects), the absence of a Swiss car industry probably enabled lightweight vehicle innovators to attract both good engineers and government resources; an excellent rail service provides an important long-distance complement to electric vehicles used in the city; and the economic success of the NEAT rail tunnel project will probably depend in part on Switzerland's ability to influence track and infrastructure construction in Italy and Germany.

7.1.4.5 Processes of Social Learning and Multiple Equilibria The above observations generate two general arguments for a nonmechanistic approach to environmental policy. First, on a micro-level, sociotechnical innovations should be analyzed as a sort of social learning. Second, on a macro-level, the dominant conception of a single equilibrium of quantities and prices should be enlarged to make multiple equilibria a guiding metaphor in analyzing options for and barriers to sociotechnical transformations.

Taken together, the concepts of economies of scale, the social construction of technologies and preferences, and social contact networks have two important implications. First, actors may have to reconceptualize their products and technologies from different points of view. They must be able to unlearn (and learn anew) propositions they have taken for granted about their product and its context. Decision options must be built up in the course of the decision procedure. Therefore, learning has to be distinguished from a mere reactive optimizing behavior in a taken-for-granted world in which the choice set is unambiguous and perfectly known to the decision maker.

Second, in the context of radical innovations, isolated learning by individual decision makers is insufficient. Atomistic learning is possible only if the key factors are under the decision maker's control (as market strategies are), or if they are beyond any decision maker's control (as is a commodity's competitive market price). If, however, some decision criteria depend on other actors' decisions (as with the definition of a product niche), learning becomes interdependent. Furthermore, in complex situations, any single decision maker's cognitive capacity may be exceeded. In such situations, contact networks and the institutional environment are indispensable factors in developing new solutions. Because of this social dependence of innovation processes, we describe this learning as a form of social learning.

By taking processes of social learning as a focus of analysis, we are able to avoid the pitfalls of a technocratic bottom-up approach. Technological opportunities are a necessary but insufficient condition for environmentally relevant sociotechnical innovations to occur. Only a careful understanding of the social processes forming around technological opportunities helps identify the potentials and barriers to transformation, which are invisible when only technical parameters are discussed. The regional dimension could prove a crucial direction of analysis for actors that want to develop environmentally benign technologies and lifestyles.

If we accept processes of social learning as a fruitful focus of analysis, we cannot assume that markets have unique equilibria. Economies of scale and the complexity of economic systems may give rise to a multitude of stable demand and supply configurations, or multiple equilibria (Jaeger and Kasemir 1996; Kreps 1990)[15]. These individual equilibria may well have similar patterns of relative prices. Compared on other dimensions, however, such as energy consumption and rates of unemployment, they may differ substantially.

The central metaphor of a single optimal equilibrium that informs much economics-based policy advice should therefore be questioned. Many scholars, among them several prominent economists, have expressed their hopes that with a correction of the mechanistic framing of neoclassical economics, technological development could be discussed and supported more appropriately (Sen 1995; Norgaard 1994; Romer 1994; Krugman 1991). However, so far no consistent analytical frame has been proposed that might replace neoclassical economics in all respects. Anyway, much more urgent than achieving this general aim is the task of developing theoretical concepts in problem areas where neoclassical economics has a blind spot, particularly innovation processes where technologies and preferences develop as endogenous to the market. Because the potential of sociotechnical innovations is a decisive factor for climate policy, we see a major need for research in this area.

7.1.5 Implications for Climate Policy

The assumed central importance of processes of social learning and multiple equilibria has major implications for both the construction of assessment models and the development of policy instruments. Here leave aside assessment models[16] and concentrate instead on climate policy that aims to reduce GHG emissions substantially. We do this in three steps. First, we spell out some implications of complex dynamics in the field of climate policy. Then we elaborate on elements of an innovation-oriented policy by focusing on the policymaker's role and identifying some guidelines for supporting innovation processes. Finally we discuss innovation-oriented climate policy in the general context of both technology policy and climate policy.

Under which conditions does an innovation-oriented climate policy make sense? The above analysis suggests that conventional measures such as market-based instruments and conventional regulations may be appropriate if policy is intended neither to encourage development of new technologies nor to change preferences.[17] If, however, one desires to transform technologies and lifestyles radically, as climate policy does, an innovation-oriented approach is called for. And this means, as has already been said, a focus on mesoscale processes linking social learning with multiple equilibria.

Processes of social learning and multiple equilibria influence the interaction of policy and technological development. If, in particular, change in technologies and preferences is seen as a market-endogenous process, this leads to fundamental uncertainties about future states of the sociotechnical system. This may make it impossible to decide a priori whether a specific policy instrument is superior to another with regard to achieving any given goal (Metcalfe 1995). Theoretical analyses therefore must inform the policy process at the level of rules for robust action (Eccles and Nohria 1992). This means that climate policy should aim at procedural rules on the basis of which a specific short-term goal may be achieved without unduly narrowing long-term options for action (Dowlatabadi 1995). Furthermore, complex systems' development trajectories tend to depend on variations of specific local parameters, so policies will probably be highly context specific. Therefore, criteria for a successful policy strategy can only be defined very broadly. Finally, any fundamental uncertainty about the relationship between means and ends also influences the way policy goals are fixed. Goals may well be transformed through experiences gained in the process of developing solutions (Weick 1995).

The innovation-oriented policymaker has basically two roles. First, he or she manipulates the degrees of freedom for and the mechanics of sociotechnical change. The move from policy built around optimal choices toward policy built around social learning modifies the policymaker's role from structuring choices to feeding the process of social learning with specific resources (Metcalfe 1995). As a consequence, efficiency is no longer policy instruments' crucial criterion but rather their ability to engender and support variability and flexibility in the innovation process (Wallace 1995).

Second, the innovation-oriented policymaker helps frame the political process, where means and ends are negotiated. Policymakers' ability to cooperate with myriads of societal actors decides their success or failure. Cooperation depends crucially on trust among the involved parties, which in turn depends on a continuous and reliable policy strategy. Such a strategy should be framed cooperatively. Cooperative framing is important because it leads stakeholders to construct and explore specific windows of opportunity rather than struggle over narrowly defined positions. However, this does not imply that policymakers should kowtow to the specific interests of industry or any other involved party. The success of a specific innovation process depends crucially on clear identities and power bases of the negotiation partners. Preventive climate policy—and the search for sustainable development—could offer a frame for such clarification (Wallace 1995).

But how can a specific innovation-oriented climate policy be conceived? We propose four general rules from our theoretical analysis. First, our discussion of social construction of technologies and preferences suggests that for identifying and valuing potentially innovative alternatives, technical and social determinants have to be considered on an equal footing. One must ask, for instance, whether social processes could substantially increase a specific technology's potential. Alternatively, the specific new product's availability could influence the pace and direction of preference change and therefore effect social change. These relationships have to be analyzed carefully at each stage of the innovation process, be it the development of a technical artifact, its large-scale production, or its market diffusion.

Second, our discussion of networks suggests that the actor setting in which processes of social learning could develop must be initiated and supported reflexively. Radical innovations often depend on the specific knowledge of a wide range of actors (Bijker 1993). The exchange between industry, policymakers, consumers, NGOs, and firms traditionally remote from the specific product sector may breed the development of new solutions (Truffer and Dürrenberger 1997; Irwin, George, and Vergragt 1994; Schot 1992). Such actor networks may crucially depend on personal contacts and shared cultural backgrounds, which form the basis for forming trust. As has been argued above, such contact networks often do exist at the regional level. A regional approach to climate policy may therefore prove most profitable. Policymakers should try to identify or facilitate such networks and support their innovative capacity. Regionally based experiments with innovations may hence be an appropriate way of establishing and stabilizing new concepts and forms of cooperation. Furthermore, cooperation and trust may grow only if voluntary agreements rather than strict command and control regulations form the default solution for resolving specific problems.

Third, our discussion of economies of scale suggests that attempts at further sociotechnical innovations should consciously try to manage the problem of a premature lock-in of specific technologies. Economies of scale are the very basis of mass-produced technologies and play a crucial role in the pace of diffusion of new technologies and products. However, policy should try to prevent them from becoming a problem for future innovative activities. It is important, where feasible, to break up the forces stabilizing inefficient technologies already locked in, and thus facilitate the transition from one equilibrium to another. For this aim, policy instruments should support such a transition but should dissolve once the task is accomplished (Erdmann 1993).[18] To manage the risks of prematurely excluding promising alternatives at an early stage, niches for new products and technologies should be consciously managed. This is the aim of a policy approach known as "strategic niche management" (Schot, Hoogma, and Elzen 1994) that might form a promising basis for an innovation-oriented policy approach.

Finally, our discussion of resource dependence reminds us that a given regional network may lack the capacity to produce or purchase all the resources needed for any given innovation. Therefore, without assistance, its history will limit it to a subset of the possible innovations. When a desired trajectory emerges through regional processes, a key role for government is to ensure, through providing resources such as educational opportunities, that those trajectories can be pursued. Our discussion of legitimacy similarly highlights the fact that although key resources may be available within the region, they may be unavailable to members of regional networks with low legitimacy. To give a glib example, banks with well-established industrial linkages may not lend money to people they perceive as ecofreaks driving souped-up sewing machines. In fact, initiators of radical innovations often have very low legitimacy (Rogers 1983). Once again, government may need to intervene actively to foster the required legitimacy as particular technologies or technological trajectories emerge.

This list of rules is certainly not exhaustive and has to be reinterpreted in any specific policy context. In this respect, much can be learned from cases where environmental innovations have been supported successfully by all sorts of policy instruments as well as by notable failures. We elaborate on this in the remaining two sections, on the basis of two extensive case studies. (Clearly, a large number of other recent examples could be regarded.)

One particular aspect of a successful political intervention should be mentioned beforehand. Voluntary agreements have great potential for supporting environmental innovations, because they may underlie both relationships of mutual trust and the formation of networks. An initiative uniting the big-three U.S. automobile manufacturers and the federal government for the environmental improvement of the conventional gasoline car, for instance, started in September 1993. It aims to develop vehicles with three times the energy efficiency of today's cars while providing comparable comfort and performance.[19] Recent proposals from Swiss industry to base energy policy in Switzerland on voluntary agreements of the involved parties—coordinated by a private energy agency—may prove similarly interesting. A decisive precondition for these initiatives' success, however, will be their ability to achieve public credibility with regard to their endeavors, in other words, to present them as something other than mere lobbying tactics.

To achieve this credibility, it will be important to include other social actors in the decision processes, such as NGOs, consumers, and representatives from other industries. These actors' specific roles will have to be carefully defined in the innovation process, and traditional identities will have to be transformed. One success story based on the participation of a nonindustrial social actor is that of the CFC-free refrigerator "greenfreeze." This refrigerator was promoted from prototype to large-scale production through the efforts of Greenpeace Germany (see Irwin, George, and Vergragt 1994).

Although this initiative may not serve as a paradigmatic example to be copied step by step into other problem fields, it represents a promising vision of how diverse actors could work together to develop environmentally benign products and technologies.

To sum up, an innovation-oriented policy approach must attach importance to the above-mentioned fundamental dimensions if it wants to contribute to a more effective preventive climate policy. But are there any incentives for a particular nation to engage in such processes of social learning, which go beyond compliance with internationally agreed environmental goals? Or are innovation-oriented policies rather stopped immediately if ever they interfere with other political goals that affect technological development? Although these questions clearly have no general answer, two aspects should be noted (Porter 1990). First, a newly developed technology likely yields an innovation rent. This "first mover advantage" could pay for the additional risks the innovating actor must bear. Second, a general innovative atmosphere may be profitable for many product sectors and technologies and improve a nation's competitive situation. Both factors should be investigated more carefully.

Finally, it has to be acknowledged that all arguments about increasing technical efficiency expounded so far are an answer to at best one-half the problem of global climate change. The other half requires solutions about how rights to act should be distributed among present societies and future generations. Sociotechnical innovations affect this aspect of the problem insofar as industrialized countries actually export their technologies and lifestyles on a global scale. They thus have a specific responsibility for the impacts associated with their exports. An additional responsibility befits those regions more able than others to mobilize the necessary capital and resources to develop a sustainable transformation of technologies and lifestyles (Homer-Dixon 1995). These innovative regions should contribute to developing alternatives that will be available on a worldwide scale and for future generations. We therefore discuss in a final subsection why the Alpine region may be particularly interesting for studying processes of social learning relevant to an innovation-oriented climate policy.

7.1.6 Innovation, Climate, and the Alpine Region

Based on the theoretical arguments above, we have concentrated on a regional approach and have chosen the alpine region as a background for analysis. But the Alpine region's contribution to global GHG emissions is essentially insignificant. Switzerland, for instance, is not typical of other countries in central Europe. It contributes by 0.2 percent to worldwide CO_2 emissions. Furthermore, the structure of its GHG emissions is not representative of most other countries in the world. Swiss GHG emissions divide as follows: CO_2 is the most significant GHG, responsible for about 80 percent of the Swiss global warming potential (OFEFP 1994). N_2O is second, with a

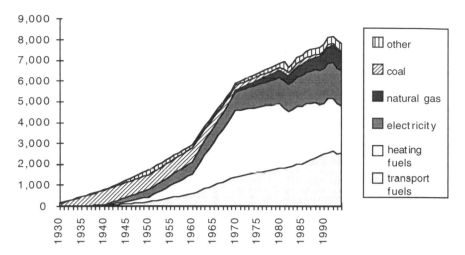

Figure 7.1 Commercial energy consumption in Switzerland (in PJ). (Source: Schweizerische Gesamtenergiestatistik 1996.)

global warming potential of about 15 percent. Agriculture accounts for 93 percent of the N_2O. Methane (6 percent) and other GHGs play only a minor role.

The causes of CO_2 emissions in the Alpine region, again, have a very particular profile. Electricity generation, for example, accounts for only a minor portion of emissions, since the mountains are a major water reservoir and are intensively exploited for electricity generation. In Switzerland, for instance, some 60 percent of electricity is generated with hydropower, another 38 percent with nuclear power (Schweizerische Gesamtenergiestatistik 1996). Burning fossil fuels for heating and transport makes a major contribution to CO_2 emissions (see figure 7.1). Heating has shown a steady decline over the last twenty years, and there is considerable potential for a further reduction in fuel consumption (Rothen 1995). With regard to transportation, the situation looks much less promising. Energy consumption attributable to transportation has increased steadily. Official scenarios predict a growth of about 32 percent from 1990 to 2030 (OFEFP 1994, 83). Transportation is hence one sector where reduction potentials will have to be exploited most urgently. So, innovation potentials associated with technologies and preferences in the field of transportation deserve particular attention.

Although the alpine region is not particularly interesting or representative with respect to its GHG emissions, it nevertheless proves highly interesting for the analysis of sociotechnical innovation processes. First, the larger territory around the Alps is one of the most wealthy and innovative regions on earth. Switzerland and Austria, the region of Rhône-Alpes in France, the German Bundesland Baden-Württemberg, and northern Italy as well are the home base of a large number of multinational firms. Besides, these regions host a range of dense networks of highly innovative small and medium-sized enterprises that hold strong positions as exporters of globally marketed technologies, goods, and services. Furthermore, the wider Alpine region offers great cultural diversity on small spatial scales accompanied by a high degree of political stability.

The Alpine region may, moreover, be well suited for developing environmentally benign technologies, especially with regard to global climate change. Environmental concern has, in general, a strong position on the region's political agenda, and willingness to act to solve environmental problems is very strong (Dunlap, Gallup, and Gallup 1993). The Alpine mountain valleys play a special role in the memories and myths of large parts of the population in the wider Alpine region. Mountains, ecosystems, and people in the Alps are therefore highly valued. Strong political majorities support measures for their protection.[20] Global climate change heavily threatens the mountain ecosystems (see chapter 5). And this is perceived all the more strongly because the Alps are an important recreational area for large parts of the urban population.

Therefore, developing new technologies, policies and lifestyles relevant for reducing GHG emissions on a national, and presumably, global scale will probably meet with a strong commitment in these regions. In addition, the necessary cultural, economic, and political resources may be available to carry out experiments and to engage in risky sociotechnical innovation processes. This is not to be misunderstood as claiming that the Alpine regions have any inherent superiority with respect to solving the problem of global climate change.

Against the background of the described emission situation, we have chosen to study two cases concerning transportation in Switzerland. We selected these cases because they may reduce GHG emissions considerably. In both cases, the perceived risks of climate change for the alpine ecosystems have significantly motivated innovation. Yet they are not the only motivating factor. A lot of other environmental and societal problems of more local and immediate character have contributed to innovative impulses. A purely climate-focused innovation policy is hardly realistic, anyway. It is much more effective to draw on all sorts of environmental concern to reduce GHG emissions radically.

The first case deals with how to bring the transalpine flow of freight goods from road to rail. The official approach is the NEAT—the project to build two baseline tunnels of about fifty kilometers each crossing the Alps north to south. It exemplifies the enormous costs that may be incurred when policy is limited to the role of providing infrastructure, taxing, and regulating. We argue transalpine freight transport will most likely not shift from road to rail unless the concerned actors are involved more explicitly and unless the buildup of new institutional interfaces is addressed explicitly. Against the background of the identified barriers to innovation, we discuss some opportunities for innovation based on the formation of networks and the social embedding of the infrastructure project.

The second case study discusses the options and barriers associated with a radical reconception of individual mobility. In Switzerland, extensive experience has been gathered with so-called lightweight vehicles. We first show how cultural and social factors have motivated and influenced the development of radically new technological concepts. Citizens' movements have played a decisive role in the unfolding of the early innovation process. Second, we argue that large-scale production of lightweight vehicles would go along with a number of fundamental uncertainties with regard to manufacturing techniques and market dynamics. Third, we postulate that market dynamics for these vehicles will depend strongly on product identities that are socially constructed. Such product identities could be a central means for creating and communicating sustainable lifestyles. We then propose policies that aim to support relevant processes of social learning in the three phases of an innovation process described.

7.2   THE PROBLEM OF TRANSALPINE FREIGHT TRAFFIC

7.2.1   Climate Change and the Transformation of Transalpine Freight Traffic

Both the ongoing economic globalization process and our everyday life depend highly on efficient freight transport. However, the unprecedented growth of freight, especially of freight transported by road, is having tremendous environmental impacts. In 1988, transport accounted for 28 percent of global CO_2 emissions. Within the transportation sector of most western countries, more than 80 percent of transportation emissions come from road transport, one third of which stems from road freight traffic (Gwilliam 1993). And the projected growth of freight transportation will make this activity an ever more important topic for climate policy (Shipper et al. 1992). Increasingly, the truck is seen as the "environmental villain" and is at the center of public attention (Cooper 1991).

However, transformation of freight transport is confronted with manifold barriers resulting from socioeconomic and political development processes. Demand for the transport of raw materials and intermediary products depends heavily on the prevailing spatial organization of production. In recent years, new organizational principles in manufacturing, relying on the reduction of intermediate stocks and the just-in-time delivery of components, have brought about a massive expansion of freight transport demand (O'Laughlin, Cooper, and Cabocal 1993; Ruijgrok and Wandel 1992). The change in the structure of transported goods from heavy, bulky materials to smaller, more expensive freight goods has clearly favored road transport, as have investments in road infrastructures. The road now provides the dominant transport paradigm, locking out railway technology. Preferences for road transport have therefore been shaped mainly by socioeconomic and political factors and not by a technological development in the haul industry alone.

One way to reduce truck usage would be to organize it more efficiently. An estimated 40 percent of truck transports run empty. A second alternative would be to reduce or restrain commodity flows. This is not politically viable in the short run, because industry would perceive it as a massive threat to its further economic growth and prosperity. Therefore, analysts have focused on transforming transport services technologically. One of the most promising and innovative alternatives would be to shift freight from road to rail by fostering combined transport, thereby reducing GHG emissions and increasing energy efficiency (IEA 1993).

Can rail maintain or even increase its market share of freight traffic? If so, by which means would a different modal split be achieved? In Switzerland, the answer to these two questions is presumed to be the provision of further railway infrastructure. Thus, the New Transalpine Railway project (NEAT), meaning two new base tunnels crossing the Alps, was proposed and adopted by a referendum. The assumption that it would be unproblematic to move freight to the rails has moreover engendered a political movement promoting the "Initiative for the Protection of the Alps." The Swiss people voted recently in favor of this initiative, which aims to ban transalpine freight traffic from the roads. Because such a ban would create severe obstacles for future negotiations between the EU and Switzerland, the general ban has recently been replaced by new price incentives to move traffic from road to rail.

However, as yet these policy measures have failed to engender any radical innovation in the freight sector. In the absence of any attempt to redefine the product radically, the policies may fail to achieve both the financial and the environmental goals for which they were conceived. In particular, the current approach ignores present market conditions and underestimates the embeddedness of the haul industry in a historically grown production and distribution system. Nevertheless, overcoming these barriers could result in substantial greenhouse gas emission reductions, but this would require a process of social learning among people involved in the contact networks related to transportation. Barriers against any change in the organization of transportation will, however, not become lower unless corresponding uncertainties for the economic actors in the transalpine freight market can be reduced.

We expound this argument in four parts. First, we describe the relationship between regional geography and European north-south freight flows, illustrating why a tremendous increase in freight traffic across Switzerland could be expected if existing regulations were immediately withdrawn. Second, we show that in the present market major—but not insuperable— barriers exist against shifting freight transport from road to rail. Third, we present some empirical data about the potential for social learning in the hauling industry, a likely leader with regard to maintaining or abandoning the dominance of road-based freight transport. Fourth, we draw conclusions about the GHG reduction potential of an innovation-oriented freight transport policy.

7.2.2   The Swiss Framing of the Transalpine Freight Problem

7.2.2.1   Growth Pattern of Freight Transport across the Alps   The flow of transalpine freight to and from Italy increased sixfold from 1965 to 1992, reaching an annual total of 65 million tons. New highways and highway tunnels in France and Austria along the traditional transalpine routes particularly facilitated this increase (Ratti and Rudel 1993). The share of this freight traveling through Switzerland has constantly diminished in the same period. Now Austria, Switzerland, and France carry roughly 25, 24, and 16 million tons, respectively. But the three countries show fundamentally different patterns of splitting the traffic between road and rail. Figure 7.2 shows the tremendous growth in road freight traffic across Austria and France. In contrast, an almost negligible amount of international road freight crosses the Swiss Alps. Railway freight transport, which still accounts for about 50 percent of transport volume, has increased modestly compared to road transport, as figure 7.3 shows. This is partially due to structural changes and

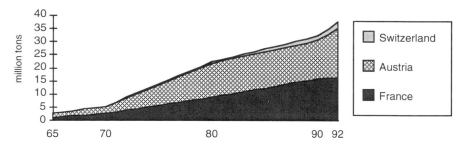

Figure 7.2   Evolution of road freight traffic across the Alps. (Source: Data collected from national statistical sources and published by LITRA (1993).)

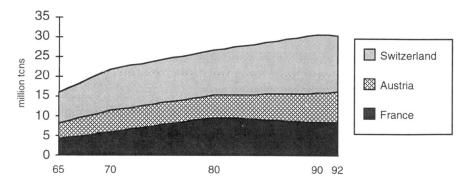

Figure 7.3   Evolution of railway freight traffic across the Alps (Source: Data collected from national statistical sources and published by LITRA (1993).)

deep crises in heavy industry, which relies exclusively on the railway. However, transalpine piggyback transport and container transport to Rotterdam, which became Italy's most important intercontinental sea port following difficulties with trade unions in various Mediterranean ports, partly compensated for these losses.

The striking differences in the modal split among Switzerland, France, and Austria can be explained mainly by Switzerland's policy to limit the weight of trucks to 28 tones, as opposed to the general 40-ton limit in Europe, and to ban night, holiday, and weekend freight traffic. This regulatory regime was designed to protect the Swiss railway company and Swiss transport industry from competition, and to preserve the environmental amenity of the alpine landscape. The policy pushed road traffic to Austria and France, often leading to longer trips and hence more environmental pollution for the alpine region as a whole.

Yet Switzerland was unwilling to abandon these restrictions, which compound the natural obstacle the Alps present with an institutional one. With the increase in freight transport in the last decade, European integration, and the deregulation and liberalization of European transport markets, industry began to lobby neighboring European governments to have this institutional obstacle removed. In particular, Switzerland's transportation policy increases transport costs considerably, because the Swiss transalpine route is the shortest one. Therefore, the European governments pressured Switzerland to open a "transit corridor" that would allow unrestricted international north-south freight traffic for the larger 40-ton trucks. This would even have helped to reduce environmental impact, reducing the number of circulating trucks. Meanwhile, the European Union decided to foster combined freight traffic as well (CCE 1993): The new railway connection linking Rotterdam and the Ruhr region exemplifies this new priority.

However, the EU and especially the neighboring countries maintained their pressure on Switzerland, forcing it to react to this external challenge. As noted above, the Swiss conceive open road freight transport along the shortest transalpine route as both an invasion and an ecological disaster. Various forecasts that predicted a doubling of the transalpine freight traffic over the next twenty years (Vittadini 1992; Graf 1995) enhanced this public perception. Therefore, the European request was met with a popular outburst. Swiss political authorities realized that it would be politically untenable to loosen the transport restrictions. Notwithstanding, Switzerland had to offer the EU an alternative.

In the transit negotiations, Switzerland, the EU, and the neighboring countries found a mutually satisfactory solution. In return for continuing the restrictive transport policy, Switzerland promised to improve its existing railway tracks in the short term and build the New Transalpine Railway infrastructure (NEAT) in the long term. The existing railway tracks across the Alps have already been improved. The Swiss railways can now satisfy the existing and expected short-term demand for combined freight traffic and piggybacking. The new railway infrastructure will create capacity for up to four times the current freight volume across Switzerland. The NEAT, or an equivalent project in Austria or France, is considered absolutely necessary to absorb the long-term demand in transalpine freight transport.

7.2.2.2 New Policy Initiatives to Regulate Transalpine Freight Traffic The NEAT aims at solving the transalpine transport problem by simply constructing new infrastructure. This is a conventional, technocratic approach based on mechanistic assumptions about preferences for transport technology. In particular, it ignores the causes of the radical changes in the transalpine freight market and the increase in road transport. The assumptions underlying the proposal are the same as they were in the 1970s, when a new base tunnel was first suggested. At that time, the Swiss railway dominated the transalpine freight traffic market (Bertschi 1985) and the alternative road infrastructure was very poor. The project, oriented toward providing additional infrastructure, could have easily attracted more freight traffic. But although the project was adapted several times for different political reasons, it was never actually implemented. Only now is there sufficient international and national pressure to force Switzerland to construct new railway infrastructure to prevent trucks from overrunning the Alps.

The project consists principally of two base tunnels across the Alps, the eastern Gotthard line and the western Lötschberg line (figure 7.4). The key element of the Gotthard line would be a bitube base tunnel of more than 54 kilometers, the longest tunnel ever built in Europe. The original project included provisions to increase the capacity of the access tracks to the tunnels. These provisions, however, extended only to the foothills on both sides of the Alps. In particular, they did not extend to the borders. Nor was capacity beyond the borders taken into account. Hence, the rail structure will face an interface problem in the near future—a major contradiction in the whole project. Although the Swiss national parliament pleaded for the construction of two new base tunnels, necessary to absorb future freight traffic, the access line tracks would be unable to carry the load to take advantage of the new base tunnels' capacities. This would lead mainly to an underutilization of the new rail capacities or allow a shift of trucks from road to rail only in the proximity of the base tunnels, that is, a distance of only about 100 kilometers, from one side of the Alps to the other. This, however, not only contradicts an economically reasonable conception of combined railroad freight transport but also implies a considerable increase in truck traffic in the alpine valleys that provide access to the tunnels, with the corresponding environmental impact.

But the construction of these tunnels also presents several tricky problems for engineers and geologists. The western Lötschberg line has a shorter tunnel and fewer technical problems, but the Gotthard line offers the greater market potential, because it is situated at the shortest north-south connection. However, neither tunnel can carry significantly more international

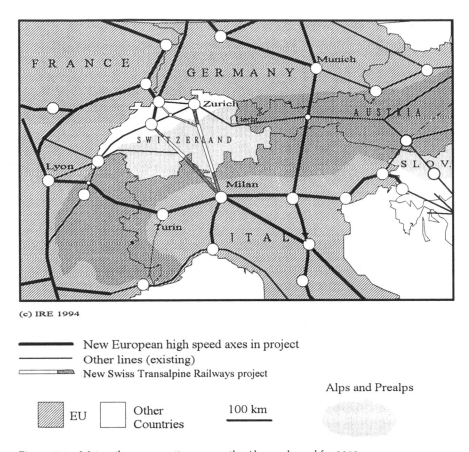

(c) IRE 1994

▬▬▬▬▬ New European high speed axes in project
───────── Other lines (existing)
▭▬▬▭ New Swiss Transalpine Railways project

Alps and Prealps

▨ EU   ▢ Other
           Countries

100 km

Figure 7.4 Main railway connections across the Alps as planned for 2010.

freight traffic without a link to extensive new infrastructures in Italy. The initial proposal called for both tunnels to be constructed simultaneously.

Concerns mainly about increasing freight transport across the Alps drove the 1992 referendum in favor of the new tunnels. This favorable vote was, however, predicated on the construction of two base tunnels, which means new public infrastructure projects in different parts of Switzerland as well as a good and easy access to either of the new baselines. The choice of two baselines must clearly be understood as a political ploy to gain the support of the majority of the Swiss population. Neither the western (French) nor the eastern (German) parts of Switzerland would have supported only one tunnel in the other half of the country. To offer an ecologically sound alternative to road freight transport, at least one tunnel must reduce travel time and operational costs significantly over the present situation. Two tunnels, however, can carry all the freight currently crossing the mountains by road or rail through France, Switzerland, or Austria (64 million tons per year).

A small group of ecologists and the Swiss left-wing political party started a second initiative. They were particularly concerned about the projected increase in road freight traffic across the Alps. They wanted to ban transit freight traffic across Switzerland totally from the roads to prevent an environmentally disastrous "traffic collapse in the Alps." This rather radical initiative focused on transport that neither started nor ended in Switzerland. With haul distances longer than 300 kilometers, this market segment would likely be the first to shift from road to rail. Without any doubt, this initiative also would have improved the profitability of the planned NEAT through its unprecedented interference in the transport market in the name of protecting alpine ecosystems.

A slim majority approved the popular initiative banning transit trucks in February 1994. This voting result reflects the potential for conflict in the transport area and illustrates the degree to which freight transport has become a topic in the public environmental debate and the symbol of an ecologically harmful activity. It is nevertheless most striking that the whole of public attention could be focused on transalpine freight traffic, almost 75 percent of which goes by rail, representing only 25 percent of the total kilometer-tons transported in Switzerland. This attention can be explained by the high concentration of traffic in a few alpine valleys, which must carry the environmental impact. Not only is the alpine region a major leisure resort, but the mountains are also the subject of a common mythological memory and contribute to the national identity. Supporters of the ban believed this natural and symbolic landscape had to be protected from the dangers of a further increase in transalpine freight traffic and an unlimited invasion by the trucks. The public clearly perceived the trucks as particularly harmful and made them a scapegoat for the environmental problems in the Alps.

In some respects, the initiative was the final logical step in a policy approach designed to shift freight in a mechanistic way from road to rail. In other respects, however, it represents a qualitative change in policy. The ban makes international transport a taboo and in so doing violates the commonly held assumption that transport operators have the freedom and right to choose their transport means. Such a shift endangered Switzerland's position in the European political landscape and meant a significant interference in the international transport market.

The European Union reacted immediately. On the heels of the transit agreement, which acknowledged Switzerland's particular situation in the transalpine freight market and its need to protect the alpine region, the approval of the Initiative for the Protection of the Alps reopened the conflict between the economic interests seeking free access to Swiss roads and the interests wanting to protect the alpine ecosystems from heavy road traffic.

Switzerland now intends to resolve this new conflict with the European Union with a transit toll, which should gradually shift international and national heavy road traffic from road to rail. A transit toll is compatible with the EU s transportation and environmental policy and is acceptable to the leaders of the Initiative for the Protection of the Alps. Furthermore, it does not discriminate between international and national operators, which is consistent with the European environmental policy principle that the target of a significant modal shift should be achieved through market forces designed to internalize external costs (Button 1993; Isenmann 1993). In fact, the transit toll can be understood as an increase in road transport costs, which signals the possibility of a mode change and makes the railway more competitive.

7.2.2.3 Questions about the Economic Feasibility of the New Transalpine Railway Project

The vote for the NEAT also expressed Switzerland's significant financial commitment in favor of a more environmentally friendly transalpine freight transportation system. However, the vote was based on the assumption that freight traffic would shift to the railway. Therefore, the public authorities and the transport ministry presented the financial risks of Switzerland's most expensive infrastructure project ever as very small. All estimates assumed that both transport in general and demand for rail services in particular would increase to yield sufficient demand to ensure an immediate and sufficient return on public and private investment.

Such a prediction conflicts with the present abundance of available road transport, which recently increased again through an invasion of truckers from eastern Europe who have further driven down road freight transport prices. In this light, the whole investment looks less economically feasible. Because of the assumed high financial return, two important questions were never addressed. First, how will the present generation and future generations share the cost of the new infrastructure? Second, should the European Union, which will doubtless be a major user of this new railway infrastructure, contribute to its cost?

Furthermore, economic feasibility hinges on successfully integrating the project into the European high speed railway network and completing a key link between southern Switzerland and northern Italy (Nijkamp and Vleugel 1994). Both these aspects have been ignored, presumably to avoid difficult problems of international cooperation in railway transport policy. International cooperation requires overcoming old and deep institutional barriers within and between countries. The historical burden of nationally shaped transport policy clearly leads to limits and difficulties in adapting to the changing institutional, organizational, and economic context. The reluctance to solve this problem causes major uncertainties for transport operators, who have to decide where to locate other transport facilities and whether to shift to combined road-rail freight transport, making sequential use of different transport modes in the transport process.

In addition, two other base route railway tunnels are planned: one under the Mont Cenis in France and one under the Brenner in Austria. Both reflect the EU's transport and environmental objectives to shift long-distance freight traffic massively from road to rail, and both would considerably change the NEAT's competitiveness. They were adopted by the European Council in June 1994 and should be finished at about the same time as the NEAT. This will inevitably lead to large overcapacities and high competition in the transalpine transport market among the national railway companies involved. A squeeze on profit margins would ultimately endanger the NEAT's financial success.

Considering the traditional, infrastructural approach, it is hardly surprising that the new infrastructure's actual potential to attract freight traffic from the roads has hardly been considered and that the success of the NEAT investment has always been taken for granted, particularly after the decision to ban transit freight traffic from the roads. Questions arose only in recent debates about whether to build both tunnels simultaneously or serially. Even in that context, the problem has still been cast independent of the NEAT's potential to connect major economic regions north and south of the Alps or to absorb a large amount of transalpine freight traffic.

In the following sections, we argue that neither providing an infrastructure nor a restrictive transport policy will shift a significant volume of freight traffic from road to rail. We assess the market potential for such a shift by analyzing causes and conditions that led to road dominance as well as organizational and institutional advantages of road over rail. This allows us to identify some barriers against a shift of transportation to rail and to point out innovative opportunities in rail transportation that might help overcome these barriers.

7.2.3   Barriers and Options Concerning a Shift of Freight Traffic from Road to Rail

7.2.3.1   Barriers to Breaking the Dominance of Road Freight Transport   Providing new infrastructure and an economic instrument to shift freight traffic from road to rail, a policy approach based on conventional economic theory, fails to consider recent changes in the transport market. The railway's steady decline and the accompanying tremendous increase in road freight traffic must be understood in light of changes in economic production and distribution in the last twenty to twenty-five years. These changes, characterized by a disintegration of large firms, an intensification of interfirm relations with corresponding new logistical requirements, a major change in the nature of the transported goods, and an extended international division of labor, in large part caused the expansion of road transport. External factors such as abundance of road capacity and low fuel prices in turn favored the expansion.

A few key concepts particularly relevant to the transport industry (Capello and Gillespie 1993) can best summarize the structural changes. The general adoption of just-in-time production principles reduced the stocks previously available in a firm, requiring a continuous flow of goods to and from the firm rather than large, infrequent shipments. In this new production system, transport and the whole logistics became a strategic production factor. Recent institutional changes such as abolishing the national barriers and liberalizing in the EU further increased freight transport and the circulation of goods. Deregulation and abolishing restrictions imposed on internal transport in third countries ("cabotage") will further increase road transport capacity in the short term.

With these changes in production and distribution, freight transport has become strategically important to manufacturers, who now require a reliable, flexible, and punctual transport service. Railway can apparently not provide this. Trucks using the highly interconnected and densely knitted road infrastructure can instead. Trucks can also respond very quickly to rapid changes in transport demand because they do not depend on rigid, centralized national schedules and scheduling systems. Material sent by truck does not need to be unloaded and reloaded from truck to train or between trains. Reloading is costly and time-consuming and generates possibilities of error. Normally, analysts assume that rail, with its high fixed costs, is cheaper than road only for distances greater than 500km. However, the current abundance of trucks has generated a price competition that has increased the break-even distance considerably (Cheung 1991).

Given these market and institutional conditions, it is difficult to believe that constructing a new infrastructure alone will induce a significant shift from road to railway. In fact, Coopers and Lybrand (1995) estimate that, even with the new infrastructure, the price of road freight would need to increase by about 35 percent to shift transalpine freight to rail. Since they do not specify their method of calculating price elasticity, it is not possible to determine the behavioral assumptions underlying this estimate. Notwithstanding, taxes sufficient to increase road freight costs by 35 percent would be politically untenable. Furthermore, Coopers and Lybrand's estimate excludes qualitative changes in the rail services provided. With appropriate changes considered, their figure could be much smaller.

Road transport dominates the market not because of simple technological superiority but because of increasing embeddedness in a highly transport-intensive production and distribution system. The dominance of road transport can be seen as the outcome of the interplay between internal dynamics in the haul industry and external selection and stabilization processes such as the economic restructuring process, the abundance of roads, the absence of a transport policy that would internalize external or social costs, and poor relationships among national transport agencies. To consider the dominance of road traffic as either the result of deterministic progress toward increasing efficiency in transportation or the outcome of a free decision process among technological alternatives would clearly be misleading.

The apparent inability of rail traffic to match the flexibility of road traffic is a major barrier to a significant shift to rail. Although infrastructure construction and changes in relative prices through tunnel tolls will generate some change, significant shifts will not occur until the railways learn to mimic the road system's flexibility and to take advantage of their potential to industrialize freight transport.

7.2.3.2 Combined Freight Transport: A Promising Sociotechnical Innovation

Given the inability of the conventional approach of providing intrastructure to deal with the problem of flexibility and industrialization, a theoretical framework is needed on which to base the following argumentation. Hughes's (1987) analysis of large technological systems provides such a framework. Hughes differentiates between the technical core—here the transport technology and the infrastructure—and the organizational and institutional context, which together form a technological system, here the freight transport system (Läpple 1993). The interplay of this technological system's different elements stabilizes a particular technological, organizational, and institutional configuration that excludes alternative conceptions. Rail's historical dominance in freight transport, combined with its centrally planned and rationally oriented organization, made it unable to respond to the threat from road. Not only was there essentially no technical innovation for a century but services were locked in rigid timetables and significant technical ruptures at international borders. It comes as no surprise that rail could not compete in an internationally organized transport market requiring flexible transport technology.

This does not mean that the road will necessarily maintain its competitive advantages. In fact, its technological development is probably constrained to the one-man, one-truck, one-cargo paradigm. This paradigm can hardly be changed, and it precludes any productivity increase. In particular, such a mode of production precludes massive automation, mechanization, and standardization in transportation, as happened with the introduction of container shipping (Bukold 1991). The railways, instead, using information and communication technologies, might provide the basis for such a transformation (Bukold 1991; Hepworth 1992). Industrialized freight transport could allow the railway to benefit from economies of scale, which would make it competitive over medium and long distances. Road transport would then be limited to short distances only. Trucks could be used to take goods to and from combined terminals and to distribute them in regional and metropolitan areas within a range of about fifty to one hundred kilometers. Such a transition would also require new loading facilities throughout the railway network and adequate organizational integration among the different actors along the transport chain. Furthermore, to guarantee smooth organizational integration, firms would have to share unprecedented amounts of information, implying a substantial redefinition of the boundary of the firm and requiring the successful establishment of trust between industry and the hauler.

Whether such opportunities can and will be seized is a question of whether processes of social learning among various actors and organizations will occur, on the one hand, and whether the necessary facilities like terminals for an efficient road-rail transport system will be constructed, on the other hand. The diffusion of lean production concepts may facilitate this learning process, because firms now tend to outsource transport functions, making it easier for transport operators to organize freight transport efficiently by bundling and grouping single shipments for medium- and long-distance transport. If a firm is to outsource long haul transport, however, it must qualitatively change its supply relationships. In particular, it has to grant its transport operator access to certain information flows. This requires that the firm be reconceived and its boundaries be blurred, which can only occur in a climate of high trust. As a consequence, the interface between the firm and the transport operator would become more permeable, and the transport offer would change from a trivial transport service to an integrated logistic service, introducing into the whole process considerable additional flexibility as to choice of transport mode. The new services would include tailor-made transport, stocking, and commissioning solutions. We studied a specific regional transport milieu to determine whether such processes of social learning are really taking place.

7.2.4 A Regional Transport Milieu and Its Innovation Potential

7.2.4.1 A Traditional Transport Milieu in the Transalpine Freight Market In this section we present the results of a study examining actual learning processes and corresponding difficulties in the economy of the Swiss canton of Ticino, a region traditionally involved in transalpine freight traffic.[21] The study focused on the role of transport and forwarding operators and explored their current and future marketing strategies in the context of the NEAT and the Initiative for the Protection of the Alps.

As a general tendency, the market for Ticino's freight forwarders (Switzerland's Italian-speaking canton) has been shrinking since the 1970s, after a long period of economic expansion and prosperity. The growth had been due to this region's location near the Swiss-Italian border along the shortest route joining Italy with northern Europe. In addition to location, Ticinese freight forwarders could boast the combination of Swiss administrative efficiency, low levels of corruption, and competence in both German and Italian language and culture. As their industry evolved, the freight forwarders developed a dense contact network of correspondents, spanning from Sicily to northern Germany. Within this network, they marketed excellent transport services, including financial arrangements. (We refer to the providers of forwarding, transport, and financial operations as the regional transport milieu.) Transport services meant chiefly rail transport, provided by the Swiss national railway company.

In the 1970s, numerous highways were built all over the alpine region; in particular, new alpine road tunnels were constructed in France and Austria. Road freight transport increased, and various new enterprises from all over Europe entered the transalpine transport market. These developments rapidly threatened the transport and forwarding industry of southern Switzerland (Ratti 1982) and also the railway company, which constantly lost business to cheaper and more flexible road transport.

European unification at the beginning of 1993 further aggravated the regional haul industry's decline. Not being part of the EU, Switzerland had to maintain its border formalities. Consequently, more commodities flowed through France and Austria. The regional haul industry lost a significant share of the international freight market. Business volumes shrank. With this institutional change, the position of the transport milieu at the border turned from an advantage into a handicap. Most firms changed their activities to distributing goods imported from Italy in Switzerland. A couple of firms moved directly to the metropolitan area of Milan to have direct access to the new hub of the transalpine freight market. With this change, many resources available in the transport milieu which had evolved over the long period of market expansion deteriorated.

An abundant offer of road transport, mainly from small and self-exploiting transport firms, currently dominates the transport market. A cutthroat competition, undermines the trust built over the years between the transport milieu and its customers in industry. This erosion of trust is probably a major barrier to the innovations needed to redefine the transport product substantially.

To assess the probable effects of the NEAT and a transit toll in the context of these recent developments in the transport market, we interviewed a carefully chosen sample of twenty firms in the Ticino region. These forwarding agents, haulers, and logistic service suppliers represented about 20 percent of the regional transport industry, were predominantly small or medium-sized, and specialized in a particular segment of the transalpine freight transport market. Five were branch establishments of large international transport and forwarding companies. In particular, we tried to understand how the different actors perceived the NEAT and transit toll and what innovation, if any, they were planning in response. The interviewees fell into three broad groups: The large majority of the firms was overtly skeptical about the short- and medium-term competitiveness of rail for freight transport. A second, smaller group saw a minor market share for combined transport systems. A third group, however, saw an opportunity for developing a new transport product.

The skeptics generally indicated they oppose the new railway infrastructure and perceive it as offering no substantially new opportunities. In general, they argued that technology organization, and institutional setting of the railway cannot provide the flexibility freight markets demand. Declining cooperation and mutual understanding within the transport milieu exacerbates the situation. Today, strongly bureaucratic interactions between the Swiss national railway and the other transport suppliers have essentially replaced cooperation, flexible pricing and a highly commercial attitude. In sum, this group's view of the situation was quite simple: A number of structural factors make it unlikely that the railway could compete with the road. Furthermore, declining relations with the railway will probably make searching for innovations both costly and unproductive. Therefore, this group has moved its activities to the road, where they can exploit excess capacity and compete on the basis of prices. Finally, they expressed opposition to the new railway project, which connects the main economic centers south and north of the Alps, because it could exclude them definitively from the transalpine transport market.

The second group of firms indicated they see the NEAT as offering distinct opportunities. In particular, because of increasing road bottlenecks and political opposition to trucks, they believe some freight must be shifted to rail. Notwithstanding, their own studies and scenario analyses suggest the share must be quite small, about 15 percent of the whole European transport market, but given stronger incentives and greater certainty, this percentage could be higher. In particular, in their opinion the viability of combined transport would increase markedly if important ancillary facilities, such as intermodal transfer terminals, were built in addition to the NEAT. Yet considering that no definite route has yet been designated and that the NEAT tunnel capacities will remain largely underutilized unless Italy builds new and improved rail connections—which it shows no signs of intending—they are skeptical that such additional facilities will be built.

7.2.4.2 Emerging Sociotechnical Innovations in the Haul Industry

In contrast to this picture of economic decline and technological abandonment, we describe the work of two firms that actually have implemented significant innovations in the freight transport product. The innovators have chosen to offer "integrated logistics" instead of traditional "box moving." Their product integrates the whole logistic chain, from the factory stock to the store shelves, which gives them a central position in the transport and information network. Managing the flows of information becomes as important as managing the flow of goods. Managing the information flows, however, requires the company to reconceive their product radically, including the coordination of interactions among different parties, such as national railway companies, wagon owners, freight forwarders, road haulers, terminal operators, shippers, and receivers.

Through this innovation, these firms have evaded the cutthroat price competition prevailing in the market and offer a service with high addedvalue. Furthermore, client relations are much more secure because they include a dimension of confidence that has elsewhere disappeared as the traditional regional milieu has deteriorated. The innovative forwarding agents perceive considerable opportunities in integrating the railway into the transport chain. Their companies provide combined road-rail services and thus increase through their own innovations the use of combined or intermodal freight transport facilities across the Alps.

In the Ticino region, major forwarding agents, transport companies, and the national railway company founded, in 1967, a new mixed private-public company for combined road-railway traffic. In the neighborhood of Milan, this company built an intermodal terminal for loading containers, trailers, and accompanied trucks (trailer and cabin) on trains. Founding this company was a first reaction to the growing competition from road transport. However, the terminal was located outside the traditional regional milieu, for reasons of ready access to the rapidly growing transport demand of Italy's Lombardy region. The product of the new company was a considerable success from the beginning and, by 1989, had reached a regional market share of about 40 percent in this market segment.

From 1989 to 1993, the company increased its market share to about 80 percent on specific routes, principally by offering an innovative transport product called "shuttle trains." These trains provide scheduled, safe, and punctual services between terminals in major industrial regions north and south of the Alps. For example, six shuttle trains a day connect Milan and Cologne (the best-served route). The company provides its own rolling material or trains. The company's unexpected success is because it offers highly reliable transport, allowing shippers to plan their transport very precisely in advance. This seems therefore the most likely sociotechnical alternative to the dominant road transport. In particular, for distances of more than 500 kilometers, the combined transport can take advantage of the economies of scale the railway can offer.

Nevertheless, many commentators believe that the product's competitiveness depends on several forms of indirect public funding. This may be true, because the Swiss national railway company subsidized the terminal's construction, although it is situated in Italy. Further, some people argue that the firm pays national railway companies a "political price" for tracking shuttle trains in their respective countries.[22] But with a projected tunnel toll, the railway will probably be competitive, even at full price. In view of the higher quality of service, price may eventually not even be the decisive aspect. The capacity of the terminal and the rail tracks, however, limits further increase in combined transport. A fourfold increase of the tunnel capacities with the NEAT requires, obviously, new road-rail terminals as well as sufficient capacity on the access rail tracks, neither of which is yet an integral component of the NEAT.

Now back to the two innovative Ticinese transport companies. Whereas the other eighteen companies could conceive only of a future with declining trust relations, fractured networks, and environmental pressure caused by trucks overrunning the alpine valleys, these two firms were looking forward to a different future. With their use of information technology for developing a new kind of service and with their establishing growing contact networks, they were able to increase dramatically the value they could add to transportation services. Using services like the shuttle trains, they could reduce costs. If more terminals for shuttle trains were built, and hence more frequent and more location-specific services were offered, they would even be able to exploit economies of scale and expand their services.

It follows that from a considerably improved railway system with reduced transport time and cost, as the NEAT could provide, two interesting opportunities arise: The first is combined road-rail transport, which can offer high punctuality and reliability. The second is integrated logistic services, which might make use of road-rail transport offers. If these opportunities were exploited, price competition might be of secondary importance compared to quality competition.

Among the representatives of the Ticinese transport industry interviewed, only two learned to seize (or rather create) one of the above-described opportunities. We have to conclude that this was not learning embedded in the contact network of the transport milieu. Hence it seems unlikely that their innovative approach will automatically spread among the milieu if only new infrastructure is provided. On the contrary, the milieu would probably oppose any solution other than a minimal one, for reasons of skepticism and fear. This potential opposition should be taken seriously. The chances connected with the NEAT should be made explicit, and it should be made clear that a minimal solution (one or two base tunnels with access rail track alone) is likely both to be very expensive and to provide very limited environmental benefits, which would make the pessimistic vision of one group of the interviewed forwarders become reality.

7.2.5 Innovation-Oriented Policies Concerning Freight Transportation

The NEAT and the Initiative for the Protection of the Alps assume that freight transport will somehow shift mechanically from road to rail, resulting in a new modal split. The NEAT and the initiative have mutually reinforced this idea, ignoring necessary sociotechnical innovation processes. Our analysis suggests that innovations in the railway sector could mobilize a substantial market potential in the freight transport sector. This would require, however, that political strategies not be confined to providing infrastructure. They must also deal with the lock-in situation of truck dominance and with the uncertainties in the transportation market. As a consequence, policymakers have to transcend their role of optimizers of individual solutions and become facilitators of innovation processes. For the freight traffic sector, this basically means fostering innovation processes in two main and mutually interrelated directions. First, combined or intermodal freight transport must be supported through technical improvements, international interconnectivity of the rail systems, and a price increase in road transport. Second, the social reconstruction of the transport product must be facilitated through new relations in the transport chain and new organizational networks. This can mainly be achieved by overcoming institutional barriers between nation states as well as between public and private firms to allow and foster new cooperative networks.

The innovation potential of combined road-rail freight transport must be identified and fostered so that it becomes a true alternative to road transport. This, however, can happen only if the railway infrastructure is linked to intermodal terminal facilities. In these terminals, standardization and automatization of transport bodies (containers or swap bodies) could reduce costs of loading and unloading considerably. More shuttle trains and fixed connections would therefore lead to economies of scale, a situation quite different from that of trucks. Economies of scale would give rail transport considerable competitive advantages over road transport. An ever enlarging network of linked intermodal terminals could further yield network benefits and induce an increasing modal shift from road to rail, involving more and more actors in the combined freight transport.

This would require that the international dimension and the macroinstitutional setting be taken into account: The NEAT infrastructure must be conceived in the context of a transnational European network. Its success depends on the construction of appropriate interfaces between the different nationally shaped infrastructures. In particular, there must be a connection to the already existing terminals in Lombardy. Until recently, analyses of the NEAT have largely neglected this international dimension.

The same argument holds for the transit toll the Initiative for the Protection of the Alps has demanded. A unilateral transit toll could simply result in more traffic going through Austria and France and hence increase GHG emission in the alpine region as a whole. Therefore, a differential price increase for transalpine road freight traffic has to be a concerted effort on the international level. This appears to be happening: The European Union is opting for the introduction of a tax on heavy road traffic and is following the Swiss example of a transit toll for transalpine freight traffic.

Developing an intermodal transport infrastructure could facilitate the second direction of innovation processes, namely the redefinition of the transport commodity. This would depend on redesigning the interfaces between public national and private international transport organizations. Furthermore, innovative firms would have to redefine the customer-transporter relationships in the transport chain. To develop new, flexible forms of interaction and new network organizations, the rhetoric of public versus private or regulation versus deregulation must be transcended. Uncontrolled deregulation could easily set the railway transport system back further. An effective policy might be to expand rail access to trains owned by third parties, such as large transport companies or cooperative ventures of national railways and private companies operating at the international level.

New trust relations among operators in the transport chain would also play a crucial role in changing the concept of transport supply to one of supply of integrated logistic services. This transformation could have lasting impacts on the strategic significance of transport for industry. The NEAT and the Initiative for the Protection of the Alps might support such a development if they were searching for innovation-oriented policies.

Notwithstanding, an efficient road-rail transport system should not be considered the definitive solution to the freight transport problem. In the long term, a shift to an environmentally less harmful transport mode is certainly insufficient to resolve the environmental problems freight transport causes. It will be necessary to reduce GHG emissions caused by freight transport by considerably reducing material flows. This means major changes in industry's spatial localization patterns and radical dematerialization of many industrial products. The transport intensity of intermediary and final products must become more apparent to consumers.

7.3 LIGHTWEIGHT VEHICLES AND THE SOCIAL CONSTRUCTION OF INDIVIDUAL MOBILITY

7.3.1 Social Learning and Individual Mobility

People are consuming increasing amounts of energy and emitting increasing amounts of GHGs to get around (ECMT 1993; Shipper, Steiner, and Myers 1993). Mobility forms, modes of transport, economic activities, land use, social status, and lifestyles have co-evolved to make social activity contingent on the availability of individualized transport (Petersen and Schallaböck 1995). Furthermore, the automobile is vital to the economies of most industrialized countries. In Germany, for instance, it is directly or indirectly associated with about 20 percent of GNP and 16 percent of the workplaces (Canzler and Knie 1994). Finally, the automobile has entered the everyday lives and myths of our current societies to the point that its role has been compared to that of cathedrals in former times (Barthes 1964). Given the current political, social, technical, and cultural environment, attempts to reduce automobile usage sharply are thus likely to fail, and attempts to substantially modify usage are likely to face considerable barriers. Criticism of the automobile has, for instance, engendered political movements that explicitly defend its free use.[23]

In the present case study, we discuss the options and barriers for a sociotechnical innovation in the field of individual transport that could substantially reduce GHG emissions; the development, manufacturing, and market diffusion of "lightweight vehicles" (LWVs). Lightweight vehicles (see figure 7.5) are extremely small (usually two or four seaters) and light (less than 600 kilograms) automobiles that consume considerably less energy than conventional cars (about two liters of gasoline per 100 kilometers, or 120 miles per gallon).

In the first part of this section, we discuss the importance of individual mobility, its impact on global climate change, and current alternatives to the conventional gasoline car. We also estimate lightweight vehicles' potential to reduce GHG emissions in Switzerland. In the remainder of the section, we illustrate how economies of scale, social construction processes, regional contact networks, and the macroinstitutional context influence the dynamics of technological change and preference formation. We focus on three stages of the innovation process: development of technical prototypes, large-scale

Length, height, width:	3.07m x 1.42m x 1.55m
Weight (empty):	650 kg
Max speed:	120 km/h
Range:	120 km electric, 600 km hybrid
Energy consumption:	2.6 lt/100km (about 100 mi/ga)

Figure 7.5   ESORO AG's H301 hybrid lightweight vehicle.

manufacturing, and diffusion within the market. In a final part, we draw some conclusions for an innovation-oriented climate policy.

7.3.2   The Impact of Lightweight Vehicles on Global Climate Change

7.3.2.1   GHG Emissions and the Transformation of the Gasoline Car   The automobile as a means of individual mobility has been subjected to widely differing interpretations throughout its history. Originally, the majority of people perceived it as a sports toy for upper-class men, with largely negative effects on the rest of society (Canzler and Knie 1994; Wachs and Crawford 1992; Flink 1988; Sachs 1984). Subsequently, the automobile diffused in all industrial societies in a couple of decades. Since the late 1960s, however, the car has become a symbol of the modern lifestyles that endanger our planet's survival. Most recently, the problem of global climate change has accentuated the global impact of the massive use of fossil fuels for individual mobility.

Individual motor traffic contributes approximately 6 percent to the overall global warming potential (Shiller, McCarthy, and Shiller 1991). However, it is one of the fastest growing contributors (Michaelis et al. 1996). The number of cars on the world's roads grew at a rate of 5.2 percent per annum between 1960 and 1989, surpassing population growth by 2.1 percent (Walsh 1993). In the same period, CO_2 emissions stemming from transportation grew at a rate of about 2.4 percent per annum (Walsh 1993).

These figures may vary quite markedly for different parts of the world. In the EU, for instance, transportation accounts for about 22 percent of CO_2 emissions (Wynne 1993) and in the US for 32 percent (Rayner 1993). In the alpine region, this percentage is even higher, because electricity is generated principally by hydro power and nuclear power. In Switzerland, for instance, transportation accounts for about 40 percent of CO_2 emissions (OFEFP 1994), mainly from gasoline cars.[24] In developing countries, growth rates are high, especially in urban areas, varying from 5 to 15 percent per annum in the 1980s (Sathaye and Walsh 1992).

The markets in industrialized countries seem insatiable.[25] Furthermore, the unequal distribution of car ownership worldwide indicates a truly massive potential market. In the United States, 1,000 inhabitants own 762 cars, in European OECD countries 495, in Asian OECD countries 589, and in the rest of the world a mere 41 (Walsh 1993). These figures suggest that cars will increase in importance as a source of GHGs in the near future (Michaelis et al. 1996).

Given this worrisome outlook, many analysts have focused on technical improvements and alternatives to the gasoline car (Berger and Servatius 1994; Bliss 1988). Industry has claimed repeatedly that specific standards for cars—for example, gasoline consumption of five liters per 100 kilometers— will be achieved automatically, or should at most be encouraged by a moderate carbon tax (Shiller, McCarthy, and Shiller 1991). An "autonomous" efficiency improvement trend can be identified subsequent to the oil crisis of the 1970s.[26] However, some commentators believe that further efficiency improvements will be ever more difficult to achieve (Dowlatabadi, Lave, and Russell 1996; Lovins and Lovins 1995; Shiller, McCarthy, and Shiller 1991).

More radical technical approaches have therefore tried to develop non-gasoline cars. After the oil crisis of the 1970s, natural and liquefied gas, methanol, ethanol, synfuels, hydrogen, and electricity all received special attention (Schot, Hoogma, and Elzen 1994; ISATA 1994; Sperling 1989). Each of these alternatives has specific advantages, but so far none can compete with gasoline in the marketplace. Problems include a lack of appropriate fuel storage and distribution systems, safety problems with respect to distribution and storage of fuel, and high production costs of the fuels themselves. In some countries, alternative fuels have, however, been quite sucessful, for example, liquefied gas in the Netherlands and ethanol in Brazil.

Many commentators consider electric vehicles the most promising solution to the environmental problems automobiles cause (Sperling 1995; MacKenzie 1994; OECD 1992). Electricity can be produced by various energy sources, including renewable ones. Many countries have well-established electricity distribution networks, and the technology—independent of its use for individual transport—is well tested. Furthermore, electric motors can be very energy efficient (OECD 1993a). The main problem is storage. Batteries available today have a low ratio of storage capacity to weight, which means that cars are very heavy or have low range.[27]

Given the problems with current batteries, hybrid propulsion systems hold a lot of promise. Lightweight construction combined with a hybrid propulsion system could exploit various synergies to reduce fuel consumption to one liter per 100 kilometers or less (Lovins and Lovins 1995; Lovins, Barnett, and Lovins 1993).

Prototypes of electric or hybrid vehicles appear regularly at international automobile fairs. However, to date electric cars have not been manufactured on a large scale.[28] To encourage the industry, a number of policy initiatives have been developed. The best known is the California Clean Air Act's requirement that by 1998, 2 percent of the vehicles sold must have no emissions at their point of use (CARB 1994). Currently, only electric vehicles can satisfy this requirement for "zero-emission vehicles," or ZEVs, so the act opens up a huge potential market for electric vehicles in the next few years. However, the success of the California initiative will largely depend on whether attractive electric vehicles are available by 1998 and whether consumers will buy them.

The automobile industry has been extremely reluctant to manufacture electric vehicles. This creates room for other initiatives that might be less restrained by sunk costs and traditional conceptions of the product. One example is the CALSTART consortium, established by the crisis-shaken defense industry, electric utilities, and public agencies in southern California (Scott and Rigby 1995; Scott 1993). Another set of promising initiatives has emerged in several European countries with regard to the development of special-purpose–design lightweight vehicles. Switzerland has been able to assemble the widest amount of expertise and gain a relatively high market share for these vehicles in the last ten years.

7.3.2.2 Lightweight Vehicles and GHG Emissions in Switzerland
Lightweight electric vehicles might have a decisive impact on reducing GHG emissions in Switzerland and worldwide. To assess their potential, one must conduct ecological life cycle analyses for individual vehicle concepts and aggregate the effects of gradually substituting lightweight vehicles for conventional cars. Life cycle assessments for electric vehicles abound. However, they yield widely varying results depending on various assumptions (Ecosens 1995; Seiler 1993; IEA 1993; OECD 1993a; Blümel 1992; Morcheoine and Chaumin 1992; Schaefer 1992)[29] Rather than try to resolve these differences, we instead sketch in rough fashion the GHG reduction potentials based on rather robust assumptions.

We assume all vehicles are the same size and provide the same comfort. We have considered the following four variants (see figure 7.6): a car with an internal combustion engine consuming five liters of gasoline per 100 kilometers (ICE-5), a similar vehicle consuming three liters per 100 kilometers (ICE-3), an electric vehicle (EV) consuming electricity from the Swiss distribution net,[30] and a corresponding vehicle with a serial hybrid power system. Although all four cars require a comparable amount of energy for the con-

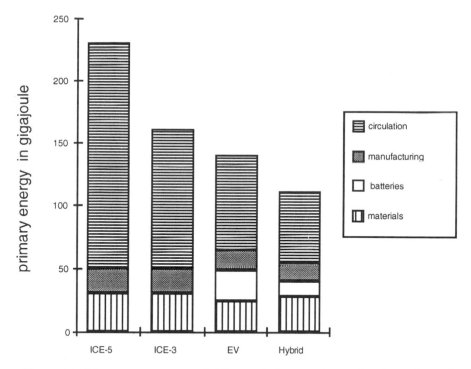

Figure 7.6 Primary energy content of different vehicle types. Data are based on Bukowiecki and Musslers 1993, Furuholt 1993, and Seiler 1993. Energy contents are calculated as primary energy equivalents (i.e., before electricity generation).

struction of bodies and motors, electric and hybrid vehicles also require batteries. In our example, hybrid vehicles have a lower battery weight, but have an additional small internal combustion engine working as a generator. The energy figures for circulation would change for the electric vehicle if based on the European electricity generation mix (instead of the Swiss one) and would lie between those of the ICE-3 and the ICE-5. Basing the hybrid's life cycle assessment on the European mix would essentially result in no change because the hybrid produces electricity on board exclusively. The data would much more favor the electric variant if air pollutants and GHG emissions were considered for Switzerland. On the basis of the European energy production mix, electric vehicles fare better with regard to certain pollutants and are comparable with regard to CO_2 emissions (Seiler 1993).

We estimated the impact on Swiss CO_2 emissions of a massive substitution of lightweight vehicles with hybrid propulsion systems for conventional cars with the following scenario: Assume that by the year 2025, two types of lightweight hybrid cars have replaced traditional cars. One-third are small city cars that consume two liters of gasoline per 100 kilometers, and the other two-thirds are somewhat larger family cars that consume three liters per 100 kilometers. If we set traffic growth rates according to the official scenarios and assume that substitution rates will increase linearly over time,

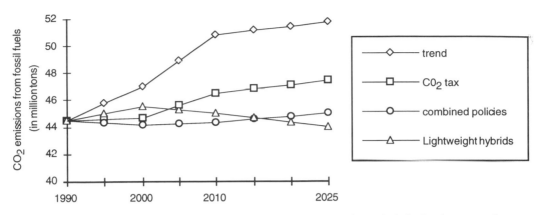

Figure 7.7 CO_2 scenarios for Switzerland including lightweight hybrids. The curve indicating the innovation of lightweight vehicles has been calculated excluding other potential measure for reducing GHG emissions. The data for "trend," "CO_2 tax," and "combined policies" are based on BUWAL 1994, p.77; the data for "lightweight hybrids" are rough estimates, assuming a complete substitution of LWVs for traditional cars by the year 2025.

CO_2 emissions in Switzerland would not only stabilize but even decrease (see figure 7.7).

We compared this result with the Swiss federal government's official scenarios, which calculated the effects of a pure carbon tax and of a specific policy mix.[31] Our comparison may be far too optimistic and simplistic[32]. Notwithstanding, it shows massive reduction potential due to radically improved automobile energy efficiency. In view of this estimated savings potential, we now discuss the social determinants of innovation in development, manufacturing, and market diffusion.

7.3.3 Outsider Networks Engaging in the Development of Prototypes

Switzerland has no domestic car industry, and its car market represents only about three million vehicles. Nevertheless, a number of initiatives there have attempted to develop radically cleaner cars. In view of the powerful automobile industry and the highly entrenched conventional concept of a car, it may be worthwile to analyze outsiders' role in the development of radically new concepts. Such outsider initiatives have been more successful in Switzerland over the last ten years than similar endeavors in most other countries (Truffer and Dürrenberger 1997)[33].

The Swiss initiatives for reconstructing the car involve networks of different actors. In general, these initiatives may be divided into two clearly distinguishable groups. The first is associated with the Swiss watch company SMH (Société suisse de Microélectronique et d'Horlogerie SA) and is known as the "Swatch-Mobile" project. This project served as the basis for an SMH-Mercedes joint venture, which is now starting the production of a car called SMART.[34] Because of the SMH initiative's industrial background, we refer to it as the "industrialists' innovative milieu." The second group comprises various individual projects originating from citizens' movements that support solar energy technologies. We call these the "pioneers' innovative milieu".[35] Both milieus have developed high-quality prototypes. Their current challenge is to turn the prototypes into an industrial reality.

The two innovative milieus show interesting similarities and differences in style and approach. Both endeavors may be termed "outsider initiatives." Neither had input from any of the major car companies, initially. Both milieus overtly distrusted the current automobile industry's ability and willingness to radically reconceive its product. Furthermore, both initiatives drew their technical competence from a more or less extended regional base of know-how.[36] Finally, both initiatives began initially to work on electric drivetrains and are now experimenting with hybrid propulsion systems or highly efficient gasoline engines.

The milieus had different motivations, however, for developing alternative cars. SMH rose from the ashes of its taken-for-granted industrial product, the Swiss watch (Karrer-Rüedi 1992). The Swiss watch industry virtually collapsed in the 1970s in part because it had not taken seriously the cultural and technical redefinition of its product. Japanese producers took up the quartz watch, which notably was invented in Switzerland, and marketed it successfully. Swiss manufacturers could reestablish their predominance in the global watch market only after a turbulent period of industrial restructuring (Maillat et al. 1995). The "Swatch" played a major role in this process. It was successful because it radically redefined the product's cultural message. Based on this experience, SMH was convinced that other consumer products could be similarly redefined. The automobile was the obvious choice. However, a high standing competence in microtechnologies, which was available in the regional milieu of the Jura region, could complement this market experience. Furthermore, the engineering school of Biel, located in the region, and several medium-sized industries specializing in lightweight construction had already accumulated very promising know-how in developing lightweight electric vehicles. These factors formed the background against which the watchmakers decided to risk challenging the automobile industry on its home ground (Linnenschmidt 1993; Meyer 1993; Niggli 1991).

The pioneers had their origins in a cultural protest movement. Members of the Swiss antinuclear and environmental movements, which were tightly related, undertook a wide variety of projects aimed at developing renewable energy. The use of solar energy for electricity generation and heating was one of their most vivid fields of development. From there they turned to the central symbol of destructive industrialized lifestyles, the car. The prospect of a vehicle driven by clean energy motivated many students and individual developers. The Tour de Sol, a one-week race of solar vehicles, first organized in 1985, became the most important event where these people met and exchanged their experiences.

However, social and cultural factors are not sufficient for developing a radical technological innovation. Both the industrialists and the pioneers could rely on highly specialized know-how in such fields as lightweight construction, accident mechanics, and electric drivetrains. The shared vision of nonpolluting individual transport built a contact network among various firms and individuals that became a resource available to both milieus. However, the two milieus drew on it in different forms. SMH hired people from within the network and sequestered them, following a strategy of tight control of information and trying to build up the necessary competence within its laboratory walls. The pioneers, in contrast, exchanged information quite openly.

Markets and development were tightly interconnected from the beginning of both initiatives. SMH received about 35,000 orders when it announced its Swatch-Mobile project (Süddeutsche, Zeitung March 5–6, 1994). SMH emphasized that the "product message" would be one of the most important bases for the Swatch-Mobile's market success. The pioneers also received very enthusiastic responses at public events and in the media. In addition, they profited from close contacts with people who wanted to drive lightweight electric vehicles and were prepared to discuss their experiences. These people generally drove very small gasoline cars that some pioneers imported from France and converted to electric propulsion. These conversions, together with the well-known City-El from Denmark, created an early market for lightweight vehicles with its own distribution channels and service stations. By 1994, Switzerland housed approximately 1,700 lightweight vehicles,[37] which equals the total number of Rolls Royces and Bentleys in Switzerland. This early LWV market was a seedbed for testing vehicle concepts and components and a major resource for the prototype developers as well.

But what did these outsider initiatives achieve? Did the Swiss really reinvent the automobile, or have they just camouflaged golf carts? There is very little reliable information on the concrete form of the SMART.[38] The pioneers, for their part, have assembled highly competitive know-how in the application of lightweight construction to automobiles.[39] The milieu formed a competitive arena for developing technical components and varying vehicle concepts. Variations were made with regard to body materials, number of seats, number of wheels, propulsion unit, and so on. However, some problems had to be resolved jointly in the milieu.

In particular, the pioneers developed designs that solved the most critical problem of safety.[40] In several crash tests, they showed that lightweight vehicles could meet the automobile industry's current safety standards (Walz 1995). The success is not easily attributable to any single actor but rather to interchanges within the network. Therefore, one could state that the pioneer milieu has established and defined a new technical artifact that could be called a "lightweight vehicle". The vehicle shown in figure 7.5 was developed in the pioneers' milieu. These results indicate that the lightweight vehicles developed by both outsider initiatives in Switzerland are far from being souped-up golf carts. However, the relevance of these vehicles for mitigating global climate change risks does not depend on technical characteristics alone: They must be mass-produced and accepted by consumers.

7.3.4 Uncertainty and Economies of Scale

7.3.4.1 Transforming Prototypes into Commodities
We have shown that LWVs offer substantial GHG-reduction potentials and that networks of technically competent individuals and small-sized enterprises forming an "innovative milieu" originated their development. Now we ask whether a mobility system based on lightweight vehicles could form the basis of an alternative sociotechnical equilibrium, and we identify the barriers to an autonomous transition from the present steel-based–car equilibrium to a lightweight-vehicle equilibrium. On the basis of this analysis, we propose a set of innovation-oriented policy instruments that would support the transition between these equilibria.

High-quality prototypes do not create a market. Cars must be produced in sufficient quantities to drop the price. However, mass production requires a fundamentally different organizational know-how and considerably more resources than developing prototypes. Hence it comes as no surprise that the industrial milieu, from the beginning, has searched for contacts with the automobile industry. Its initial joint venture with Volkswagen lasted about two years but was terminated shortly after a change in VW's senior management. Subsequently, SMH has formed a joint venture with Mercedes Benz. The pioneers have followed various strategies. Some have started a small-scale production of several hundred vehicles per year and seem to make a living. Others have tried to develop relationships with firms outside the automobile industry, including several firms in California (Vester 1995).

To assess crucial barriers and opportunities which might present themselves if someone wanted to produce lightweight vehicles in large quantities, we interviewed representatives of relevant Swiss industries and the finance sector (Dürrenberger, Jaeger, and Truffer 1995).[41] We asked the interviewees whether they considered the production of lightweight vehicles to be an opportunity for their firm and/or Switzerland and what they saw as the most important barriers regarding such a project.

In general, those interviewed saw few technical obstacles to production. Swiss industry could mobilize large parts of the technological know-how for mass production of lightweight vehicles. However, without a well-established and experienced partner, risk capital would be essentially unobtainable. Furthermore, all the interviewees doubted strongly that their own firms would get involved. They judged the risks as excessive. Two sources of risks were mentioned: first, the automobile industry's market power and the regulatory density in the automobile sector would make gradual learning extremely difficult. However, this highly innovative sector would undoubtedly need the opportunity to learn. Secondly, they wondered strongly about the marketability of lightweight vehicles.

Given this assessment of innovation barriers, the strategy the industrial milieu followed seems quite understandable. Outsider initiatives may well be able to develop highly innovative prototypes with relatively minor investments. Large-scale production, however, requires the know-how of an experienced industrial partner. Small innovative firms often lack the capacity to make the transition from small-scale production to industrial manufacturing (Nell 1993). In the following sections, we propose some instruments of an environmentally motivated innovation policy that would support social learning in the field of manufacturing alternative cars and help diminish some of the corresponding uncertainties.[42]

7.3.4.2 Policies That May Reduce Uncertainties

Our interviewees emphasized that substantial economies of scale could be mobilized for lightweight vehicle production. Assuming mass production, prices for lightweight vehicles could be grossly comparable to those of steel-based cars. Major uncertanties are, however, associated with a sudden massive expansion of the production capacity, not only with respect to manufacturing techniques but also with regard to aspects like liabilities for product safety.[43] These risks would be substantially lower if production lots were small to start with, but corresponding production prices would then push the vehicles into the luxury segment, where their market prospects would be extremely limited.

How could this deadlock be broken? Policymakers could exploit the fact that lightweight vehicles are more energy efficient than their conventional counterparts. If fuel prices were high enough, people would switch to LWVs. Yet because demand for gasoline tends to be highly inelastic, cars usually last for more than ten years, and fuel is a relatively minor component of the cost of owning and running a car, such a tax would have to be unacceptably high and would have to be in effect for an unacceptably long time to accomplish its purpose. Hence we believe this policy would be politically unfeasible.

Marketable gasoline permits are a second possibility.[44] Permits could be issued equaling a fixed total quantity of gasoline—let's call it a "gasoline cap." Market processes would determine the prices of the permits. The cap size could be tied directly to the FCCC's CO_2 emission stabilization objectives. In the beginning, a region's gasoline cap could be set near its current gasoline consumption level. With growing mobility demand and/or a cap reduction to the FCCC stabilization level, permit prices would rise from an initially low value and would favor the more efficient lightweight vehicles. In this way, an expanding mobility demand could be satisfied without transgressing the gasoline cap. The stabilization goal for GHG emissions could thus be achieved.

The demand for travel, the size of the cap and the difference between life cycle costs of conventional cars and lightweight vehicles would essentially determine permit prices. Let us start from the simplifying heuristic assumption that permit prices affect the mix of vehicles only (and not the demand

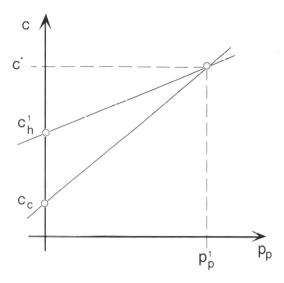

Figure 7.8 Life cycle cost for lightweight hybrids and conventional cars depending on permit prices under the hypothesis of small-scale production of the lightweight hybrids.

for travel, for instance) and that people base their purchase decisions on life cycle costs of the car only.[45] For LWVs to be competitive in such a situation, the permit price would have to rise to a level where both kinds of vehicles had equal life cycle costs. Figure 7.8 shows how the permit price would have to change to equalize the life cycle costs. The vertical axis, c, represents the costs for running a vehicle for its entire life. The point c_c represents the costs of owning and running a conventional car, excluding permit purchases. The point (c_h^1 represents the cost of owning and running a hybrid manufactured in small-scale production, again excluding the expenditures for permits. An equilibrium total cost, c^*, is determined where the vehicle costs plus the expenditures for gasoline permits are equal for conventional cars and hybrids. The amount paid for permits equals the product of the price of a permit per liter, p_p^1,[52] times the vehicle's fuel consumption over its entire life. The slope of the curve measures the respective cars' fuel consumption.

Given such a permit price, the expansion of mobility demand or a reduction of the cap would favor lightweight vehicles.[47] This is an incentive for technological learning. Experience with production techniques and day-to-day use might continually lower the risks of manufacturing LWVs in greater quantities. Therefore, economies of scale, which lower the production costs, could be realized. If everything else remained constant, the price of the permits would fall as the lightweight vehicle market expanded. In figure 7.9, lightweight vehicles are now mass-produced and hence have lower production costs per unit, c_h^2. The price of permits, p_p^2, and the overall costs to the consumers, c^{**}, are considerably lower than in figure 7.8 (p_p^1 and c^*, respectively).

As the market for light weight vehicles grew and economies of scale were realized, however, the price of the permits and therefore the price of gasoline

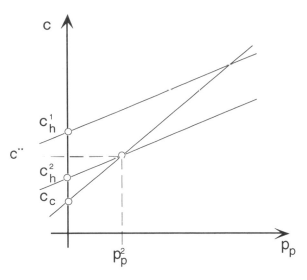

Figure 7.9 Life cycle cost for lightweight hybrids and conventional cars depending on permit prices under the hypothesis of mass production for lightweight hybrids.

would fall. Mass production costs of lightweight vehicles might very well eventually be lower than those of conventional cars. If so, lightweight vehicles could replace conventional cars completely and, assuming constant demand for transportation, the price of the permits would fall to zero.

The crucial element in this strategy is the gradual transformation of one existing equilibrium, which seems to be economically stable, to another as stable as the first and providing mobility services at comparable prices to consumers. The two equilibria, however, have different energy consumption characteristics. The transition from one equilibrium to the other depends on learning processes on the part of both technicians and users. During the transition, emissions would remain constant or even decline. Hence, regulators could foster energy-efficient cars without falling into the trap of promoting dubious new forms of mobility.

7.3.4.3 Acceptability of Policy Measures The initial rise in permit prices under the scenario presented indicates that permits are not a magic policy tool, either.[48] At its peak, the price would probably be grossly comparable to that of a carbon tax implemented to achieve the same goals.[49] Voters would probably not accept a doubling or tripling of current gasoline prices in the name of permit schemes any better than they would accept a tax. Other mechanisms and incentives, such as "feebate" schemes, to which we will now turn, could, however, compensate for this disadvantage.

Permits' equilibrium price could be lowered considerably if inefficient cars cross-subsidized energy-efficient cars. Inefficient cars would be charged a fee at the point of sale. The corresponding returns would be redistributed to efficient vehicles in the form of a rebate, hence the term "feebates" (Davis,

Levine, and Train 1995; Sperling 1995; Lovins and Lovins 1995). In the beginning, when few LWVs would be produced, a small fee on conventional cars would generate rebates that would drop LWV prices to very low levels. As LWV popularity rose, the fee returns would decline and would be distributed among more and more LWVs. Therefore, the instrument would slowly dissappear. However, feebates alone would give policymakers no control over total fuel consumption in the regulated area. As a consequence, feebates might actually encourage the purchase of lightweight vehicles as an additional, rather than a primary, means of transportation and thus contribute to an increase in total gasoline consumption.

The relative disadvantages of permits and feebates could be significantly reduced if the two instruments were combined. Under the above assumptions, the price of the gasoline permits would be directly proportional to the difference in life cycle costs of conventional cars and lightweight vehicles. A positive permit price would therefore be achieved only if the costs (excluding permit expenditures) of lightweight vehicles were higher than the costs of gasoline cars, that is, if $c_h > c_c$. Therefore, the fee on conventional gasoline cars, f, should initially be set to a level where the following relation holds:

$$c_h - \frac{q_c \cdot f}{q_h} > c_c + f,$$

that is,

$$f < (c_c - c_h)\frac{q_h}{q_h + q_c}.$$

Current sales numbers for conventional cars are given by q_c and the corresponding quantity of hybrids is given by q_h. From the second inequality we see that the fees would not exceed the difference in life cycle costs between the two vehicle types times lightweight vehicles' market share. The growth of the lightweight vehicle market and the decline of the conventional car market would slowly reduce the rebate q_c^*/q_h for any single lightweight vehicle. The economies of scale mobilized through the market's expansion would reduce the production cost of lightweight vehicles, c_h. All in all, selling prices, that is, production prices minus rebates, would be held more or less constant throughout. Permit prices would then be positive and considerably lower than if no feebates existed. Furthermore, permit prices would remain more or less constant over the whole period of market expansion of LWVs. By this means, the market would be transformed to a new, more efficient equilibrium without stressing customer frugality. More important, by attaining mass production levels the new technology could compete with conventional cars on world markets without any further supporting policy intervention.[50]

Combined permits and feebates as policy instruments designed to foster innovation processes have some important strengths: first of all, policymakers would interact with the market at the point where they have the most reliable information and where political decisions are most transparent: with respect to quantities of emissions or energy use. Promoters of a GHG-reduction strategy would not have to defend taxes on the basis of doubtful assessments of external costs but could argue in terms of emission quantities. Moreover, stable emissions do not mean stagnant mobility. Besides, fees would likely meet only minor resistance and do not induce distribution conflicts. Finally, and what is most important from the point of view of evolutionary instruments influencing the transition from one equilibrium to another, feebates and permits would end up diluting themselves in proportion to their success.[57]

The above analysis assumes that lightweight vehicles and conventional cars would provide the consumer with largely equivalent mobility. They are assumed to differ only in production price and energy efficiency. If, however, consumers were to perceive a substantial difference between the two alternatives,[52] permit prices could still rise to a level that would probably generate high political resistance against the measure. In the following subsections, we therefore investigate how preference dynamics for lightweight vehicles might lower resistance to the promotion of energy-efficient alternatives to the conventional car.

7.3.5 The Social Construction of a Consumer Product

7.3.5.1 Processes of Social Learning and Market Dynamics A new consumer product's market prospects depend on its ability to match consumer preferences and on its price relative to that of its competitors.[53] We have discussed prices above. In this section, we examine preferences for lightweight vehicles compared to conventional gasoline cars. We assume that prices are more or less comparable.

If preferences emerge as exogenous to the market, as conventional analysis assumes, market diffusion can be analyzed in two ways. First, if preferences are given, consumers decide about their purchases in a largely autistic way (i.e., completely independent of others). The market potential may then be estimated by segmenting markets according to preference structure and income. Second, if preferences are given but information about the new product's existence and utility depends on its presence in the consumer's social environment, the topology of social contact networks shapes the product's diffusion (Rogers 1983).

If preferences develop endogenous to the market, a third possiblity emerges that is especially probable for innovative products: Processes of social learning might drive market dynamics. This might occur at two levels: Either the individual consumer's preferences may change in the course of adopting the innovative product—sociologists have called this "domestication" (Sørgaard and Sørensen 1994)—or alternatively, learning processes may occur within a collective of consumers, which will lead to new socially accepted rules of interpreting the new artifact. These interpretative rules may relate to the vehicles' characteristics or the social images of their drivers or producers (Michman 1991; Nyström 1990; Umiker-Sebeok 1987). The interpretative rules then combine to create product identities that often develop concomitantly with the market for the new products. It is therefore quite difficult to determine the exact product identity of innovative products in advance,[54] which implies that the overall innovation process (Dolan, Jeuland, and Muller 1986) and hence the future market (de Bont 1992) are both highly indeterminate.

When examining endogenous preference formation, it is useful to distinguish between the emergence of a product identity and its later diffusion. We refer to the early markets, where symbolic attachments and use patterns develop, as "protomarkets." These early markets are often based on social contact networks, or "consumer milieus" (see Truffer, Dürrenburg, and Jaeger 1994), which have a regional basis. Consumer milieus consist of consumers who share (at least some) cultural views and values relevant for purchasing a specific innovative product. The product may then serve as a marker for membership in the consumer milieu (Featherstone 1991).

To accomplish this function, the product will probably have characteristics that contradict the product characteristics and symbolic significance of comparable established commodities (Featherstone 1991). The design of the mountain bike, for instance, contradicted the secular development path of making lighter and faster bikes. Snowboards show a similar departure from traditional paths. Initially, the major ski manufacturers perceived snowboards as strange, impractical, and dangerous devices with very limited market prospects. In both cases, however, it was exactly these startling product characteristics which established the protomarket. The new customers did not want to be confused with "ordinary" consumers.

Relevant markers are not necessarily restricted to the innovative product's technological characteristics. Characteristics of use (such as social behavior in handling the commodity) and users (such as lifestyle attitudes, e.g., hedonistic values) may also be important. For instance, snowboarders identify each other by their clothes, hats, glasses, hairstyle, attitudes toward risk, idols, and so forth. Some years ago, a product technically quite similar to the snowboard, the "monoski," was introduced to the market. The innovation was unsuccessful, though, because no consumer milieu formed around it.

Protomarkets for environmental technologies are of economic interest only if wider parts of the population develop use patterns, lifestyles, and symbolic attachments with regard to the innovative product. Diffusion and transformation of models of consumption may follow different paths. The automobile provides a classic example of one such development of a consumption model.[55] Here, protomarkets emerged in higher-income and high-status strata. Product characteristics, symbolism, and use patterns were socially constructed and demonstrated correspondingly. The innovative product served then as a marker for wealth and affluence. To the extent that lower-income segments were attracted to the new commodity and producers were able to adjust production and prices, the innovation became a mass-produced good.

Among innovative products that have been successfully launched recently, however, consumption patterns have developed differently (Featherstone 1991). Initially, producers and consumers of the innovative product are all kinds of creative people searching for new modes of self-expression and new lifestyles. They are united by either the search for fun and individuality or specific concerns and beliefs, such as environmentalism. Consumption then transmits identity, not status. The Swatch provides perhaps the most well-known and successful example.[56]

We now discuss the implications of these different kinds of market dynamics for lightweight vehicles by relating recent Swiss experience. In particular, we ask whether lightweight vehicles could play a role in the social construction of environmentally benign mobility forms.

7.3.5.2 Exogenous Preferences and Market Niches
No mass-produced commodity called "lightweight vehicle" yet exists. It is therefore difficult to assess its market potential. Market projections must draw on analogies to similar products recently commercialized. Because hybrids are so new—none has been commercialized to date—we focus on lightweight electric vehicles (LEVs), which constitute the majority of LWVs in Switzerland.[57]

The literature presents a mixed picture of electric vehicles' market prospects. A very technical assessment states that 20–50 percent of current U.S. car trips (see Turrentine and Sperling 1992) and 90 percent of current Swiss car trips (see BEW 1990) could be made in electric vehicles. If electric vehicles were accepted widely as an environmentally benign technology, these figures would suggest rather high market shares, so long as people could avail themselves of other means of transport for the remaining trips.[58] However, more detailed analyses of attitudes suggest that only a tiny proportion of the population would accept high prices and inferior product characteristics in return for environmental superiority (Meinig 1994). This shows up most clearly in studies that confront consumers with profiles of product alternatives (Urban, Hauser, and Dholakia 1987).

Investigations into preferences related to a car's limited range, its recharging conditions, and similar aspects suggest market potentials of only 1–2 percent (Segal 1995) and high disutilities for these product characteristics (Bunch, Bradley, and Golob. 1993). The reduction in environmental harm does not compensate in consumers' minds for the disadvantages in performance and convenience.

As noted above, the exogenous preference approach assumes that the market can be segmented into niches. To examine market niches in Switzerland, we conducted a series of 300 telephone interviews with a sample of the German-speaking Swiss population.[59] Switzerland is especially interesting because LEVs have gained considerable public attention and can be assumed to be well-known. In our study, 95 percent of the respondents reported

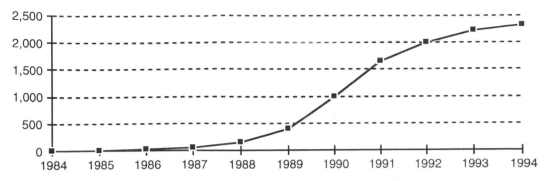

Figure 7.10 Cumulative sales of electric vehicles in Switzerland, 1984–94.

having seen such a vehicle in the street, 57 percent said they had discussed the topic of LEVs extensively with relatives or friends, 36 percent indicated they personally knew a LEV driver, and more than 25 percent said they had already considered buying a LEV for their own transportation needs. Furthermore, approximately 2,500 electric vehicles have been sold in Switzerland in the past ten years (see figure 7.10). This represents about one-quarter of the electric vehicles registered for on-road use worldwide.

To assess LEVs' potential market niches, we defined a series of items related to mobility behavior; the perception of LEVs and their drivers; general concerns and attitudes related to environment, technology and society; and sociodemographic variables. We analyzed dependence by carrying out a categorical logit regression analysis.[60]

The analysis was based on two models. The first model represents the identification of consumers with LEV drivers as the dependent variable. The independent variables refer to various attitudes and other properties of the respondents. The distribution of these variables in the total population then defines a hypothetical market niche that an appropriate product could fill. Because market niches require that customers actually buy, not just identify with, people who use the respective product, we have estimated a second model that relates the independent variables to the willingness to buy a LEV.

For the first model, we offered respondents several images that may be attributed to LEV drivers and asked them to indicate whether they considered "green fundamentalists," "car enthusiasts," "elderly persons," "mothers and housewives," and so forth to be typical LEV drivers. Immediately following, we asked the respondents whether they would consider themselves typical LEV drivers: 37 percent of the respondents answered "yes," whereas 32 percent answered "no." The remaining 31 percent were "undecided."

In the resulting model (see figure 7.11), the independent variables distinguish quite clearly between the population segments that identify themselves with LEV drivers and the ones that do not.[61] As a general pattern, the variables measuring LEV-related characteristics have the strongest parameter

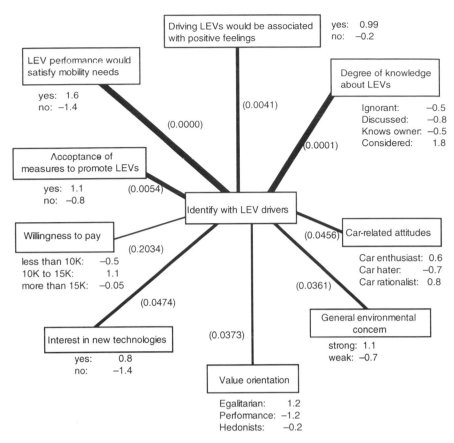

Driving LEVs would be associated with positive feelings
yes: 0.99
no: −0.2

LEV performance would satisfy mobility needs
yes: 1.6
no: −1.4

Degree of knowledge about LEVs
Ignorant: −0.5
Discussed: −0.8
Knows owner: −0.5
Considered: 1.8

(0.0041)

(0.0000)

(0.0001)

Acceptance of measures to promote LEVs
yes: 1.1
no: −0.8

(0.0054)

Identify with LEV drivers

Willingness to pay
less than 10K: −0.5
10K to 15K: 1.1
more than 15K: −0.05

(0.2034)

(0.0456) Car-related attitudes
Car enthusiast: 0.6
Car hater: −0.7
Car rationalist: 0.8

(0.0361)

(0.0474)

(0.0373)

Interest in new technologies
yes: 0.8
no: −1.4

General environmental concern
strong: 1.1
weak: −0.7

Value orientation
Egalitarian: 1.2
Performance: −1.2
Hedonists: −0.2

Goodness of fit = 0.29

Figure 7.11 Logit model describing the proportion of respondents who identify with LEV drivers. Numbers following the response categories stand for the estimated logit-parameter values. Numbers in brackets indicate the overall level of significance of the categorical variable, related to the hypothesis that the parameter values are equal to zero.

values and significance levels. General concerns and beliefs have somewhat weaker levels of significance but still quite strong parameter values. However, sociodemographic and mobility-related variables are poor predictors.

These statistical results suggest that LEVs could find a considerable market niche if consumers decided on the basis of the positive images they associate with LEVs. Information campaigns would improve LEVs' prospects, because the degree of acquaintance has a positive impact on identification. The product message would have to be based on environmental concern, fascinating technology, and positive feelings associated with driving LEVs. The identification might be weak, however, if potential users conceived of LEVs as a snobbish gadget. The egalitarian value orientation together with the high acceptance of specific LEV policy measures found in the interview indicates that product identification depends on a positive image in society at large. Therefore, attempts to market LEVs as a good that the upper-income classes can use to demonstrate environmental concern publicly could prove a bad strategy.

However, identification does not mean buying. To get closer to decision-relevant criteria, we asked the respondents whether they would consider using an LEV as a first or a second car: 17 percent of the sample indicated that they would buy such a vehicle as a first car, and 37 percent indicated they would buy such a vehicle as a second car. The willingness to consider LEVs as a first car did not show any consistent explanation pattern in our analysis. We therefore concentrate on the responses associated with the second-car market.

Willingness to buy an LEV as a second car is best explained by expected satisfaction with its performance and by the positive feelings expected from driving it. General environmental concern and technical interest play a significant but secondary role. Not surprisingly, the willingness to pay is positively related to the stated intent to purchase a LEV. Finally, current car ownership status explains potential second-car ownership significantly but not strongly: nondrivers respond more positively than single-car owners, who, in turn, respond more favorably than multicar owners.

We can therefore conclude that a marketing strategy built on current preference patterns would have to rely on the second-car market. General concerns and attitudes play a much weaker role than identification with LEV drivers. A market niche could be opened if environmental concern, interest in technology, and fun of driving could be combined into a culturally accepted product message. Vehicle prices between $15,000 and $20,000 would then be considered fair. Most probably, however, those parts of the population that consider expanding their mobility options would purchase LEVs. A promotion effort for LEVs would hence lead to a mere increase in rates of vehicle ownership.

7.3.5.3 Endogenous Preference Dynamics and the Social Construction of New Mobility Forms

The results of the previous section suggest that LEVs would reduce GHG emissions only by improving the efficiency of second cars. However, the above results assume that consumers have static preferences. In this section, we discuss the possibilities of preference dynamics concerning lightweight vehicles. Using data from the Swiss LEV consumer milieu, we ask what kind of processes of individual and social learning might go along with the use of LEVs and whether they could be a vector for the social construction of environmentally benign mobility forms.

The early commercialized Swiss LEVs did not meet any broadly accepted standard of performance or reliability. During the pioneering years, performance and reliability improved. Yet the LEV market must still be seen as a protomarket.[62] The early users stayed in contact with each other. They were, in fact, part of the pioneers' milieu. The prototype developers profited from consumers' experiences with everyday use of LEVs. The LEV market itself profited from the symbolic potential of highly performing prototypes, which was interpreted as a promise regarding electric vehicles' future characteristics. The users formed a sort of consumers' milieu. Many users met at fairs, races, and clubs. This milieu had a regional dimension, too. The vast majority of the users lived in the three biggest urban regions of the German-speaking part of Switzerland.

But, who are the pioneer users of LEVs? Are they just a small group of affluent people with high environmental concern? Does the diffusion curve in figure 7.10 indicate that the market is saturated? To answer these questions, we conducted a standardized survey among approximately one-fourth of the Swiss LEV owners in July 1995.[63]

Our sample included users of all ages, from eighteen to eighty-five, with a median of forty-three, predominantly males (88 percent), and predominantly living in two-person households (60 percent). About half had children. In general, their income and education was above average. Typical professions were engineer (architect, computer specialist, electrical engineer), pilot, doctor, teacher, manager, and consultant. They drove a mean of eighteen kilometers a day in their LEV. For only 19 percent of the respondents, the LEV was their first and only car. Environmental concern played a major role in the decision to buy and drive a lightweight electric vehicle. Almost 90 percent of the respondents mentioned environmental concern as an important decision factor, followed by fascination with a new technology (43 percent) and better parking possibilities (31 percent).

Sociodemographically, LEV owners resemble the typical early user described in the literature on innovation diffusion (Rogers 1983). Therefore, such early users may well be the segment that tests and proposes future rules for using and interpreting lightweight vehicles. Moreover, the LEV owners interviewed mentioned being a pioneer as an important motivating factor. About 87 percent perceived themselves as pioneers who support an environmentally sensible technology.

However, this self-understanding did not imply an adverse attitude toward conventional cars. Instead, people incorporated LEVs into the existing array of transport alternatives in their household. This is consistent with the purchase intentions the interviewees representing the German-speaking Swiss population declared. The relatively high rates of interest in and ownership of LEVs in Switzerland might therefore be connected to the existence of an extremely frequent, high-quality public railway service that provides a noncar alternative for long-distance trips.[64] LEVs might actually be used to diversify households' transportation options. It must be mentioned, though, that LEV owners report having changed their initial idea about the vehicles' role when they learned to use them.

The literature on LEVs describes how LEV drivers learn to adapt to electric vehicles' specific handling conditions and discusses their driving behavior, traffic perception, mobility, energy use and perception of spatial structures in the surroundings. As one would expect, more involvement with the technology helps consumers adapt to electric vehicles' performance characteristics

(Sperling 1995). In particular, those product characteristics of electric vehicles that are obviously inferior, namely, limited range and specific recharging conditions, decrease in importance once people have acquired experience (Turrentine and Sperling 1992).

Besides these adaptation processes, the literature on LEVs also reports that learning takes places with respect to driving behavior and traffic perception. Several studies state that LEV drivers drive more cautiously and adapt their driving better to the traffic circumstances (Risser et al. 1993; Streicher 1994). In our survey, 55 percent of the respondents said that they have been driving more cautiously since they purchased their LEV.

With an LEV, drivers become more sensitive to mobility issues. Although most survey respondents bought the LEV as a second car, many LEV owners indicated in preliminary interviews that they were nonetheless using their LEV more often than their gasoline car and that in principle, they could do without the gasoline car altogether. In the survey, 21 percent of the respondents who purchased the LEV as a second car claimed that they could dispense with their gasoline car without difficulty. Furthermore, the LEV owners' total mobility demand (LEVs and gasoline cars) was about 11,000 km a year, slightly lower than the mean found in our survey of the German-speaking Swiss population.[65]

Finally, drivers surveyed had a growing awareness of environmental issues, energy consumption, and the spatial structures in which they lived. Most (70 percent) reported increased consciousness of energy consumption through everyday activities. The vehicles' limited range illustrates tangibly what energy consumption means. Some respondents said that since they have driven their LEV they have differently perceived the structures of urban land use in their surroundings. They have begun to consider it a problem if they have to move over greater distances for their everyday activities. For them, short-distance land use patterns have clearly become more attractive. In the standardized survey, 41 percent responded positively to a corresponding question.

In general, people reported having learned to satisfy their mobility needs more flexibly and to reduce the environmental impact their moving around caused. They met specific transportation needs with the most sensible means of transport according to environmental considerations. LEVs were thought ideal for daily mobility and short trips. Longer trips and heavy transportion could be carried out by public transport or the temporary use of a gasoline car.[66] These learning processes point to an important potential of LEVs. The use of lightweight vehicles on the one hand enlarges a household's range of mobility choices, but at the same time it induces the household's members to satisfy their needs more flexibly. These learning processes could form a basis for a break with the automobile's currently dominant role as a generic means of individual transport.[67]

Hence data from our interviews suggest that an endogenous development of increasing preference for lightweight vehicles would very probably not make them a direct substitute for conventional cars. But LEVs could play a decisive role in the social construction of flexible mobility forms—so-called integrated mobility services—in which a wide range of technologies could replace the gasoline car's hegemonic position.

7.3.6   Innovation-Oriented Policies for Individual Mobility

We have now followed the innovation of the lightweight vehicle all the way from development through manufacturing to market diffusion. We have been eager to illustrate several dimensions of a sociotechnical innovation associated with economies of scale, the social construction of technologies and preferences, the importance of regional contact networks, and the macroinstitutional environment. We must now summarize the theoretical arguments and relate them to policy implications that have arisen in the discussion of options for and barriers to this particular innovation.

With regard to the theoretical dimensions, we have shown that the interaction between social and technical factors is fundamental to understanding the emergence of this radical technological innovation. A cultural protest movement, on the one hand, and an experience with the social redefinition of an industrial product, on the other hand, motivated the development of the technical artifact "lightweight vehicle." The initiatives drew on regionally and historically specific cultural images of clean energy. Both networks profited therefore from a positive response in the media and early users' accumulated experience. Without an understanding of these social processes, it would be difficult to explain the specific technological trajectories followed and the sustained existence of both initiatives over several years.

However, technical competence and cultural background alone may not explain the impressive achievements on the part of technology development. The regional contact networks connecting individuals, firms, and other organizations formed a necessary resource for the innovation process. No individual firm or inventor would have been able to find solutions to the full range of technical problems the technology posed. As to large-scale manufacturing, the pioneers proved unsuccessful in mobilizing a stabilized regional contact network outside the automobile industry. Finally, regional contact networks also played a role in establishing a new product identity for lightweight vehicles. They provided early users with an important opportunity to discuss the specific difficulties they encountered. Similar contact networks could play a role in the development of integrated mobility forms.

Finally, we have seen that increasing returns to scale play a crucial role in all stages of the innovation process and stabilize the current car-dominated mobility system. We have extensively discussed the role of returns in manufacturing. Here, the realization of economies of scale depends on the extent of the market, which in turn depends on production prices for lightweight vehicles, which in turn depend on the scale of production. Without specific policy tools, no change in this deadlock will be possible.

A number of lessons regarding innovation-oriented policies may be learned from the case of lightweight vehicles. From the very beginning, the pioneers profited from public authorities' subsidizing efforts. The sustained existence of the pioneering projects would not have been feasible without these financial inputs. However, public funding was mostly conditional on a considerable proportion of funds' coming from third parties.[68] Swiss policymakers thus wanted to avoid the danger of getting too closely involved with specific projects or of promoting self-exploitation by the pioneers.

However, funds were only a necessary but not a sufficient input for the pioneers' success. Much more important was the provision of expertise and networking services, which public policy officials and students and faculties of universities and engineering schools brought into the pioneers' milieu. As a consequence, boundaries between innovators whose activities relied on public funding and those who had private capital at their avail became blurred in the course of time. Some former university students, for instance, founded their own commercially successful engineering firms and became important players in the network of innovators. The careful management of these networks is probably one of the most important success factors in the case of the Swiss lightweight vehicle.

Networking activity has been more successful in the development of technical prototypes than in the launching of large-scale production, though. As has been argued above, full-scale manufacturing is associated with fundamental uncertainties and investment levels so high as to leave little room for nonprofessional players. In sectors like car production, where economies of scale are pervasive, reducing specific uncertainties could, therefore, be a productive target for an innovation-oriented policy. For instance, defining and supporting particular niches where promising technological alternatives could develop and where learning could occur with regard to large-scale production would be an important strategy. Furthermore, market-based instruments or conventional regulations could play a role insofar as they may break lock-ins of technologies. Our proposal to combine feebates with permits suggests one direction in which such endeavors should move.

The strategic reduction of uncertainties, however, implies identifying specific technologies as worth being promoted. There is no a priori reason why any state-controlled institution should be more competent in doing this than a private enterprise. Decisions about what technologies to promote should therefore be made in open dialogs involving various actors. Voluntary agreements between government and industry would be an appropriate form of intervention at this stage. Narrowly defined technology forcing is likely to meet with manifold difficulties, not least because policymakers will always lack the appropriate information, and trust relationships will be eroded. To build and maintain trust, environmental organizations and consumers should be involved in defining such innovative products from the beginning. In the near future, for instance, it remains to be seen whether the rather closed industry-internal strategy of the joint venture between SMH and Mercedes will be successful.

As stated above, one of the major vectors of uncertainty in the innovation process is market dynamics. To transform individual mobility sustainably, it will be crucial to break consumers' highly stabilized preferences for the present form of gasoline cars. A technical alternative alone, such as lightweight vehicles, cannot initiate this break: Actual consumers will have to change their preferences in everyday contexts. Experiments aiming at establishing new mobility forms may therefore prove an essential tool of an innovation-oriented policy. The lightweight vehicle milieu in Switzerland exemplifies such an experiment that emerged with virtually no state support. However, the recent decline in numbers of vehicles sold indicates the difficulties of wider diffusion. Regionally more concentrated experiments could prove fruitful here.

An interesting example of a such a regionally concentrated experiment has recently started in Switzerland (Schwegler, von Orsouw, and Wyss 1994). It aims to foster processes of social learning regarding the use of lightweight vehicles. In a community with 10,000 inhabitants in the Italian-speaking part of the country, LEVs will be substituted for 8 percent of the cars during the next six years. A series of promotional activities, subsidies, and user advantages (such as preferential parking) will foster the development of new mobility forms. The experiment may show whether lightweight vehicles are actually a promising means of advancing integrated mobility services.

ACKNOWLEDGMENTS

The chapter relies on research financed in the CLEAR framework, supplemented by research financed by the Swiss National Research Program 31 under grant no. 4031-033525 and by EAWAG under grant L+F no. 94-45. The current version of the chapter has profited a great deal from discussions with a large number of people. Special thanks are due to Urs Dahinden, Ottmar Edenhofer, Christopher Flavin, Franco Furger, Gabriela Hüsler, Bernd Kasemir, Amory Lovins, Steve Rayner, Meinrad Rohner, Benno Seiler, Thomas Stadler, Dimitri Sudan, Richard Volz, Matthias Wächter, Arnd Weber, Bernhard Wehrli, and three anonymous referees. The usual disclaimers apply.

NOTES

1. In this chapter, we do not discuss the many possibilities of adaptive regional responses to climate impacts, like building dams to confront sea level rise, and so on. Although this topic no doubt warrants careful research, the solutions such adaptation processes generate would have hardly any impact on the GHG emissions of the respective country or at the global scale. Here we are interested in innovations that yield technologies, products, and institutions that could diffuse to other parts of the world and that along these lines could contribute to a massive reduction of GHG emissions globally.

2. As Norgaard (1994, 92ff) argues, this is rooted in a broader, "mechanistic" world view.

3. This proposition is known as the "first theorem of welfare economics" (Kreps 1990).

4. This has been described as a "seamless web" that connects social and technical realities (see Bijker 1995). Some authors (Callon 1987) have gone even further and pleaded that human and nonhuman elements of an innovation project should be seen as "acting" in a similar way. An innovation will be successful only if all elements behave, from the every beginning, in the way its promoters predict. A sequential model of innovation in which technical problems are solved first and social and economic problems enter only in the second-line place should, therefore, be refused.

5. For a detailed comparison between these approaches see McKelvey 1991.

6. E.g., Johansson and Swisher (1994, 55) estimate that 10 to 30 percent of total GHG emissions could be reduced in most industrialized countries at little or no cost to society.

7. For Switzerland, a CGE model (Stephan, van Nieuwkoop, and Wiedmer 1991, 1992) and a couple of optimization models have been developed in recent years (Kypreos 1990; Pillet et al. 1993; Rothen 1993, 1994).

8. For an overview of integrated assessment models, see Parson 1994.

9. These services include the production of biomass; the absorption of anthropogenic emissions in soils, atmosphere and water circulation; the provision of low ultraviolet light and clean water; and so on.

10. The most elaborate proposals suggest raising the level of the tax continually over a given period of time (e.g., von Weizsäcker 1989).

11. The EU, in particular, has been unable to find a scheme on which all member states can agree. Also, Switzerland recently postponed plans for a CO_2 tax.

12. Arrow's foreword in Arthur 1994 gives a short overview of the history of economies of scale.

13. Jaeger (1990) has emphasized the relevance of innovative regional milieus for the resolution of environmental problems.

14. We call the early contact networks "user milieus" that trade the innovative products in "protomarkets" (cf. case study on lightweight vehicles).

15. Dosi and Orsenigo (1988) use the term "evolutionary equilibria"; these equilibria have similar properties.

16. Assessment models that are able to represent nonmechanistic dynamics of technologies and preferences could take advantage of recent advances in computer simulation and the mathematical modeling of nonlinear systems. For an overview of these methods see, for example, Casti 1992. However, these models are not well suited for the standard "technocratic" exchange between economic experts and policy makers. See chapter 8 as well as Edenhofer 1996.

17. In fact, Cebon (1995) argues on the basis of the literature on persistent market failures of efficient technology retrofits that this is the case only for particular technologies.

18. In the case study on lightweight vehicles, we explicitly develop a policy instrument with an inherent mechanism of self-dissolution.

19. Germany recently undertook a similar initiative, whereby the big-three German car producers have promised to develop a three-liter per 100 kilometer (80 miles per gallon) car by the year 2000. In exchange, the three southern German Länder will refrain from any strong regulation that would endanger the selling of cars in Germany.

20. The Initiative for the Protection of the Alps is a good case in point.

21. The Ticino is located in the southern part of Switzerland close to the Italian border and has about 300,000 inhabitants.

22. The price of running a train or truck can (at least analytically) by broken down into three components: the direct cost of loading and running the train or truck itself, the cost of tracking the train (providing signaling, station, control services) or truck (police, emergency services), and cost caused by congestion (or capacity provision) as well as other social externalities. The company bears the first component directly. The other two components are very difficult to price, both for road and rail.

23. In Switzerland, the "Autopartei" (car party) has garnered several percent of the votes in recent elections and is represented in the Swiss parliament (see Künzi 1994).

24. Gasoline accounted for 62 percent of the total transportation fuels used in 1993, aviation fuels for 20 percent, and diesel for 18 percent (Schweizerische Gesamtenergiestatistik 1993).

25. Enquete-Kommission 1994 gives forecasts and actual growth data for Germany over the last fifty years.

26. In the United States fuel efficiency nearly doubled between 1974 and 1988 (Rayner 1993). However, the transport sector's overall energy consumption and its share in national energy consumption have still risen.

27. Furthermore, the environmental consequences of using and recycling lead batteries in mass-produced vehicles have been questioned (Lave, Hendrickson, and McMichael 1995).

28. Production plans are announced regularly, though. For instance, General Motors announced the production of an electric car for the California market in 1998 but has given no indication that it will be a particularly lightweight car. The French automobile industry also pursues development of electric vehicles actively. The SMART car of SMH and Mercedes will initially be equipped with an internal combustion engine. Electric drives are promised for the distant future.

29. One major bone of contention is how electricity generation should be translated into GHG emissions. Should local or global emissions be taken into account, and should the marginal or average averted emissions be considered? Further issues are the construction materials real-world properties (for instance, the recycling characteristics of aluminium; see Dowlatabadi, Lave, and Russell 1996) and batteries' actual performance.

30. As noted above, about 60 percent of Swiss electricity is generated by hydropower, about 38 percent by nuclear sources, and about 2 percent from fossil fuels.

31. In March 1994, the Swiss government proposed the introduction of a carbon tax to stabilize CO_2 emissions. The tax should amount to 12 Swiss francs (SFr) per ton of CO_2 in 1996, to 24 SFr in 1998, and finally to 36 SFr in the year 2000. This would mean a price increase per liter gasoline of 0.03 SFr in 1996, of 0.06 Sfr in 1998, and of 0.09 Sfr in the year 2000. Current gasoline prices amount to 1.20 Sfr (about one U.S. dollar) per liter.

32. Similar scenarios that essentially support our claims have been calculated for Switzerland (BEW 1992) and for other countries such as the United States (DeCicco and Ross 1994; DeCicco et al. 1993).

33. An interesting comparison case for the Swiss initiatives was the electric vehicle development program of EDF (Eléctricité de France) in the early 1970s (see Callon 1987). However, this project was executed by large, centralized, bureaucratic organizations. The Swiss initiatives, in contrast, started as bottom-up endeavors and developed in a highly unplanned and unpredictable way.

34. SMART is defined as an independent vehicle brand and stands for SMH-Mercedes-ART.

35. See Truffer and Dürrenberger 1997. Maillat, Quévit, and Senn 1993 have elaborated the notion of innovative milieus in detail.

36. The factors that have made Switzerland the seed bed of this kind of initiative have been elaborated elsewhere (see Truffer and Dürrenberger 1994).

37. This number excludes electric vehicles limited to the local area of some Swiss ski resorts.

38. Press releases from the SMH-Mercedes joint venture state that the SMART will be a two-seater that may be equipped with several versions of the propulsion unit. In the first phase, however, it will be produced with a small internal combustion engine consuming about four liters of gasoline per 100 kilometers (60 miles per gallon). More specific product characteristics remain somewhat in the mist, though (Tagesanzeiger, May 18, 1995, 33).

39. This assessment was offered in a workshop with international lightweight vehicle experts and Swiss developers (see Truffer and Dürrenberger 1994), but various authors (Vester 1995; ISATA 1994; Flavin 1993) have also corroborated it.

40. Several pioneers reported that they had achieved their safety goals, although the automobile industry's safety experts repeatedly claimed that lightweight construction and vehicle safety were irreconcilable.

41. We conducted a series of about 70 focused interviews early in 1994.

42. To concentrate on a direct substitution process between lightweight vehicles and conventional cars, we assume that the former will be equipped with a hybrid drive and that consumers will accept it as providing largely the same product services as conventional cars.

43. LWVs protect passengers by bouncing instead of crunching. This technology works well, but it has not yet been juridically acknowledged as meeting safety requirements.

44. We omit the details of how a permit market might be designed. Furger and Truffer (1996) elaborate the potentials of new information technologies in more detail. For an extensive discussion of greenhouse permits in general, see Hahn and Stavins 1995 or Hourcade and Baron 1993.

45. This means that purchase decisions would not be biased toward the vehicle's purchase price, recent changes in fuel price, or the social message conveyed by ownership, for instance. None of these assumptions are trivial.

46. Technically, the price of the gasoline permits may be called a rent. The equilibrium price for permits, which equals the total life cycle costs of conventional cars and hybrids, is known as an intensive rent (Sraffa 1960).

47. Tingtening the gasoline cap would essentially have the same effect. Regulators could hence control the emissions from individual motor traffic quite directly.

48. In fact, permits as an isolated measure could even adversely affects the promotion of radical innovations and thus stabilize existing technologies (see Walter 1989).

49. In fact, following economic theory, the permit price should exactly equal the level of the tax (see Baumol and Oates 1988).

50. Therefore, the new technology would also immediately be relevant for developing countries. It would not remain a luxury technology to calm the bad consciousness of the north but would contribute efficiently to reducing global emission levels.

51. Therefore, feebates and permits demonstrate one of the most important characteristics of an evolutionary policy instrument, as identified by Erdmann (1993), for instance.
Besides exhibiting these strengths, the actual implementation of such a policy would certainly meet with a lot of problems. The deliberate selection of a specific equilibrium is fraught with the danger of creating a host of future problems associated with possible new lock-ins. This is a momentous argument which cannot be dealt with in the context of this chapter. As a general rule, however, fundamental uncertainties related to the choice of any specific equilibrium

should be handled in an open political dialog where all societal representatives may have their say.

52. The permit-feebate scheme would be able to outweigh any minor differences perceived in the respective mobility services, as has already been pointed out.

53. More appropriately, preferences are defined with regard to product characteristics. Consumer products may then be analyzed as bundles of specific characteristics (Lancaster 1991).

54. The literature on consumer behavior, where interpretative approaches have gained considerable attention in recent years, offers this same assessment. See, e.g., Hirschman and Holbrook 1992; Holbrook and Hirschman 1993.

55. Flink (1988) and Sachs (1984) tell the history of the automobile's protomarket dynamics. A more general analysis of taste formation and social differentiation is found in Bourdieu 1979.

56. The Swatch's success was tightly associated with its "classless' image. (Linnenschmidt 1993)

57. In fact, LEVs also represent the majority of electric vehicles in Switzerland. Converted conventional gasoline cars account for only a small proportion of electric vehicles. Our argument, however, does not depend substantially on the technical form of the propulsion unit. Similar concepts have become known as "neighborhood electric vehicles" (Sperling 1995) and "Stadtauto" (urban vehicles; see Hautzinger 1995).

58. In a recently conducted national survey on environmental concern in Switzerland (Diekmann and Franzen 1995), 25 percent of the interviewees stated that they strongly favor a massive transition from gasoline cars to electric vehicles; 45 percent said they are rather in favor; and only 30 percent indicated they would oppose such a transition.

59. The sample was restricted to urban regions of the German-speaking part of Switzerland, where about 60 percent of Swiss population live. The majority of LEVs are driven in these regions.

60. Individual items were combined into categorical variables. These variables were normally formed so as to have two clear categories of pros and cons. Respondents who were not easily attributable to either of the two extremes were classified as "undecided." Dependence among variables was calculated by using fully categorical logit models. The parameters presented here correspond to the "best fit" models as determined in series of evaluations with various combinations of explanatory variables.

61. The goodness of fit value of 0.29 is comfortably high for a categorical logit model.

62. In the early years of the pioneers' milieu, about 2,500 electric vehicles had been sold (see figure 7.10) and another 1,700 were officially registered in 1994 (BfS 1995). In Germany, approximately 1,900 electric cars were registered in 1993 (Knie and Berthold 1995).

63. The questionnaire was sent to 702 LEV owners in the urban regions of the German-speaking part of Switzerland. About 400 questionnaires were returned. The respondents included not only current LEV owners but also people who have given up their vehicle after some time (about 20 percent of the sample). The representativity of the sample was controlled for the vehicle type.

64. Half-hourly or hourly service joins virtually all Swiss cities and towns. Major Swiss cities have service to the rest of Europe approximately every two hours.

65. The travel distances estimated both by LEV owners and by the sample representing the German-speaking Swiss urban population in general may, of course, be subjectively biased.

66. In most cases, this was their own gasoline car. However, the latter became less important to the respondents in the course of time. More flexible forms of gasoline car use could actually

lower the expenditures of a household. Recent experience with car sharing in Switzerland has pointed to this potential (Muheim and Inderbitzin 1992).

67. The automobile's role is that of a vehicle providing high speeds, strong acceleration, and the ability to cover long distances. Canzler and Knie (1994) have coined the term *"Renn-/ Reiselimousine"* (race and travel sedan) for this paradigmatic form of the car.

68. In such projects, federal funds cover no more than 30 percent of the project costs.

GLOSSARY

economies of scale: decreasing average costs in the production of a good or service with the expansion of produced quantities. They constitute a pervasive empirical reality in industrial (mass) production but have hardly received sufficient attention at a theoretical level so far. They lead to specific technological development paths and lock-ins.

feebates: a dynamic and market-conformable policy tool suitable for introducing environmentally benign products. They basically rely on a cross-subsidizing mechanism combining fees charged on harmful products with rebates redistributed on benign products.

Initiative for the Protection of the Alps: a referendum adopted in February 1994 by a slim majority of the Swiss population. The constitutional initiative bans from the road all freight traffic transiting Switzerland across the Alps. It establishes an unprecedented and highly conflictual transport regime in the Swiss Alps.

institutions: provide actors with resources and legitimacy and exclude others from access to these resources. They also empower actors to behave in a certain way, stabilize social behavior in routines, and define relations between different actors.

lightweight vehicles (LWVs): extremely small two or four seaters weighing less than 600 kilograms and therefore consuming considerably less energy than conventional automobiles (about two liters of gasoline per 100 kilometers). An innovative construction design has solved critical safety problems.

multiple equilibria: contrast with the dominant conception of a single equilibrium of quantities and prices of goods demanded and offered in exchange markets. This concept is used as a guiding metaphor in the theoretical analysis of options for and barriers to sociotechnical transformation processes, offering an alternative to the mechanistic approach often found in the climate policy debate.

New Transalpine Railway project (NEAT): consists of two base tunnels crossing the Alps (about 50 kilometers in length) and represents the pillar of an environmentally oriented Swiss transport policy. The new infrastructure aims to shift freight transport from road to rail.

social construction: a concept that hints at the role of social processes in forming and perceiving crucial characteristics of technical artifacts. Preference dynamics may be subject to similar processes of social construction. These processes hence influence innovative products or technologies' success or failure.

social contact networks: a basic trait of social relationships, including interacting individuals as well as firms or collective actors. They may range in scale from local to global, but in everyday life they are generally confined to a regional dimension. Social contact networks play a crucial role in defining and transforming values, attitudes, and preferences. Regional contact networks based on trust relations display a particular innovative capacity in the globalizing world economy.

social learning processes: developments in which actors learn to alter their conception of a specific technology fundamentally . This may happen in the course of either construction or use of the technical artifact. The respective new understanding must, however, transcend the individual level, be made available to other actors, and be accepted by them.

REFERENCES

Abegg. B., and R. Froesch. 1994. Climate change and winter tourism: Impact on transport companies in the Swiss canton of Graubünden. In *Mountain Environments in Changing Climates*, ed. M. Beniston. London: Routledge, pp. 328–341.

Aglietta, M. 1976. *Régulation et crises du capitalisme. L'experience des États-Unis*. Paris: Calmann-Levy.

Amendola, M., and J.-L. Gaffard. 1995. Complémentarité intertemporelle de la production, coordination de l'activité économique et croissance hors de l'équilibre. *Revue Economique* 46(6):1527–41.

Amendola, M., and J.-L. Gaffard. 1988. *La dynamique economique de l'innovation*. Paris: Economica.

Anderson, P. W., K. J. Arrow, and D. Pines. 1988. *The economy as an evolving complex system*. Redwood City, CA: Addison-Wesley.

Appelbaum, E., and R. Batt. 1994. *The New American Workplace: Transforming Work Systems in the United States*. Ithaca, NY: ILR Press.

Arthur, B. W. 1994. *Increasing returns and path-dependence in the economy*. Ann Arbor: University of Michigan Press.

Arthur, B. W. 1989. Competing technologies, increasing returns to scale, and lock-in by historical events. *The Economic Journal*, 99(March):116–31.

Aydalot, P., ed. 1986. *Milieux innovateurs en Europe*. Paris: Groupe de recherche européen sur les milieux innovateurs.

Barthes, R. 1964. *Die Mythen des Alltags*. Frankfurt am Main: Edition Suhrkamp.

Baumol, W. J., and W. E. Oates. 1988. *The Theory of Environmental Policy*. Cambridge: Cambridge University Press.

Baumol, W. J., and E. N. Wolff 1995. Les dynamiques de déséquilibre et le mécanisme de croissance de la productivité. Les implications quant au rôle de la rationalité limitée. *Revue Economique* 46(6):1391–404.

Belk, R. W. 1995. Studies in the new consumer behaviour. In *Acknowledging Consumption*, ed. D. Miller. London: Routledge, pp. 58–95.

Berger, R., and H.-G. Servatius, 1994. *Die Zukunft des Automobils hat erst begonnen*. Munich: Piper.

Bertschi, H.-J. 1985. Der alpenüberquerende Verkehr. Ph.D. diss., Handelshochschule St.Gallen.

BEW. 1992. *CO_2-Perspektiven Verkehr. Sozio-Ökonomische Energieforschung*. Bern, Switzerland: Bundesamt für Energiewirtschaft.

Bundesamt für Energiewirtschaft (BEW). 1990. *Elektromobile—Ein Leitfaden für Interessenten*. Bern, Switzerland: Bundesamt für Energiewirtschaft.

BfS (Bundesamt für Statistik). 1995. *Motorfahrzeugbestand in der Schweiz am 30. September 1994*. Bern, Switzerland: Bundesamt für Statistik.

Bijker, W. E. 1995. Sociohistorical technology studies. In *Handbook of Science and Technology Studies*, ed. S. Jasanoff, G. E. Markle, J. C. Petersen and T. Pinch. London: Sage Publications, pp. 229–257.

Bijker, W.E. 1993. Do not despair. There is a life after constructivism: Science. *Technology & Human Values* 18:113–138.

Bijker, W. E., T. P. Hughes, and T. J. Pinch. 1987. *The Social Construction of Technological Systems*. Cambridge, MA: MIT Press.

Bijker, W. E., and Law, J. 1992. *Shaping Technology/Building Society: Studies in Socio Technical Change*. Cambridge, MA: MIT Press.

Bimber, B. 1994. Three faces of technological determinism. In *Does Technology Drive History?* ed. M. R. Smith and L. Marx. Cambridge, MA: MIT Press, pp. 49–101.

Bliss, D. L. 1988. *The New Oil Crisis and Fuel Economy Technologies: Preparing the Light Transportation Industry for the 1990's*. New York: Quorum Books.

Blümel, H. 1992. CO_2 and pollutant emissions of catalyst-equipped, battery-powered and hybrid cars: A comparison. In *The Urban Electric Vehicle: Policy Options, Technology Trends, and Market Prospects*, ed. OECD. Paris: OECD.

Boehmer-Christiansen, S. 1994a. Global climate protection policy: The limits of scientific advice. Part 1. *Global Environmental Change* 4(2):140–59.

Boehmer-Christiansen, S. 1994b. Global climate protection policy: The limits of scientific advice. Part 2. *Global Environmental Change* 4(3):181–200.

Bourdieu, P. 1979. *La distinction. Critique sociale du jugement*. Paris: Les éditions de minuit.

Breiling, M., and P. Charamza. 1994. Localizing the threats of climate change in mountain environments. In *Mountain Environments in Changing Climates*, ed. M. Beniston. London: Routledge, pp. 341–366.

Brown, P. G. 1992. Climate change and the planetary trust. *Energy Policy* 20(3):208–22.

Brown, P. G. 1991. Why climate change is not a cost/benefit problem. In *Global Climate Change: The Economic Costs of Mitigation and Adaptation*, ed. J. C. White, W. Wagner, and C. N. Beal. New York: Elsevier, pp. 33–45.

Bukold, S. 1991. Die Industrialisierung des Güterverkehrs durch den Container. *Internationales Verkehrswesen* 43(1):264–75.

Bukowiecki, A., and P. Mussler. 1993. Ressourcenbedarf des Leichtmobils. Diplomarbeit ETH-Z (Abt. VIII), ETH Zürich.

Bunch, D., M. Bradley, and T. Golob. 1993. Demand for clean-fuel vehicles in California: A discrete choice stated preference pilot project. *Transportation Research* 27A(3):237–54.

Burt, R. S. 1987. Social contagion and innovation. Cohesion versus structural equivalence. *American Journal of Sociology*, 92(May):1287–335.

Burton, A. 1989. Human dimensions of global climate change. In *Greenhouse Warming: Abatement and Adaptation*, ed. N. C. Rosenberg. Washington D.C.: Resources for the Future.

Button, K. J. 1993. *Transport the Environment and Economic Policy*. Aldershot, Hauts: Elgar.

BUWAL. 1994. *Die globale Erwärmung und die Schweiz: Grundlagen einer nationalen Strategie*. Bern, Switzerland: Bundesamt für Umwelt, Wald, and Landschaft.

Callon, M. 1987. Society in the making: The study of technology as a tool for sociological analylsis. In *The Social Construction of Technological Systems*, ed. W. E. Bijker, T. P. Hughes, and T. J. Pinch, 83–103. Cambridge, MA: MIT Press.

Canzler, W., and A. Knie. 1994. *Das Ende des Automobils. Fakten und Trends zum Umbau der Autogesellschaft*. Heidelberg: C.F. Müller.

Capello, R., and A. Gillespie. 1993. Transport, communications and spatial organisation: Conceptual frameworks and future trends. In *Europe on the Move*, ed. P. Nijkamp. Amsterdam: Nectar, p. 43–67.

CARB (California Air Resources Board) 1994. Low-Emission Vehicle and Zero-Emission Vehicle Program Review. Staff report of the Mobile Sources Division of the Californian Air Resources Board, California Environmental Protection Agency, April 1994.

Casti, J. L. 1992. *Reality Rules: Picturing the World in Mathematics*. New York. John Wiley & Sons.

CCE. 1993. *Croissance, compétitivité, emploi. Les défis et les pistes pour entrer dans le 21ème siècle. Livre blanc*. Brussels: Bulletin des Communautés Européennes (supplément).

Cebon, P. B. 1995. The role of location in industrial environmental behavior. Paper presented to the First Open Meeting of the Human Dimensions of Global Environmental Change Community at Duke University, Durham, NC, June 1–3.

Cebon, P. B. 1992. Twixt cup and lip: Organizational behavior, technical prediction and conservation practice. *Energy Policy* 20(9):802–14.

Cheung, B. 1991. Determing the prospect for a shift in modal split in freight transport. In *Freight Transport and the Environment*, ed. M. Kroon, R. Smit, and J. van Ham. Amsterdam: Elsevier Science Publishers, pp. 223–233.

Chiaromonte, F., G. Dosi, and L. Orsenigo. 1993. Innovative learning and institutions in the process of development: On the micro-foundation of growth regimes. In *Learning and Technological Change*, ed. R. Thomson. New York: St. Martin's Press, pp. 117–150.

Clark, K. B., and T. Fujimoto. 1991. *Product Development Performance: Strategy, Organization, and Management in the World Auto Industry* Boston: Harvard Business School Press.

Cline, W. R. 1994. Costs and benefits of greenhouse gas abatement: A guide to policy analysis. In *The Economics of Climate Change*, ed. OECD. Paris: OECD.

Cohen, S. J. 1993. Climate change and climate impacts: Please don't confuse the two!" *Global Environmental Change* 3(2):2–6.

Collins, H. M. 1992. *Changing Order: Replication and Induction in Scientific Practice*. London: University of Chicago Press.

Commission on Global Governance. 1995. *Our Global Neighbourhood: The Report of the Commission on Global Governance*. Oxford: Oxford University Press.

Cooper, J. 1991. Innovation in logistics: The impact on transport and the environment. In *Freight Transport and the Environment*, ed. M. Kroon, R. Smit, and J. van Ham. Amsterdam: Elsevier Science Publishers, pp. 235–253.

Coopers and Lybrand Co. 1995. *Financial Review of the Neue Alpen-Transversale Project*. Bern, Switzerland: Report for the Swiss Bundesamt für Verkehr.

Cowan, G. A., D. Dines, and D. Meltzer. 1994. *Complexity: Metaphors, Models and Reality*. Reading, MA: Addison-Wesley.

Crevoisier, O. 1991. Functional logic and territorial logic and how they inter-relate in the region. In *Technological Change in a Spatial Context*, ed. E. Ciciotti, N. Alderman, and A. Thwaites. New York: Springer, pp. 17–37.

David, P. 1985. Clio and the economics of QWERTY. *American Economic Review* 75:332–7.

Davis, W. B., M. D. Levine, and K. Train 1995. Analysis of Revenue Neutral Feebates as a Policy Tool for Increasing Auto Fuel Economy. Energy Analysis Program 1994: Annual Report. Energy and Environment Division, Lawrence Berkeley Laboratory, Berkeley, California.

Dean, A. 1994. Costs of Cutting CO_2-Emissions: Evidence from "Top-down" Models. In *The Economics of Climate Change*, ed. OECD. Paris: OECD, pp. 25–42.

de Bont, C. J. P. M. 1992. *Consumer evaluation of Early Product-Concepts*. Delft, the Netherlands: Delft University Press.

DeCicco, J., S. S. Bernow, D. Gordon, D. B. Goldstein, J. W. Holtzclaw, M. R. Ledbetter, P. M. Miller, and H. M. Sachs. 1993. Transportation on a greenhouse planet: A least-cost transition scenario for the United States. In *Transportation and Global Climate Change*, ed. D. L. Greene and D. J. Santini. Washington, DC: American Council for an Energy Efficient Economy, pp. 283–318.

DeCicco, J., and M. Ross. 1994. Improving automotive efficiency. *Scientific American* (December): 30–5.

Diekmann, A., and A. Franzen. 1995. Schweizer Umweltsurvey 1994, Codebuch. Bern, Switzerland: Soziologisches Institut der Universität.

Dierkes, M., and U. Hoffman, eds. 1992. *New Technology at the Outset: Social Forces in the Shaping of Technology*. Frankfurt: Campus.

Dolan, R. J., A. P. Jeuland, and E. Muller. 1986. Models of new product diffusion: Extension to competition against existing and potential firms over time. In *Innovation Diffusion Models of New Product Acceptance*, ed. V. Mahayan and Y. Wind. Cambridge, MA: Ballinger Publishing Company.

Dornbusch, R., and J. M. Poterba. 1992. *Global Warming: Economic Policy Responses*. Cambridge, MA: MIT Press.

Dosi, G. 1988. The nature of the innovative process. In *Technical Change and Economic Theory*, ed. G. Dosi, C. Freeman, R. Nelson, G. Silverberg, and L. Soete. London: Pinter, pp. 221–239.

Dosi, G., C. Freeman R. Nelson, G. Silverberg, and L. Soete. 1988. *Technical Change and Economic Theory*. London: Pinter.

Dosi, G., and L. Orsenigo. 1988. Coordination and transformation: An overview of structures, behaviors and change in evolutionary environments. In *Technical Change and Economic Theory*, ed. G. Dosi, C. Freeman, R. Nelson, G. Silverberg, and L. Soete. London: Pinter, pp. 13–38.

Dower, R. C., and B. M. Zimmerman. 1992. *The Right Climate for Carbon Taxes: Creating Economic Incentives to Protect the Atmosphere*: New York: World Resources Institute.

Dowlatabadi, H. 1995. Factors affecting climate change policies: Uncertainty, subjectivity, inertia and opportunity. In *An Economic Perspective on Climate Change Policies*, ed. M. Thorning. Washington DC: Amercian Council for Capital Foundation Center for Policy Research.

Dowlatabadi, H., L. B., Lave, and A. G. Russell. 1996. A free lunch at higher CAFE? A review of economic, environmental and social benefits. Energy Policy 24:253–264.

Dunlap, R., G. H. Gallup Jr., and A. M. Gallup. 1993. Of global concern: Results of the health of the planet survey. *Environment* 35:7–39.

Dürrenberger, G., C. Jaeger, and B. Truffer. 1995. Potentiale und Hindernisse einer Industrialisierung von Leichtmobilen in der Schweiz. Internal report, EAWAG, Dübendorf., Switzerland.

Eccles, R. G., and N. Nohria. 1992. *Beyond the Hype: Rediscovering the Essence of Management*. Boston: Harvard Business School Press.

European Conference of Ministers of Transport (ECMT). 1993. *Transport Policy and Global Warming*. Paris: OECD.

Ecosens. 1995. *Energetische und gesamtökologische Beurteilung verschiedener Antriebssysteme für Leichtmobile*. Bern, Switzerland: Bundesamt für Energiewirtschaft.

Edenhofer, O. 1996. Das Management globaler Allmenden. In *Sonderheft 36 "Umweltsoziologie" der Kölner Zeitschrift für Soziologie und Sozialpsychologie*, ed. A. Diekmann and C. C. Jaeger. Opladen, Germany: Westdeutscher Verlag, pp. 390–419.

Edge, D. 1995. Reinventing the wheel. In *Handbook of Science and Technology Studies*, ed. S. Jasanoff, G. E. Markle, J. C. Petersen, and T. Pinch. London: Sage Publications, pp. 3–23.

Eliasson, G. 1992. Business competence, organizational learning, and economic growth: Establishing the Smith-Schumpeter-Wicksell connection. In *Entrepreneurship, Technological Innovation and Economic Growth*, ed. F. M. Scherrer and M. Perlman. Ann Arbor: University of Michigan Press, pp. 251–275.

Elzinga, A., and A. Jamison. 1995. Changing policy agendas in science and technology. In *Handbook of Science and Technology Studies*, ed. S. Jasanoff, G. E. Markle, J. C. Petersen, and T. Pinch. London: Sage Publications, pp. 572–626.

Enquete-Kommission "Schutz der Erdatmosphäre" des Deutschen Bundestages, eds. 1994. *Mobilität und Klima. Wege zu einer klimaverträglichen Verkehrspolitik.* Bonn, Germany: Economica Verlag.

Erdmann, G. 1993. Innovation-oriented environmental policy. Zurich: Institut für Wirtschaftsforschung, ETH (no. 93/117).

Ernste, H., and C. Jaeger, eds. 1989. *Information Society and Spatial Structure.* London: Belhaven Press.

Farmer, M. K., and M. L. Matthews. 1991. Cultural difference and subjective rationality: Where sociology connects with the economics of technological change. In *Rethinking Economics*, ed. G. M. Hodgson and E. Screptani. Aldershot, England: Edward Elgar, pp. 63–117.

Featherstone, M. 1991. *Consumer Culture and Postmodernism.* London: Sage.

Flavin, C. 1993. Die neue Auto-Revolution. *World-Watch* 2(4):28–36.

Flavin, C., and N. Lenssen. 1995. *Power Surge: Guide to the Coming Energy Revolution.* New York: W. W. Norton & Company.

Flink, J. J. 1988. *The Automobile Age.* Cambridge, MA: MIT Press.

Frenzen, J., P. M. Hirsch, and P. C. Zerillo. 1994. Consumption, preferences and lifestyles. In *Handbook of Economic Sociology*, ed. N. J. Smelser and R. Swedberg. Princeton, NJ: Princeton University Press, pp. 403–423.

Furger, F., and B. Truffer. 1996. Umweltzertifikate und die ökologische Umgestaltung moderner Gesellschaften. In *Energiepolitik*, ed. H. G. Brauch. Heidelberg, Germany: Springer.

Furuholt, E. 1993. Life cycle analysis of PIV and conventional small cars. Mimeo graphed. Oslo, Norway.

Geller, H. S. 1991. Saving money and reducing the risk of climate change through greater energy efficiency. In *Global Climate Change: The Economic Costs of Mitigation and Adaptation*, ed. J. C. White, W. Wagner, and C. N. Beal. New York: Elsevier, pp. 175–211.

Goldemberg, J. 1987. *Energy for a Sustainable World.* Washington, DC: World Resources Institute.

Gore, A. 1992. *Earth in Balance—Ecology and Human Spirit.* Boston: Houghton Mifflin.

Goulder, L. H., and S. H. Schneider. 1995. The costs of averting climate change. A technological bias in standard assessments. Paper presented at the First Open Meeting of the Human Dimensions of Global Environmental Change Community at Duke University, Durham, NC, June 1–3.

Graf, H. G. 1995. *Perspektiven des schweizerischen Güterverkehrs 1992–2015.* Bern, Switzerland: Generalsekretariat Eidgenössisches Verkehrs-und Euergiewirtschafts departement, Dienst für Gesamtverkehrsfragen.

Granovetter, M. 1991. The social construction of economic institutions. In *Socio-Economics: Towards a new synthesis*, ed. A. Etzioni and P. R. Lawrence. New York: Sharpe Inc, pp. 75–85.

Granovetter, M. 1985. "Economic action and social structure: The problem of embeddedness. *American Journal of Sociology* 91(3):481–510.

Gwilliam, K. M. 1993. On reducing transport's contribution to global warming. In *Transport Policy and Global Warming*, ed. ECMT. Paris: OECD, pp. 7–25.

Hahn, R. W., and R. N. Stavins. 1995. Trading in greenhouse permits: A critical examination of design and implementation issues. In *Shaping National Responses to Climate Change: A Post-Rio Guide*, ed. H. Lee. Washington, DC: Island Press.

Hansen, N. 1992. Competition, trust, and reciprocity in the development of innovative regional milieus. *Papers in Regional Science* 71:95–105.

Harvey, D. 1989. *The Condition of Postmodernity*. Cambridge, MA: Basil Blackwell.

Hautzinger, H. 1995. Neue Ergebnisse zur PKW-Nutzung: Konsequenzen für innovative Fahrzeugkonzepte. In *Stadtauto: Mobilität, Ökologie, Ökonomie, Sicherheit*, ed. H. Appel. Brunswick, Germany: Vieweg, pp. 70–84.

Hepworth, M. 1992. *Transport in the Information Age: Wheels and Wires*. London: Belhaven Press.

Hirschman, E. C., and M. B. Holbrook. 1992. *Postmodern Consumer Research: The study of Consumption as Text*. London: Sage Publications.

Hodgson, G. M., and B. Screptani, eds. 1991. *Rethinking Economics: Markets Technology and Economic Evolution*. Aldershot, England: Edward Elgar.

Holbrook, M. B., and E. C. Hirschman. 1993. *The Semiotics of Consumption: Interpreting Symbolic Consumer Behavior in Popular Culture and Works of Art*. Berlin: Mouton de Gruyter.

Homer-Dixon, T. 1995. The ingenuity gap: Can poor countries adapt to resource scarcity? *Population and Development Review* 21(3):587–612.

Hourcade, J. C., and R. Baron. 1993. Tradeable permits. In *International Economic Instruments and Climate Change*, ed. OECD. Paris: OECD.

Hughes, T. P. 1987. The evolution of large technological systems. In *The Social Construction of Technological Systems*, ed. W. E. Bijker, T. P. Hughes, and T. J. Pinch. Cambridge: MIT Press, pp. 51–83.

International Energy Agency (IEA). 1993. *Cars and climate change*. Paris: OECD/IEA.

Irwin, A., S. George, and P. Vergragt. 1994. The social management of environmental change. *Futures* 26(3):323–34.

International Symposium on Automative Technology and Automation (ISATA), ed. 1994. *Proceedings of the dedicated conferences on electric, hybrid and alternative fuel vehicles and supercars*. Aachen, Germany: ISATA.

Isenmann, T. 1993. Marktwirtschaftliche Massnahmen im Verkehr. Report no. 65 of the Swiss National Science Foundation Research Programme 25: City and Traffic., Bern.

Jacobs, M. 1994. The limits of neoclassicism: Towards an institutional environmental economics. In *Social Theory and the Global Environment*, ed. by Redclift, M. and Benton, T. London: Routledge, pp. 67–92.

Jaeger, C. C. 1990. Innovative milieus and environmental awareness. *Sociologia Internationalis* 28(2):205–16.

Jaeger, C. C., and B. Kasemir. 1996. Climatic risks and rational actors. *Global Environmental Change* 6(1): 23–36.

Johansson, T. B., and J. N. Swisher. 1994, pp. 43–58. Perspectives on "bottom-up" analyses of the costs of CO_2. In *The Economics of Climate Change*, ed. OECD. Paris: OECD.

Karrer-Rüedi, E. 1992. *Der Trend zum Wirtschaftsstil der flexiblen Spezialisierung: Eine Diskussion am Beispiel der Schweizer Uhrenindustrie*. Bern, Switzerland: Peter Lang.

Knie, A. 1994. *Wankel-Mut in der Autoindustrie. Aufstieg und Ende einer Antriebsalternative.* Berlin: Edition Sigma.

Knie, A. 1992. Yesterday's decisions determine tomorrow's options: The case of the mechanical typewriter. In *New Technology at the Outset,* ed. M. Dierkes and U. Hoffman. Frankfurt: Campus.

Knie, A., and O. Berthold. 1995. Das ceteris paribus Syndrom in der Mobilitätspolitik. Internal discussion paper, Wissenschaftszentrum für Sozialforschung, Berlin.

Kreps, D. M. 1990. *A Course in Microeconomic Theory.* New York: Harvester Wheatsheaf.

Krugman, P. 1995. Technological change in international trade. In *Handbook of the Economics of Innovation and Technology,* ed. P. Stoneman. Oxford. Blackwell, pp. 342–366.

Krugman, P. 1991. *Geography and Trade.* Cambridge, MA: MIT Press.

Künzi, D. 1994. Das Automobil—Kathedrale der Moderne. Kulturelle Dimension der Verkehrsthematik. Research paper, Soziologisches Institut, Universität Zürich.

Kypreos, S. 1990. Energy scenarios for Switzerland and emission control, estimated with a normative model. *PSI-Bericht* 70(Juni):1–98.

Lancaster, K. 1991. *Modern Consumer Theory.* Brookfield, VT: Edward Elgar Publishing Company.

Läpple, D., ed. 1993. *Güterverkehr, Logistik und Umwelt.* Berlin: Sigma.

Lave, L. B., C. T. Hendrickson, and F. C. McMichael. 1995. Environmental implications of electric cars. *Science* 268: 993–5.

Lee, H., ed. 1995. *Shaping National Responses to Climate Change: A Post-Rio Guide.* Washington, DC: Island Press.

Linnenschmidt, B. 1993. Die Swatch-Strategie. Eine Chance für die deutsche Automobilindustrie? Diplomarbeit, Wirtschafts- und Sozialwissenschaftliche Fakultät, Universität Nürnberg.

Schweizerische Liga für rationelle Verkehrswirtschaft (LITRA). 1993. *Vademecum 1993: Data and facts on traffic.* Bern, Switzerland: Information Services for Public Transport.

Lorenz, E. H. 1992. Trust, community and cooperation: Toward a theory of industrial districts. In *Pathways to Industrialization and Regional Development,* ed M. Storper and A. J. Scott. New York: Routledge, pp. 195–205.

Lovins, A. B. 1977. *Soft Energy Paths: Toward a Durable Peace.* Cambridge, MA: Friends of the Earth.

Lovins, A. B., J. W. Barnett and L. H. Lovins. 1993. *Supercars: The Coming Light-Vehicle Revolution.* Snowmass, CO: The Rocky Mountains Institute.

Lovins, A. B., and L. H. Lovins. 1995. Reinventing the wheel. *The Antlantic Monthly* (January):75–93.

Lundvall, B., ed. 1992. *National Systems of Innovation: Towards a Theory of Innovation and Interactive Learning.* London: Printer Publishers.

MacKenzie, J. J. 1994. *The Keys to the Car: Electric and Hydrogen Vehicles for the 21st Century.* New York: World Resources Institute.

Maillat, D., B. Lecoq, F. Nemeti, and M. Pfister. 1995. Technology district and innovation: The case of the Swiss Jura Arc. *Regional Studies* 29(3):251–63.

Maillat, D., M. Quévit, and L. Senn, eds. 1993. *Réseaux d'innovation et milieux innovateurs: un pari pour le développement régional.* Neuchâtel, Switzerland: EDES, presse universitaire.

McCracken, G. 1988. *Culture and Consumption.* Bloomington: Indiana University Press.

McKelvey, M. 1991. How do national systems of innovation differ? A critical analysis of Porter, Freeman, Lundvall and Nelson. In *Rethinking Economics*, ed. G. M. Hodgson and E. Screptani. Aldershot, England: Edward Elgar, pp. 117–138.

Meinig, W. 1994. Marktforschung und Marketing für die Mobilität. In *Zweites Europäisches Symposium Solar- und Elektromobile* Universität Regensburg, Regensburg, Germany, pp. 43–53.

Metcalfe, S. 1995. The economic foundation of technology policy: Equilibrium and evolutionary perspectives. In *Handbook of the Economics of Innovation and Technology*, ed. P. Stoneman. Oxford: Blackwell, pp. 409–513.

Meyer, J. H. 1993. Wer oder was steckt hinter dem Verwirrspiel um das mysteriöse Swatch-Mobil? *Tagesanzeiger* (4 März):73.

Michaelis, L. D. Bleviss, J. P. Orfeuil, and R. Pischinger. 1996. Mitigation options in the transportation sector. In *Climate Change 1995: Impacts, adaptations and mitigation of climate change*, ed. R. T. Watson, M. C. Zinyowera, R. H. Moss, and D. J. Dokken. Cambridge: Cambridge University Press, pp. 679–712.

Michman, R. D. 1991. *Lifestyle Market Segmentation*. New York: Praeger.

Miller, D. 1995. *Acknowledging Consumption: A Review of New Studies*. London: Routledge.

Mintzer, I. M., and J. A. Leonard. 1994. *Negotiating Climate Change. The Inside Story of the Rio Conference*. Cambridge: Cambridge University Press.

Mokyr, J. 1990. *The Lever of Riches: Technological Creativity and Economic Progress*. New York: Oxford University Press.

Morcheoine, A., and G. Chaumin. 1992. Energy efficiency, emissions and costs: What are the advantages of electric vehicles? In *The Urban Electric Vehicle: Policy Options, Technology Trends, and Market Prospects*, ed. OECD, Paris: OECD, pp. 115–124.

Muheim, P., and J. Inderbitzin. 1992. *Das Energiesparpotential des gemeinschaftlichen Gebrauchs von Motorfahrzeugen als Alternative zum Besitz eines eigenen Autos*. Bern, Switzerland: Bundesamt für Energiewirtschaft.

Nell, E. J. 1993. Transformational growth and learning: Developing craft technology into scientific mass production. In *Learning and Technological Change*, ed. R. Thomson. New York: St. Martin's Press, pp. 217–243.

Nelson, R. R. 1995. Recent theorizing about economic change. *Journal of Economic Literature* 33:48–90.

Nelson, R. R. 1993. Technical change as cultural evolution. In *Learning and Technological Change*, ed. R. Thomson. New York: St. Martin's Press, pp. 9–24.

Nelson, R. R., and S. G. Winter. 1982. *An Evolutionary Theory of Economic Change*. Cambridge, MA: Bellknap Press.

Niggli, P. 1991. Das Swatchmobil. Unpublished discussion paper, EAWAG, Dübendorf Switzerland.

Nijkamp, P., and J. M. Vleugel, eds. 1994. *Missing Transport Networks in Europe*. Avebury, U.K.: Aldershot.

Nordhaus, W. D. 1994. *Managing the Global Commons: The Economics of Climate Change*. Cambridge: MIT Press.

Nordhaus, W. D. 1991. To slow or not to slow: The economics of the greenhouse effect. *Economic Journal* 101:920–37.

Norgaard, R. B. 1994. *Development Betrayed. The End of Progress and a Coevolutionary Revisioning of the Future*. London: Routledge.

Nyström, H. 1990. *Technological and Market Innovation: Strategies for Product and Company Development*. Chichester, U.K.: John Wiley and Sons.

OECD. 1992. *The Urban Electric Vehicle: Policy Options, Technology Trends, and Market Prospects*. Proceedings of an International Conference, Stockholm, Sweden, May 25–27. Paris. OECD.

OECD. 1993a. *Electric Vehicles: Technology, Performance and Potential*. Paris: OECD/IEA.

OECD. 1993b. *International Economic Instruments and Climate Change*. Paris: OECD.

OECD. 1994. *The Economics of Climate Change*. Paris: OECD.

OFEFP (Office fédéral de l'environnement, des forêts et du paysage). 1994. *Rapport de la Suisse 1994. Convention-cadre des Nations Unies sur les changements climatiques.* Bern, Switzerland: OFEFP.

Okken, P. A., R. J. Swart, and S. Zwerver. 1989. *Climate and Energy: The Feasibility of Controlling CO_2 Emissions*. Dordrecht, the Netherlands: Kluwer Academic Publishers.

O'Laughlin, K. A., J. Cooper, and E. Cabocal, 1993. *Reconfiguring European Logistics Systems*. Oak Brook: Council of Logistics Management.

Oravetz, M., and H. Dowlatabadi. 1995. Is there autonomous energy efficiency improvement? Internal discussion paper, Carnegie-Mellon University, Pittsburgh, PA.

O'Riordan, T., and J. Cameron, eds. 1995. *Interpreting the Precautionary Principle*. London: Earthscan.

Orr, D. W. 1993. Agriculture and global warming: Agriculture. *Ecosystems and Environment* 46(1):81–8.

Parson, E. A. 1994. Searching for integrated assessment: A preliminary investigation of methods and projects in the integrated assessment of global climate change. Paper presented at the third meeting of the CIESIN-Harvard Commission on Global Environmental Change Information Policy.

Petersen, R., and K. O. Schallaböck. 1995. *Mobilität für morgen. Chancen einer zukunftsfähigen Verkehrspolitik*. Basel Switzerland: Birkhäuser Verlag.

Pfeffer, J., and G. R. Salancik. 1978. *The External Control of Organizations: A Resource Dependence Perspective*. New York: Harper and Row.

Pillet, G., W. Hediger, S. Kypreos, and C. Corbaz. 1993. The economics of global warming. National and international climate policy. The requisites for Switzerland. *PSI-Bericht No. 93–02* (Mai).

Piore, M., and C. Sabel. 1984. *The Second Industrial Divide: Possibilities for Prosperity*. New York: Basic Books.

Porter, M. 1990. *The Competitive Advantage of Nations*. London: Macmillan.

Porter, M. E., and C. van der Linde, 1995. Towards a new conception of the environment-competitiveness relationship. *Journal of Economic Perspectives* 9(4):97–118.

Powell, W. W. 1991. Expanding the scope of institutional analysis. In *The New Institutionalism in Organizational Analysis*, ed. W. W. Powell and P. J. DiMaggio. Chicago and London: University of Chicago Press, pp. 183–204.

Powell, W. W., and P. J. DiMaggio, eds. 1991. *The New Institutionalism in Organizational Analysis*. Chicago and London: University of Chicago Press.

Ratti, R. 1982. Les relations commmerciales européennes à travers les Alpes: l'espace de marché du St-Gothard. *DISP* 68:23–31.

Ratti, R., and R. Rudel. 1993. Tableau de l'évolution des transports dans l'arc alpin. *Revue de géographie alpine* 4:11–26.

Rayner, S. 1993. Prospects for CO_2 emissions reduction policy in the USA. *Global Environmental Change 3(1)*:12–31.

Risser, R., C. Chaloupka, K. Liedl, and M. Stockinger. 1993. Studie über das Fahrverhalten beim Lenken von Leichtmobilen. Studie im Auftrag der Universität Zürich.

Rogers, E. M. 1983. *The Diffusion of Innovations.* New York: The Free Press.

Rohner, M., and O. Edenhofer. 1996. Kann sich Klimapolitik auf die Nutzen-Kostenanalyse verlassen? In *Klimapolitik*, ed. by H.-G. Brauch. Berlin: Springer Verlag, pp. 153–168.

Romer, P. M. 1994. The origins of endogeneous growth. *Journal of Economic Perspectives 8(1)*:3–22.

Rosenberg, N. 1994. *Exploring the Black Box: Technology, Economics and History.* Cambridge: Cambridge University Press.

Rosenberg, N. J., P. R. Crosson, K. D. Frederick, W. E. Easterling III, M. S. McKenney, M. D. Bowes, R. A. Sedjo, J. Darmstadter, L. A. Katz, and K. M. Lemon. 1993. The MINK methodology: Background and baseline. *Climatic Change 24(1–2)*:7–22.

Rothen, S. M. 1995. Hemm- und Förderfaktoren für Energiesparinvestitionen im Gebäudeberreich. Internes Diskussionspapier, EAWAG, Dübendorf.

Rothen, S. M. 1994. Carbon dioxide reduction in an optimization model: A case study for Switzerland. *Swiss Journal of Economics and Statistics 130(2)*:145–70.

Rothen, S. M. 1993. *Kohlendioxid und Energie. Eine Untersuchung für die Schweiz.* Chur, Zürich: Verlag Rüegger.

Rothman, D. S., and D. Chapman. 1991. A critical analysis of climate change policy research. In *Global Climate Change: The Economic Costs of Mitigation and Adaptation*, ed. J. C. White, W. Wagner, and C. N. Beal. New York: Elsevier, pp. 285–303.

Ruijgrok, C., and S. Wandel. 1992. *Spatial and Structural Changes in Logistics.* Amsterdam: International Symposium, March 18–21, Nectar.

Sabel, C. S. 1989. Flexible specialisation and the re-emergence of regional economies. In *Reversing Industrial Decline? Industrial Structure and Policy in Britain and Her Competitors*, ed. P. Hirst and J. Zeitlin. Oxford: Berg.

Sachs, W. 1984. *Die Liebe zum Automobil. Ein Rückblick auf die Geschichte unserer Wünsche.* Reinbek bei Hamburg: Rowohlt.

Sanstad, A. H., and R. B. Howarth. 1994. "Normal" markets, market imperfections and energy efficiency. *Energy Policy 22(10)*:811–8.

Sathaye, J., and M. Walsh. 1992. Transportation in developing nations: Managing the institutional and technological transition to a low-emission future. In *Confronting Climate Change*, ed. I. M. Mintzer. Cambridge: Cambridge University Press, pp. 195–217.

Schaefer, H. 1992. Elektrostrassenfahrzeuge im Spannungsfeld zwischen Gesellschaft und Umwelt. In *Elektrostrassenfahrzeuge*, ed. VDI (Verband Deutscher Ingenieure). Düsseldorf, Germany: VDI Verlag, pp. 1–16.

Schmidheiny, S., and the Business Council for Sustainable Development. 1992. *Changing Course—A Global Perspective on Development and the Environment.* Boston: MIT Press.

Schot, J. 1992. Constructive technology assessment and technology dynamics: The case of clean technologies. *Science, Technology & Human Values* 17:36–56.

Schot, J., R. Hoogma, and B. Elzen. 1994. Strategies for shifting technological systems: The case of the automobile. *Futures 26(10)*:1060–76.

Schwegler, U., M. von Orsouw, and W. Wyss, 1994. *Grossversuch mit Leicht-Elektromobilen. Vorstudie.* Bern, Switzerland: Bundesamt für Energiewirtschaft.

Schweizerische Gesamtenergiestatistik. 1993. Bern, Switzerland, Bundesamt für Engergiewirtschaft.

Schweizerische Gesamtenergiestatistik. 1996. Bern Switzerland: Bundesamt für Energiewirtschaft.

Scott, A. J. 1993. *Electric Vehicle Manufacturing in Southern California: Current Developments, Future Prospects.* Los Angeles: The Lewis Center for Regional Policy Studies, University of California.

Scott, A. J. 1988. *Metropolis: From the Division of Labor to Urban Form.* London: Pion.

Scott, A. J., and D. L. Rigby. 1995. Economic geography in action: Building an electric vehicle complex in Southern California. *Environment and Planning A* 27(6):831–4.

Scott, W. R. 1992. *Organizations: Rational, Natural, and Open Systems.* 3d ed. Englewood Cliffs, NJ: Prentice-Hall.

Sebenius, J. K. 1995. Overcoming obstacles to a successful climate convention. In *Shaping National Responses to Climate Change: A Post-Rio Guide*, ed. H. Lee. Washington, DC: Island Press.

Segal, R. 1995. Forecasting the market for electric vehicles in California using conjoint analysis. *Energy Journal* 16(3):89–112.

Seiler, B. 1993. Das Elektromobil. Eine sinnvolle Alternative zum Auto? Diplomarbeit ETH Zürich.

Sen, A. 1995. Rationality and social choice. *The American Economic Review* 85(1):1–24.

Shiller, J. W., P. D. McCarthy, and M. A. Shiller. 1991. Dealing with the economic costs of climate change mitigation: A perspective from the automobile industry. In *Global Climate Change: The Economic Costs of Mitigation and Adaptation*, ed. J. C. White, W. Wagner, and C. N. Beal. New York: Elsevier, pp. 77–111.

Shipper, L., S. Meyers, R. B. Howarth, and R. Steiner. 1992. *Energy Efficiency and Human Activity: Past Trends, Future Prospects.* New York: Cambridge University Press.

Shipper, L., R. Steiner, and S. Meyers. 1993. Trends in transportation energy use, 1970–1988: An international perspective. In *Transportation and Global Climate Change*, ed. D. L. Greene and D. J. Santini. Washington, DC: American Council for an Energy Efficient Economy, pp. 51–90.

Smelser, N. J., and R. Swedberg, eds. 1994. *The Handbook of Economic Sociology.* Princeton, NJ: Princeton University Press.

Sørensen, K. H., ed. 1994. *The Car and Its Environments: The Past, Present and Future of the Motorcar in Europe.* Brussels, Belgium: COST A4.

Sørensen, K. H., and N. Levold. 1992. Tacit knowledge, heterogeneous engineers and embodied technology. *Science, Technology & Human Values* 17:13–35.

Sørgaard, J., and K. H. Sørensen. 1994. Mobility and modernity: Towards a sociology of the car. In *The Car and Its Environments: The Past, Present and Future of the Motorcar in Europe*, ed. K. H. Sørensen. Brussels, Belgium: COST A4, pp. 1–32.

Sperling, D. 1995. *Future Drive: Electric Vehicles and Sustainable Transportation.* Washington, DC: Island Press.

Sperling, D., ed. 1989. *Alternative Transportation Fuels: An Environmental and Energy Solution.* New York: Quorum Books.

Sraffa, P. 1960. *The Production of Commodities by the Means of Commodities: Prelude to a Critique of Economic Theory.* Cambridge: Cambridge University Press.

Stein, D. L., and L. Nadel. 1990. *Lectures in Complex Systems*. Redwood City, CA: Addison-Wesley.

Stephan, G., R. van Nieuwkoop, and T. Wiedmer. 1992. Social incidence and economic costs of carbon limits. *Environmental and Resource Economics* 2:569–91.

Stephan, G., R van Nieuwkoop, and T. Wiedmer. 1991. Kohlendioxid-Abgabe und Auswirkungen auf die Schweiz. Eine wirtschaftliche Analyse. *Mitteilungsblatt für Konjunkturfragen* (December).

Storper, M., and R. Walker. 1989. *The Capitalist Imperative: Territory, Technology, and Industrial Growth*. Cambridge, MA: Basil Blackwell.

Stram, B. N. 1995. A carbon tax strategy for global climate change. In *Shaping National Responses to Climate Change. A Post-Rio Guide*, ed. H. Lee. Washington DC: Island Press.

Streicher, W. 1994. *Begleitforschung zur Förderaktion von Elektromobilen*. Vienna: Bundesministerium für Wissenschaft und Forschung.

Teubal, M. 1979. On user needs and need determination: Aspects of a theory of technological innovation. In *Industrial Innovation: Technology, Policy, Diffusion*, ed. M. J. Baker. London: MacMillan.

Thomsen, E. F. 1992. *Prices and Knowledge: A Market-Process Perspective*. London: Routledge.

Thomson, R., ed. 1993. *Learning and Technological Change*. New York: St. Martin's Press.

Tietenberg, T., and D. G. Victor. 1994. *Combating Global Warming: Possible Rules, Regulations and Administrative Arrangements for a Global Market in CO_2 Emission Entitlements*. New York: United Nations.

Tobey, J., J. Reilly, and S. Kane. 1992. Economic implications of global climate change for world agriculture. *Journal of Agricultural and Resource Economics* 17(1):195–204.

Truffer, B., and G. Dürrenberger. 1997. Outsider initiatives in the reconstruction of the car: The case of light weight vehicle milieus in Switzerland. In *Science, Technology, & Human Values* 22:207–34.

Truffer, B., G. Dürrenberger, and C. C. Jaeger. 1994. Protomarkets for light weight vehicles. In *Dedicated Conference on Electric, Hybrid & Alternative Fuel Vehicles and Supercars (Advanced Ultralight Hybrids)*, ed. ISATA. Croydon, U.K.: Automotive Automation Limited, pp. 785–796.

Turrentine, T., and D. Sperling. 1992. How far can the electric vehicle market go on 100 miles? In *The Urban Electric Vehicle*, ed. OECD. Paris: OECD, pp. 259–269.

Tushman, M. L., and P. Anderson. 1986. Technological discontinuities and organizational environments. *Administrative Science Quarterly* 31:439–65.

Umiker-Sebeok, J., ed. 1987. *Marketing and Semiotics: New Directions in the Study of Signs for Sale*. Berlin: Mouton de Gruyter.

Urban, G. L., J. R. Hauser, and N. Dholakia. 1987. *Essentials of New Product Management*. Englewood Cliffs, NJ: Prentice-Hall.

Utterback, J. M. 1994. *Mastering the Dynamics of Innovation: How Companies Can Seize the Opportunities in the Face of Technological Change*. Boston: Harvard Business School Press.

Varian, H. R. 1992. *Microeconomic Analysis*. (3d ed.) New York: W. W. Norton & Company.

Vester, F. 1995. *Crashtest Mobilität: Die Zukunft des Verkehrs*. Munich: Willhelm Heyne Verlag.

Victor, D. G., and J. E. Salt. 1995. Keeping the climate treaty relevant. *Nature* 373:280–2.

Vittadini, M. R. 1992. *Il quadro della domanda di trasporto. Il trasporto di merci e di persone attraverso le alpi. Situazione e prospettive di evoluzione*. Milan: Istituto regionale di ricerca della Lombardia, pp. 7–127.

von Hippel, E. 1988. *The Sources of Innovation*. New York: Oxford University Press.

von Weizsäcker, E. U. 1989. *Erdpolitik. Ökologische Realpolitik an der Schwelle zum Jahrhundert der Umwelt*. Darmstadt, Germany: Wissenschaftliche Buchgesellschaft.

von Weizsäcker, E. U., A. B. Lovins, and L. H. Lovins. 1995. *Faktor Vier. Doppelter Wohlstand— hulbierter Naturverbrauch*. Munich: Droemer Knaur.

Wachs, M., and M. Crawford, eds. 1992. *The Car and the City*. Ann Arbor: University of Michigan Press.

Waldorp, M. M. 1992. *Complexity: The Emerging Science at the Edge of Order and Chaos*. New York: Simon and Schuster.

Wallace, D. 1995. *Environmental Policy and Industrial Innovation: Strategies in Europe, the U.S. and Japan*. London: The Royal Institute of International Affairs, Energy and Environment Programme, Earthscan.

Walsh, M. P. 1993. Motor vehicle trends and their implications for global warming. In *Transport Policy and Global Warming*, ed. ECMT. Paris: OECO, pp. 69–95.

Walter, J. 1989. *Innovationsorientierte Umweltpolitik bei komplexen Umweltproblemen*. Heidelberg: Physica Verlag.

Walz, F. 1995. Das Sicherheitspotential von kleinen Leichtfahrzeugen. In *Stadtauto. Mobilität, Ökologie, Ökonomie, Sicherheit*, ed. H. Appel. Vieweg, Braunschweig.

Weick, K. E. 1995. *Sensemaking in Organizations*. London: Sage Publications.

Williamson, O. E. 1975. *Markets and Hierarchies. Analysis and Antitrust Implications*. New York: The Free Press.

Witt, U. 1991. Reflections on the present state of evolutionary economic theory. In *Rethinking Economics*, ed. G. M. Hodgson and E. Screptani. Aldershot, U.K.: Edward Elgar, pp. 83–103.

Womack, J. P., D. T. Jones, and D. Roos. 1990. The machine that changed the world. New York: Rawson.

Wynne, B. 1993. Implementation of greenhouse gas reductions in the European Community. Institutional and cultural factors. *Global Environmental Change* 3(1):101–28.

8 Regional Integrated Assessment and the Problem of Indeterminacy

Claudia Pahl-Wostl, Carlo C. Jaeger, Steve Rayner, Christoph Schär, Marjolein van Asselt, Dieter M. Imboden, and Andrej Vckovski

8.1 INTRODUCTION

The previous chapters have discussed key processes relevant to an assessment of climate change and its impacts on the Alpine region. Chapter 2 has provided an overview of what is known about today's climate in the Alpine region where, as in any mountainous area, climate displays high variability on small spatial scales and microclimate is contingent on local conditions and topography. At the same time, larger-scale features of the climate system, the *Grosswetterlagen*, strongly influence the Alpine climate's overall characteristics. This combination makes forecasting climate change and its impacts a difficult task. The Alpine region has already experienced considerable changes in climate over the past few millenia. Chapter 3 has shown how paleo-climatological tools can be used to extend our knowledge about climate, its variations, and its impacts on ecological systems, enabling us to put anticipated climate change in perspective, relating it to changes experienced in the past. Convincing evidence suggests that anthropogenic climate change will occur faster than any previous changes. However, chapter 4 has explained why making specific predictions is exceedingly difficult. Plausible scenarios of future climate illustrate the range of expected changes. In the coming decades, precipitation over the Alpine region may, for example, in general increase, although snowfall may decrease. At the same time, some areas within the region may actually experience a drier climate. The uncertainties in future Alpine climate are of major importance for the discussions in chapters 5 and 6, which have examined ecosystems' response to climate change. These examinations imply that present Alpine ecosystems would have great difficulties adapting beyond critical rates of climate change. At the same time, quantifying such critical rates for specific Alpine ecosystems, let alone for ecosystems in general, seems very difficult indeed. On the other hand, some forest ecosystems may just as likely benefit from a warmer climate and display increased diversity and productivity. How should the inhabitants of the Alpine region respond to potential threats arising from global and regional climate change? Because of climate change's global nature, regional climate policies cannot be based on the argument that regional action will prevent regional impacts. However, chapter 7 has argued that no-regret policies may have considerable scope for reducing emissions by innovative action. Because of their innovative character, however, such policies' effectiviness is even harder to estimate than that of policy measures in general.

By assembling more and more knowledge, we become more and more aware of the major and presumably irreducible uncertainties inherent in any predictions of regional climate change, in its impacts on ecological and socioeconomic systems, and in the precise effects of possible instruments of climate policy. This chapter, therefore, addresses the question of how policymakers and others can use the kind of knowledge assembled in the preceding chapters to make sound decisions with regard to climate change on a regional scale. It emphasizes the need to develop innovative approaches for regional assessments that take into account regional decision-making processes' specific nature. The regional focus reflects an increasing awareness within the climate change integrated assessment community that many decisions will depend on the individual choices of millions of organizations and individual citizens and will be driven by local interests and conditions (Morgan and Dowlatabadi 1996), one of several intricate aspects of the climate change problem.

Climate change throws a number of issues of environmental decision making into high relief:

• Although the enhanced greenhouse effect's basic processes are well known, the science of climate change also contains areas of high uncertainty, some aspects of which may, for all practical purposes, be irreducible.

• Climate change is pervasive; that is, it cannot be resolved by simply prescribing the use of a single product or process.

• Climate change cannot be contained within any individual political or economic jurisdiction; it requires coordination and solidarity across geographical and institutional borders.

• Climate change's effects will be practically irreversible and will unfold over a period of decades to centuries, perhaps with substantial time lags between actions (emissions) and impacts; it requires coordination and solidarity across generations.

• the benefits and costs of the impacts of climate change and the effects of actions taken to mitigate it will fall unevenly across the world's population; it requires coordination and solidarity across the world's regions.

Hence climate change represents a class of issues that raise tough challenges for decision makers. To help meet these challenges, scientists have started to design methods and tools for integrated assessment. In a comprehensive review, Rotmans and Dowlatabadi (1998) define the overall process of integrated assessment (IA) of environmental issues as "an interdisciplinary process of combining, interpreting and communicating knowledge from diverse scientific disciplines in such a way that the whole cause-effect chain of a problem can be evaluated from a synoptic perspective with two characteristics: (i) it should have added value compared to single disciplinary oriented assessment; (ii) it should provide useful information to decision-makers". The current emphasis, particularly with respect to climate change, is on pursuing global assessments using IA computer models. Some of these models combine physical, ecological and socioeconomic processes in much detail to represent the full feedback loop from human action through the climate system back to human reaction. Data collection, theoretical understanding, and computer technology have now reached the stage where this is feasible to some degree, but only with a tremendous number of ad hoc decisions and rules of thumb. Another procedure involving subjective judgment by teams of human experts is therefore unavoidable. Relying on a single expert's judgment would introduce unacceptable arbitrariness into the outcomes. However, even collective expert judgments are by necessity substantially subjective.

The long timescale involved compounds the scientific uncertainties associated with complex natural systems. Human systems' nondeterministic behavior is even harder to predict over periods of decades to centuries (see also Funtowicz and Ravetz 1990; Morgan and Henrion 1990). Important differences in value systems among members of the present generation and the prospect of values' changing over time further complicate climate change decision making. Climate change's pervasiveness also means that the issue touches many aspects of life. Climate policy is not just environmental policy, but industrial, agricultural, social, and developmental policy as well. Irreducible uncertainty, plural values, and multiple goals mean that analysis can discover and reveal no unique best solution to the challenge of climate change. Real-world solutions must be negotiated.

This uncertainty becomes even more important as we explicitly take into consideration the fact that numerous natural and social phenomena are indeterminate. They are not just hard to predict, they are not predictable at all, irrespective of how much we know about them. Regarding social phenomena, surprises in political or economic developments are quite common. A prediction's failure is often blamed, in hindsight, on having neglected some essential facts; or to put it another way, it is often claimed that a correct prediction would in principle have been possible had better knowledge been available. The failure is thus attributed to uncertainty in knowledge rather than to indeterminacy of the phenomena under consideration. Even when it may not be practically possible or even meaningful to disentangle aspects related to imprecise or insufficient knowledge about a phenomenon from the phenomenon's indeterminate nature, the distinction has major importance for the whole decision process. Accounting for indeterminacy shifts integrated assessment's emphasis away from achieving a preconceived goal toward sustaining an ongoing goal-seeking adaptive process comprising numerous distributed and sequential decisions on regional scales.

Climate change is a global phenomenon. This implies that regional impacts and regional actions have to be considered in a global context.

However, integrated assessment of climate change on a global level is insufficient, because it neglects regional characteristics and does not account for regional decision-making processes' dynamic nature. In this respect it important to point out the following:

• Global integrated assessment modeling has developed in isolation from a longer tradition of interdisciplinary assessment focused on local issues. This tradition's primary value may be its experience in explicitly confronting the issue of stakeholder perceptions, interests, and actions in the assessment process, which requires that the investigation begin with stakeholder scoping of the issues to be examined.

• At the practical level, climate change is essentially a regional issue. The nature of the human activites contributing to climate change, the level of societal resilience to climate change's potential impacts, and the industrial, agricultural, and economic capacity to act in the face of potential climate change all vary from region to region.

To address the needs of regional integrated assessments, we set out to explore ways of developing IA in general and IA models in particular both to yield a better reflection of our pluralistic world and to provide information to aid the processes of negotiation that occur within it. The approach we present here shifts integrated assessment away from the unidirectional communication of model results to decision makers toward a framework for the coproduction of knowledge, where modeling is embedded as an activity within a larger societal process of assessment and negotiation. Such a process reflects stakeholder perceptions, priorities, and regional orientation—a decision-making style very strongly rooted in the traditions of the Alpine peoples.

We begin by elaborating the concepts of irreducible uncertainty and value diversity that can be involved and the implications of the absence of unique solutions that can be implemented in climate change decision-making processes. We suggest that the integrated assessment process must incorporate local knowledge and intuitive insights not captured in current modeling processes. We follow with an assessment of the achievements and limitations of three current approaches to IA modeling: best-guess, probabilistic, and pluralistic strategies. Each has a certain appeal as the basis for developing an IA approach for the Alpine region, but each achieves its advantage at a cost to another aspect of the IA process that we value from a societal decision-making perspective.

8.2   DECISION MAKING, UNCERTAINTY IN KNOWLEDGE, AND INDETERMINATE PHENOMENA

Humans, be it at the individual or the collective level, organize their knowledge of phenomena (natural and socioeconomic) with the help of mental models. The term "mental model" here refers to a particular concept and the perception of phenomena derived from it. On one hand, we may now identify heuristically uncertainties that reside where phenomena, mental models, and knowledge overlap, uncertainties that arise from limitations in our knowledge about the phenomenological world. On the other hand, we have to deal with uncertainties that reside more where knowledge, mental models, and social processes overlap, uncertainties that arise from the fact that knowledge is produced within a social process and is thus to a certain extent socially constructed. To give a trivial example, in modern societies people have mental models of cars. If one hears only a noise that sounds like a car's, one may be uncertain about whether it is a car or something else. There exist, however, factual criteria for determining whether it is a car. The implications of having a certain kind of car and the social prestige that may be associated with it are normative issues. Similarly, factual criteria exist for determining and measuring climate change and its impacts. However, what this implies for human societies, how risks are perceived, and what action should be taken depends on individual and collective normative judgments. Let us now explore these issues in more detail with regard to climate change.

8.2.1 Uncertainties in Knowledge about the Phenomenological World

Conventionally, uncertainty is viewed as the grey area between the darkness of ignorance (or the absence of knowledge) and the light of certainty (or the presence of demonstrated knowledge). This view may be quite appropriate at the level of experimental bench science, where the focus is on margins of error in experimental measurements under highly controlled conditions. This view of uncertainty as imprecision has another source in engineering design, where margins of safety are allowed around estimates of materials' or structures' behavior because of actuarial experience of past failure rates.

The representation of uncertainty as a range around a given measurement or estimate and the view of knowledge as the progressive discovery of objective truth both suggest that uncertainty is always remediable, at least in principle, by improved measurement or further investigation. Clearly, science will always be concerned with increasing the precision of measurements or estimates of important transformations in the environment. However, understanding environmental systems and the impacts of human activities and management efforts involves qualitatively different types of uncertainty, some of which may prove irreducible (see also Funtowicz and Ravetz 1990; Morgan and Henrion 1990). We adopt here the typology of Funtowicz and Ravetz (1985), who identify three types of uncertainty: technical, methodological, and conceptual. These types of uncertainty exist not on a smooth continuum between low and high uncertainty, but on a broken one in which qualitative discontinuities appear as the scale and source of uncertainty change from issues of precision to issues of conception.

Box 8.1 Uncertainties in Measurements

Error models are frequently used in the assessment of measurement uncertainties. That is, measurement uncertainties are interpreted as errors. One assumes that there exists a well-defined "true value" and that the actual measurements represent slight deviations from this value. This assumption allows a measurement value X to be expressed as the sum of a fixed (true) value x_0 and a random error component $E : X = x_0 + E$. The error component E is a random variable with a corresponding distribution often approximated by a normal distribution. If the error's distribution has zero mean the measurement is said to be unbiased (no systematic error). The true value x_0 is then the expected value of the measurement X, which can be estimated as the mean value of a data set. Using this model, a measurement is characterized by the value x_0 and by the distributional characteristics of the error E.

In Alpine regions, measurements in remote locations often pose serious technical problems. Accurately measuring precipitation falling as snow is notoriously difficult because of the high spatial inhomogeneity involved. This is especially pronounced in high mountain regions, where winds are often strong and the topography is rugged. Measuring temperatures is not free of technical problems either. Daily mean temperatures are commonly estimated as the mean of the daily minimum and daily maximum temperatures, quite easily obtained from minimum-maximum thermometers. In most cases these estimates are reasonably accurate. However, if the temperature values are distributed asymmetrically about the mean, this method introduces an additional measurement error. It may even result in a systematic bias if, for example, the distribution's shape has changed systematically because of climate change. A difference in mean temperatures of two to three degrees may result simply from a shift in observation time from morning to afternoon (Schaal and Dale 1977). Such effects, difficult to detect, are at the interface between technical uncertainties, which can be represented by error bars, and methodological uncertainties, which may be detected by comparing the results obtained using different independent methods.

8.2.1.1 Technical Uncertainty

Technical uncertainty corresponds to the degree of precision in the results of an inquiry. Error bars on histograms, for instance can represent this sort of uncertainty appropriately, and improving techniques of measurement and estimation can reduce it. Box 8.1 gives further details and discusses as examples the measurements of crucial climatic variables such as temperature and precipitation which, despite considerable sophistication in measurement devices, are still not free of technical problems.

Technical uncertainty creates conditions often referred to as "decision making under risk." Although there may be disagreements over data quality, analysis methods are not subject to serious dispute. Because an event's probabilities and the magnitude of its consequences are well known from sources such as actuarial data, applying standardized methods of analysis, including probabilistic risk analysis and cost-benefit analysis, is both feasible and appropriate. Issues of technical uncertainty, by definition, belong to well-known, recognizable classes of events that can all be treated similarly based on previous experience. Conceptual and methodological analysis is not required for each new issue. Both scientific analysis and decision making are reducible to routine procedures and the application of rules.

8.2.1.2 Methodological Uncertainty Methodological uncertainty arises in connection with questions about the adequacy of the tools used to measure and estimate specific variables, or even, sometimes, to detect or identify them. Error bars can appropriately represent uncertainty here, but their usefulness may be rather in illustrating that different measurement or modeling techniques may produce results that do not even overlap. Let us explore such a case using temperature "measurement" as an example. Because of the long timescales associated with climate dynamics, which range from decades to millennia, information about climate variations must be derived from the past, in particular from paleoecological investigations of environmental archives. Chapter 3 discussed the methodology underlying such approaches. Such long-term data sets covering thousands of years are essential for improving understanding of the climate system's internal dynamics and of climate's influence on ecological phenomena (especially in terrestrial systems). Because temperature values must be inferred from indirect evidence, methodological issues may be far more important here than technical difficulties as illustrated, for example, the reconstruction of sea surface temperatures from paleoecological data. Box 8.2 shows that the bounds on the uncertainty derived for different methodological approaches do not even overlap!

Making decisions under conditions of methodological uncertainty corresponds closely to decision making under uncertainty, where the probabilities or the consequences of an event are uncertain. Disputes here center on selecting the right tools for obtaining better measurements or estimates of probabilities and impacts. Both scientific and political judgments involve a high level of skill in diagnosing the situation based on a mixture of knowledge of parts of a problem and informed guesswork about others. For this reason, decision making here has been described as being in a "clinical mode" of practical judgment.

8.2.1.3 Conceptual Uncertainty Perhaps the least tractable type of uncertainty, conceptual uncertainty, involves wide disagreement about whether we understand a problem or a system at all appropriately. A closer look at some conceptual controversies related to the forecasting of regional climate and its impacts on ecosystems is perhaps illustrative.

Mountainous regions often pose difficult weather forecasting problems, and they may also amplify larger-scale uncertainties. Precipitation processes are more complex and more intense in mountainous regions than in surrounding areas, and they are also likely to be more sensitive to global change. Consider for instance a warming of 2 K, a warming that might well occur within the next century. Such a warming could increase air's potential to transport water vapor by as much as 25 percent. Such a moistening of the atmosphere might have no major effect on precipitation over flat terrain but is likely to exert a very significant influence in regions such as the Alps, which extract a substantial fraction of the ambient atmospheric moisture

Box 8.2 Uncertainties in Representing Uncertainties

Differences between Glacial and present sea surface temperatures (SSTs) in the tropics indicate the sensitivity of this region to climate change and provide an important test for the accuracy of climate models. Anderson and Webb (1994) compared a number of different approaches to the reconstruction of SSTs from paleoecological data:

Method	Change in SST [in K]	Uncertainty [in K]
Coral Sr/Ca	−5	±0.5
Coral $\delta^{18}O$	−5	±0.5
CLIMAP	−2	±1.5
Foraminifera $\delta^{18}O$	−2	±1
Algae alkenones	−1	±0.5

The fractional incorporation of oxygen isotopes $^{18}O/^{16}O$ and of strontium and calcium and the saturation of long-chain alkenones are temperature-dependent processes. Consequently, one use these temperature sensitivities to reconstruct historical temperatures from fossil remnants, where these parameters are preserved (cf. section 3.2.5.6 Climap (the Climate Long-Range Investigation and Mapping Programme) derived SSTs based mainly on the fossils of several plankton groups whose distributions in the tropical ocean have not changed since Glacial times. Judging from modern distributions, a cooling would bring an entirely different species assemblage. What strikes the eye is the discrepancy between the uncertainties derived from measurement errors (estimated from between-sample variability) and the overall scatter of the results from the different methods. Only conceptual errors in the basic assumptions not captured by merely accounting for measurement errors can explain such a discrepancy. Quantifying conceptual sources of uncertainty may well be essentially impossible. Using redundant but independent sources of information can at least increase the level of confidence in such reconstructions or reveal obvious shortcomings, as in the example above. The discrepancy observed has triggered further research. De Villiers, Nelson, and Chivas (1995), for instance, reported that the underestimation of biological control of corals' calcium and strontium uptake may have introduced a major error into temperature reconstructions.

---

content through topographically controlled precipitation mechanisms. The atmospheric flow may either be over the mountain barrier or around it, leading to two entirely different flow regimes and precipitation patterns (see chapter 2; Trüb 1993; and Schär and Durran 1997). Box 8.3 explains why the Alps are located just at the transition zone between one regime and the other. Subtle effects that may lead to a shift toward one regime or the other may determine which of the two regimes comes to dominate in a particular situation. Forecasts of precipitation patterns are thus highly contingent on the type of model chosen, in particular with respect to spatial resolution.

This is not the only conceptual uncertainty inherent in forecasting future Alpine climate. Chapter 4 discussed in more detail the advantages and limitations of different approaches for producing regional climate scenarios. If we wish to assess impacts on ecological systems, the uncertainty in climate scenarios are amplified by conceptual issues which arise within the context of choosing the appropriate level of description for ecosystems.

Box 8.3 Flow-over and Flow-around Regimes

Figure 8.1 shows an idealized regime diagram applicable to the atmospheric flow past the Alps. The two flow regimes are separated by a regime boundary (bold line), which has some of the characteristics of a bifurcation. For "small" nondimensional mountain heights the "flow-over" regime dominates, while beyond a critical mountain height the incident airmass is deflected laterally and detours the obstacle without experiencing significant ascent, giving rise to a "flow-around" regime. Which of the two regimes comes to dominate in a particular situation may well be determined by subtle effects, in particular in the vicinity of the regime boundary.

Two inferences can be drawn from this for the current discussion. First, the typical range of the control parameters (grey area in figure 8.1) applicable to the alpine setting is located near the regime boundary. It follows—in agreement with observational studies—that the flow can be either over or around the Alps, depending on the ambient atmospheric circulation. This indicates that the precipitation response in the Alps is highly sensitive to the continental-scale atmospheric environment and might change significantly in climate change through a modified distribution of the so-called *Grosswetterlagen* (cf. chapter 2). Second, representing this sensitivity is a subtle task. Because the effect observed depends on mountain height and the horizontal extension of the mountain barrier, an appropriate representation requires high spatial resolution. For illustration purposes, the diagram indicates the typical range of flow parameters represented in a global climate model with a low horizontal resolution of approximately 300 kilometers (hatched area in figure 8.1). The low resolution results in an erroneous predominance of the flow-over regime, and the above-mentioned sensitivity of the alpine precipitation response is hence not appropriately represented. On a technical level, representing sensitivity in numerical models requires much higher computational resolution than that required for simulating global circulation. Even if these difficulties can be resolved through high-resolution numerical methods (cf. chapter 4), a probabilistic analysis of future Alpine precipitation is a daunting task.

Species are often assumed to respond individualistically to changes in important environmental parameters, irrespective of the community of which they are a member. However, ecosystems' functional properties are more robust than their structural properties (such as species composition, for instance) with regard to the impacts of disturbance or stress (Pahl-Wostl 1995a; Schindler 1987). Functional properties may be stabilized by structural changes and thus be maintained over a long period of time, only to decrease drastically once a critical threshold is exceeded. Ecosystem response differs from the response of individual species, which may be highly contingent on the current state of the ecosystem as a whole. What frame of reference is then appropriate for investigating and modeling ecological systems and deriving criteria for evaluating climate change's impacts? Our projections may differ, depending on the level of ecological organization chosen. Chapter 5 addressed this question within the context of climate change's impact on Alpine vegetation. Box 8.4 discusses the issue of different levels of ecological organization in more detail. This topic, still a major problem for ecological research, involves a number of essentially irreducible uncertainties.

Conceptual uncertainty necessitates decision making under conditions of indeterminacy, where even the sign of a change may not be known, let alone

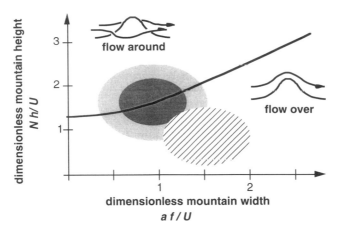

Figure 8.1 Regime diagram of flow past an Alpine-scale topographic obstacle with a sharp regime boundary (bold line) separating the flow-over from the flow-around regime. The grey area indicates the observed range of typical control parameters for the Alpine setting; the hatched area shows the same area but for the Alps' numerical representation in a low-resolution climate model. The regime boundary is from Trüb 1993, and the variables defining the dimensionless control parameters are the mountain height h and the horizontal scale a; the upstream Brunt-Väisälä frequency N (a measure for the vertical stability and hence the vertical density gradient); the Coriolis parameter f (a frequency measure of the latitude-dependent effect of the earth's rotation with the units of frequency); and the wind velocity U.

its probabilities and consequences. Disputes here revolve essentially around world views, including basic beliefs about nature as being fragile or resilient and beliefs about what constitutes prudence in the face of unknown futures.

Thinking about uncertainty in this way provides an important clue as to why more information does not necessarily reduce uncertainty in complex systems, such as climate, which depend on the interaction of a series of interrelated biological, geophysical, and socioeconomic systems. There are instances of irreducibility within each type of uncertainty. Moreover, climate change is not a unitary phenomenon but a complex of highly interdependent, specific risks. Reducing specific uncertainties about the behavior of part of the system at the technical, methodological, or conceptual level may increase rather than decrease our uncertainty about the larger system by revealing that we know less about another level of the system than we thought. Ignorance, uncertainty, and knowledge are not stages on the path to enlightenment but omnipresent factors interacting dynamically within chaotic indeterminate systems such as climate.

In addition, all knowledge is collected, valued, and selected within certain conceptual and societal frameworks. The whole scientific enterprise is based on a number of conventions, rules of good practice, and expertise. Such practice is a necessity when dealing with complex issues. At the same time it introduces another aspect of uncertainty, one that already starts at a very basic level of knowledge generation, namely at the level of data production.

Box 8.4 Ecological Effects' Dependence on Level of Organization

Ecosystems are complex systems with several levels of organization (cf. figure 8.2). Functional properties at the macroscopic level of the system as a whole depend on the diversity of the component species and functional groups (Pahl-Wostl 1995a). The system components are embedded within dynamic networks. Such networks, with numerous internal "degrees of freedom," convey to ecosystems a high degree of flexibility. At the same time, this causes unexpected behavior and surprise. If boundary conditions (e.g., climate) change, multiple responses are possible, introducing a major source of uncertainty into any evaluation of climate change's ecological impacts.

Little is known about the relationship between ecosystem function and its structural organization, mainly because ecological systems have traditionally been investigated according to a dual hierarchy that may be distinguished as the process-functional versus the organismic-structural perspective:

Process-functional perspective:

Level of organization	Scientific discipline
cell	biochemistry
organism	physiology
ecosystem	ecosystems ecology

Organismic-structural perspective:

Level of organization	Scientific discipline
individual	organismic ecology
species	population ecology
community	community ecology

Obviously the different subdisciplines overlap somewhat. However, in general the communication has not been very intense. Conflicting results may be obtained with respect to climate change's impacts depending on the level of organization focused on. A prominent example of this is the potential effect of an increased atmospheric concentration of CO_2 on the primary production of terrestrial ecosystems. One may observe an enhancement of production on the level of the individual plant over short timescales, but no effect, or even a reduction in production, on the level of the ecosystem as a whole (Körner and Arnone 1992). This example emphasizes process-functional issues. In addition to these, however, species-specific differences in response may change the plant community's species composition. Changes in the plant community may in turn affect, for instance, the soil's nutrient balance. The effects of such changes may be communicated to other parts of the ecological web, affecting, for example, insects or pests. Similar reflections can be made on the effects of climate change, such as an increase in temperature or precipitation. Chapter 5 has also discussed this topic and emphasized that climate change's effects are different for different levels of organization in plant communities.

8.2.2 Generation of Knowledge as a Social Process

One might assume that subjective judgment would have little chance of biasing measurements made using highly sophisticated technical devices. However, expectations shaped by experience and prior knowledge influence to a large extent what is observed and whether the observations are regarded as reasonable. Measurements not regarded as reasonable are often discarded,

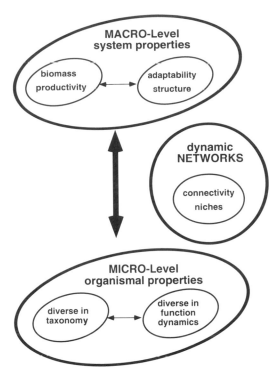

Figure 8.2 Different levels of ecosystem organization and their interdependence. Changes in the ecological network comprising the component species and functional groups stabilize functional properties at the level of the ecosystem as a whole and maintain them in a variable environment. A high diversity in functional pathways thus endows an ecosystem with a high degree of flexibility and adaptability. At the same time, such organization renders ecosystem development difficult to predict and control.

sometimes with serious negative consequences. The discovery of the Antarctic ozone hole was delayed because data analysts at NASA decided to exclude from the data sets all values lower than predictions derived from computer models that had produced reliable results over many years. Based on these modified data sets, they generated "corrected" maps showing values in the expected range (MacEachren 1995). Only after other scientists measured ozone concentrations independently and found large discrepancies was the ozone hole finally detected.

This failure should not be attributed primarily to any interest in concealing information, but rather to scientists' tendency to distrust measurements outside the range they consider plausible. Such outliers are usually attributed to technical failure and measurement errors. The increased availability and widespread use of large data sets has raised more and more questions regarding data quality control. Quality control procedures, such as the use of plausibility rules to filter out "erroneous measurements," have been automated over the past few years. The example of the ozone hole shows that problems may arise in using data with little quality control but may also arise if the quality control leads to "purified" (and therefore biased) data that make certain scientific findings impossible. This lowers confidence in the supposedly firm observational base fundamental to any modeling effort and provides another source of disagreement and contradictory interpretations.

In the light of these difficulties, no "stopping rule" can tell a decision maker what amount of knowledge justifies policy action. Furthermore, characterization of the types of uncertainty present is itself often subject to dispute and negotiation. Even the polar states of certainty and ignorance are more problematic than they at first seem. Thompson (1982) has described conditions under which "contradictory certitudes" may hold sway, not simply as a result of (at least) one party's being in error but of situations where differences in the guiding assumptions of a question and the framework for answering it lead the questioners to incommensurable, but equally valid, conclusions. Because each questioner confronts the problem from a different angle, he sees quite clearly aspects not visible from another angle. It is impossible, however, to look at the problem from all possible angles simultaneously.

The notion of contradictory certitudes recognizes that people may have different world views, fundamentally different perspectives on how the world is, or ought to be, ordered. From these different starting points, individuals and institutions can define problems, assess opportunities and costs, and arrive at very different conclusions about appropriate actions. Each conclusion is completely credible to those who share the world view in which it originates, yet may not be at all credible to those who do not. This situation reflects not a level of uncertainty that further research can resolve, but conflicting certainties with significantly different implications for both research and policy agendas.

A particular conflict centering on an Alpine hydropower project provides an instructive example of conflicting certainties arising from a differing world view. The project involved a plan to flood a specific mountain landscape (the Greina plateau). A strong environmental movement, based mainly in Swiss urban regions, fought the project because it perceived that landscape as displaying a unique beauty that should be preserved. Most inhabitants of the mountain region surrounding the landscape in question, however, strongly favored the project. They perceived the landscape as inhospitable and emphasized that they and their forebears had perceived the natural environment as hostile for centuries. The environmentalist movement won, and the Swiss government agreed to pay compensation to the communities and the canton concerned for their willingness to preserve the mountain landscape (SGS 1988). Likewise, any evaluation of the potential impacts of climate change must take into account subjective judgments based on moral, esthetic, and economic considerations.

The flip side of contradictory certitude is the "social construction of ignorance" (Rayner 1986), in which available information is excluded implicitly but systematically from a discourse. In principle, the information could be admitted, but there are strong institutional pressures to forget or ignore it. The social construction of ignorance can have many causes, one of which is boundary maintenance, when the excluded information or expertise would frustrate action or weaken institutions. For example, the history of Alpine pastures provides a wealth of experience about how to combine the common ownership of land with individual responsibility for its use that would be highly relevant to strategies for dealing with the "tragedy of the commons" (Hardin 1968). It might also be helpful in designing new institutional arrangements for land use in urban areas. Both scientific and public discourse about Alpine pastures, however, usually treat the sophisticated arrangement of traditional common property rights merely as a curiosity. Its relevance to contemporary environmental problems is systematically neglected, probably because it would challenge contemporary notions about the unique efficiency of private property rights.

The problems of contradictory certitudes and the social construction of ignorance exist within scientific inquiry. They are compounded, however, when universal scientific knowledge clashes with or ignores the local experience of nonscientists. The gap between universal and local knowledge is not merely spatial but instead raises fundamental issues of what kinds and sources of knowledge analysts value. Contrasting sixteenth-century iconography with contemporary satellite photographs, Ingold (1993) illustrates the situation by tracing the change in the standpoint of human inquiry from the Enlightenment to the present. Maffei's *Scala Naturale* (1564) shows the scholar at the center of the environment, consisting of fourteen concentric spheres envisaged to form a giant stairway, the ascent of which, step by step, affords more comprehensive knowledge of the world through experience within it. In modern satellite imagery, an observer viewing the world from without displaces the scholar experiencing the world from within. There is nothing intrinsically objectionable in this standpoint, except that we privilege observation over experience. We devalue local knowledge, originating in experience, as partial, parochial, and ultimately unreliable, whereas we treat global knowledge as universal, total, and real. Against this view, Ingold argues that the difference between local and global perspectives is not one of hierarchical degree, in scale or comprehensiveness, but one of kind. In other words, the local is not a more limited or narrowly focused apprehension than the global, it is an altogether different mode of apprehension based on an active perceptual engagement with components of the dwelt-in world in the practical business of life, rather than on the detached, disinterested observation of a world apart (Ingold 1993).

In examining global climate change's potential impacts at the Alpine regional level, local knowledge of varied microclimatic conditions and ecological systems and of socioeconomic systems may prove an important input to the integrated assessment process. Local initiative may be highly relevant for triggering innovative action. Such local knowledge is likely to be missed or excluded where it is not consistent with universal assumptions. Including local knowledge is all the more important because universal knowledge may prove quite insufficient for developing viable strategies for dealing with an indeterminate future.

8.2.3   Decision Making amid Uncertainty

Integrated assessment of climate change should assist society in taking responsible action despite all the uncertainties involved (Weyant et al. 1996). The type of strategy chosen will depend very much on how the decision-making process is perceived—in terms of who should participate, what the important arguments are, how the decision problem is framed and how decisions are made. To study the problem of decision making under uncertainty in IA, it is useful to assume, at least in a preliminary fashion, the perspective of decision theory. This perspective, which to a large extent frames current work in economics, management, political science, and other social sciences, offers powerful tools for analyzing decision-making problems quantitatively (Kreps 1988). It is widely used in dealing with decision-making problems related to climate change on a global scale (Arrow et al. 1996). The framework proves quite useful for discussing the strengths and limitations of different strategies for IA and their usefulness in making regional assessments.

Decision theory starts with the simple distinction between an individual decision maker's actions and preferences. When trying to solve a problem, a decision maker is well advised to ask systematically what actions she can perform and what preferences she wishes to pursue with these actions. For example, the government of an Alpine country like Switzerland or Austria may deal with climate change by simply choosing to wait and see, it may engage in international negotiations on a systematically coordinated climate policy, or it may take the initiative by reducing emissions on the national level. As for preferences, the government in question may be seeking voter approval, an international reputation as a cooperative player, or the economic advantages accruing from sociotechnical innovations.

In some cases, a careful elaboration of possible actions and relevant criteria is sufficient for reaching a sound decision. With regard to regional climate change, however, this is simply not possible because of the numerous uncertainties involved. The decision sciences deal with this problem by distinguishing between actions and consequences. A wait-and-see policy may lead to catastrophic damage from climate change, but it may also lead to a successful muddling through in the face of minor changes in climatic conditions. For each action, then, the decision maker should try to determine the range of possible consequences. Obviously, ranking different consequences according to their likelihood is also important, and the mathematical concept of probability is often used for this purpose. As soon as the varying likelihood of different consequences enters the picture, however, the problem of risk aversion arises. The decision maker then must develop, in addition to decision criteria referring to different outcomes, criteria for choosing between different risks. Usually, he will be willing to accept small risks but reluctant

(a)

0.5 — $ 0

0.5 — $ 10

0.5 — $ 90

0.5 — $ 100

(b)

0.5 — $ 0

0.5 — $ 10,000

0.5 — $ 4,000

0.5 — $ 6,000

Figure 8.3 Examples for lotteries that differ in (a) expected value and (b) risk.

to accept large risks. Determining the boundary between the two, however, is often a thorny problem. Suppose we knew that with a probability of 10 percent, snowfall in certain areas of the Alpine region would decrease by 50 percent fifty years from now: Is this risk small or large? An integrated assessment of regional climate change, then, must provide reasonable answers to the following questions: What actions can be taken? What possible consequences does each action have? What criteria are relevant for comparing different consequences? How likely are the different consequences? What criteria are relevant for comparing different risks? Criteria and consequences must correlate in some fashion: decision makers can safely ignore consequences irrelevant according to their criteria but usually need to gather at least some information about those consequences that do matter according to their criteria.

Choice amidst uncertainty can be analyzed in a systematic fashion using the metaphor of a lottery. An action and a range of possible consequences characterize a lottery. Consider for example a choice between two lotteries, both based on flipping a coin. Figure 8.3 depicts the set of alternative outcomes, which are monetary payoffs. The first lottery (a, left), has two possible payoffs: [$10; $0]. In the second lottery (a, right), the payoffs are [$100; $90]. Most people would no doubt prefer the second lottery to the first. A convenient way to describe this situation is to say that the first lottery's expected value is $5, whereas the second lottery's expected value is $95. Sometimes, therefore, a single figure may describe a lottery, namely its expected value, and the criteria for choice may be obvious. Suppose, however, that an individual is again facing a choice between two lotteries based on flipping a coin. In the first lottery (b, left), the two possible payoffs are [$10,000; $0] In the second lottery (b, right), the payoffs are [$6,000; $4,000]. Both lotteries have an expected value of $5,000. The choice of which to play depends now on preferences between outcomes rather than on preferences based on expected values. Risk seekers will find the possibility of winning $10,000 attractive enough to choose the first lottery despite the risk of winning nothing. Those who are risk averse will find the possibility of winning nothing very threatening and choose the second lottery, which guarantees a gain. Which lottery is preferred now depends on individual attitudes towards risk, and in such situations, one needs to know more than just the the different lotteries' expected values. Consider, for example, the risks associated with potential catastrophic damage due to extreme events, which are of particular relevance in Alpine regions. Extreme events are exceedingly difficult to predict and are therefore often excluded from formal decision analysis. However, the presence of incalculable risks may be the relevant criterion on which a risk-averse decision maker bases his choice from among different lotteries.

Actors have to choose an action before they can find out which consequence will actually ensue. In this sense, choice amidst uncertainty can be represented as a choice among lotteries. In the context of climate change, the different lotteries may, for example, correspond to different IPCC scenarios (cf. Arrow et al. 1996). The decision about which action to take is based on a knowledge of possible outcomes. Indicating a consistent preference ordering for the various subsets of potential outcomes from different actions is assumed possible. This implies a rank ordering of the different actions, which can be represented by any utility function that allows numbers to be assigned consistently to the different actions. In practice, utility functions are often expressed in monetary terms.

Both probabilities and preferences may express purely subjective dispositions. Together, they constitute a disposition to act in a particular way under conditions of uncertainty. Experience, social norms or even the inherited features of a human organism may shape both preferences and probabilities. In some cases, this may lead to uniform preferences or probabilities for different decision makers. In many cases, no convergence will occur. For our present purposes, it is important to notice how closely decision theory links questions of uncertainty with questions of valuation. This link is in fact crucial in developing sensible strategies to deal with uncertainty in IA.

IA may now be viewed as a process in which different lotteries are established and means are provided to discuss preference orderings to come to a decision on what action to take. The research strategies discussed in the following differ in how they deal with preference orderings and probabilities and how they take into account the perception of risks and the importance of subjectivity. More specifically, one might conceive of the following three strategies:

1. The *best-guess strategy* bases decisions on the expected value of a lottery. Here, the rational choice corresponds to the lottery with the highest expected value (where this value may be expressed as some index of utility).

2. The *probabilistic strategy* takes into account the fact that decisions depend on the likelihood of specific outcomes and on the risk associated with a

lottery. In such a case, the rational choice depends explicitly on a decision maker's attitude toward risk.

3. The *pluralistic strategy* takes the argument a step further by acknowledging that a range of different subjective probabilities are often associated with a corresponding range of different subjective preferences. This means that different agents may frame a decision problem in very different ways, because both probabilities and preferences vary. The problem may be framed so differently because several equally valid but complementary perspectives exist on the decision problem at large. In particular, assessments on the regional scale have to face such a situation. Regional action can hardly be linked unequivocally to regional consequences, because this link is contingent on cooperation on the global scale. Expectations about the likelihood of such cooperation may be closely linked to preferences, or more broadly speaking, to value systems. Regional climate lotteries are thus shaped to a considerable extent by both subjective probabilities and subjective preferences.

The next three sections are devoted to discussing these strategies in detail.

8.3   THE BEST-GUESS STRATEGY

8.3.1   The Unavoidability of Best Guesses

Any effort to make an integrated assessment of anthropogenic climate change must deal with the problem of uncertainty. A best-guess strategy tries to tackle an uncertain situation by developing a best guess of what the situation might look like. The well-known claim that for ecosystems, adaptation to climate change becomes exceedingly difficult when average global temperature increases by more than 0.1 degrees Kelvin per decade, is an example of a best guess.

The integrated assessment model MERGE 2 (see box 8.5) provides an instructive example of the best-guess strategy. This model computes the costs and benefits of various climate policies, including a no-action policy. In the model, best guesses handle all the formidable uncertainties and indeterminacies involved in the interactions between humankind and the climate system over the next two centuries. With this strategy, a model structure is defined and a long series of parameters is set. The outcome is quite astonishing. Two figures can unambiguously characterize whatever emission path one may wish to enforce: The first indicates the benefits gained by reducing both market and nonmarket impacts of climate change as compared with a business-as-usual scenario. The second indicates the costs required to implement the policy in question. Both figures are obtained by using a discount rate to reduce the expected time series of costs and benefits to present values of costs and benefits. Defining a preference ordering of the various possible actions is then easy, and the optimal choice then corresponds to the decision resulting in the lowest ratio of costs to benefits.

Box 8.5 MERGE 2

Recently, Manne and Richels (1995) have presented a revised version of MERGE (a Model for Evaluating the Regional and Global Effects of GHG reduction policies; see Manne, Mendelsohn, and Richels 1995). MERGE 2 represents the interactions between humankind and the climate system by distinguishing five geopolitical regions: the United States, the other OECD nations, the former Soviet Union, China, and the rest of the world (ROW).

MERGE 2 computes GHG concentrations as functions of emissions over time by a model of the carbon cycle combined with exponential decay models for methane and nitrous oxides. It then computes mean temperatures in temperate and tropical countries as functions of GHG concentrations. These two steps represent the climate system in MERGE 2. In particular, a doubling of CO_2 concentrations leads to an increase of 2.5 K in global equilibrium temperature.

Next, MERGE 2 computes two kinds of impacts as functions of mean temperature. First, market impacts are computed. These represent changes in GDP as a result of anthropogenic climate change. Following Nordhaus (1991), MERGE 2 assumes that an increase of 2.5 K would reduce the U.S. GDP by 0.25 percent. This estimate is generalized to the other OECD countries and to the former Soviet Union. Moreover, it is generalized across temperature changes by assuming that GDP losses are proportional to temperature increases. For China and for ROW, MERGE 2 assumes that the fraction of GDP lost as a consequence of a given temperature increase is twice that of other regions. Second, MERGE 2 computes estimates of nonmarket impacts, including impacts on biodiversity, human health, and environmental quality. These impacts are valued by estimating how much consumers in each region would be willing to pay to avoid them. Obviously, available income operates as a powerful constraint here. The rich may be willing to pay quite a lot to avoid environmental degradation, whereas the poor are unable to do so even if they wish. This situation raises painful moral issues, but it certainly represents an important aspect of contemporary reality. MERGE 2 assumes that in high-income countries, consumers would be willing to pay about 2 percent of their income to avoid a 2.5 K increase in temperature. Manne and Richels (1995, 7) compare this with the fact that the United States spends about 2 percent of its GDP on all kinds of environmental protection.

The model reprersents GHG emissions as a consequence of economic activities aiming at GDP production. Basically, only reducing GDP can reduce emissions, because the model yields a unique intertemporal economic equilibrium for given exogenous conditions. Given exogenous conditions leave no choice between different trajectories that would yield different equilibria characterized by different baskets of commodities yielding different GHG emissions (Jaeger and Kasemir 1996). MERGE 2 represents a one-commodity world with consumer goods, capital goods, electricity, and other forms of commercial energy all representing different uses of the same single good.

MERGE 2 computes GDP as a function of capital, labor, electrical energy, and non-electrical energy. If the price of energy increases, capital and labor can substitute for it. However, as far as one can tell from the documentation in Manne and Richels 1995, energy shortages induce only substitution effects among available technologies; they do not induce technical change in their own right.

Whereas the model measures labor, electrical, and nonelectrical energy in physical units, it measures GDP and capital in monetary units. It is assumed that somehow this creates no problems, even when computing the partial derivative of GDP with regard to capital in situations where the prices needed to evaluate both variables may be ill-defined. This partial derivative is assumed to exist and to be positive. The second partial derivative is again assumed to exist but to be negative. Such assumptions are essential for the working of the model because they guarantee that with given exogenous variables, only one equilibrium can be reached.

Box 8.5 (Cont.)

Population growth is considered exogenous, and so is the growth of labor force productivity. Moreover, various cost constraints are assumed for energy technologies and for their improvement over time. Economic agents' behavior is conceptualized as intertemporal utility maximization. Accordingly, each of the five regions maximizes the present value of the future stream of consumption, which may or may not be adjusted for the impacts of climate change. Future consumption is discounted at a rate of 5 percent. This is justified with present rates of return on capital, which are interpreted as marginal productivities of capital.

This brief description shows that a large set of parameters is built into the model. The exogenous variables represent various climate policies. More precisely, carbon quotas coupled with tradable emission rights are defined for the time horizon considered (the next two centuries). A business-as-usual scenario is defined as a model run without emission constraints. Alternatively, the model can be run with various predetermined emission constraints. Finally, an optimal emission path can be computed by comparing the market and nonmarket impacts of climate change with the costs of emissions reduction. The various functional forms used in the model are defined to yield a unique optimal emission path.

A great deal has been written about the merits and demerits of cost-benefit analysis (e.g., Munasinghe et al. 1996), and we do not enter into this discussion here. For our present purposes, what matters is using a best-guess strategy for finding an optimal decision. MERGE 2 is by no means the only optimization model based on a best-guess strategy. A well-known model with similar features is DICE (Nordhaus 1994).

The best-guess strategy, however, is not restricted to optimization models. The Pacific Northwest National Laboratory (PNNL) has used the best-guess strategy in developing GCAM, the Global Change Assessment Model (PNNL 1994; Edmonds Pitcher, and Rosenberg 1996). GCAM is not an optimizing model but rather a modeling framework that combines four modules for integrated assessment. The various modules are the result of cooperative efforts involving PNNL and a series of other institutions. The four modules deal with human activities, atmospheric composition, climate and sea level, and ecosystems. GCAM comes in two formats. The "Mini-CAM" is highly parameterized with little process detail, whereas the "process level GCAM" (or PGCAM) incorporates significant process detail into the framework. Each module relies on specific detailed models, each a full-blown model that can also be used on a stand-alone basis to investigate specific problems. Examples include Edmonds, Wise, and MacCracken 1994 on advanced energy technology and climate change and Wigley, Richels, and Edmonds 1996, showing that different emission paths can lead to the same CO_2 concentrations. As this kind of work shows, the best-guess strategy can be used both for analyzing optimization problems and for studying a wide variety of other problems related to anthropogenic climate change.

The difference between a best guess and a statement of fact is simple: The latter claims that something is the case, whereas the former explicitly acknowledges that it refers to an uncertain situation. Science usually aims at statements of fact, and such statements can become important tools of policy making. It would be silly to say that scientists are uncertain about the existence of the ozone hole. However, many scientific activities dealing with anthropogenic climate change produce best guesses, and best guesses are a major tool of policy making. In IA, statements of facts are highly unlikely, which often makes a best-guess strategy unavoidable. Does this mean that IA is necessarily an unsatisfactory exercise that is not really adequate for the problem at hand?

Such an attitude still characterizes many reactions to current IA modeling efforts. However, a best-guess strategy is in fact unavoidable in most scientific research for a very simple reason: The issue of the quantity and quality of available data is rarely trivial. The discussion in chapter 2 emphasized the pressing need to develop more comprehensive, accurate, and homogenized Alpine-wide data sets of the basic climate elements. Technical and methodological uncertainties (cf. boxes 8.1 and 8.2) further compound the situation, which is even worse with respect to biological data. Time and again, best guesses are the only way to make progress in empirical research.

Similar situations arise in developing of any computer model that involves a wide variety of decisions leading to best guesses. Because the best-guess strategy is often unavoidable in scientific research, it would be strange to reject IA models building on that research on the grounds that they, too, involve that strategy. The real task is to distinguish situations in which such a strategy offers an appropriate framework for decision making from situations in which it becomes misleading.

8.3.2 Climate Lotteries and the Scope of the Best-Guess Strategy

Integrated assessment models are designed as tools to support decision making. Decision theory relies to a considerable extent on the metaphor of a lottery. How does this metaphor relate to the best-guess strategy in climate change research? One may say that a choice faces humankind among various paths of future GHG emissions. Each path might lead to a variety of consequences, ranging from sea level rise to improved agricultural yield. A specific emission path together with the set of its possible consequences is then analogous to a lottery, and the choice among different emission paths is analogous to a choice among different lotteries.

Decision making under uncertainty involves two kinds of preferences: preferences among single outcomes and preferences among lotteries. Probability provides a bridge between these two kinds of preference. As stated in section 8.2.3, in certain situations this bridge enables one to describe a decision problem under conditions of uncertainty simply by indicating the expected values of different lotteries. These are the cases where the best-guess strategy is appropriate.

If the decision maker is not risk neutral, the expected value of monetary consequences (i.e., the difference between monetary costs and benefits) is no longer an appropriate basis for decision making. As long as the degree of risk aversion is given, this problem may be resolved by transforming monetary values into utility indices that take risk aversion into account. Under these circumstances, it is still possible to characterize each lottery in terms of a single figure representing that lottery's utility for a specific decision maker. In this case, utility is judged subjectively, depending on the decision maker's attitude toward risk. Although it is well known how to perform this transformation in theory, it is rarely done in making an integrated assessment of climate change. In particular, the widespread practice of using monetary GDP values to characterize different emissions scenarios ignores the problem of risk aversion.

A best-guess strategy is therefore appropriate only under quite restrictive conditions. It must be possible to describe lotteries using an empirically meaningful measure that simultaneously expresses preferences for outcomes, probabilities of outcomes, and degrees of risk aversion. In practice, however, the existence of differing degrees of risk aversion is one of the major reasons for debates about climate policy. And if different actors differ in their degree of risk aversion, it will be impossible to base the choice among lotteries on a single figure characterizing each lottery.

Characterizing climate lotteries in terms of a single figure neglects some major issues in the climate debate. A risk-averse and a risk-seeking person will simply not be able to agree on a utility measure for different climate lotteries. Moreover, it is by no means sure that preferences for particular outcomes in climate lotteries depend on a measure of GDP. It may be perfectly reasonable for some people to prefer a situation with a lower GDP to one where the GDP is higher. An adjustment that corrects for this effect with regard to one actor will inevitably be inappropriate with regard to another.

One might try to find another empirically based measure for climate lotteries that would make a best-guess strategy more appropriate. So far, however, no such measure seems to be in sight, and as long as risk aversion issues remain relevant to climate policy, we cannot expect one single measure to be appropriate. Avoiding using monetary values as descriptors of climate lotteries is often an appropriate approach. In addition to avoiding the limitations of the best-guess strategy, such an approach can actually enhance the strategy's usefulness.

8.3.3 Achievements and Limitations

The best-guess strategy in IA has already been remarkably successful in two respects. First, it has played a crucial role in framing the policy debate on climate change. The IPCC process's success in informing policy makers has depended largely on a careful use of the best-guess strategy; and if in the future, attempts at regional integrated assessments should become more common in the field of climate change, such attempts will not be feasible without using the same strategy. Up to now, the scope of the best-guess strategy for the Alpine region has been quite limited. It has, for example, been used to assess the potential consequences of specific policy measures, such as the introduction of a carbon tax.

Moreover, this strategy has highlighted crucial questions of research and policy. For example, MERGE 2 implies that climate policy has no influence on technical change. Under such conditions, model simulations show quite plausibly that only small emission reductions pay. In other words, the expected value of a climate lottery with constant GHG emissions is lower than the expected value of a climate lottery with rising emissions. Accounting for induced technical change may completely alter model simulations' results. This is of crucial importance, in particular for assessments on the regional scale.

Another example of how the best-guess strategy can highlight important questions is the use of official data in GCAM. Many statistical figures are quite reliable for industrialized countries but less so for developing countries. To a large extent, this results from the fact that the institutions collecting specific data have more resources in rich countries than in poorer ones. But there is also a more subtle aspect. Many data are defined in ways that make sense in some countries but not in others. Especially when it comes to assessing the causes of GHG emissions, issues of data quality gain crucial importance.

Although the best-guess strategy is well established, and rightly so, it also has obvious limitations. Morgan and Henrion (1990) show the need to indicate at least some kind of confidence interval, as in the IPCC's best guesses of future global mean temperature. A best guess has the advantage of offering a simple description, but quite often this description may actually be too simple.

Confidence intervals can be used to describe probability distributions. A normal distribution, for example, is fully described once its expected value and its standard deviation (or some multiple of it) are known. If this is the meaning of a confidence interval, then we have moved well beyond the best-guess strategy into the realm of the probabilistic strategy, which will be discussed below. But the IPCC is not claiming that future global mean temperature is a random variable whose probability distribution is identified with the help of a confidence interval. In actuality several best guesses are being combined here: one for the highest global mean temperature that it seems reasonable to expect, say, at double the current CO_2 levels; another for its lowest counterpart; and a third for the temperature level that seems most likely. It is not clear under what conditions a combination of best guesses is an appropriate approach to accounting for a range of possible outcomes, but it is certainly rather problematic whenever outcomes are sufficiently uncertain to make issues of risk aversion essential. In this case, we need to explore the whole range of possible outcomes, even when they are very unlikely. In many aspects of the regional assessment of climate change, this is a relevant issue.

A crucial property of regional climate change assessments is the impossibility of including a direct feedback from actions influencing regional emissions to the regional impacts of climate change. Inevitably, impact models are distinct from models of policy measures. Therefore, we must specify judgments about the appropriateness of the best-guess strategy for these two kinds of models. A best-guess strategy may be sufficient for impact models as long as extreme events do not matter. For impact assessments in Alpine regions, however, extreme events may well be one of the most important factors. Such events include floods from heavy rainfall, the devastation of mountain forests from intense storms, and landslides from melting permafrost. A best-guess strategy may be sufficient for models of policy measures whenever the relevant actors share the same degree of risk aversion. This is hardly, if ever, the case. In conclusion, for regional integrated assessments of climate change, the best-guess strategy is no doubt a necessary, but rarely a sufficient, approach.

8.4 THE PROBABILISTIC STRATEGY

Manne and Richels (1995) use the best-guess model MERGE 2 to perform the following thought experiment. Imagine that the coupled economy-climate system has just two different possibilities for reacting to increases in GHG concentrations. The first possibility, which is very likely, involves limited damage; the second, which is rather unlikely, involves major damage. Thinking of long-term impacts of climate change on the Alpine region, one could consider as an example the danger that, as a consequence of persistently rising global emissions, the oceanic circulation could shift in such a way as to give rise to a new glaciation period in the Alpine region. The probability of such a shift would be somewhat smaller if at least the industrialized countries would reduce their emissions to some extent.

This decision problem can be framed by performing two runs of a best-guess model for each possible action. Each action is represented as a climate lottery with two possible outcomes (minor and major damage). To compare ten different emissions scenarios, for example, we need twenty model runs. Each pair of model runs yields two figures for the present value of the action in question that can be blended into a unique present value by calculating a weighted average. Suppose that the two possible outcomes have probabilities of 5 percent and 95 percent. If these figures are used as weights, the weighted average represents the present value of the climate lottery for a risk-neutral decision maker. Along these lines, the best-guess model serves as a decision support tool by providing a ranking order for different actions based on the values derived from model simulations.

The procedure described above actually uses a best-guess model for representing climate lotteries. As soon as more than a few possible outcomes are considered, however, it gets very clumsy. A probabilistic modeling strategy therefore seems appropriate for representing decision problems in which a diversity of attitudes toward risk aversion matters. Such a strategy is much harder to design and implement than the best-guess strategy. Moreover, its links to various features of the climate system require careful attention.

8.4.1 Chaotic Systems and Climate Forecasts

The handling of uncertainty in IA is not restricted to the best-guess strategy. On the contrary, one may be quite dissatisfied with reducing a whole range of uncertain outcomes to an overall estimate of an expected value. Such dissatisfaction has primarily driven the development of the probabilistic strategy, now well established in the design of IA models. Using this strategy for model development implies that the full plausible range of a model's parameter space has been explored and that model results yield probability distributions rather than average values. The PAGE model (Hope, Anderson, and Wenman 1993) made probably the first attempt to use the probabilistic strategy in the integrated assessment of climate change. Another attempt to include some probabilistic features in IA modeling has been made with the development of IMAGE (Rotmans 1990), which nonetheless is still basically a best-guess model. It seems fair to say that the ICAM marked the breakthrough of the probabilistic strategy in IA (see box 8.6).

How then can researchers and decision makers dealing with climate change assess the relevant probabilities? This question has very different answers for the various components of the comprehensive system formed by humankind and climate. In some cases, scientific research can provide probabilities for specific events; in other cases this is impossible. Here we consider the former case, taking the problem of climate forecasts as an example.

There is a conceptual difference between weather forecasting and climate forecasting. To make weather forecasts, we need models depicting the atmosphere as a dynamic system. We are interested in predictions such as "Tomorrow scattered showers can be expected, in particular in the afternoon." To speak in the language of dynamic systems, we try to predict the state of the dynamic system, namely the atmosphere, in as much detail as possible. Weather forecasts may be entirely wrong, and even if correct they are reliable only over a short period of time, say from one day to several days. Such behavior is characteristic of nonlinear chaotic systems, in which the predictability is limited and the forecasting horizon is not constant. Special weather situations, in particular during transition phases, render forecasting exceedingly difficult.

How does climate forecasting differ from weather forecasting? In forecasting climate, we are not interested in predicting whether it is going to rain on the August 1, 2049. Instead, we want to assess, for instance, the overall probability of precipitation during summer fifty years from now. In contrast to weather, which is defined in terms of individual events, climate is defined statistically. In the case of climate, we need to forecast the probability distri-

Box 8.6   Integrated Climate Assessment Model

ICAM (Integrated Climate Assessment Model) was developed at Carnegie-Mellon University (Dowlatabadi and Morgan 1993; Dowlatabadi 1995). Like other IA models, ICAM represents a set of interdependent processes ranging from economic growth and carbon emissions through changes in atmospheric CO_2 concentrations, temperature changes, and sea level rise to impacts on ecosystems and human societies. Whereas other models start from the basic structure they want to represent and deal with the inevitable uncertainties later on, ICAM carefully attempts to represent the uncertainties relevant to climate policy explicitly.

Take the example of the Antarctic ice shield, parts of which may break away as a consequence of global warming, possibly leading to a sudden rise in global sea level. There is no way of telling today at what future level of CO_2 concentration such an event might take place. A modeler following a best-guess strategy can either ignore this event or make an arbitrary decision about when it might occur. The probabilistic strategy implemented in ICAM, however, leads to a different treatment, because ICAM includes a probability distribution for the event in question. Moreover, ICAM users can open windows that inform them about the problem's main features and about relevant literature. This general philosophy is applied to all processes represented in ICAM.

ICAM makes users aware of the many uncertainties involved in climate change in three different ways. First, it represents no major process without explicitly using probability distributions. Second, it documents the sources that have led the modeler to choose a given representation. Third, not even the probability distributions are fixed. The user herself defines her own specific characteristics based on information ICAM provides. Along these lines, the user defines a set of stochastic model components.

The strength of computer simulation comes into play when one tries to figure out the interactions among the probability distributions defined at a model's various junctures. In principle, an analytic treatment might lead from these probability distributions to distributions for a model's output variables. In practice, this is usually impossible even for a trained mathematician with plenty of time. ICAM, however, enables users without mathematical training to produce a model output consistent with their definition of the various distributions. If the model is run a second time, it again respects the definitions the user has set, but it usually yields a different output. In this respect, ICAM behaves more like dice than the model DICE. By performing a series of runs, the user becomes acquainted with the output variables' probability distributions.

Great efforts have been made to design ICAM in a user-friendly way. The multitude of uncertainties represented in ICAM, however, may confuse the user considerably. To some extent, this may even be an appropriate reaction to the problem of climate change. On the other hand, ICAM also enables the user to perceive aspects of the problem that are not a matter of subjective guesswork. By running the model several times, it is possible to get a reasonable feeling for, say, the danger of sea level rise as a consequence of global warming.

The probabilistic strategy has the additional advantage of enabling the modeler to deal with important but poorly understood processes, such as induced technical change, in a reasonable way. Contemporary economic theory has a hard time accommodating induced technical change. The difficulty runs deep, because economic theory is based on the idea of relative scarcities. Scarcity, however, is defined with respect to a given set of technologies, and so it is hard to discuss the choice of incentives for technical progress within a conceptual framework based on relative scarcities. This has led many modelers simply to neglect the problem of induced technical change. Climate policy then can only induce substitution effects among given technologies but cannot induce the generation of new technologies. Clearly, this is a very unsatisfactory situation. ICAM, on the other hand, simply includes a parameter for induced technical change together with relevant documentation. The uncertainties associated with the parameter in question can then be treated exactly as all the other uncertainties the model represents.

bution over all the climate system's possible states corresponding to the various weather patterns. Predictions from weather and climate forecasting are often referred to as predictions of the first and the second kind, respectively. Predictions of the first kind forecast a dynamic system's state at a specific time. Predictions of the second kind speculate on the likelihood of finding the dynamic system in a certain state. Predictions of the second kind are possible even for chaotic systems. Box 8.7 explores these issues in more detail for a simple model, the Lorenz model, one of the most famous models of a chaotic system.

Because we are interested in probability distributions and statistical properties, we might at first sight assume that the inherent limitations in predicting a chaotic system's defined state affect climate forecasting less than weather forecasting. However, we need to take some important points into consideration here.

Climate is experienced and defined on regional, or even local, spatial scales. IA in the field of climate change aims not only to provide information on a global scale (e.g., on the evolution of the global mean surface air temperature) but, in addition, to give reasonable forecasts of regional climate (e.g., on the future of the hydrological cycle in the Alpine region). Such information is required both to assess the risks of climate change and to guide appropriate response strategies. In section 8.2.1, we discussed the problems associated with regional climate forecasts, in particular in mountainous regions. From these considerations, we may conclude that even the probability distributions of certain climatic parameters are uncertain and need to be replaced by a whole family of probability distributions. This is particularly important for forecasting the tails of such distributions, which are relevant for determining the likelihood of extreme events.

The above considerations strongly suggest that a probabilistic strategy may be essential for many IA efforts regarding the problem of climate change, in particular on regional scales. To a considerable extent, such a strategy can be based on statistical data and on an understanding of key processes characterized by nonlinear dynamics. However, our discussion so far has assumed that research can provide the probability distributions to be used. We now turn to the problems that arise when this is not feasible.

8.4.2   Climate Variability and Subjective Probabilities

Even predictions of the second kind have their limits. The climate system itself exhibits variations on a variety of different temporal scales that may actually change the shape of its probability distributions, possibly on timescales much longer than those usually employed in the definiting climate. Because climate is not just averaged weather, which poses a serious problem, we need to account for a whole array of processes that can largely be ignored when making weather forecasts. Figure 8.4 gives an overview of the timescales involved in weather and climate forecasting and the important

Box 8.7 Forecasting Weather and Climate

To make the difference between weather and climate forecasting more explicit, let us use the Lorenz attractor as a simple model for the climate system. The system of equations that governs this attractor originally stems from a meteorological problem and has been extensively used, ever since Lorenz (1963) developed it, as "a poor man's climate model" and a prototype of a model displaying chaotic behavior (Sparrow 1982; Palmer 1993). The model describes climate in terms of three relevant variables referred to in general as X, Y, and Z.

In a dynamic model, the first and most straightforward task is to describe individual variables' evolution in time. Figure 8.5 provides an example for the X variable obtained in simulations with the Lorenz model. X exhibits chaotic fluctuations, jumping irregularly between positive and negative values. Weather forecasting may be looked upon as analogous to predicting X's temporal evolution as shown in figure 8.5.

Sensitivity to initial conditions, sometimes referred to as the butterfly effect, limits such forecasting. To demonstrate this sensitivity, figure 8.5 shows the result of two simulation runs (solid and dashed curves) conducted under exactly the same conditions with the exception of the starting values of X, which differed by 10^{-6}. After about 13 time units, the curves obtained during the two simulation runs begin to differ visibly. Two initial states that differ only slightly diverge exponentially to become completely independent after a finite length of time. This is where the metaphor of the butterfly comes in. At times, a fluctuation as small as the one caused by the flutter of a butterfly's wing may lead to an entirely different outcome in the weather's macroscopic pattern. It would obviously be pointless to establish a cause-and-effect relationship between the presence of a butterfly and the subsequent occurrence of a thunderstorm: Any attempt to train butterflies to trigger rainfall when required would be doomed to failure! Nonlinear chaotic systems by nature cannot be predicted or controlled in important respects in a world where small fluctuations are practically unavoidable.

For illustrative purposes, it is sometimes useful to represent the simulation results differently. Instead of the individual variables, the state of the system itself, characterized by a specific combination of all relevant dynamic variables, can be plotted as a function of time. A coordinate system comprising these relevant variables is referred to as the dynamic system's phase space. At any moment, the state of the system can be described as a point in this phase space, much as the location of a point in three-dimensional space can be described by its spatial coordinates. Rather than X, Y, and Z, as in the Lorenz model, we have air pressure, humidity, and temperature as relevant variables for weather forecasting.

We depict the temporal evolution of the system in the phase space of our model in panel (a) of figure 8.6. Each point in the phase space is taken to represent one particular state of the atmosphere. The figure sketches the atmosphere's evolution over five days as a phase space trajectory. Over a few days, the planetary-scale atmospheric circulation is deterministically predictable, meaning that the effect on the forecast of a small uncertainty in the initial conditions (the location of the trajectory at day 0) is still tolerable within this time period. We call this *predictability of the first kind*. The degree of divergence, and hence the reliability of a forecast over several days, depends on the specific weather situation under consideration. Over longer timescales (say, one month), however, deterministic predictability is entirely lost. Phase space trajectories that start from slightly different initial conditions diverge and may end up in completely different regions of the phase space (i.e., may represent completely different states of the atmosphere). To counteract this effect, we may try to improve our knowledge of the initial conditions (by improving the meteorological observations), but the gain in predictability is limited. As illustrated in figure 8.5, an error in the initial conditions, no matter how small, can result in an arbitrarily large error later. This is a key property of chaotic sys-

Box 8.7 (Cont.)

tems. In relation to weather forecasting, a broad consensus exists nowadays that this effect makes predictions of individual low-pressure systems practically impossible beyond about two weeks.

Nevertheless, a different kind of forecast may be possible beyond these horizons. Panel (b) of the figure shows the phase space representation of the Lorenz model in the X-Z plane. The plane is not covered uniformly, as one might expect from a totally random system. Despite the dynamics' chaotic nature, only certain combinations of X-Z values are realized. The set of all possible states in a dynamic system's phase space is also called its attractor. Though one cannot forecast in detail when the climate system will be in the left or right wing of the Lorenz attractor, it might be possible to predict how often it is in each. Information of this type has the character of a probability distribution (cf. density of points in the phase space in panel (b) of the figure). Predictability based on such information is often referred to as *predictability of the second kind*. Climate refers thus to statistical properties rather than to single events.

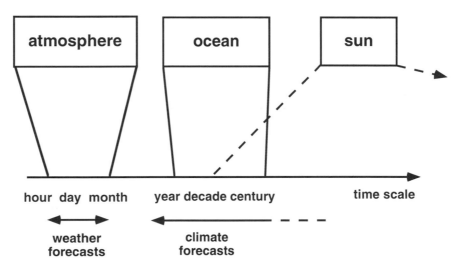

Figure 8.4 Approximate range of timescales for important processes affecting weather and climate forecasts. A process's typical timescale refers to the time interval over which significant variations take place.

processes influencing them. In making weather forecasts, one may safely ignore changes on annual or decadal timescales associated with the variability of the oceans or the sun. Such effects enter the models as boundary conditions derived from numerous measured data. However, in making climate forecasts, these slow processes need to be predicted accurately over decades. We thus face similar problems to those in weather forecasting, albeit on much longer timescales. Box 8.8 offers an example of long-term climatic variations derived from climate records and also discusses in more detail the problems arising for regional climate forecasting.

Let us disregard anthropogenic effects for a moment and consider the natural variability of the climate system alone. Even with this simplification,

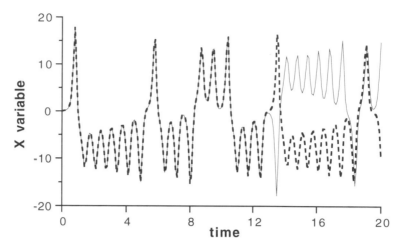

Figure 8.5 The dynamics of the Lorenz model: variable X as a function of time obtained in two (solid and dashed curve) simulation runs under exactly the same conditions except for the starting values of X at $t = 0$, which differed by 10^{-6} in the two simulation runs. Time is given in relative units.

climate forecasting over several decades is still not feasible today, mainly because we lack sufficient ability to simulate slow natural fluctuations on this timescale deterministically. For instance, some of the variability with periods exceeding approximately 10 years may result from internal fluctuations in the ocean. These processes and their effect on the atmosphere have only just recently begun to be simulated realistically (e.g., Mikolajewicz and Maier-Reimer 1990; Delworth, Manabe, and Stouffer 1993), and such simulations still have a very low deterministic predictive potential. Furthermore, either adequate understanding of other processes acting in the same band of frequencies is still lacking (e.g., solar variability), or these processes appear to be intrinsically unpredictable within the relevant time frame (e.g., volcanic activity).

Based on current understanding of the climate system, we cannot tell whether it will become possible to forecast the climate's natural variability, even on decadal timescales. An illustrative example of this is the ENSO phenomenon, which has a timescale of 2.1 years and arises from a complex interplay between Pacific Ocean and atmospheric circulations. On the one hand, several El Niño events have been predicted successfully during the last decade (Zebiak and Cane 1987); on the other hand, recent work suggests that these events can also become chaotic as a result of an interaction with the annual cycle (Fei-Fei, Neelin, and Ghil 1994; Tziperman et al. 1994).

On the whole, we currently have a rather limited capability to predict natural variations of climate. Nevertheless, current climate models can estimate anthropogenic climate change. Here "climate change" is defined as the change that would take place in the absence of natural variability, assuming that all natural forcing factors remain constant. However, there is no guarantee that the climate system behaves in the manner this expectation presupposes. Cli-

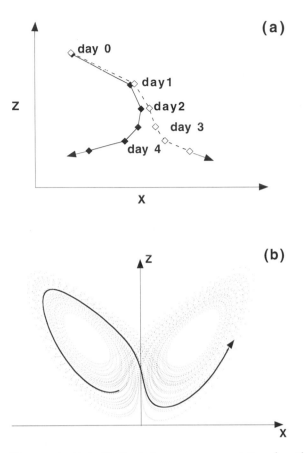

Figure 8.6 (a) An idealized phase space representation of weather forecasting. (b) Comparison of weather forecasting and climate prediction in the Lorenz phase space. Weather forecasting corresponds to the short trajectory, drawn as an arrow. Climate predictions, on the other hand, are based on a probability density distribution for long-run trajectories, indicated by the scattered points.

mate variability itself may experience major modifications in the event of climate change. Ice core records indicate that large and rapid climate fluctuations have occurred in the earth's past, whereas the climate of the current Interglacial has been remarkably stable (Dansgaard et al. 1993; GRIP Members 1993).

From a decision perspective, however, the probabilistic strategy can make sense even where predictions of the second kind fail, because the strategy can be based on subjective probabilities (see box 8.9). If one uses probability distributions to represent the degree of confidence that a given actor holds with regard to a variety of statements, the probabilistic strategy can be related to these distributions. In a model like ICAM, users can try to express their subjective probabilities by defining suitable input probabilities. The model then delivers outputs consistent with the user's subjective probabilities.

Box 8.8   The Spectrum of Climate Variability

The climatic record presented in figure 8.7 shows the mean annual temperature anomaly in northern Switzerland since 1525. This display demonstrates modes of variability characterized by different timescales. First, interannual variability is huge. On several occasions since 1525, the annual mean temperature between two consecutive years has changed by as much as 2 K. Secondly, if these rapid fluctuations are removed (solid line), another much slower mode of variability becomes apparent.

The situation illustrated in figure 8.7 is qualitatively comparable to that of the Lorenz attractor. The latter can also be interpreted as arising from several types of fluctuation: a rapid fluctuation that takes place within each of the two wings interrupted by an occasional transition from one wing to the other.

Climate records such as those shown in figure 8.7 may be used to compute a spectrum of climate variability like the one presented in panel (a) of figure 8.8. The panel presents the variance of the global surface air temperature as a function of the period of oscillation. Daily and seasonal oscillations are due to external influences, as are some of the slower oscillations. Some oscillations, however, are related to the climate system's internal variability. For instance, the peak around three to seven days results from the successive generation and decay of low-pressure systems, whereas slower processes with periods exceeding twenty years are presumably the result of solar, oceanic, and volcanic variability.

Despite differing in details, spectra of climate variability show one significant common feature that is important for what follows: the variance increases with increasing length of timescale. This feature presents a serious difficulty for climate forecasting. Basically, the slow processes, which are the least understood and for obvious reasons, difficult to observe, might determine the nature of future climates.

Which of the modes of variability must be taken into consideration for climate forecasting? Consider for a moment the simpler case of medium-range weather forecasting (cf. panel (b) of figure 8.8). Here it is possible to neglect the slow portion of the spectrum of variability. Fluctuations with periods exceeding approximately twenty days are too slow to affect the weather significantly at validation time. Next, consider a climate forecast for the years 2020–50. It follows analogously that variability on timescales exceeding about 100 years can be disregarded (cf. panel (c) of figure 8.8). However, the phase of fluctuations with periods between approximately 10 and 100 years must be predicted precisely.

8.4.3   Achievements and Limitations

The probabilistic strategy offers powerful tools for dealing with various kinds of uncertainty. Moreover, it can help reduce the impacts of a modeler's individual subjectivity on model outcomes. For this purpose, one can rely on collective subjective judgment to derive probability distributions. These distributions still represent subjective probabilities; the subject in question, however, is not an individual modeler but a scientific community of experts.

Representing relevant parameters by their probability distributions is not a panacea for the problem of uncertainty in IA modeling. In particular, for models with a huge number of more or less independent parameters, it becomes increasingly difficult to explore the whole parameter space systematically because of the large computational demands of varying many

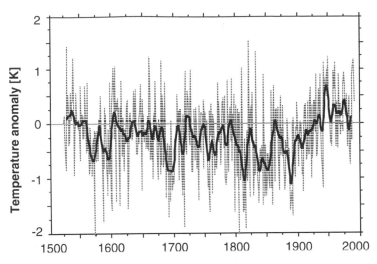

Figure 8.7 Mean annual air temperature anomaly in Basel according to the record of Pfister (1984) given with respect to the mean 1900–60 temperature. The dashed line shows the record at yearly resolution; the solid line is based on an eleven-year weighted mean. The record is based on the combined use of instrumental data, historical documentary records, and proxy information (e.g., harvest dates, tree rings).

parameters simultaneously. Hence, it would be practically impossible to account for the uncertainties associated with the results of model simulations even if the parameter uncertainties were well defined. In addition, whenever there is a disparity in methods and a broad range of opinions about a chosen approach's appropriateness, one would be well advised to doubt seriously the technical probabilities obtained using any particular method (cf. box 8.2).

Although one should certainly not neglect the probabilistic strategy in the integrated assessment of climate change, one should avoid the trap of "naive subjectivism" (Funtowicz and Ravetz 1985), which considers the probabilistic strategy as not only necessary but also sufficient for dealing with the problem of uncertainty. This is certainly not the case for at least two related reasons. First, the probabilistic strategy has little to say about the important processes in which ignorance may actually be socially constructed, a topic discussed in section 8.2.2. Second, and partly related to the first point, the probabilistic strategy remains silent about value orientations.

The attempt to keep value judgments in the background is quite appropriate as long as one sticks to a frequentist interpretation of probability (cf. box 8.9 for explanations of different approaches in probability theory). If it is possible to determine the objective probability, say, of heavy rainfall in areas of the Alpine region, the result is a statement of fact that should be largely independent of value judgments. A frequentist approach to anthropogenic climate change, however, is hopeless, for it would require humankind to perform a large number of climate experiments. The personalist approach, on the other hand, leads to the question of how subjective probabilities are related to preferences and to larger preference patterns, such as value systems.

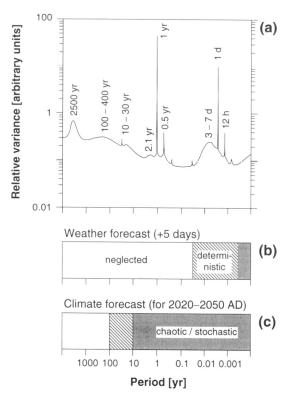

Figure 8.8 Panel (a) shows the spectrum of climate variability and the relative contribution of various oscillations that contribute to the observed variability of the global surface air temperature as a function of period (in years). The other panels highlight those parts of the spectrum that must be predicted when forecasting (b) weather and (c) climate. The part of the spectrum requiring deterministic prediction is hatched; the chaotic or stochastic part is shaded. The part left white can be neglected. (Panel (a) is adapted from Mitchell 1976).

If subjective probabilities are taken into account, it becomes quite natural to consider subjective preferences on the same footing. In particular, if an IA model or procedure invites users to specify their subjective probabilities, it would be quite sensible to ask them to specify preferences, too. Obviously, one would not be equally interested in all kinds of preferences. It seems natural to ask whether certain value orientations shape whole systems of preferences relevant to the problem of climate change. Such value orientations would be linked to subjective probabilities, but the problem of uncertainty would now appear closely linked to a seemingly very different issue: the existence of incompatible value orientations.

Under these circumstances, the privileged position that even avowedly subjective judgments made by scientific experts often enjoy must be carefully examined. Even atmospheric scientists studying the Alpine region may be unable to calculate an objective probability for increased fog as a result of climate change. Nevertheless, their subjective probability on this issue carries more weight than the subjective probability that a decision maker or an

Box 8.9 A Note on Mathematics and Uncertainty

Probability theory is certainly the most widely and successfully used technique for dealing with uncertain information. Probability theory has its origins in the sixteenth century, when scholars tried to gain a better understanding of games of chance. It was codified in an axiomatic format by Kolmogorov (1933). As is often the case with mathematical structures, probability theory can be related in various ways to the experiences of everyday life and of scientific inquiry. In other words, various interpretations of probability theory are possible.

When a measurement with several possible outcomes can be repeated many times, the relative frequencies of the various observed states tend to approach numbers that obey the same rules as the probabilities Kolmogorov introduced. This leads to the classical frequency interpretation of probabilities, which allows the theoretical concept of probability to be related to measurable real-world observations. In this sense, the frequency interpretation deals with objective probabilities. It assumes that the state of a system can be analyzed (i.e., measured) many times whereas a system's distributional properties remain the same. This concept is expressed mathematically using the idea of random variables and a corresponding set of independent and identically distributed realizations, called a sample. If we have many different realizations of this random variable, the relative frequency of a certain outcome approaches its probability.

Box 8.7 indicates that important features of a chaotic system can be described in terms of probability distributions. On the other hand, for a given probability distribution, it may be possible to find a chaotic process that could generate it. This duality fits nicely with a frequentist interpretation of probabilities.

Applications of probability theory, however, need not rely on a frequentist interpretation. Probability can also be interpreted as a measure of the degree of belief that some person associates with a given statement (Morgan and Henrion 1990). This is the personalist interpretation of probability, and it leads to the idea of subjective probabilities as developed in Bayesian statistics. It can be further elaborated with notions like upper and lower probabilities and belief functions.

We need not discuss all the interpretations of probability theory that the literature offers. The contrast between objective and subjective probability is no doubt the most salient feature of the various interpretations proposed so far. In the personalist camp, Savage (1954) made an important step when he suggested dealing with probability and utility within a unified mathematical framework. His approach is related to the one von Neumann and Morgenstern (1944) chose for the foundation of game theory. Whereas game theory deals with interdependences among decision makers, however, Savage considers a single decision maker faced with a choice among lotteries (see section 8.2.3). Each lottery is characterized by a set of possible outcomes, and the decision maker is characterized by preferences among outcomes, subjective probabilities of outcomes, and preferences among lotteries. This formalization of decision under uncertainty has become a major tool in contemporary decision theory. It is important for IA in the field of climate change because it emphasises the links between uncertainty and diverging value orientations.

The techniques discussed so far are based on the idea that a particular system under consideration can assume various states and that each state is characterized by different values of "observables." For example, a person's nationality may be considered a state variable and the color of his passport an observable. However, some systems do not display the same observables in every state. This feature has been studied extensively with regard to quantum-mechanical systems, but it is a normal ingredient of everyday life, too. When a person assumes the role of a tennis player, scores are an observable; when she assumes the role of a tennis umpire, the corresponding decisions are an observable. Because the two roles are incompatible, the two observables cannot arise simultaneously.

Box 8.9 (Cont.)

Mathematical representations of such systems rely on nonstandard logics (e.g., Blau 1977). A nonstandard logic may be combined with probability theory, as is the case in quantum mechanics. For an application to the problem of climate change, see Jaeger and Kasemir 1996.

ordinary citizen may hold with regard to the same question. But clearly we would not consider the preferences the scientists in question may hold about climate issues, and even less so those they may hold about other issues, to be more appropriate than the preferences entertained by decision makers. And in a democratic setting, there are strong reasons to take the preferences of ordinary citizens as yardsticks for formulating the value judgments that define the decision rules for climate policy.

In other words, the probabilistic strategy introduces the indispensable subjective moment needed to deal with most problems of decision making under indeterminacy. But it does so only halfway, by acknowledging subjective probabilities while ignoring subjective preferences. Therefore it seems quite sensible to combine subjectively defined probability distributions with equally subjective utility functions. This step leads to the pluralistic strategy in integrated assessment.

8.5   THE PLURALISTIC STRATEGY

8.5.1   The Importance of Values

We have seen that the probabilistic strategy, despite its undoubtable strengths, involves major difficulties, the core problem being that in many cases the concept of probability can be interpreted only in relation to a specific observer (De Finetti 1974). Probability is then used as a measure of uncertainty in belief (Morgan and Henrion 1990), which is hardly avoidable in climate change research. Thus, by probability's very nature, a strong normative component in implementing the probabilistic strategy is unavoidable.

This view of probability has two important consequences for integrated assessment. First, it implies that probabilities are subjective in the same way that tastes are. However, this does not mean that probabilities are abitrary or that people can never agree on them. Under normal conditions, people agree on some probabilities. People usually believe that the probability that the sun will rise tomorrow is pretty close to one. In many cases, however, people differ in their estimates of probability. In the context of the climate issue, the work of Morgan and Keith (1994) is of great interest. They interviewed sixteen leading climate experts and asked them to draw probability distributions for certain key variables, such as climate sensitivity. The answers varied significantly with respect to the size of the confidence intervals and to the best estimates of average values. A sound approach to integrated assessment might therefore be to say, "Instead of telling you my assessment, I'd prefer to tell you how I arrived at that assessment, because I think that this will help you come up with an assessment of your own" (Kreps 1988, pp. 151–152.). Second, it can be very misleading to treat the issue of probability separately from that of subjective value systems. As Funtowicz and Ravetz (1985) emphasized, such a separation becomes completely inappropriate when both decision stakes and systems uncertainties are high. This is certainly the case with the integrated assessment of climate change.

Humans conduct scientific research, whether natural or social, and it is thus subject to human choice. Humans choose what to study and what to ignore, what methods to use in their analysis and what criteria to apply in determining the validity of the data gathered. If there is no standpoint outside of history and society from which life can be observed and described, we must make the underpinnings of our analysis as explicit as we can. We must be reflexive. Reflexivity is simply the self-conscious examination of the implicit assumptions inevitably embedded in any analytical approach. By making these assumptions explicit, the researcher has an opportunity to question them, rather than taking them as given. In making these assumptions, we inevitably make value judgments. However, when the value-based assumptions disappear into the background, they come to be seen as intrinsic and are uncritically accepted, often without any conscious thought about either their presence or their implications.

The value that we place on anything is an assertion of its importance to us, often expressed in terms of what else we would give up to acquire, use, or preserve it: in other words, we prefer one to the other. That people have different preferences or values is an indisputable commonplace. Different branches of the social sciences teach that biased interpretations are not arbitrary but relate to value systems, conceptual frames, and different perspectives (Tversky and Kahnemann 1974, 1980, 1981). Obviously one could argue that only one perspective exists on nature: the right one. Much of the discussion about the impacts of climate change is centered on such topics, which is unfortunate because it seems inherent in the nature of complex systems that complementary, even contradicting perspectives are legitimate. The fact that people hold fundamentally different perspectives on how the world is or ought to be ordered may be expressed as "contradictory certitudes" (see section 8.2.2). Considering the question of evaluating the impacts of climate change illustrates that such contradictory certitudes are not simply a matter of deplorable insufficiencies in knowledge.

8.5.2 Complex Systems, Extreme Events, and the Need for Pluralistic Approaches

The United Nations Framework Convention on Climate Change states as its goal the stabilization of GHG concentrations in the atmosphere at a level that would prevent dangerous anthropogenic interference with the climate system. What must be considered dangerous depends on the impacts expected on ecological and social systems. Value system considerations undoubtedly enter at the level of whether specific possible impacts are perceived as threatening. However, the complexity of the questions under consideration leaves room for different complementary descriptions, the choice among which depends on subjective judgment.

The discussion about the potential impacts of low-probability, high-risk events provides an instructive example. A discussion on extreme events has been notably absent in IPCC reports because scientific evidence concerned with this topic is fragmented and contradictory. Sir John Houghton, chairman of the IPCC's scientific working group, wrote in a letter (Shackley and Wynne 1995c):

There are those who home on surprises as their main argument for action. I think that is a weak case. No politician can be expected to take on board the unlikely though possible event of disintegration of the west Antarctic ice sheet. What the IPCC scientists have been doing is providing a best estimate of future climate under increased greenhouse gases—rather like a weather forecast is a best estimate. Within the range of possibility, no change of climate is very unlikely. Sensible planning, I would argue needs to be based on the best estimate, not on fear of global catastrophe or collapse. (p. 226)

Dealing with such topics responsibly is doubtless a delicate issue: however, an integrated assessment of climate change cannot simply ignore them. The possibility that an increase in average global temperature may increase extreme climate events on regional scales is of major and immediate relevance for regional assessments. It is a crucial topic because of the catastrophic damage that natural disasters such as flooding or landslides can cause. Regional climate models indeed predict an increase in average precipitation in the Alpine region that will manifest itself mainly in an increase in heavy precipitation events (Frei et al. 1997). These are changes that happen on timescales people care about; in addition, the associated economic costs may be considerable, a possibility of major concern to insurance companies (Brauner 1994).

A regional decision maker who considers the risks engendered by an increase in extreme climate events to be an important issue faces several problems in coming to any decision at all. The catastrophic effects of extreme climate events have more than one cause: Changes in demographic structure and land use, for instance, may be of more immediate relevance than climate change. The impacts of climate change are experienced on a regional or even local scale. However, actions taken on regional or local scales have no direct mitigating influence on these impacts. Any consequences of regional action are contingent on cooperation at the global level. Actors involved in the decision process have different subjective preferences and risk perceptions, which may lead to quite different interpretations of the same set of limited and uncertain information. The degree to which investments in abating climate change are perceived as costs that must be justified by the prevention of major damage also shapes discussion about possible options. An alternative point of view might emphasize the role of innovative responses to climate change. Such responses might even stimulate economic development. The discussion in chapter 7 shows that such expectations are not simply wishful thinking but may be based on empirical facts.

In terms of decision theory, one might describe the situation as one in which different perspectives lead to the construction of different, complementary sets of climate lotteries. We assumed that the decision problem could be described as a choice among different climate lotteries (cf. section 8.2.3). However, if we have to deal with complementary perspectives, outcomes essential for some major decision criterion in one set of climate lotteries may not even be accounted for in others (for example, catastrophic damage, innovative responses), although the different actors involved may argue consistently within their own sets of decision criteria. If integrated assessment does not address the issue of constructing incompatible sets of lotteries based on subjective probabilities and preferences, communication among the different actors involved in the decision process may grind to a complete halt.

8.5.3   Capturing Values and Complementary Perspectives in Integrated Assessment

If attempts at an integrated assessment of climate change cannot avoid subjectivity and value judgments, such attempts run a serious risk of becoming completely subjective and arbitrary exercises without scientific content. This danger should not be taken lightly. Scientific institutions cannot maintain credibility if they back assessments of climate risks that do little more than express the individual subjectivity of the researchers involved. On the other hand, both for the needs of policymakers and as a response to public opinion, scientists simply must develop integrated assessments of climatic risks. It would be very irresponsible for scientific institutions to refrain from any serious attempt to contribute to such assessments. Methodologies that would enable scientists to do this reasonably and acceptably are definitely needed.

Because a modeling approach currently dominates integrated assessment, we first focus on strategies to address subjectivity in integrated assessment modeling. It is important to realize how inherent subjective value judgments are to any serious effort at computer modeling in the domain of climate change (e.g., Shackley and Wynne 1995a; Shackley et al. in press). No modeling is possible without a long sequence of small decisions based on the modeler's subjective judgment. Subjectivity is already present in the conceptual phase in deciding which elements the model will include and which it will leave out. Decisions in the modeling phase involving value judgments range from parameterizations to the choice of algorithms, from the treatment of problematic data sets to the definitions of functional forms. For example, economists' attempts to estimate climate change's impact in monetary terms have been denounced as immoral and misleading because they use the market cost of labor as the basis for valuing lost productivity through mortality. This, the critics contend, amounts to placing a differential value on human life in the industrialized and less-industrialized worlds in violation of the basic principle of human worth and equality.

Models are shaped by value choices on the part of their creators, including preferences for objectivity, parsimony, reproducibility, and efficiency. To the extent that modelers share value systems, peer review reinforces that set of preferences. However, this means that the values of other social groups that will participate in the societal negotiation about policy may be missed entirely, rendering the models at best irrelevant and at worst pernicious in their eyes.

8.5.4 Implementing Complementary Perspectives in Model Routes

As a complementary methodology to the currently used methodologies of uncertainty treatment described above, Van Asselt and Rotmans (1995a, 1996) propose the model route strategy. This is an attempt to supply tools that enable modelers to be explicit about their subjectivity and value judgments by relating the whole sequence of decisions to socially relevant value orientations. Any integrated assessment model designed to describe aspects of real-world functioning can be thought of as a "map" of a landscape. Both modelers and users have to deal with the fact that no landscape can be known completely in advance and that interpretations of what can actually be seen may vary significantly. Model routes can then be thought of as representations of the different pathways, reflecting different assessments of the landscape's unknown features. In other words, model routes are different mathematical representations of state-of-the-art knowledge, including varying interpretations of uncertainties and indeterminacies. The model route strategy enhances the notion brought forward by Morgan and Henrion (1990) of assimilating different plausible model structures into one single model that "contains" these various structures as special cases.

The model route strategy was applied in developing the integrated assessment model TARGETS (box 8.10). Model routes consist of chains of alternative formulations of model quantities and model relationships. This means that alternative values for model quantities have to be determined. In the initial implementation of model routes in the TARGETS model, this is achieved by simply changing parameter values. When seeking a more sophisticated implementation, one may consider subjective probability distributions, as discussed in the previous section (Morgan and Henrion 1990; Morgan and Keith 1994), or alternatively, nonprobabilistic concepts. Where uncertainty and indeterminacy affect model relationships, the model's functional structure needs to be reformulated. This means either changing the

Box 8.10   TARGETS

The National Institute of Public Health and the Environment (RIVM) in the Netherlands launched the Global Dynamics and Sustainable Development Program in 1992. This research program's main objective is to clarify the concept of sustainable development from a global perspective. To achieve this, Rotmans and his group developed the global integrated assessment model TARGETS (Rotmans and De Vries 1996), intended to yield insights into the complex interrelations in time and space among the increasing worldwide pressure on the environment, human health, and society as a whole, in particular in the long term (the next 100 years).

The TARGETS model comprises submodels covering human health and demographic dynamics, economy, energy resources, land and water resources, and global cycles. Adopting a systems dynamics approach as a guiding principle, the biosphere can, in global terms, be considered a system of reservoirs, with natural and human-induced processes connecting the reservoirs. The interrelated cause-and-effect chains are described in terms of pressure (the driving forces), state (the changes in environmental and human resources), impact (the consequences for people and environment) and response (policy options).

One of the research's major aims is to use the TARGETS model to improve communication among scientists from different disciplines as well as between the scientific and policy-making communities. To be explicit about uncertainties and value judgments is in this respect a necessity. The model addresses this challenge by offering various perspective-based model routes. Furthermore, a user-friendly interface enables people to explore both facts and uncertainties with respect to global change as well as a limited variety of perspectives. Finally, it allows the user to design and evaluate policy strategies for a hypothetical world governor. Such exercises might help users to get a feel for the complex interrelations between environmental and human processes and their consequences.

However, in TARGETS the model route strategy has been applied within an existing modeling framework that may be characterized as hierarchical. Adopting a pluralistic attitude toward models would involve building from scratch using different models originating from different perspectives.

function by changing the constants or the function's form (e.g., exponential instead of linear or vice versa), or deleting, adding, or changing the function's arguments.

In principle, an enormous number of potential model routes could be implemented by providing alternatives for multiple parts and randomly clustering them. However, this makes no sense, because it introduces a bewildering variety of possible choices instead of attempting to provide a method of ordering complex issues. In addition, it is reasonable to assume that not all possible model structures are equally plausible and consistent with respect to the underlying assumptions. Modelers therefore need to search for an organizing principle or heuristic concept that would enable them to provide a limited number of coherent model routes. Subjectivity must be rendered explicit in such a way that both modelers and users understand that value systems (perspectives) are involved in any attempt to describe real complex systems. Van Asselt and Rotmans (1995a, 1996) therefore introduce the concept of a "perspective-based model route," a coherent chain of biased interpretations of the crucial uncertainties present in an integrated assessment model.

Incorporating perspective into the model route strategy implies that it is the modelers role to erect signposts in the landscape and to indicate them on the map to provide routes that reliably lead to a specific end. However, instead of providing merely one description of the functioning of aspects of the real world, thereby adopting a cavalier attitude toward uncertainty and indeterminacy and leaving aside the wide variety of alternative perspectives, modelers need to illuminate the major ambiguities and to provide reasonable alternative interpretations. Erecting signposts is thus equivalent to identifying crucial uncertainties and indeterminacies in state-of-the-art knowledge and indicating where these affect model choices, then providing mathematical translations consistent with a particular perspective's bias and preferences. Less experienced walkers, namely, users of the model, can then choose to follow one of these routes to explore the landscape instead of blindly hoping to stumble across a suitable path.

In elaborating on perspectives, the methodology of ideal types developed by Weber (1906, 1913; see also Burger 1987; Morishima 1990; Jaeger 1994; and Hirsch Hadorn 1997) can be helpful. Ideal types are often identified with scientific abstractions in general. Ideas of perfect competition, frictionless motion, or ideal gases are then all described as ideal types. Weber, however, developed a methodology specifically for the social sciences. In a nutshell, he tried to organize a scientific search for moral choices faced by a given social system. Such choices refer to historical possibilities framed in moral terms. An ideal type connects historical facts and moral norms in such a way as to enable the researcher to understand crucial causal connections in human history. The moral norms and values to be taken into consideration, however, are not those that concern any particular researcher, but rather those relevant within the society under study. In this vein, a researcher studying ancient Greece might try to develop ideal types of aristocracy, tyranny, and democracy. Ideal types, then, can be understood as socially legitimate perspectives on social issues—legitimate, that is, in specific, relevant social settings.

The methodology of ideal types can be used to move from the individual subjectivities of researchers involved in integrated assessment to the collective subjectivities present in the global society of which those researchers are members. The so-called cultural theory of risk (Douglas 1985; Rayner 1988; Thompson, Ellis, and Wildavsky 1990) offers an interesting way of introducing a parsimonious typology of such ideal types. This theory can be used to define three different ideal types: the egalitarian, the hierarchist, and the individualist (see box 8.11).

Van Asselt and Rotmans (1995a, 1996) decided to use these three ideal types as a heuristic to arrive at a limited set of sound model routes. In the present situation of social theorizing, this choice seems interesting. We wish to emphasize, however, that the model route strategy makes sense independently of the cultural theory of risk.

Box 8.11   Cultural Theory

Cultural theory (e.g., Douglas and Wildavsky 1982; Douglas 1985; Rayner 1988, 1991, 1992; Thompson, Ellis, and Wildavsky 1990) generally distinguishes four perspectives from which people perceive the world and behave in it: the hierarchist, the egalitarian, the individualist, and the fatalist. The types generated are derived from two common social dimensions: one that measures social restrictions on individual autonomy ("grid") and one that signifies solidarity versus egocentrism ("group"). Hierarchism, individualism, egalitarianism, and fatalism are not novel concepts. In some sense, cultural theory duplicates conventional science. Hierarchism and individualism, for example, reflect a conventional duality that has been capitalized upon in various forms under headings like "collectivism versus individualism" and "state versus market" (Grendstad 1995).

Individualism defines roles as little as possible, restricts authority continuously, keeps institutions small and leaves the individual to himself on the basis of personal ability and fortune, and thus represents a life without boundaries. Emphasizing freedom, individualists want to carry out their enterprises without any interference. However, individualism does not imply abstention from exerting control over others; the individualist's success is often measured in terms of the size of the following the person can command. The individualistic driving force can be characterized as growth, that is, increasing needs and extension of activities. In arguing that a perspective achieves viability by perfectly matching social relations to cultural biases, cultural theory holds that individualism, which implies that anything is obtainable exclusively, matches the myth of nature benign. In perceiving nature as cornucopian, the individualists clearly have the scope to manage both needs and resources. Therefore, they choose to manage needs and resources upward to the very limits of their skills. Individualists can be considered risk-seeking in advocating an adaptive management style.

In hierarchism, strong group boundaries and binding prescriptions characterize an individual's social environment. In other words, an individual interacts with others, and is separated from them, by strong rules and tight role definitions. Because the parts are supposed to sacrifice themselves for the good of the whole, most matters are of collective concern. Therefore, a central role is ascribed to collective institutions, such as the government, to secure the collective, and thereby the individual's private, interests. The exercise of control (and more generally the very existence of inequality) is justified on the grounds that different roles for different people enable people to live together more harmoniously than alternative arrangements would. Because nature perverse/tolerant requires strong control to ensure stability, cultural theory has it that hierarchical relationships match a myth of nature as tolerant and perverse. The managing institution must take steps to prevent unusual dangerous occurrences. The hierarchists therefore advocate the exercise of control as a management style.

Egalitarianism is a shared life of voluntary consent that makes status insignificant. Because egalitarian groups lack internal role differentiation, no individual is granted the authority to exercise control over others. Egalitarians emphasize equality and collective benefits. The encroachment of hierarchy (which brings status differences) and individualism (which all too easily introduces inequalities of wealth, power, and knowledge) constantly threatens this desired state of affairs. Nature, for egalitarians, is strictly accountable. Cultural theory thus holds that an ephemeral nature suits egalitarianism very well. Because the egalitarians perceive resources as being fixed and believe that people can do nothing about them, the only available strategy is to decrease their needs so as to ensure a nonnegative overlap. The myth of nature ephemeral tells us that radical change must come, before it is too late. The egalitarians thus advocate a management style that can be characterized as preventive and risk-aversive.

Fatalism in cultural theory terms is associated with the view of reality as capricious. Fatalists cannot be ascribed a particular management style or attitude toward risk. These survivalists merely try to cope with the situation with which they are confronted.

8.5.5 Achievements and Limitations

Because we are dealing with complex systems, proponents of any particular value system can always find arguments for their own perspectives. We therefore need a plurality of complementary perspectives that acknowledges the heterogeneity of value systems in any IA effort. We consider it an important challenge to find ways to incorporate complementary perspectives into IA modeling. The model route strategy described in the previous section at least enables us to consider more than one (hidden) perspective. The existence of divergent perspectives usefully illustrates the difficulty of attempting to account for a policy phenomenon such as climate change; whether the "problem" exists at all, and how to tackle it if it does, will always be contested and will be subject to competing definitions (Shackley and Wynne 1995a). A model that allows the implications of the various perspectives underlying its structure to be explored possesses certain advantages. Such models allow constructive discussion of the topic of subjectivity in model-building between model developers and model users.

Obviously, the model route strategy also has limitations. A first problem occurs because translating qualitative descriptions into model quantities and relationships is a hard and tough job. This means that selective compromises are made in arriving at perspective-based model routes. In other words, it is not clear whether different people would assign a set of model characteristics to the same perspective. This subjectivity in assignment necessitates explicitness with regard to ambiguities. In the case of recognized ambiguities, one needs to assess how sensitive the outcomes are to different model choices. Furthermore, this reveals the need to view the implementation of model routes as an iterative process, going back and forth between interpreting qualitative descriptions and implementing model choices. Communication with social scientists and a broad range of modelers is at least one step toward arriving at intersubjective model routes. Confronting participants engaged in the actual debates with the model routes would be a promising subsequent step toward improving the latter.

Perspectives may be derived from within a specific theoretical framework used as a tool to arrive at coherent descriptions. Limitations reside in the fact that any theoretical framework such as cultural theory is generally considered merely one possible scheme of classification. Social scientists' involvement in IA is desperately needed to explore and illuminate the range of perspectives and their implicit assumptions about policy making, society, and science (Van Asselt and Rotmans 1995b; Shackley and Wynne 1995b). However, even if it were possible to come up with a generally accepted and empirically validated typology of perspectives, the problem remains that each type is necessarily an idealization and generalization. The value systems of people engaged in actual debates are generally agreed to be more hybrid than stereotypes can account for. Therefore, model routes can be no more than heuristic tools for exploring the consequences of differences in ideal-typical value systems.

However, adopting a pluralistic strategy cannot be limited to implementing new types of IA models. The challenge for practitioners in the field of IA is to design institutional settings that would enable us to cope with plural perspectives in the broadest sense. We need to find ways to address those aspects and nuances of human values that modeling frameworks can never incorporate. In other words, we need to assess new avenues in integrated assessment that are not focused primarily on improving IA modeling but address the institutional dimension as well. Searching for new avenues to realize pluralistic strategies in the IA process is a major challenge for integrated assessment on regional scales.

8.6 THE COPRODUCTION OF KNOWLEDGE

We have now compared three different strategies for the integrated assessment of climate change. Currently, these are mainly pursued with an IA model as nucleus. The three strategies and the corresponding models differ in what types of uncertainty they account for, how they account for them, and what criteria they consider relevant to the decision process.

The best-guess strategy assumes that IA models can be used as tools to derive a set of forecasts based on the current state of all available knowledge—namely, a set of best guesses. It accounts for uncertainties only implicitly, by admitting that a best guess is not a statement of fact. In addition, model results' sensitivity to specific assumptions can be investigated. However, uncertainties as such are not perceived as important criteria in the decision process. It is assumed that each action can be mapped on to an expected value of a well-defined utility and that all possible outcomes can be ranked according to a consistent preference ordering. The decision criteria are then clearly structured: Any rational person would choose the action that leads to the outcome with the highest preference ranking.

However, decision makers may differ in their degree of risk aversion. In this case, unique solutions based on best guesses cannot be found. The probabilistic strategy takes this into consideration by taking uncertainties explicitly into account. Models are used as tools to quantify both objective and subjective uncertainties. As a result, actions can be mapped onto possible outcomes' probability distributions. The emphasis shifts from choosing the most desirable option to managing the whole set of risks in a responsible fashion. "Responsible" in this sense means exploring and making transparent the whole range of available knowledge and its uncertainties. Decisions are based on preferences for specific outcomes, which may imply avoiding undesirable outcomes even when these are highly unlikely.

Despite its emphasis on uncertainties, the probabilistic strategy limits its attention largely to uncertainties that can be quantified and handled by probabilities. However, uncertainties may also relate to different preferences, beliefs, and perspectives of individuals and different social groups. The pluralistic strategy takes into account that values and different perspectives enter at many stages of an IA process, starting with the very first step of formulating the problem. In the pluralistic strategy, models may serve as tools for approaching a problem by entirely different routes. As a result, quite different sets of actions and outcomes may be taken into consideration within different perspectives, and they may not even overlap. Each proponent of a certain perspective may argue consistently within the framework of his or her own beliefs. However, as long as the proponents of different perspectives do not transcend their own framework, communication remains difficult. Responsible risk management now means, first of all, communicating the importance of a diversity in values and facilitating a dialogue on complementary perspectives. The decision process is thus viewed as a process of social learning instead of an optimizing choice made by a single rational decision maker.

Neither of these strategies offers explicit suggestions about how IA models should be integrated into an overall IA process. The most evident integration is given in the best-guess strategy, where models are used mainly to derive decision proposals for politicians. Solutions may be sought in regulatory action based on best guesses derived from model simulations. The situation starts to become more involved in the probabilistic strategy. Here models are perceived as tools for informing the policymakers during the policy-making process quantitatively and systematically about uncertainties. The decision process itself, however, remains ill defined. It has already been mentioned that the pluralistic strategy is by no means clear how models can serve as efficient tools to convey qualitative insights into the importance of different perspectives. Also it is not clear into what type of IA process such models should be integrated.

We assert that it is important to adopt a pluralistic strategy for the overall IA process into which an IA model is embedded. This assertion follows logically from our argument that accounting for indeterminacy shifts the integrated assessment's emphasis away from achieving a preconceived goal toward sustaining an ongoing goal-seeking adaptive process. This shift is not only pragmatic, it is fundamental. We therefore now outline an approach for adopting a pluralistic perspective that we call the "coproduction of knowledge." It emphasizes the need for fostering processes of social learning on regional scales by involving scientists and stakeholders from different social groups in an intense dialogue.

8.6.1 The Societal Role of Science and Scientific Information

Practitioners in the field of IA modeling agree that models are not the be all and end all of integrated assessment but rather a component of a broader assessment process (e.g., Parson 1995; Rotmans and Dowlatabati 1997). However, there is no consensus about what the nature of that process should be and how models should fit into it. To date, the models have been designed and employed principally to provide insights to high-level policy makers. Models often appeal to policymakers because of their immediate potential for instrumental use: "for speaking truth to power." Quantitative analyses of responsiveness to tax rates, for instance, can in principle be directly translated into a set of policy choices about whether to implement a carbon tax, and even at what level taxes should be set. In other words, modelers usually manage to come up with a bottom line, or if not a bottom line, something that can be appropriated and used as one in a policy debate.

Yet our discussion of irreducible uncertainty and the role of values in integrated assessment indicates that such bottom lines are often illusory. Most modelers carefully deny that their models (even the best-guess variety) are "truth machines," describing them rather as aids to systematic thought. Nevertheless, in practice, any model that produces a single result (even with an uncertainty range) tends to seduce even the sophisticated modeler into regarding it as at least a plausible or possible unitary reality (Wynne and Shakley 1994). Political decision makers can hardly be blamed for ignoring the warnings about how to use the information that models provide. Our argument implies that we must abandon the idea of speaking truth to power. But if we do, what would replace it? The question is a serious one; the answer, profoundly unsettling.

First, we can recognize that policymakers' use of scientific information seems inconsistent with the explicit norms of the speaking truth to power approach. Empirical research in the United States (where the instrumentalist emphasis on the bottom line is probably most strongly emphasized) shows that, despite a generally positive attitude to such analysis, it is seldom acted upon in any directly identifiable fashion (Rich 1978; Starling 1979; Weiss and Bucuvalas 1980; Whiteman 1985). Once a course of action is chosen, modeling data are frequently invoked for the purposes of rationalization and persuasion (Patton 1978; Whiteman 1985), but this does not mean that those data shaped the decision.

Weiss (1982) suggests that policy analysis's real usefulness may be less as a tool for solving specific problems than as a way of orienting policymakers toward issues: "And much of this is not deliberate, direct, and targeted, but a result of long-term percolation of social science concepts, theories, and findings into the climate of informed opinion" (Weiss 1982 p. 534). As Heineman et al. (1990, p. 43) put it: "In summary, instrumental use of policy analysis is not as widespread as analysts would like; however, there is a more diffuse use of policy analysis which can be significant. This use for "enlightenment" is often underplayed in the literature on utilization of policy analysis.... If the analyst cannot easily shape a specific policy, his or her findings may still have an impact on the broader policy agenda...." (p. 43) An important answer to "What will we do if we abandon speaking truth to power?" is "We will do what we have always done, but with greater awareness of what we are doing." Abandoning illusions may be uncomfortable, but it is the first step on the path to authentically empowering individuals and communities.

Once that step is taken, we find that useful concepts do exist for the relationship between science researchers and policymakers. For example, Robinson, in a series of articles (1982, 1988, 1991, 1992a, 1992b), has criticized the misconception underlying standard views of the science-policy relationship: that it is a one-way flow of objective information from science to policy (1982). He proposes instead a relationship in which researchers, policy-makers, and the public form "mutual learning systems" (1992a, 1992b) that use modeling tools to explore alternative futures ("backcasting") rather than trying to predict the future ("forecasting"). Elements of this model include the explicit recognition that policy questions are questions not mainly of fact but to a large extent of value, and that both a "physical flows" perspective and an "actor-system" perspective are needed to provide a usefully inte-grated approach to policy questions (1991). This expresses the attitude toward model development and use the pluralistic approach implies.

The controversies about transAlpine traffic discussed in chapter 7 are a case in point. The fact that the Swiss population has actually decided to force international transapline freight traffic to use railways cannot be understood within a "physical flows" perspective. The ratio of international to intrana-tional traffic implies that such a decision is rather ineffective at the level of physical flows. From an "actor-system" perspective, however, the same deci-sion is perfectly intelligible as a powerful expression of the Alps' meaning for the regional and national culture. Chapter 7 also suggests that a similar approach would be needed to analyze the thorny decision-making process dealing with baseline tunnels through the Alps. Regional integrated assess-ment of climate change, then, cannot ignore the cultural orientations and conflicting social interests that exist in the regions in question.

In this respect, the experience of three decades of environmental impact analysis and research into technological siting issues has some potential lessons for IA. Although environmental impact analysis has been routine and perfunctory in many cases, studies of the process in controversial cases reveal three primary lessons pertinent to economic-ecological assessment. First, the basis for disagreement on policy issues is frequently social rather than technical. Second, a focus on improved technical analysis may polarize rather than resolve controversy. And third, differences among the various parties to a policy debate extend beyond differences over preferred technical solutions; frequently, these differences represent fundamental differences in world views or rationalities that affect the interpretation and framing of the issues.

Nelkin's early case study research (1979) demonstrated both the distribu-tive, equity issues that arise in ostensibly technical disputes and also the lim-itations of technical expertise in bringing about solutions. Her analyses showed that while many of the debates were about technical questions, in the end they involve political choices from among competing social values (Nelkin 1979). Socolow, reflecting on the controversy surrounding construc-tion of a proposed dam on the Delaware River, similarly argued that the public debate, which focused purely on the technical issues, was cloaked in a formality that excludes a large part of what people most care about (Socolow 1976).

Wynne's (1982) analysis of the Windscale Inquiry, held in Britain in 1977 to investigate plans to build a thermal oxide reprocessing plant for spent nuclear fuel, points out the need to acknowledge the rationality of and address the arguments advanced by the very different frames of reference of the parties to the debate. For supporters of the plant, who constituted the "dominant rationality," the issue was framed as one of "rational and objective discussion" by which emotional arguments can be reduced to an objective and intelligent level (Wynne 1982). For opponents of the plant, however, the issue was one of belief in the impartiality of the social institutions responsible for managing nuclear technology. Further expansion of this technology was opposed on the grounds that these institutions could not be trusted. The opponents' frame of reference was, in actuality, no less rational than that of the supporters, if one defines rationality as a self-conscious process of using reasoned argument to make and defend advocative claims (Dunn 1981), yet their views were characterized as factually wrong or irrational.

Research on nuclear waste disposal in the United States confirms the necessity to go beyond considering technical issues only. Passage of the Nuclear Waste Policy Act of 1982 followed many years of controversy and criticism of the Department of Energy's (DOE) focus on technical issues and failure to recognize the political and social implications of nuclear technology and nuclear waste. Senior managers of the DOE program most frequently cited lack of public acceptance as the most significant barrier to successful policy implementation. Nevertheless, research has documented a continuing pattern of giving lip service ... to the importance of non-technical factors ... and almost nothing ... invested in the analysis or evaluation of these factors (Hewlett 1978). In particular, the agency limited stakeholder participation to formal provision of information and opportunities for comment. The prevailing pattern of rationality resulted in the production of a never-ending series of technical analyses that, given the inherent uncertainty of disposing of waste that would remain radioactive for thousands of years, could not prove the project's feasibility to the critics' satisfaction. This failure to address the full scope of the problem in the light of irreducible uncertainty and conflicting values undermined the program's legitimacy and resulted in a major revision of the act (Bradbury 1989).

Ironically there is almost no overlap or contact between the researchers who have investigated stakeholders' role in environmental controversies and IA modelers. One attempt to embed modeling in a broader framework is that of Parson (1995), who has used the MiniCAM (see section 8.3.1) to generate information during live simulation of climate negotiations. However, Parson's simulation focused on the level of high policy (negotiations among nation states), and the participants were essentially role playing. The challenge remains to embed the process of IA in a pluralistic institutional setting. Such an IA dialogue would thus comprise stakeholders from many walks of life as well as politicians and scientists.

8.6.2 Different Voices in the Climate Debate

Our argument so far shows that decision making on a regional scale may be viewed as a process of social learning in which scientific arguments must supply information to a process of sociocultural and economic change. This implies shifting the focus on rationality in decision making from substantive to procedural rationality, from evaluating the result of a decision irrespective of how it is made to evaluating the decision-making process itself (Faucheux, Froger, and Noel 1995; Simon 1982). The metaphor of a dialogue among different voices in society may describe societal decision-making processes that involve social learning. Voices are not single individuals but may be social institutions such as national or regional parliaments or NGOs. The role of science and scientific information in such a dialogue must evolve to meet the challenges a comprehensive process of integrated assessment poses.

Voices in a society may be viewed as expressing a certain cultural perspective and value system. Among the plethora of voices that characterize democratic pluralistic societies, we may identify at least three voices that shape the climate debate: We call them "alarmism," "skepticism," and "administration." These voices may be linked to three perspectives distinguished in cultural theory: namely, egalitarianism, individualism, and hierarchism, respectively (cf. box 8.11). Let us explore now in more detail these voices' meaning and how they have shaped the current debate about climate change.

While being interviewed about his GHG policy, former White House chief of staff John Sununu stated (Sununu 1992, p. 24): "Yet there is this tremendous rush to have the world tie its hands today, instead of waiting for the results of that research to come through in the next few years.... This idea that we have to tie our hands now, on the basis of bad data and bad models, when good data and good models will be available in a relatively short time, is part of what makes me urge people to look a little bit below the surface and examine what is really the agenda here." Sununu argues that policy toward reducing fossil fuel consumption ought to be delayed because more reliable climate change scenarios are imminent (see also Schlesinger and Jiang 1991). The rapidly growing power of high-speed computers, which progressively increase climate models' computational resolution, feeds the expectations for their improvement. Nevertheless, more reliable scenarios are by no means guaranteed (cf. Risbey Handel, and Stone 1991; Risbey and Stone 1992).

It would be a dangerous kind of arrogance, however, to dismiss the voice of people like Sununu as plain nonsense dictated by vested interests. This voice is as much part of the dialogue as the voice of those who claim that scientific evidence about climatic risks warrants immediate action of a very dramatic kind. Actually, it is this voice of alarmism that the media have taken up: The notion of a climate catastrophe has replaced the notion of climate change. However, this voice seems to influence the policy process only minimally. Such conflicts among different voices characterize all complex societies. Current debates over climate change certainly involve at least the two conflicting perspectives in which perceptions are clustered at two extremes, with the middle ground only sparsely occupied. Such a situation embodies the risk of an extreme polarization, where processes of social learning become virtually impossible.

However, the scientific assessments of the Intergovernmental Panel on Climate Change published over recent years (IPCC 1990, 1992, 1996) seem to show a different picture. The participants in this enterprise agreed that some climate change seems inevitable during the 21st century, but significant uncertainty remained about how large and what its impacts would be on human systems and the environment and when they would be felt. This third voice represented itself as that of reason, between the outlying extremes of alarmism and skepticism.

Both extreme perspectives have criticized this picture of an Aristotelian golden mean of scientific reasonableness as self-serving. Alarmists accuse the scientific community of fiddling while Rome burns and insist that in any case there are perfectly sound reasons to embark on radical actions to reduce emissions apart from climate change. Skeptics accuse the scientific community of playing off the scaremongering of environmental Jeremiahs to expand their research budgets.

Is the truth about climate change really to be found between these extremes, as the alleged voice of reason would have us believe? This voice itself seems to embody a technocratic conviction that precise description of natural and social systems will enable us to manage our way to a preferred future. In a sense, this viewpoint reduces a broad range of global social, economic, demographic, and political changes to the issue of controlling GHG emissions by political means. As is well known, fiscal means play a crucial role here, ranging from the proposal for a carbon tax to that one of tradeable permits issued by some international authority. The third voice in the climate debate, then, may be labeled as the voice of administration: It argues that appropriate government decisions must deal with the problem of global climate change, and it tries to justify these decisions using scientific evidence. Contrasting with alarmism, which calls for stringent political regulations, the voice of administration argues for managing climate change rationally by evaluating any political measure's costs and benefits. Elaborate models for integrated assessment, providing a means to find an optimal strategy, support this voice.

The multiplicity of different perspectives and their dynamic interplay is an important element of societal dynamics. The three voices discussed above define a framework for a complex discourse. Each can define itself in opposition to the other two. These voices are the proactive participants in the discourse. As solidarities of each sort are established, participants in the discourse distinguish themselves from those creating alternative modes. The

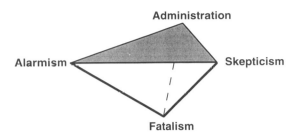

Figure 8.9 Interplay of voices in the climate debate.

conflict among different forms of solidarity and the struggle of the different voices to become dominant provide crucial dynamics of social change.

To understand such conflicts it is important to account for another voice, namely that of fatalism. This is the passive voice of those who lack direct access to market or political power or are unable to mobilize a group of equals. Figure 8.9 represents the resulting discourse space. The triangle of the three voices considered above becomes the roof of an inverted pyramid. The voice of fatalism, expressing the inability to believe any one of the three voices competing for general agreement, forms the apex of the pyramid.

Strategists of the voices at each corner of the top triangle compete to pull the inverted pyramid's apex into their own corner by claiming to speak on the fatalists behalf. The very existence of competing and contradicting voices bewilders the fatalists. From their point of view, the existence of different ways of framing the many uncertainties involved in climate change only amplify those uncertainties. The voice of fatalism expresses the feeling that each of the three other voices can advance a warranted claim to truth, but that none can claim a monopoly of truth for itself. We have discussed in detail that to dealing adequately with the uncertainties of climate change requires pluralistic approaches. The inconsistencies of human knowledge, then, are not deplorable imperfections. They may actually be essential to humankind's ability to deal with natural and social systems' indeterminacy. The voice of fatalism expresses the current inability to deal with the ambiguities the existence of complementary descriptions engenders and also expresses a feeling of powerlessness in the face of the issue's overwhelming complexity. If we claim that complementary descriptions are essential for dealing with complexity, we require a new voice that takes account of this fact and mediates a pluralistic dialogue in society.

8.6.3 Indeterminacy and a New Voice for Regional Integrated Assessment

Social learning can take place only when different social groups engage in a fruitful dialogue. A new voice, a new institutional role, may be required that would explicitly try to foster a conversation among several complementary voices. We call this voice the voice of facilitation (figure 8.10). It accepts

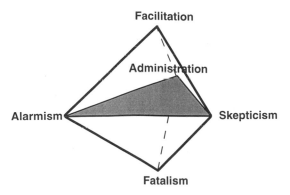

Figure 8.10 The additional voice of facilitation fosters communication among the other voices. Such communication is a prerequisite for any social learning process to take place at all.

nature and society as complex and sees indeterminacy as an opportunity for dialogue. It is not a reactive but a future-oriented perspective that tries to mediate innovative action by pointing out options for the future. The facilitators seek as practical conclusions joint actions by various actors rather than simple political measures by one controlling actor.

This conceptualization agrees with our perception of regional integrated assessment as a dynamic adaptive process of sociocultural change that acknowledges a diversity of approaches and different sources of knowledge. It is based on a view of robust action and sequential decision making in which short-term goals are realized without sacrificing degrees of freedom for decisions on longer timescales. The voice of facilitation in such processes should be strong and must not be identified with a single person or group of persons. Specifically designed regional IA processes could equally well take this role.

This suggestion addresses issues Morgan and Dowlatabadi (1996) raise in their recent review of lessons learned from the integrated assessment of climate change. In their view, the biggest challenges in IA are philosophical and methodological. How do we deal with both manageable and extreme uncertainties? How do we deal with the adaptive process of a wide range of values and expectations that shape decision making and are not accessible by a view of decision making based on a single, rational, utility-maximizing decision maker? An approach as outlined in this chapter, focusing on implementing appropriate and legitimate decision frameworks, with a specific focus on uncertainties, should provide a major step forward in meeting these challenges.

Models may play an important role in such an endeavor. So far, highly aggregated IA models have mainly been used in a hierarchical mode and as means of a one-way communication to inform policymakers about the current state of scientific knowledge. However, one may perceive a new role for such computer models as tools of communication to exchange qualitative and quantitative insights into the origins and the evaluation of uncertainty

Regional Integrated Assessment and the Problem of Indeterminacy

Box 8.12 Models and IA Focus Groups

Social learning is essential for decision making in conditions of complexity. A promising way to integrate a microcosm of social learning into IA is establishing IA focus groups. The idea of a microcosm used here is drawn from one stream of risk-assessment research in microbiology. The behavior of an organism introduced into a natural environment is hard to predict from isolated laboratory experiments. Therefore, biologists sometimes study this behavior by inserting the organism into a biological microcosm of an ecosystem established within an experimental setting. In a similar vein, a social microcosm enables researchers to study the dynamics of a social group exposed to specific information. The social microcosm processes expert knowledge, especially in the form of computer models, and gauges its policy relevance. Although computer models work best for mathematically well-defined systems, human groups are well suited to dealing with complex, ambiguous, open-ended problems. Such a microcosm should express the ordinary citizens' points of view, but in a dialogue with scientists, policymakers, and representatives of NGOs and business. Monitored discourse groups of laypersons are currently used with considerable success in social research and in public participation processes. Current procedures for using such groups include focus groups and citizen juries.

IA focus groups use IA models in their discussion process, drawing on elements of both public opinion research and marketing studies. They differ from both these fields, however, in their explicit goal of providing ordinary citizens as well as various other stakeholders with an opportunity to articulate their voice in the climate debate, and to do so in a way that draws on the state of the art in scientific research and that is suitable for shaping actual policy making. IA focus groups include people from one or more of the following categories: laypersons, experts, policymakers, and representatives of businesses. Recently, experimental projects have been launched in several countries to conduct and analyze IA focus groups.

Within the ULYSSES project,[1] IA focus groups have been established in seven European countries. They use different existing IA models that vary in scope and aggregation. Expected results include a framework for IA focus groups and experience with features of IA models and their usefulness within the context of such groups.

and indeterminacy. Such a role is in line with our emphasis on procedural rationality. The models and the results they produce cannot be separated from the IA processes in which they are embedded. Different types of model frameworks may serve the purpose of communication tools in a social process. One can, for example, design phenomena-oriented models that do not integrate all available knowledge but focus instead on important phenomena of complex indeterminate systems and how these phenomena are perceived (cf. Pahl-Wostl 1995b, 1996). The overall integration of knowledge and perceptions can then take place in a discursive dialogue in specifically designed IA focus groups (see box 8.12 for further details). The input to such groups may consist of different types of models together with other scientific and nonscientific information of relevance. Participants may be citizens, politicians, scientists, or representatives of business and nongovernmental organizations. By providing patterns of subjective preferences and equally subjective probabilities, the common sense of laypersons contributes to identifying the perspectives needed to synthesize the available scientific results into an integrated assessment. Such an assessment does not provide regional decision

Box 8.13   Regional IA and the Second Phase of CLEAR

The CLEAR project's second phase (1996–99) focuses on using research on advanced IA, involving IA focus groups, to integrate different disciplinary projects ranging from regional climate models to ecosystem dynamics and economic innovations. In this process, a specific model platform is being designed that includes global and regional scenarios and simple dynamic models expressing essential elements of the dynamics of complex systems. Such modeling tools are important for conveying the meaning of irreducible uncertainties and indeterminacy. The model platform is attuned to the specific requirements of laypersons using the model interactively.

Scientists are actively involved in developing integrated regional scenarios for the modeling platform to be used in the focus groups. These scenarios are based on the subjective judgments of the experts involved. This approach has the advantage of allowing the uncertainty gap pointed out by Wynne (personal communication) to be bridged. Wynne emphasizes that awareness of uncertainty in scientific results is high in the scientific expert community because of scientists' detailed knowledge of the systems under investigation. According to Wynne, there is a danger that the awareness of uncertainties might decrease when IA analysts incorporate scientific results into IA models. Uncertainty increases again when results from IA modeling efforts are communicated to the public because citizens may distrust expert knowledge.

Specifically designed model platforms may now facilitate a dialogue between scientists and citizens for the coproduction of knowledge and for the building of a shared awareness of uncertainties, risks, and options. In this respect, uncertainties may be viewed as both a threat and an opportunity, the latter because uncertain prospects arising from potential climate change may trigger innovative action for socioeconomic and technological change.

makers with an algorithm that would lead to the optimal decision but rather provides insights about the mental space in which the decision has to be made, including different actors' conflicting viewpoints. It mediates what we call the coproduction of knowledge, involving laypersons and scientists. Such an iterative process combines scientific expert knowledge and local knowledge to identify options for societal action that are desirable and feasible from both a scientific and a societal perspective. The second phase of the CLEAR project is centered on such an approach for regional IA (cf. box 8.13). Producing this volume has been essential for providing a framework and a basis for this assessment project. The assessment is built on the assumption that climate change does not pose a major economic threat to the Alpine region over the next few decades. However, the public is concerned about the risks associated with extreme events and potential catastrophic consequences—even if these, like a major shift in global climate, are very unlikely to occur. This increasing socioeconomic uncertainty may not hinder, but trigger, innovative societal action. A number of sequential climate-related decisions will be made primarily driven by local nonclimate considerations. Therefore, an integrated assessment of climate change cannot be separated from questions related to ecological and economic sustainability in general. The discussion about climate change must be embedded in an overall discussion about future scenarios for socioeconomic development.

The approach chosen in CLEAR is only one possibility for meeting the specific challenges facing science in building new institutional structures, which can be summarized as follows:

• search for effective methods on how to reach agreement on some common tasks rather than searching for an elusive optimal goal

• shape environments where open discussion and fruitful dialogue can take place, such as IA focus groups

• define and communicate new roles of models in such participatory IA processes

• develop a plurality of models and new approaches to incorporate a diversity of values and perspectives into such processes

Obviously, approaches to such integrated assessments will differ from region to region. It would be contradictory to argue in favor of pluralistic strategies for regional integrated assessment without acknowledging the importance of accounting for regional characteristics in adopting such an approach. In this vein, we have sketched an attempt to organize IA as an open dialogue. The whole process of integrated assessment must be integrated into a pluralistic political and social framework that reflects the regional or national diversity of cultural, political, and technological perspectives. This suggests a diversity of approaches, including a variety of regional perspectives on global problems. Science's role in such assessments, in which the laypersons' participation becomes an integral part of the research process, may be described as "postnormal" (Funtowicz and Ravetz 1993, 1994).

Respect for laypersons' judgment is an important element of political traditions in the Alpine region. It is also a crucial feature of the Swiss system of "cooperative democracy" (Imboden 1962). This institutional arrangement combines popular elections of political authorities at the local and national level with regular votes on a very large class of political decisions. It is rooted in long-standing traditions of cooperative use of the local commons. In a historical situation marked by the need to develop a responsible use of the global commons, it seems worthwhile to explore cooperative democracy's relevance for the future. *The Economist* (1996) has even argued that one of the most important steps in the further development of democracy might be a widespread adoption of similar institutional schemes in highly industrialized countries.

The experience of Swiss democracy provides impressive proof that heavily involving laypersons in political decision making may produce quite valuable results. There is little reason to think that political decisions made mainly by civil servants and specialized experts yield systematically superior results according to any sensible criteria. With regard to integrated assessment, this suggests that attempts at "speaking truth to people" may be a promising complement to the well-established rituals of speaking truth to power.

Two major difficulties arise, however. First, the righteous claim to speak "the truth" is often questionable. In the case of climate change, questions and doubts scientists entertain may be as useful to public debate as the consolidated knowledge that scientists may be able to offer. Second, it is far from obvious whether a common language is readily available to ensure that lack of intelligibility and systematic misunderstandings does not cripple communication between scientific experts and the general public. To a considerable extent, integrated assessment's task is precisely to foster such a common language. In its absence, cooperative democracy degenerates into a farce, with citizens voting on issues they do not understand and government bureaucracies carrying out hidden agendas behind smokescreens of expert knowledge. Obviously, the same problem arises in conventional representative democracy. But this makes its resolution all the more urgent.

Integrated assessment cannot be based on a dreamworld without power relations and without serious communication barriers. In such a world, the need for integrated assessment would hardly arise. The challenge, then, is to find ways of limiting the abuse of power, not by eliminating power relations, but by embedding them in relations of political responsibility (Jaeger 1994), and of overcoming communication barriers not by dismantling scientific knowledge, but by embedding its production in a wider social debate. This task arises with regard to myriad local and national issues, but it has acquired a special importance with regard to global problems like climate change. Developing procedures of participatory integrated assessment like IA focus groups seems a promising way of tackling this task. And the Alpine region may offer a particularly promising setting for renewing institutions of cooperative democracy by enriching them with procedures of regional integrated assessment.

Integrated assessment of climate change on a global scale might be sufficient for decision making were there a world government that could take care of the problem by itself. Fortunately, perhaps, no such government exists. Climate policy will unfold as a patchwork of national policies, international agreements, initiatives by NGOs, business strategies, and decisions by many other agents. For all these actors, climate change is relevant not only on a global scale, but also on various regional scales. And for most of them, action can be globally effective only in an indirect way. The immediate range of action is again given by some regional scale. For these reasons, the integrated assessment of regional climate change will be a major task of the scientific community in the years to come. The Alpine region seems very well suited to play a fruitful role in this respect. It has a strong tradition in involving the public in decision-making processes. It is one of the world's most prosperous and dynamic economic regions. It comprises a wide range of different cultural traditions. If an integrated assessment succeeds in mediating the discussion about seeming and real conflicts among social, economic, and ecological goals, the chances are good for this region to become one of the pioneers in adopting a viable path into a sustainable future.

NOTE

1. ULYSSES is financed by *Framework Programme IV for Research and Technological Development* of the European Community under the specific program *Environment and Climate* and the research theme *Human Dimensions of Environmental Change*. Modelers and social scientists from throughout Europe contribute to this project. C. Jaeger coordinates ULYSSES.

REFERENCES

Anderson, D. M., and R. S. Webb. 1994. Ice-age tropics revisited. *Nature* 367:23–24.

Arrow, K. J., J. Parikh, G. Pillet, M. Grubb, E. Haites, J. Hourcade, K. Parikh, and F. Yamin. 1996. Decision-making frameworks for addressing climate change. In *Economic and Social Dimensions of Climate Change, IPCC*, ed. J. P. Bruce, H. Lee, and E. Haites, chap. 2. Cambridge, UK: Cambridge University Press.

Blau, U. 1977. *Die dreiwertige Logik der Sprache*. Berlin: de Gruyter.

Bradbury, J. A. 1989. *The Use of Social Science Knowledge in Implementing the Nuclear Waste Policy Act*. Pittsburgh, PA: University of Pittsburgh, Graduate School of Public and International Affairs.

Brauner, C. 1994. *Global Warming: Element of Risk*. Zürich: Swiss Reinsurance Company.

Burger, T. 1987. *Max Weber's Theory of Concept Formation: History, Laws, and Ideal Types*. Durham; NC: Duke University Press.

Dansgaard, W., S. J. Johnson, H. B. Clausen, D. Dahl-Jensen, N. S. Gundestrup, C. U. Hammer, C. S. Hvdberg, J. P. Steffensen, A. E. Sveinsbjörnsdottir, J. Jouzel and 6 Bond 1993. Evidence for general instability of past climate from a 250-kyr ice-core record. *Nature* 364:218–20.

De Finetti, F. 1974. *Theory of Probability*. New York: Wiley.

Delworth, T., S. Manabe, and R. J. Stouffer. 1993. "Interdecadal variations of the thermohaline circulation in a coupled ocean-atmosphere model." *Journal of Climate* 6:1993–2011.

De Villiers, S., B. Nelson, and A. Chivas. 1995. Biological control on coral Sr/Ca and $\delta^{18}O$ reconstructions of sea surface temperatures. *Science* 269:1247–49.

Douglas, M. 1985. *Risk Acceptability According to the Social Sciences: Social Research Perspectives*. New York: Russell Sage Foundation.

Douglas, M., and A. Wildavsky. 1982. *Risk and Culture: An Essay on the Selection of Technical and Environmental Dangers*. Berkeley and Los Angeles: University of California Press.

Dowlatabadi, H. 1995. *Integrated Climate Assessment Model 2.0: Technical Documentation*. Pittsburgh, PA: Carnegie-Mellon University, Department of Engineering and Public Policy.

Dowlatabadi, H., and G. Morgan 1993. A model framework for integrated studies of the climate problem. *Energy Policy* (March):209–21.

Dunn W. N. 1981. *Public Policy Analysis: An Introduction*. Englewood Cliffs, NJ: Prentice-Hall.

Edmonds, J., H. Pitcher, and N. Rosenberg. 1996. *The Global Change Assessment Model (GCAM): An Integrated Capability to Analyze Greenhouse Gas Emissions, Atmospheric Accumulation, and Impacts on Human Systems*. Washington, DC: Pacific Northwest National Laboratory, Global Climate Change Group, Technology Planning and Analysis Center.

Edmonds, J., M. Wise, and C. MacCracken. 1994. *Advanced Energy Technologies and Climate Change: An Analysis Using the Global Change Assessment Model (GCAM)*. Richland, WA: Pacific Northwest National Laboratory.

Faucheux, S., G. Froger, and J.-F. Noel. 1995. What forms of rationality for sustainable development? *The Journal of Socio-Economics* 24:169–209.

Fei-Fei, J., D. Neelin, and M. Ghil. 1994. El Niño on the devil's staircase: Annual subharmonic steps to chaos. *Science* 264:70–2.

Frei, C., M. Widmann, D. Lüthi, H. C. Davies, and C. Schär. 1997. Response of the regional water cycle to an increase of atmospheric moisture related to global warming. Proc. 7th Conf. on Climate Variations, American Meteorological Society, February 2–7, Long Beach, CA.

Funtowicz, S., and J. Ravetz. 1994. The worth of a songbird: Ecological economics as a post-normal science. *Ecological Economics* 10:197–208.

Funtowicz, S., and J. Ravetz. 1993. Science for the post-normal age. *Futures* 25:735–55.

Funtowicz, S., and J. Ravetz. 1990. *Uncertainty and Quality in Science for Policy.* Dordrecht, the Netherlands: Kluwer.

Funtowicz, S., and J. Ravetz. 1985. Three types of risk assessment: A methodological analysis. In *Risk Analysis in the Private Sector,* ed. C. Whipple and V. T. Covello. New York: Plenum Press, pp. 217–231.

Grendstad, G. 1995. *Classifying Cultures.* PhD diss., LOS report University of Bergen, Norway.

GRIP Members. 1993. Climate instability during the last interglacial period recorded in the GRIP ice core. *Nature* 364:203–8.

Hardin, G. 1968. The tragedy of the commons. *Science* 162:1243–8.

Heineman, R. A., W. T. Bluhm, S. A. Peterson, and E. N. Kearny. 1990. *The World of the Policy Analyst: Rationality, Values, and Politics.* Chatham, NJ: Chatham House Publishers, p. 43.

Hewlett, R. G. 1978. *Federal Policy for the Disposal of Highly Radioactive Wastes from Commercial Nuclear Power Plants.* DOE/MA-0153. Washington, DC: U.S. Department of Energy.

Hirsch-Hadorn, G. 1997. Webers Idealtypus als Methode zur Bestimmung des Begriffsgehalts theoretischer Begriffe in den Kulturwissenschaften. *Journal for General Philosophy of Science* 28:275–296.

Hope, C., J. Anderson, and P. Wenman. 1993. Policy analysis of the Greenhouse Effect: An application of the PAGE model. *Energy Policy* 21:327–38.

Imboden, Max. (1962). *Die politischen Systeme.* Basel: Helbing & Lichtenhahn.

Ingold, T. 1993. Globes and spheres: The topology of environmentalism. In *Environmentalism: The View From Anthropology,* ed. K. Milton. London: Routledge.

IPCC (Intergovernmental Panel for Climate Change). 1990. *Climate Change: The IPCC Scientific Assessment.* Cambridge, UK: Cambridge University Press.

IPCC. 1992. *Climate Change 1992.* Cambridge, UK: Cambridge University Press.

IPCC. 1996. The science of climate change, Vol. 1 of *Climate Change 1995: IPCC Second Assessment Report.* Cambridge, UK: Cambridge University Press.

Jaeger, C. C. 1994. *Taming the Dragon: Transforming Institutions in the Face of Global Change.* Amsterdam: Gordon and Breach.

Jaeger, C. C., and B. Kasemir. 1996. Climatic risks and rational actors. *Global Environmental Change* 6:23–36.

Kolmogorov, A. N. 1933. *Grundbegriffe der Wahrscheinlichkeitsrechnung.* Berlin: Springer Verlag.

Körner, C., and J. Arnone. 1992. Responses of elevated carbon dioxide in artificial tropical ecosystems. *Science* 257:1672–1675.

Kreps, D. 1988. *Notes on the Theory of Choice*. Boulder, CO: Westview Press.

Lorenz, E. N. 1963. Deterministic nonperiodic flow. *Journal of Atmospheric Science* 20:130–41.

MacEachren, Alan M. 1995. Approaches to truth in geographic visualization. Pages 110–118 of *Proceedings of the Autocarto 12 Conference*, Feb 27–March 2, 1995, Charlotte, North Carolina. ACSM/ASPRS.

Manne, A., R. Mendelsohn, and R. Richels. 1995. MERGE: A model for evaluating regional and global effects of GHG reduction policies. *Energy Policy* 23:1.

Manne, A., and R. Richels. 1995. The Greenhouse debate: Economic efficiency, burden sharing and hedging strategies. *The Energy Journal* 16:1–37.

Mikolajewicz, U., and E. Maier-Reimer. 1990. Internal secular variability in an ocean general circulation model. *Climate Dynamics* 4:145–56.

Mitchell, J. M. 1976. An overview of climatic variability and its causal mechanisms. *Quarternary Research* 6:481–93.

Morgan, G. M., and H. Dowlatabadi. 1996. Learning from integrated assessment of climate change. *Climatic Change* 34:337–68.

Morgan, G. M., and M. Henrion. 1990. *Uncertainty: A Guide to Dealing with Uncertainty in Quantitative Risk and Policy Analysis*. Cambridge, Mass.: Cambridge University Press.

Morgan, G. M., and D. Keith. 1994. Climate change: Subjective judgments by climate experts. *Environmental Science & Technology* 29:468–76.

Morishima, M. 1990. Ideology and economic activity. *Current Sociology* 38:51–78.

Munasinghe, M., P. Meier, M. Hoel, S. Hong, and A. Aaheim. 1996. Applicability of techniques of cost-benefit analysis to climate change. In *Economic and Social Dimensions of Climate Change, IPCC*, ed. J. P. Bruce, H. Lee, and E. F. Haites, chap. 5. Cambridge: Cambridge University Press.

Nelkin, D. 1979. *Controversy: Politics of Technical Decisions*. Beverly Hills, CA: Sage.

Nordhaus, W. 1994. *Managing the Global Commons: The Economics of Climate Change*. Cambridge, MA: MIT Press.

Nordhaus, W. 1991. To slow or not to slow: The economics of the Greenhouse Effect. *Economic Journal* 101:407–15.

Pahl-Wostl, C. 1997. *Integrated Assessment of Regional Climate Change and a New Role for Computer Models at the Interface between Science and Society*. Proceedings of the Symposium "Prospects for Integrated Environmental Assessment: Lessons Learnt from the Case of Climate Change. Toulouse, October 1996. Eds. A. Sors, A. Liberatore, S. Funtowicz, J. C. Hourcade, and J. L. Fellous, pp. 156–160. European Commission DG XII, Report No. EUR 17639.

Pahl-Wostl, C. 1995a. *The Dynamic Nature of Ecosystems: Chaos and Order Entwined*. Chichester, U.K.: Wiley.

Pahl-Wostl, C. 1995b. *Complexity, Irreducible Uncertainties and Climate Change*. Statement for the Workshop on Integrated Environmental Assessment, Brussels, Belgium, May 18–20, 1995.

Palmer, T. N. 1993. Extended range atmospheric prediction and the Lorenz model. *Bull. Amer. Meteor. Soc.* 74:49–65.

Parson, E. 1995. Integrated assessment and environmental policy making: in pursuit of usefulness. *Energy Policy* 23(4/5):463–476.

Patton, M. Q. 1978. *Utilization-Focused Evaluation*. Beverly Hills, CA: Sage.

Pfister, C. 1984. *Das Klima der Schweiz von 1525–1860 und seine Bedeutung in der Geschichte von Bevölkerung und Landwirtschaft*. Bern und Stuttgart: Verlag Paul Haupt.

PNNL (Pacific Northwest National Laboratory). 1994. *Global Studies: Preparing for the 22nd Century*. Richland, WA: Pacific Northwest National Laboratory.

Rayner, S. 1992. Cultural theory and risk analysis. In *Social Theory of Risk*, ed. S. Krimsky and D. Golding. Praeger: Westport CT, 83–115.

Rayner, S. 1991. A cultural perspective on the structure and implementation of global environmental agreements. *Evaluation Review* 15:75–102.

Rayner, S. 1988. Risk communication in the search for a global climate management strategy. In *Risk Communication*, ed. H. Jungermann, R. E. Kasperson, and P. M. Wiedemann. Jülich, Germany: KFA Jülich, pp. 169–176.

Rayner, S. 1986. "Commentary on J. Ravetz, 'Usable knowledge, usable ignorance: Incomplete science with policy implication'". *Sustainable Development of the Biosphere*, ed. W. C. Clark and R. E. Munn. Cambridge: Cambridge University Press, pp. 432–434.

Rich, R. F. 1978. Uses of social science information by federal bureaucracies. In *Using Social Research in Public Policy Making*, ed. C. H. Weiss, 199–212. Lexington, MA: Lexington Press.

Risbey, J. S., M. D. Handel, and P. Stone. 1991. Should we delay responses to the greenhouse issue? *EOS Trans. Amer. Geophys Union* 72:593.

Risbey, J. S., and P. Stone. 1992. Sununu's optimism. *Technology Review* 95:9.

Robinson, J. B. 1992a. Of maps and territories: The use and abuse of socioeconomic modeling in support of decision-making. *Technological Forecasting and Social Change* 42:147–64.

Robinson, J. B. 1992b. Risks, predictions, and other optical illusions: Rethinking the use of science in social decision-making. *Policy Sciences* 25:237–54.

Robinson, J. B. 1991. Modeling the interactions between human and natural systems. *Global Environmental Change* 2:629–48.

Robinson, J. B. 1988. Unlearning and backcasting: Rethinking some of the questions we ask about the future. *Technological Forecasting and Social Change* 33:325–38.

Robinson, J. B. 1982. Apples and horned toads: On the framework-determined nature of the energy debate. *Policy Sciences* 15:23–45.

Rotmans, J. 1990. *IMAGE: An Integrated Model to Assess the Greenhouse Effect*. Dordrecht, the Netherlands: Kluwer.

Rotmans, J., and B. De Vries, eds. 1996. *TARGETS: Insights, No Answers*. Amsterdam: Baltzer.

Rotmans, J., and H. Dowlatabadi. 1998. Integrated Assessment Modeling. In *Human Choice and Climate Change: An International Social Science Assessment*, ed. S. Rayner and E. Malone. Columbus, Ohio: Battelle Press.

Savage, L. 1954. *The Foundations of Statistics*. New York: Wiley.

Schaal, L. A., and R. F. Dale. 1977. Time of observation temperature bias and climatic change. *Journal of Applied Meteorology* 16:215–22.

Schär, C., and D. R. Durran. 1997. Vortex formation and vortex shedding in continuously stratified flows past isolated topography. *Journal of the Atmospheric Sciences* 54:534–54.

Schindler, D. W. 1987. Detecting ecosystem responses to anthropogenic stress. *Canadian Journal of Fisheries and Aquatic Sciences* 44(suppl):6–25.

Schlesinger, M. E., and X. Jiang. 1991. Revised projection of future greenhouse warming. *Nature* 350:219.

SGS (Schweizerische Greina Stiftung zur Erhaltung der alpinen Fliessgewässer). 1988. *Greina und der Landschaftsrappen*. Zürich: SGS.

Shackley, S., and B. Wynne. 1995a. Integrating knowledges for climate change: Pyramids, nets and uncertainties. *Global Environmental Change* 5:113–26.

Shackley, S., and B. Wynne. 1995b. Response to Boehmer-Christiansen, and Van Asselt and Rotmans. *Science and Public Policy* 22:5–6.

Shackley, S., and B. Wynne. 1995c. Global climate change: The mutual construction of an emergent science-policy domain. *Science and Public Policy* 22:218–30.

Shackley, S., P. Young, S. Parkinson, and B. Wynne. In press. Uncertainty, complexity and concepts of good science in climate change modeling: Are GCMs the best tools? *Climatic Change*.

Simon, H. 1982. *Models of Bounded Rationality*. Vol. 2. Cambridge, MA: MIT Press.

Socolow, R. H. 1976. Failures and discourse. In *Boundaries and Analysis*, ed. H. A. Fieveson, F. W. Sinden, and R. H. Socolow, 9–40. Cambridge, MA: MacMillan.

Sparrow, C. 1982. *The Lorenz Equations: Bifurcations, Chaos, and Strange Attractors*. New York: Springer.

Starling, G. 1979. *The Politics and Economics of Public Policy*. Homewood, IL: Dorsey Press.

Sununu, J. 1992. The political pleasures of engineering. *Technological Review* 95:22–8.

The Economist. 1996. The Future of Democracy (special insert to the Christmas issue).

Thompson, M. 1982. Among the energy tribes: The anthropology of the current energy debate. Working Paper no. 82–59, International Institute for Applied Systems Analysis, Laxenburg, Austria.

Thompson, M., R. Ellis, and A. Wildavsky. 1990. *Cultural Theory*. Boulder, CO: Westview Press.

Trüb, J. 1993. Dynamical aspects of flow over Alpine-scale orography: A modeling perspective. Ph.D. diss. no. 10339, Swiss Federal Institute of Technology, Zürich.

Tversky, A., and D. Kahnemann. 1981. The framing of decisions and the psychology of choice. *Science* 211:453–8.

Tversky, A., and D. Kahnemann. 1980. Causal schemas in judgments under uncertainty. In *Progress in Social Psychology*, ed. M. D. Fishbein. Hillsdale, NJ: Lawrence Erlbaum Associates.

Tversky, A., and D. Kahnemann. 1974. Judgment under uncertainty: Heuristics and biases. *Science* 185:1124–31.

Tziperman, E., L. Stone, M. Cane, and H. Jarosh. 1994. El Niño chaos: Overlapping of resonances between the seasonal cycle and the Pacific ocean-atmosphere oscillator. *Science* 264:72–4.

Van Asselt, M. B. A., and J. Rotmans. 1996. Uncertainty in perspective. *Global Environmental Change* 6:121–57.

Van Asselt, M. B. A., and J. Rotmans. 1995a. *Uncertainty in Integrated Assessment Modeling: A Cultural Perspective-Based Approach*. Bilthoven, The Netherlands: National Institute of Public Health and Environmental Protection (RIVM).

Van Asselt, M. B. A., and J. Rotmans. 1995b. Letter to the editor. *Science and Public Policy* 22:414–5.

von Neumann, J., and O. Morgenstern. 1944. *Theory of Games and Economic Behavior*. Princeton, NJ: Princeton University Press.

Weber, M. 1913. Über einige Kategorien der verstehenden Soziologie. In *Weber, Max: Gesammelte Aufsätze zur Wissenschaftslehre*, (2nd edition 1951) J. Mohr, 427–74. Tübingen, Germany: Teubner.

Weber, M. 1906. Kritische Studien auf dem Gebiet der kulturwissenschaftlichen Logik. In *Weber, Max: Gesammelte Aufsätze zur Wissenschaftslehre*, (2nd edition 1951) J. Mohr, 215–90. Tübingen, Germany: Teubner.

Weiss, C. H. 1982. Policy research in the context of diffuse decision-making. In *Social Science Research and Public Policy-Making*, ed. D. B. Kallen, 534. Windsor: NFER-Nelson, p. 534.

Weiss, C. H., and M. J. Bucuvalas. 1980. Truth tests and utility tests. *American Sociological Review* 45:302–13.

Weyant, J., O. Davidson, H. Dowlatabadi, J. Edmonds, M. Grubb, E. A. Parson, R. Richels, J. Rotmans, P. Shukla, R. S. Tol, W. Cline, and S. Frankenhauser. 1996. Integrated assessment of climate change: An overview and comparison of approaches and results. In *Economic and Social Dimensions of Climate Change, IPCC*, ed. J. P. Bruce, H. Lee, and E. F. Haites, chap. 10. Cambridge: Cambridge University Press.

Whiteman, D. 1985. The fate of policy analysis in congressional decision-making. *Western Political Quaterly* 38:294–311.

Wigley, T. M. L., R. Richels, and J. Edmonds. 1996. Economic and environmental choices in the stabilization of atmospheric CO_2 concentrations. *Nature* 379:240–3.

Wynne, B. 1982. *Rationality and Ritual: The Windscale Inquiry and Nuclear Decisions in Britain.* Chalfont, St. Giles, U.K.: British Society for the History of Science.

Wynne, B., and S. Shackley. 1994. Environmental models: Truth machine or social heuristics. *The Globe* 21:6–8.

Zebiak, S. E., and M. A. Cane. 1987. A model of El Niño-Southern Oscillation. *Mon. Weather Rev.* 115:2262–78.

Index